2011 69th Annual Device Research Conference

(DRC 2011)

Santa Barbara, California, USA
20-22 June 2011

IEEE Catalog Number: CFP11DRC-PRT
ISBN: 978-1-61284-243-1

Copyright © 2011 by the Institute of Electrical and Electronic Engineers, Inc
All Rights Reserved

Copyright and Reprint Permissions: Abstracting is permitted with credit to the source. Libraries are permitted to photocopy beyond the limit of U.S. copyright law for private use of patrons those articles in this volume that carry a code at the bottom of the first page, provided the per-copy fee indicated in the code is paid through Copyright Clearance Center, 222 Rosewood Drive, Danvers, MA 01923.

For other copying, reprint or republication permission, write to IEEE Copyrights Manager, IEEE Service Center, 445 Hoes Lane, Piscataway, NJ 08854. All rights reserved.

***This publication is a representation of what appears in the IEEE Digital Libraries. Some format issues inherent in the e-media version may also appear in this print version.**

IEEE Catalog Number: CFP11DRC-PRT
ISBN 13:   978-1-61284-243-1
ISSN: 1548-3770

Additional Copies of This Publication Are Available From:

Curran Associates, Inc
57 Morehouse Lane
Red Hook, NY 12571 USA
Phone: (845) 758-0400
Fax: (845) 758-2633
E-mail: curran@proceedings.com
Web: www.proceedings.com

Conference at a Glance

Monday, June 20, 2011

8:30 AM - 12:00 PM	**Plenary Session**	Corwin Pavilion
1:30 PM – 5:10 PM	**Session II.A.** III-V FETs **Session II.B.** Graphene	Corwin Pavilion East Corwin Pavilion West
7:00 PM – 9:30 PM	**Session III.** Poster Session	Lagoon Plaza

Tuesday, June 21 2011

8:20 AM – 12:00 PM	**Session IV.A.** Spin/Memory **Session IV.B.** Alternative Transistor Concepts	Corwin Pavilion East Corwin Pavilion West
1:30 PM – 5:10 PM	**Session V.A** Tunnel FETs **Session V.B** Wide Bandgap Devices	Corwin Pavilion East Corwin Pavilion West
5:45 PM – 8:00 PM	**Conference Picnic**	Goleta Beach
8:30 – 10:30 PM	**Rump Sessions** R-1: Large Area OFETs: Organic or Oxide? R-2: What is the Ultimate Low Power Device? R-3: Graphene--What is it good for?	 Corwin Pavilion East Corwin Pavilion West MultiCultural Center Theater

Wednesday, June 22, 2011

8:20 – 9:20 AM	**Joint DRC/EMC Plenary Session**

DRC SESSIONS – Wednesday AM

10:00 AM – 12:00 PM	**Session VI.A** pFETS **Session VI.B** Thin-Film Devices	Corwin Pavilion East Corwin Pavilion West

EMC SESSIONS – Wednesday AM

Session A	III-Nitrides: MBE Growth
Session B	Thermoelectrics and Thermionics I
Session C	Nanoscale Characterization
Session D	Plasmonics and Metamaterials
Session E	Organic, Printed and Flexible Electronics
Session F	Devices Utilizing Low Dimensional Structures
Session G	Photovoltaics: New Materials and Characterization

DRC SESSIONS – Wednesday PM

1:30 PM – 4:50 PM	**Session VII.A** Optoelectronic Devices **Session VII.B** High Speed Devices	Corwin Pavilion East Corwin Pavilion West

EMC SESSIONS – Wednesday PM

Session H	III-Nitrides: Electronics I
Session I	Thermoelectrics and Thermionics II
Session J	Nanowire Transport and Devices
Session K	Silicon Carbide Growth, Characterization and Devices
Session L	Graphene Fabrication and Devices
Session M	III-Nitrides: Defects and LEDs
Session N	Next Generation Solar Cell Materials and Devices

June 20 – 22, 2011
The University of California
Santa Barbara, California

DEVICE
RESEARCH
CONFERENCE

OFFICERS

Patrick Fay...General Program Chair
Miguel Urteaga ....Technical Program Chair
Suman Datta...Technical Program Vice Chair
Tom Jackson .... Treasurer

Technical Program Committee

Ian Appelbaum, University of Maryland
Seth Bank, University of Texas Austin
Brian Doyle, Intel Corporation
Dan Green, ONR (Short Course Organizer)
Yongtaek Hong, Seoul National University
Mantu Hudait, Virginia Tech
Adil Karim, Johns Hopkins APL
Andy Kent, NYU
Siyuranga Koswatta, IBM Corporation
Ioannis Kymissis, Columbia University
Erik Lind, Lund University
Souvik Mahapatra, IIT Bombay
Kingsuk Maitra, Global Foundries
Kirsten Moselund, IBM Zurich
Jeong-Sun Moon, HRL
Mikael Östling, KTH
Dimitris Pavlidis, Technical University Darmstadt
Jamie Phillips, University of Michigan
Eric Pop, University of Illinois at Urbana-Champaign
Michael Schlechtweg, Fraunhofer IAF
Yanning Sun, IBM Corporation (Short Course Organizer)
Mitsuru Takenaka, University of Tokyo

Sponsored by the IEEE Electron Devices Society

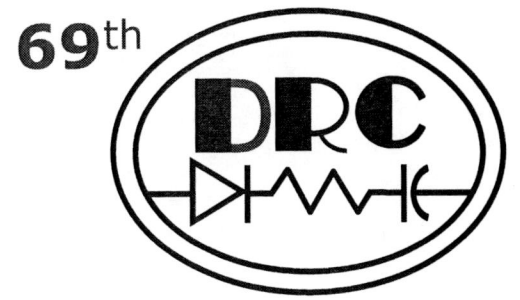

June 20 – 22, 2011
The University of California
Santa Barbara, California

DEVICE
RESEARCH
CONFERENCE

Schedule of Events

SUNDAY PM, JUNE 19TH, 2011

Registration .... 5:00 – 8:00 PM
Location .... Corwin Pavilion Lobby

Welcoming Reception .... 6:30 PM – 8:30 PM
Location .... Lagoon Plaza

MONDAY AM, JUNE 20TH, 2011

Registration .... 7:30 AM – 5:00 PM
Location .... Corwin Pavilion Lobby

Plenary Session .... 8:30 AM
Location .... Corwin Pavilion

MONDAY PM, JUNE 20TH, 2011

Session II.A. III-V FETs .... 1:30 PM
Location .... Corwin Pavilion East

Session II.B. Graphene .... 1:30 PM
Location .... Corwin Pavilion West

Session III. Poster Session .... 7:00 – 9:30 PM
Location .... Lagoon Plaza

TUESDAY AM, JUNE 21ST, 2011

Registration .... 7:30 AM – 5:00 PM
Location .... Corwin Pavilion Lobby

Session IV.A. Spin/Memory .... 8:20 AM
Location .... Corwin Pavilion East

Session IV.B. Alternative Transistor Concepts .... 8:20 AM
Location .... Corwin Pavilion West

TUESDAY PM, JUNE 21ST, 2011

Session V.A. Tunnel FETs .... 1:30 PM
Location .... Corwin Pavilion East

Session V.B. Wide Bandgap Devices .... 1:30 PM
Location .... Corwin Pavilion West

Conference Picnic .... 5:45-8:00 PM
Location .... Goleta Beach

Rump Sessions .... 8:30-10:30 PM
I. Large Area OFETs: Organic or Oxide?
Location .... Corwin Pavilion East

II. What is the Ultimate Low Power Device?
Location .... Corwin Pavilion West

III. Graphene--What is it good for?
Location .... MultiCultural Center Theater

WEDNESDAY AM, JUNE 22ND, 2011

Registration.. 7:30 AM – 11:00 AM
Location .... Corwin Pavilion Lobby

Joint EMC/DRC Plenary Session...8:20 AM
Location... Corwin Pavilion

Session VI.A. pFETs .... 10:00 AM
Location... Corwin Pavilion East

Session VI.B. Thin-Film Devices ....10:00 AM
Location..Corwin Pavilion West

WEDNESDAY PM, JUNE 22ND, 2011

Session VII.A. Optoelectronic Devices............................... ....1:30 PM
Location... Corwin Pavilion East

Session VII.B. High Speed Devices ....1:30 PM
Location... Corwin Pavilion West

Table of Contents

Session I. PLENARY SESSION — 1-10

I.-1
8:50 AM
Soft, Curvilinear Semiconductor Devices for Bio-Integrated Electronics 003
J. Rogers, Department of Materials Science and Engineering University of Illinois Urbana/Champaign, Champaign, Illinois, USA

I.-2
10:10 AM
Devices for high performance computing beyond 14nm node- Is there anything other than Si? 005
W. Haensch, IBM TJ Watson Research Center, Yorktown Heights, New York, USA

I.-3
11:10 AM
The latest performance of GaN-based nonpolar and semipolar emitting devices 009
Shuji Nakamura, Materials Department, University of California Santa Barbara, California, USA

Session II.A. III-V FETs — 11-28

II.A-1
1:30 PM
Ultra-thin Compound Semiconductors on Insulator (XOI) for MOSFETS and TFETS 013
R. Kapadia[1,2,3], K. Takei[1,2,3], A. C. Ford[1,2,3], H. Fang[1,2,3], S. Chuang[1,2,3], M. Madsen[1,2,3], S. Krishna[4,] and A. Javey[1,2,3], [1]Electrical Engineering and Computer Sciences, University of California, Berkeley, California, [2]Materials Sciences Division, Lawrence Berkeley National Laboratory, Berkeley, California, [3]Berkeley Sensor and Actuator Center, University of California, Berkeley, California, and [4]Center for High Technology Materials, University of New Mexico, Albuquerque, New Mexico, USA

II.A-2
2:10 PM
Experimental Investigation of Scalability and Transport in $In_{0.7}Ga_{0.3}As$ Multi-Gate Quantum Well FET (MuQFET) 017
L. Liu, V. Saripalli, V. Narayanan and S. Datta, The Pennsylvania State University, University Park, Pennsylvania, USA

II.A-3
2:30 PM
60 nm gate length Al_2O_3 / $In_{0.53}Ga_{0.47}As$ gate-first MOSFETs using InAs raised source-drain regrowth 019
A. D. Carter, J. J. M. Law, E. Lobisser, G. J. Burek, W. J. Mitchell, B. J. Thibeault, A. C. Gossard, and M. J. W. Rodwell, ECE Department, University of California, Santa Barbara, California, USA

II.A-4
2:50 PM
15 nm diameter InAs nanowire MOSFETs 021
A. W. Dey[1], C. Thelander[2], M. Borgström[2], B. Mattias Borg[2], E. Lind[2], and L.-E. Wernersson[1], [1]Dept. of Electrical and Information Technology and [2]Dept. of Solid State Physics, Lund University, Lund, SWEDEN

II.A-5
3:30 PM
The Nanoelectric Modeling Tool NEMO5: Capabilities, Validation and Applications to Sb-Heterostructures 023
S. Steiger[1], M. Povolotskyi[1], H.-H. Park[1], T. Kubis[1], G. Hegde[1], B. Haley[1], M. Rodwell[2] and G. Klimeck[1,3], [1]Network for Computational Nanotechnology, Purdue University, West Lafayette Indiana, USA and [2]University of California, Santa Barbara, California, USA

II.A-6
4:10 PM
Experimental Determination of Dominant Scattering Mechanisms in Scaled InAsSb Quantum Well 027
A. Agrawal[1], A. Ali[1], R. Misra[2], P. E. Schiffer[2], B. R. Bennett[3] , J. B. Boos[3] and S. Datta[1], [1]Department of Electrical Engineering, The Pennsylvania State University, University Park, Pennsylvania, USA [2]Department of Physics, The Pennsylvania State University, University Park, Pennsylvania, USA, and [3]Naval Research Laboratory, Washington, District of Columbia, USA

Session II.B. Graphene 29-44

II.B-1
1:30 PM
Integration of High Quality Top-Gated Graphene Field Effect Devices on 150 mm Substrate 031
J. Heo, H.-J. Chung, S.-H. Lee, H. Yang, J. Shin, U-I. Chung, S. Seo, Semiconductor Device Lab, Samsung Advanced Institute of Technology, Yongin-si, KOREA

II.B-2
1:50 PM
Complementary-Type Graphene Inverters Operating at Room-Temperature 033
H.-Y. Chen and J. Appenzeller, ECE Department and Birck Nanotechnology Center, Purdue University, West Lafayette, Indiana, USA

II.B-3
2:10 PM
Gate capacitance scaling and graphene field-effect transistors with ultra-thin top-gate dielectrics 035
B. Fallahazad, K. Lee, S. Kim, C. Corbet, E. Tutuc, Microelectronics Research Center, University of Texas at Austin, Austin, Texas, USA

II.B-4
2:30 PM
Synthesis and Applications of Graphene for Flexible Electronics 037
B. H. Hong, Department of Chemistry & SKKU Advanced Institute of Nanotechnology, Sungkyungkwan University, Suwon, KOREA

II.B-5
3:30 PM
Sub-10 nm Epitaxial Graphene Nanoribbon FETs 039
K. Tahy[1], W. S. Hwang[1], J. L. Tedesco[2], R .L. Myers-Ward[2], P.M. Campbell[2], C. R. Eddy[2] Jr., D. K. Gaskill[2], H. Xing[1], A. Seabaugh[1], and D. Jena[1], [1]Department of Electrical Engineering, University of Notre Dame, Notre Dame, Indiana, USA and [2]U.S. Naval Research Laboratory, Washington, District of Columbia, USA

II.B-6
3:50 PM
Effect of Oxide Thickness Scaling on Self-Heating in Graphene Transistors 041
S. Islam, M.-H. Bae, V. Dorgan, and E. Pop, Dept. of Electrical and Computer Engineering, University of Illinois, Urbana-Champaign, Illinois, USA and Micro and Nanotechnology Lab, Urbana, Illinois, USA

II.B-7
3:50 PM
Graphene Quantum Capacitance Varactors for Wireless Sensing Applications 043
S. J. Koester, University of Minnesota-Twin Cities, Minneapolis, Minnesota, USA

Session III. Poster Session 45-152

III-1
Effects of Heavily Doped Source on the Subthreshold Characteristics of Nanowire Tunneling Transistors 051
M. A. Khayer and R. K. Lake, Laboratory for Terascale and Terahertz Electronics (LATTE), Department of Electrical Engineering, University of California, Riverside, California, USA

III-2
Spatially resolved photovoltaic performance of axial GaAs nanowire pn-diodes 053
A. Lysov, C. Gutsche, M. Offer, I. Regolin, W. Prost, F.-J. Tegude, Center for Nanointegration Duisburg-Essen, University of Duisburg-Essen, Duisburg, GERMANY

III-3
Low loss AlInN/GaN Monolithic Microwave Integrated Circuit Switch 055
A. Sattu[1], D. Billingsley[1], J. Deng[1], J.Yang[1], R. Gaska[1], M. Shur[2], G. Simin[3], [1]Sensor Electronic Technology, Inc., Columbia, South Carolina, USA, [2]Electrical and Computer Science Engineering, Rensselaer Polytechnic Institute, Troy, New York, USA, and [3]Electrical Engineering, University of South Carolina, Columbia, South Carolina, USA

III-4
High-Mobility Organic Thin-Film Transistors with Photolithographically Patterned Top Contacts 057
U. Zschieschang[1], N. H. Hansen[2], J. Pflaum[2], T. Yamamoto[3], K. Takimiya[3], H. Kuwabara[4], M. Ikeda[4], T. Sekitani[5], T. Someya[5], and H. Klauk[1], [1]Max Planck Institute for Solid State Research, Stuttgart, GERMANY, [2]University Würzburg and ZAE Bayer e.V., GERMANY, [3]Hiroshima University, Higashi-Hiroshima, JAPAN, [4]Nippon Kayaku Co., Ltd., Kita-ku, Tokyo, JAPAN, and [5]University of Tokyo, Tokyo, JAPAN

III-5   **High Breakdown Voltage ZnMgO/In-Ga-Zn-O Heterostructure Transistors** 059
J. Yamaguchi, I. Soga, and T. Iwai, Fujitsu Laboratories Ltd., Atsugi, Kanagawa, JAPAN

III-6   **High-Resolution Temperature Sensing with Source-Gated Transistors** 061
R. A. Sporea, J. M. Shannon, and S. R. P. Silva, Advanced Technology Institute, FEPS, University of Surrey, Guildford, Surrey, UNITED KINGDOM

III-7   **Voltage-Controlled Spin-Wave-Based Logic Gate** 063
T. Liu and G. Vignale, Department of Physics and Astronomy, University of Missouri, Columbia, Missouri, USA

III-8   **Effect of Disorder on Superfluidity in Double Layer Graphene** 065
B. Dellabetta and M. J. Gilbert, Department of Electrical and Computer Engineering and Micro and Nanotechnology Laboratory, University of Illinois, Urbana, Illinois, USA

III-9   **Dual Pillar Spin Transfer Torque MRAM with tilted magnetic anisotropy for fast and error-free switching and near-disturb-free read operations** 067
N. N. Mojumder, S. K. Gupta and K. Roy, School of Electrical and Computer Engineering, Purdue University, West Lafayette, Indiana, USA

III-10   **Giant Magnetoelectric Effect in Nanofabricated $Pb(Zr_{0.52}Ti_{0.48})O_3$-$Fe_{85}B_5Si_{10}$ Cantilevers and Resonant Gate Transistors** 069
F. Li[1], Z. Fang[1], R. Misra[3], S. Tadigadapa[1], Q. Zhang[1,2] and Suman Datta[1], [1]Electrical Engineering, [2]Materials Research Institute, and [3]Physics Department, The Pennsylvania State University, University Park, Pennsylvania, USA

III-11   **Observation of Trap-Assisted Steep Sub-threshold Swing in Schottky Source/Drain Al_2O_3/InAlN/GaN MISHEMT** 071
Q. Zhou[1], H. Chen[1], C. Zhou[1], Z. Feng[2], S. Cai[2], K. J. Chen[1], [1]Department of Electronic and Computer Engineering, Hong Kong University of Science and Technology, Kowloon, Hong Kong and [2]National Key Laboratory of Application Specific Integrated Circuit, Hebei Semiconductor Research Institute, Shijiazhuan, CHINA

III-12   **Towards Electronics at 1000 °C** 073
D. Maier[1], M. Alomari[1], N. Grandjean[2], J.-F. Carlin[2], M.-A. Diforte-Poisson[3], C. Dua[3], S. L. Delage[3], and E. Kohn[1], [1]Institute of Electron Devices and Circuits, University of Ulm, Ulm, GERMANY, [2]École polytechnique fédérale de Lausanne, Lausanne, SWITZERLAND, and [3]III/V Lab, Marcoussis, FRANCE

III-13   **Bias Temperature Stress Analysis of ZnO Thin Film Transistors with HfO_2 Gate Dielectrics** 075
J. J. Siddiqui[1], J. D. Phillips[1], K. Leedy[2], and B. Bayraktaroglu[2], [1]EECS Department, University of Michigan, Ann Arbor, Michigan, USA and [2]Air Force Research Laboratory, Sensors Directorate, Wright-Patterson AFB, Ohio, USA

III-14   **Carbon Nanotube Purified Ink-Based Printed Thin Film Transistors: Novel Approach in Controlling the Electrical Performance** 077
N. Rouhi, D. Jain, and P. J. Burke, Electrical Engineering and Computer Science Department, University of California-Irvine, Irvine, California, USA

III-15   **Monolayer MoS_2 Transistors – Ballistic Performance Limit Analysis** 079
K. Ganapathi, Y. Yoon, and S. Salahuddin, Department of Electrical Engineering and Computer Sciences, University of California, Berkeley, California, USA

III-16   **RF performance projections for 2D Graphene Transistors: Role of Parasitics at the Ballistic transport limit** 081
P. Zhao[1], D. Jena[1], and, S. O. Koswatta[2], [1]Department of Electrical Engineering, University of Notre Dame, Notre Dame, Indiana, USA and [2]IBM Research Division, T. J. Watson Research Center, Yorktown Heights, New York, USA

III-17   **High Performance N- and P-Type Gate-All-Around Nanowire MOSFETs Fabricated on Bulk Si by CMOS-Compatible Process** 083
Y. Song, H. Zhou, Q. Xu, J. Luo, C. Zhao and Q, Liang, Key Laboratory of Microelectronics Devices & Integrated Technology, Institute of Microelectronics, Chinese Academy of Sciences, Beijing, CHINA

III-18   **The effect of field effect device channel dimensions on the effective mobility of graphene** 085
A. Venugopal[1], J. Chan[1], W. P. Kirk[1], L. Colombo[2] and E. M. Vogel[1], [1]University of Texas at Dallas, and [2]Texas Instruments Incorporated, Dallas, Texas, USA

III-19   **Top-gated single-electron transistor in germanium nanowires** 087
S.-K. Shin, S. Huang, N. Fukata, and K. Ishibashi, Advanced Device Laboratory, RIKEN, Wako, Saitama, JAPAN

III-20   **Reliability of Ambipolar Switching Poly-Si Diodes for Cross-Point Memory Applications** 089
M. H. Lee[1], C.-Y. Kao[1], C.-L. Yang[1], Y.-S. Chen[2], H. Y. Lee[2], F. Chen[2], and M.-J. Tsai[2] [1]Institute of Electro-Optical Science and Technology, National Taiwan Normal University, Taipei, TAIWAN and [2]Electronics and Optoelectronics Research Laboratories, Industrial Technology Research Institute, Hsinchu, TAIWAN

III-21   **Electrochemical supercapacitor based on flexible pillar graphene nanostructures** 091
J. Lin[1], J. Zhong[1], D. Bao[2], J. Reiber-kyle[3], W. Wang[4], V. Vullev[2], M. Ozkan[3], C. S. Ozkan[1,4], [1]Department of Mechanical Engineering, [2]Department of Bioengineering, [3]Department of Electrical Engineering, and [4]Materials Science and Engineering Program, University of California, Riverside, California, USA

III-22   **"Zero" Drain-Current Drift of Inversion-Mode NMOSFET on InP (111)A Surface** 093
C. Wang, M. Xu, R. Colby, E. A. Stach and P. D. Ye, School of Electrical and Computer Engineering and Birck Nanotechnology Center,Purdue University, West Lafayette, Indiana, USA

III-23   **Improvement of efficiency in inverted bottom-emission white OLEDs by doping the hole transport layer** 095
H. Lee[1], J. Kwak[2], J. Lim[3], K. Char[3], S. Lee[4] and C. Lee[1], [1]School of Electrical Engineering and Computer Science, Inter-University Semiconductor Research Center, Seoul National University, Seoul, KOREA, [2]Department of Electronics Engineering, Dong-A University, Busan, KOREA, [3]School of Chemical and Biological Engineering, Intelligent Hybrids Research Center, Seoul National University, Seoul, KOREA, and [4]Department of Chemistry, Seoul National University, Seoul, KOREA

III-24   **Vertical Organic Field-Effect Transistor Array Fabrication Based on Laser Holography Lithography Process** 097
D. Kim and Y. Hong, Department of Electrical Engineering and Computer Science, Seoul National University, Seoul, KOREA

III-25   **Ambipolar Nano-crystalline-silicon TFTs with Submicron Dimensions and Reduced Threshold Voltage Shift** 099
A. Subramaniam, K. D. Cantley, R. A. Chapman, B. Chakrabarti, and E. M. Vogel, Department of Electrical Engineering, The University of Texas at Dallas, Richardson, Texas, USA

III-26   **Monolithically Grown $In_xGa_{1-x}As$ Nanowire on Silicon Tandem Solar Cells with High Efficiency** 101
J. C. Shin[1], K. H. Kim[2], H. Hu[2], K. J. Yu[1], J. A. Rogers[2,1], J.-M. Zuo[2], and X. Li[1,2], [1]Department of Electrical and Computer Engineering and [2]Department of Materials Science and Engineering, University of Illinois, Urbana, Illinois, USA

III-27   **3D Simulation of Electrical Characteristic Fluctuation Induced by Interface Traps at Si/high-κ Oxide Interface and Random Dopants in 16-nm-Gate CMOS Devices** 103
H.-W. Cheng, Y.-Y. Chiu, F.-H. Li, and Y. Li, Department of Electrical Engineering, National Chiao Tung University, Hsinchu, TAIWAN

III-28   **Creating dynamic nanowire devices using wrapped gates** 105
K. Storm, G. Nylund, M. Borgström, J. Wallentin, C. Fasth, C. Thelander and L. Samuelson, Solid State Physics, the Nanometer Structure Consortium, Lund University, Lund, SWEDEN

III-29   **InAlAs/InGaAs Metamorphic HEMT and MOS-HEMT with Regrown Source/Drain by MOCVD** 107
X. Zhou, Q. Li, and K. M. Lau, Department of Electronic and Computer Engineering, Hong Kong University of Science and Technology, Clear Water Bay, Kowloon, HONG KONG

III-30 **Numerical Study of Electronic Transport Through Bilayer Graphene Nanoribbons** 109
K. M. M. Habib and R. K. Lake, Department of Electrical Engineering, University of California, Riverside, California, USA

III-31 **Tunnel-FET Architecture with Improved Performance due to Enhanced Gate Modulation of the Tunneling Barrier** 111
L. De Michielis[1], L. Lattanzio[1], P. Palestri[2], L. Selmi[2], and A. M. Ionescu[1], [1] Nanoelectronic Devices Laboratory (Nanolab), Ecole Polytechnique Fédérale de Lausanne, SWITZERLAND and [2]Department of Electrical, Managerial and Mechanical Engineering, University of Udine, Udine, ITALY

III-32 **Orientation dependent complex bandstructure of $Si_{1-x}Ge_x$ alloys** 113
A. Ajoy[1], K. V. R. M. Murali[2], S. Karmalkar[1], and S.E. Laux[3], [1]Indian Institute of Technology Madras, INDIA, [2]IBM SRDC, Bangalore, INDIA, and [3]IBM SRDC, T. J. Watson Center, Yorktown Heights, New York, USA

III-33 **Towards Planar GaAs Nanowire Array High Electron Mobility Transistor** 115
X. Miao, and X. Li, Department of Electrical and Computer Engineering, Micro and Nanotechnology Laboratory, University of Illinois at Urbana-Champaign, Urbana, Illinois, USA

III-34 **Interface States at high-κ/InGaAs interface: H_2O vs. O_3 based ALD Dielectric** 117
H. Madan[1,2], D. Veksler[1], Y.T. Chen[1,3], J. Huang[1], N. Goel[1], G. Bersuker[1] and S. Datta[2], [1]SEMATECH, Albany, New York, USA, [2]The Pennsylvania State University, University Park, Pennsylvania, USA, and [3]University of Texas at Austin, Texas, USA

III-35 **C-V Measurements of Single Vertical Nanowire Capacitors** 119
P. Mensch, K. E. Moselund, S. Karg, E. Lörtscher, M. T. Björk, H. Schmid and H. Riel, IBM Research – Zurich, Rüschlikon, SWITZERLAND

III-36 **Barrier Height, Interface Charge & Tunneling Effective Mass in ALD Al_2O_3/AlN/GaN HEMTs;** 121
S. Ganguly, J. Verma, G. Li, T. Zimmermann, H. Xing, D. Jena, University of Notre Dame, Department of Electrical Engineering, Notre Dame, Indiana, USA

III-37 **Graphene Field-Effect Transistors Using Large-Area Monolayer Graphene Grown by Chemical Vapor Deposition on Co Thin Films** 123
M. E. Ramón[1], A. Gupta[1], C. Corbet[1], D. A. Ferrer[1], H.C.P. Movva[1], G. Carpenter[2], L. Colombo[3], G. Bourianoff[4], M. Doczy[4], D. Akinwande[1], E. Tutuc[1], and S.K. Banerjee[1], [1]Microelectronics Research Center, The University of Texas at Austin, Austin, Texas, USA, [2]IBM Research, Austin, Texas, USA, [3]Texas Instruments Incorporated, Dallas, Texas, USA, and [4]Intel Corporation, Austin, Texas, USA

III-38 **Modeling of Dielectric Breakdown-Induced Time-Dependent STT-MRAM Performance Degradation 125**
G. Panagopoulos, C. Augustine and K. Roy, School of ECE, Purdue University, West Lafayette, Indiana, USA

III-39 **Introduction of ALD Beryllium Oxide Gate Dielectric for III-V MOS Devices** 127
T. Akyol[1], J. H. Yum[1,5], D. A. Ferrer[1], M. Lei[2], M. Downer[2], C. W. Bielawski[3], T. W. Hudnall[4], G. Bersuker[5], J.C. Lee[1], S. K. Banerjee[1], [1]Microelectronics Research Center – Dept. of Electrical and Computer Eng., [2]Dept. of Physics, [3]Dept. of Chemistry, University of Texas at Austin, Texas, USA, [4]Texas State University at San Marcos, San Marcos, Texas, USA and [5]Sematech Austin, Texas, USA

III-40 **Transport Properties of CVD-Grown Graphene Nanoribbon Field-Effect Transistors** 129
A. S. Lyons, A. Behnam, E. K. Chow, and E. Pop, Dept. of Electrical and Computer Engineering, Micro and Nanotechnology Lab, University of Illinois, Urbana-Champaign, Illinois, USA

III-41 **Protein Nanopore-gated Bio-transistor for Membrane Ionic Current Recording** 131
T.-S. Lim, D. Jain, and P. J. Burke, Integrated Nanosystems Research Facility, Department of Electrical Engineering and Computer Science, University of California Irvine, Irvine, California, USA

III-42 **Tunnel FET-Based Pass-Transistor Logic for Ultra-Low-Power Applications** 133
S. H. Kim, Z. A. Jacobson, P. Patel, C. Hu, and T.-J. K. Liu, EECS Department, University of California, Berkeley, California, USA

III-43 **Metal/III-V Effective Barrier Height Tuning using ALD High-κ Dipoles** 135
J. Hu, K. Saraswat, and H.-S. P. Wong, Department of Electrical Engineering, Stanford University, Stanford, California, USA

III-44 **Intrinsic DC Operation and Performance Potential of 50nm Gate Length Hydrogen-terminated Diamond Field Effect Transistors** 137
D. A. J. Moran[1], O. J. L. Fox[2], H. McLelland[1], S. Russell[1], P. W. May[2], [1]The School of Engineering, The University of Glasgow, Glasgow, United Kingdom and [2]The School of Chemistry, The University of Bristol, Bristol, United Kingdom

III-45 **Improvement of f_T in InAl(Ga)N barrier HEMTs by Plasma Treatments** 139
R. Wang[1], G. Li[1], T. Fang[1], O. Laboutin[2], Y. Cao[2], J. W. Johnson[2], G. Snider[1], P. Fay[1], D. Jena[1], and H. Xing[1], [1]Department of Electrical Engineering, University of Notre Dame, Notre Dame, Indiana, USA and [2]Kopin Corporation, Tauton, Massachusetts, USA

III-46 **Scaling behavior and velocity enhancement in Self-aligned N-polar GaN/AlGaN HEMTs with maximum f_T of 163 GHz** 141
N. Nidhi[1], S. Dasgupta[1], D. F. Brown[2], J. S. Speck[1] and U. K. Mishra[1], [1]ECE Department, University of California Santa Barbara, Santa Barbara, California, USA and [2]Hughes Research Laboratories, Malibu, California, USA

III-47 **Fermi-level Pinning at Metal/Antimonides Interface and Demonstration of Antimonides-based Metal S/D Schottky pMOSFETs** 143
Z. Yuan[1], A. Nainani[1], J.-Y. Lin[1], B. R. Bennett[2], J. B. Boos[2], M. G. Ancona[2] and K. C. Saraswat[1], [1]Dept. of Electrical Engineering, Stanford University, California, USA and [2]Naval Research Laboratory, Washington, District of Columbia, USA

III-48 **Uniaxially Tensile Strained Accumulation-Mode Gate-All-Around Si Nanowire nMOSFETs** 145
M. Najmzadeh, D. Bouvet, W. Grabinski, and A. M. Ionescu, Nanoelectronic devices lab., Swiss Federal Institute of Technology (EPFL), Lausanne, SWITZERLAND

III-49 **Spintronics Search Engines** 147
H. Dery[1], H. Wu[1], B. Ciftcioglu[1], M. Huang[1], Y. Song[2], R. Kawakami[3], J. Shi[3], I. Krivorotov[4], I. Zutic[5], and L. J. Sham[6], [1]Department of Electrical and Computer Engineering, University of Rochester, Rochester, New York, USA, [2]Department of Physics, University of Rochester, Rochester, New York, USA, [3]Department of Physics, University of California Riverside, Riverside, California, USA, [4]Department of Physics, University of California Irvine, Irvine, California, USA, [5]Department of Physics, State University of New-York at Buffalo, Buffalo, New York, USA, and [6]Department of Physics, University of California San Diego, San Diego, California, USA

III-50 **Low Frequency Transconductance and Output Resistance Dispersion of Epitaxial Graphene Nanoribbon-based Field Effect Transistors** 149
G. Aroshvili[1], N. Meng[2], D. Vignaud[2], D. Pavlidis[1] and H. Happy[2], [1]Department of High Frequency Electronics, Darmstadt University of Technology, Darmstadt, GERMANY and [2]Institute of Electronics, Microelectronics and Nanotechnology, CNRS and Univ.p Lille 1, Villeneuve d'Ascq, FRANCE

III-51 **InAs/SiGe on Si Nanowire Tunneling Field Effect Transistors** 151
C. Kshirsagar and S. J. Koester, University of Minnesota-Twin Cities, Minneapolis, Minnesota, USA

Session IV.A. Spin/Memory 153-174

IV.A-1
8:20 **Electrical measurement of the spin Hall effects in Fe/In$_x$Ga$_{1-x}$As heterostructures** 155
E. S. Garlid[1], Q. O. Hu[2], C. Geppert[1], M. K. Chan[1], C. J. Palmstrøm[2,3], and P. A. Crowell[1], [1]School of Physics and Astronomy, University of Minnesota, Minneapolis, Minnesota, USA, [2]Dept. of Electrical and Computer Engineering, University of California, Santa Barbara, California, USA, and [3]Dept. of Materials, University of California, Santa Barbara, California, USA

IV.A-2
9:00 **Simultaneous Spin and Charge Transport in Gated Si Devices** 159
J. Li and I. Appelbaum, Center for Nanophysics and Advanced Materials and Department of Physics, University of Maryland, College Park, Maryland, USA

IV.A-3
9:20 AM
Unidirectional information transfer with cascaded All Spin Logic devices: A Ring Oscillator 161
S. Srinivasan[1,2], A. Sarkar[1,2], B. Behin-Aien[1,2], and S. Datta[1,2], [1]School of Electrical and Computer Engineering, Purdue University, W. Lafayette, Indiana, USA and [2]NSF Network for Computational Nanotechnology (NCN), W. Lafayette, Indiana, USA

IV.A-4
9:40 AM
Proposal for piezoelectric-ferromagnet bilayer based microwave Oscillators without any external magnetic field or spin transfer torque 163
D. Bhowmik and S. Salahuddin, Department of Electrical Engineering and Computer Sciences, University of California, Berkeley California, USA

IV.A-5
10:20 AM
Orthogonal Spin Transfer MRAM 165
D. Bedau[1], D. Backes[1], H. Liu[1], J. Langer[2], P. Manandhar[3] and A. D. Kent[1], [1]Department of Physics, New York University, New York, New York, USA, [2]Singulus Technologies AG, Kahl am Main, GERMANY, and [3]Spin-Transfer Technologies, Quincy, Massachusetts, USA

IV.A-6
10:40 AM
Thermal Effects and Instability in Unipolar Resistive Switching Devices 167
A. Chen and M.-R. Lin, Strategic Technology Group, GLOBAL FOUNDRIES, Sunnyvale, California, USA

IV.A-7
11:00 AM
A Hybrid Ferroelectric and Charge Nonvolatile Memory 169
S. R. Rajwade, K. Auluck, J. Shaw, K. Lyon and E. C. Kan, School of Electrical and Computer Engineering, Cornell University, Ithaca, New York, USA

IV.A-8
11:20 AM
Spin-torque switchable perpendicular magnetic junctions for solid-state memory 171
J. Z. Sun[1,2], R. P. Robertazzi[1], J. J. Nowak[1], P. L. Trouilloud[1], G. Hu[1], M. C. Gaidis[1], S. L. Brown[1], D. W. Abraham[1], E. J. O'Sullivan[1], W. J. Gallagher[1], D. C. Worledge[1], and A. D. Kent[2], [1]IBM-MagIC MRAM Development Alliance, IBM T. J. Watson Research Center, Yorktown Heights, New York, USA and [2]Dept of Physics, New York University, New York, USA

Session IV.B. Alternative Transistor Concepts   175-190

IV.B-1
8:20 AM
Combinational and Sequential Logic with Transistors based on Individual Carbon Nanotubes 177
H. Ryu[1], D. Kälblein[1], U. Zschieschang[1], O. G. Schmidt[2], and H. Klauk[1], [1]Max Planck Institute for Solid State Research, Stuttgart, GERMANY and [2]Faculty of Electrical Engineering and Information Technology, Chemnitz University of Technology, GERMANY

IV.B-2
8:40 AM
Comparative Study of Fabricated Junctionless and Inversion-mode Nanowire FETs 179
C.-H. Park[1], M.-D. Ko[1], K.-H. Kim[1], C.-W. Sohn[1], C. K. Baek[1], Y.-H. Jeong[1,2], and J.-S. Lee[1,2], [1]Dept. of Electronic and Electrical Engineering, and [2]Division of IT-Convergence Engineering, POSTECH, Pohang, Gyeongbuk, KOREA

IV.B-3
9:00 AM
Fabrication of Vertical InAs-Si Heterojunction Tunnel Field Effect Transistors H. Schmid, K. E. 181
Moselund, M. T. Björk, M. Richter, H. Ghoneim, C. D. Bessire and H. Riel, IBM Research – Zurich, Rüschlikon, SWITZERLAND

IV.B-4
9:20 AM
Challenges for Post-CMOS Devices & Architecture 183
J. Welser[1,2] and K. Bernstein[2], [1]SRC-NRI,Durham, North Carolina, USA and [2]IBM Research, Yorktown Heights, New York & San Jose, California, USA

IV.B-5
10:20 AM
Correlated Oxide Phase Transition Switch: A Paradigm in Electron Devices 187
Z. Yang, C. Ko, V. Balakrishnan, and S. Ramanathan Harvard School of Engineering and Applied Sciences, Harvard University, Cambridge, Massachusetts, USA

IV.B-6
11:00 AM
Lateral Gate Suspended-Body Carbon Nanotube Field-Effect-Transistors with Sub-100nm Air Gap by Precise Positioning Method 189
J. Cao and A. M. Ionescu, Nanoelectronic Devices Laboratory (Nanolab), Ecole Polytechnique Fédérale de Lausanne, Lausanne, SWITZERLAND

Session V.A   Tunnel FETs   191-208

V.A-1
1:30 PM
Si-based Tunnel Field-Effect Transistors for Low-Power Nano-Electronics 193
A. S. Verhulst[1,] W. G. Vandenbergh[1,2], D. Leonelli[1,2], R. Rooyackers[1], A. Vandooren[1], J. Zhuge[4], K-H. Kao, B. Sorée[1,5], W. Magnus[1,5], M. V. Fischetti[6], G. Pourtois[1], C. Huyghebaert[1], R. Huang[4], Y. Wang[4], K. De Meyer[1], W. Dehaene[1,2], M. M. Heyns[1,3], and G. Groeseneken[1,2], [1]imec, Leuven, BELGIUM, [2]Department of Electrical Engineering, [3]Department of Metallurgy and Materials Engineering, K.U.Leuven, Leuven, BELGIUM, [4]Institute of Microelectronics, Peking University, Beijing, CHINA, [5]Department of Physics, Universiteit Antwerpen, , Wilrijk, BELGIUM; [6]Department of Materials Science and Engineering, University of Texas Dallas, Richardson, Texas, USA

V.A-2
2:10 PM
Compact Model and Performance Estimation for Tunneling Nanowire FET 197
P. M. Solomon, D. J. Frank, and S.O. Koswatta, IBM, SRDC, T.J. Watson Research. Center, Yorktown Heights, New York, USA

V.A-3
2:30 PM
Using Dimensionality to Achieve a Sharp Tunneling FET (TFET) Turn-On 199
S. Agarwal and E. Yablonovitch, University of California, Berkeley, California, USA

V.A-4
2:50 PM
Investigation on Superlattice Heterostructures for Steep-Slope Nanowire FETs 201
E. Gnani, P. Maiorano, S. Reggiani, A. Gnudi and G. Baccarani ARCES and DEIS, University of Bologna, Bologna, ITALY

V.A-5
3:30 PM
Self-aligned Gate NanoPillar In$_{0.53}$Ga$_{0.47}$As Vertical Tunnel Transistor 203
D. K. Mohata, R. Bijesh, V. Saripalli, T. Mayer and S. Datta, The Pennsylvania State University, University Park, Pennsylvania, USA

V.A-6
3:50 PM
Self-aligned InAs/Al$_{0.45}$Ga$_{0.55}$Sb vertical tunnel FETs 205
G. Zhou[1], Y. Lu[1], R. Li[1], Q. Zhang[1], W. Hwang[1], Q. Liu[1], T. Vasen[1], H. Zhu[2], J. Kuo[2], S. Koswatta[3], T. Kosel[1], M. Wistey[1], P. Fay[1], A. Seabaugh[1], and H. G. Xing[1], [1]Department of Electrical Engineering, University of Notre Dame, Notre Dame, Indiana, USA, [2]IntelliEPI, Richardson, Texas, USA, and [3]IBM T. J. Watson Research Center, Yorktown Heights, New York, USA

V.A-7
4:10 PM
P-type Tunneling FET on Si (110) Substrate with Anisotropic Effect 207
M. H. Lee[1], C.-Y. Kao[1], C.-L. Yang[1], and C.-H. Lee[2], [1]Institute of Electro-Optical Science and Technology, National Taiwan Normal University, Taipei, TAIWAN and [2]Graduate Institute of Electronics Engineering (GIEE) and Department of Electrical Engineering, National Taiwan University, Taipei, TAIWAN

V.A-8
4:30 PM
Late News

V.A-9
4:50 PM
Late News

Session V.B   Wide Bandgap   209-224

V.B-1
1:30 PM
Anomalous output conductance in N-polar GaN-based MIS-HEMTs 211
M. H. Wong[1], U. Singisetti[1], J. Lu[1], J. S. Speck[2], and U. K. Mishra[1], [1]Electrical and Computer Engineering and [2]Materials Departments University of California, Santa Barbara, Californa, USA

V.B-2
1:50 PM
Normally-off Gate-Recessed AlGaN/GaN-on-Si Hybrid MOS-HFET with Al$_2$O$_3$ Gate Dielectric 213
A. L. Corrion, M. Chen, R. Chu, S. D. Burnham, S. Khalil, D. Zehnder, B. Hughes, and K. Boutros, HRL Laboratories LLC, Malibu, California, USA

V.B-3	**N-Polar AlGaN/GaN MIS-HEMTs on SiC with a 16.7 W/mm Power Density at 10 GHz Using an Al$_2$O$_3$**
2:10 PM	**Based Etch Stop Technology for the Gate Recess 215**

S. Kolluri, S. Keller, S. P. DenBaars and U. K. Mishra, Department of ECE, University of California, Santa Barbara, California, USA

V.B-4	**Total GaN Solution for Electircal Power Conversion 217**
2:30 PM	Y.-F. Wu, R. Coffie, N. Fichtenbaum, Y. Dora, C.S. Suh, L. Shen, P. Parikh and U.K. Mishra, Transphorm Inc., Goleta, California, USA

V.B-5	**First AlN/GaN HEMTs power measurement at 18 GHz on Silicon substrate 219**
3:30 PM	F. Medjdoub, M. Zegaoui, D. Ducatteau, N. Rolland and P.A. Rolland, IEMN, Villeneuve d'Ascq, FRANCE

V.B-6	**Enhanced mobility for MOCVD grown AlGaN/GaN HEMTS on Si substrate 221**
3:50 PM	S. L. Selvaraj, A. Watanabe and T. Egawa, Research Center for Nano-Device and System, Nagoya Institute of Technology, Gokiso-cho, Showa-ku, Nagoya, JAPAN

V.B-7	**High Performance GaN-on-Si Power Switch: Role of Substrate Bias in Device Characteristics 223**
4:10 PM	R. Chu, D. Zehnder, B. Hughes, and K. Boutros, HRL Laboratories LLC, Malibu, California, USA

Rump Sessions   225-226

R.1	**Large Area OFETs: Organic or Oxide? 225**
8:30 PM	Session Organizers: Ioannis Kymissis, Columbia University and Yongtaek Hong, Seoul National University

R.2	**What is the Ultimate Low Power Device? 225**
8:30 PM	Session Organizers: Kirsten Moselund, IBM, Siyuranga Koswatta, IBM, and Erik Lind, Lund University

R.3	**Graphene--What is it good for? 226**
8:30 PM	Session Organizers: Eric Pop, University of Illinois Urbana-Champaign, Dimitris Pavlidis, Technical University Darmstad, and Jeong Moon, HRL

Joint DRC/EMC Plenary Session   227-228

8:30 AM	**New Concepts and Materials for Solar Power Conversion 227**
	Wladyslaw Walukiewicz, Lawrence Berkeley National Laboratory

Session VI.A pFETs   229-238

VI.A-1	**High Mobility Strained P-Channel Germanium Quantum Well Field Effect Transistor for Low Power**
10:00 AM	**(V_{cc} = 0.5 V) III-V CMOS Applications 231**

R. Pillarisetty, Intel Corporation, Components Research, Technology and Manufacturing Group, Hillsboro, Oregon, USA

VI.A-2	**Performance enhancement of GaAs UTB pFETs by strain, orientation and body thickness**
10:40 AM	**engineering 233**

A. Paul[1], S. Mehrotra[1], G. Klimeck[1] and Mark Rodwell[2], [1]ECE Department and NCN, Purdue University, West Lafayette, Indiana, USA and [2]ECE Department, University of California, Santa Barbara, California, USA

VI.A-3
11:00 AM
Highly-Strained SGOI p-Channel MOSFETs Fabricated by Applying Ge Condensation Technique to Strained-SOI Substrates 235
J. Suh, R. Nakane, N. Taoka, M. Takenaka and S. Takagi, School of Engineering, The University of Tokyo, Tokyo, JAPAN

VI.A-4
11:20 AM
Hole Mobility Enhancement in Uniaxially Strained SiGe FINFETs: Analysis and Prospects 237
R. Bijesh[1], I. Ok[2], M. Baykan[2], C. Hobbs[2], P. Majhi[2], R. Jammy[2], and S. Datta[1], [1]The Pennsylvania State University, University Park, Pennsylvania, USA and [2]Sematech, Austin, Texas, USA

Session VI.B Thin- Film   239-250

VI.B-1
10:00 AM
Defect Analysis of Roll-to-Roll SAIL Manufactured Flexible Display Backplanes 241
C. Taussig[1], R. E. Elder[1], W. B. Jackson[1], A. Jeans[1], M. Jam, E. Holland[1,] H. Luo[1], J. Maltabes[1], C. Perlov[1], S. Trovinger[1], M. Almanza-Workman[2], R. A. Garcia[2], H. Kim[2], O. Kwon[2], and F. Jeffrey[2], [1]HP Labs, Palo Alto, California, USA and [2]Phicot Inc., Palo Alto, California, USA

VI.B-2
10:20 AM
Indium-free Transparent Thin Film Transistors Based on Nanocrystalline ZnO 245
B. Bayraktaroglu[1], K. Leedy[1] and R. C. Scott[2], [1]Air Force Research Laboratory, Sensors Directorate, AFRL/RYDD, Wright Patterson AFB, Ohio, USA and [2]Arizona State University, Tempe, Arizona, USA

VI.B-3
10:40 AM
Circuit applications based on solution-processed zinc-tin oxide TFTs 247
C.-G. Lee, T. Joshi, K. Divakar, and A. Dodabalapur, Microelectronics Research Center, University of Texas at Austin, Austin, Texas, USA

VI.B-4
11:00 AM
Aluminum Top-Gate ZnO Nanowire Transistors with Improved Transconductance 249
D. Kälblein[1], B. Fenk[1], K. Hahn[2], U. Zschieschang[1], K. Kern[1,3], H. Klauk[1], [1]Max Planck Institute for Solid State Research, Stuttgart, GERMANY, [2]Max Planck Institute for Metals Research, GERMANY, and [3]Ecole Polytechnique Fédérale de Lausanne, Lausanne, SWITZERLAND

Session VII.A Optoelectronic Devices   251-266

VII.A-1
1:30 PM
Monolithic Integration of CMOS and Nanophotonic Devices for Massively Parallel Optical Interconnects in Supercomputers 253
S. Assefa[1], W. M. J. Green[1], A. Rylyakov[1], C. Schow[1], F. Horst[2] and Y. A. Vlasov[1], [1]IBM Thomas J. Watson Research Center, Yorktown Heights, New York, USA and [2]IBM Zurich GMBH, Rueshlikon, SWITZERLAND

VII.A-2
2:10 PM
Electrical pumped integrated III/V laser lattice-matched to a Silicon substrate 257
B. Kunert[1], S. Liebich[2], M. Zimprich[2], A. Beyer[2], S. Ziegler[1], K. Volz[2], W. Stolz[2], N. Hossain[3], S. R. Jin[3], and S. J. Sweeney[3], [1]NAsP III/V GmbH, Marburg, GERMANY and [2]Material Sciences Center and Faculty of Physics, Philipps-University Marburg, Marburg, GERMANY, and [3]Advanced Technology Institute and Department of Physics, University of Surrey, Guildford, Surrey, UK

VII.A-3
2:30 PM
Lateral Carrier Injection with n-type Modulation-doped Quantum Wells in VCSELs 259
C.-H. Lin[1], Y. Zheng[1], M. Gross[3], M. J. W. Rodwell[1], and L. A. Coldren[1,2], [1]Department of Electrical and Computer Engineering, University of California, Santa Barbara, California, USA, [2]Department of Materials, University of California, Santa Barbara, California, USA, and [3]Ziva Corporation, San Diego, California, USA

xv

VII.A-4
2:50 PM
Near UV AlGaN-Cladding Free Nonpolar InGaN/GaN Laser Diodes 261
D. A. Haeger[1], C. Holder[1], R. M. Farrell[1], P. S. Hsu[1], K. M. Kelchner[2], K. Fujito[3], D. A. Cohen[2], S. P. DenBaars[1,2], J. S. Speck[1], and S. Nakamura[1,2], [1]Materials Department, University of California, Santa Barbara, California, USA, [2]Electrical and Computer Engineering Department, University of California, Santa Barbara, California, USA, and [3]Optoelectronic Laboratory, Mitsubishi Chemical Corporation, Ushiku, Ibaraki, JAPAN

VII.A-5
3:30 PM
RIE Lag Directional Coupler based Integrated InGaAsP/InP Ring Mode-locked Laser 263
J. S. Parker[1], P. R .A. Binetti[1], Y.-J. Hung[2], Erik J. Norberg[1], and L. A. Coldren[1], [1]Electrical and Computer Engineering Department, University of California, Santa Barbara, California, USA and [2]Dept. of Electronic Engineering, National Taiwan University of Science and Technology, Taipei, TAIWAN

VII.A-6
3:50 PM
Integrated Non-III-Nitride/III-Nitride Tandem Solar Cell 265
N. G. Toledo[1], S. C. Cruz[2], C. J. Neufeld[1], M. A. Scarpulla[2], T. Buehl[2], A. C. Gossard[1,2], S. P. Denbaars[1,2], J. S. Speck[2] and U. K. Mishra[1], [1]Department of Electrical and Computer Engineering, University of California, Santa Barbara, and [2]Materials Department, University of California, Santa Barbara, California, USA

Session VII.B High Speed Devices   267-282

VII.B-1
1:30 PM
N-polar GaN HEMTs with f_{max} > 300 GHz using high-aspect-ratio T-gate design 269
D.J. Denninghoff[1], S. Dasgupta[1], D.F. Brown[3], S. Keller[1], J. Speck[2], and U.K. Mishra[1], [1]Department of Electrical and Computer Engineering, University of California, Santa Barbara, California, USA, [2]Department of Materials, University of California, Santa Barbara, California, USA, and [3]HRL, Malibu, California, USA

VII.B-2
1:50 PM
1.0 THz f_{max} InP DHBTs in a refractory emitter and self-aligned base process for reduced base access resistance 271
V. Jain[1], J. C. Rode[1], H.-W. Chiang[1], A. Baraskar[1], E. Lobisser[1], B. J. Thibeault[1], .M. Rodwell[1], M. Urteaga[2], D. Loubychev[3], A. Snyder[3], Y. Wu[3], J. M. Fastenau[3], W. K. Liu[3], [1]ECE Department, University of California, Santa Barbara, California, USA, [2]Teledyne Scientific & Imaging, Thousand Oaks, California, USA, and [3]IQE Inc., Bethlehem, Pennsylvania, USA

VII.B-3
2:10 PM
Effect of optical phonon scattering on the performance limits of ultrafast GaN transistors 273
T. Fang, R. Wang, G. Li, H. Xing, S. Rajan, and D. Jena, Electrical Engineering, University of Notre Dame, Notre Dame, Indiana, USA

VII.B-4
2:30 PM
Device Scaling Technologies for Ultra-High-Speed GaN-HEMTs 275
K. Shinohara[1], D. Regan[1], I. Milosavljevic[1], A. L. Corrion[1], D. F. Brown[1], S. Burnham[1], P. J. Willadsen[1], C. Butler[1], A. Schmitz[1], S. Kim[1], V. Lee[2], A. Ohoka[2], P. M. Asbeck[2,] and M. Micovic[1], [1]HRL Laboratories LLC, California, USA. and [2]Department Electrical and Computer Engineering, University of California, San Diego, California, USA

VII.B-5
3:30 PM
Trap-related Delay analysis of self-aligned N-polar GaN/InAlN HEMTs with record extrinsic g_m of 1105 mS/mm 279
Nidhi, S. Dasgupta, J. Lu, F. Wu, S. Keller, J. S. Speck and U. K. Mishra, ECE Department, University of California Santa Barbara, Santa Barbara, California, USA

VII.B-6   **130nm InP DHBTs with f_t>0.52THz and f_{max}>1.1THz 281**
3:50 PM   M. Urteaga[1], R. Pierson[1], P. Rowell[1], V. Jain[2], E. Lobisser[2], M.J.W. Rodwell[2], [1]Teledyne Scientific
           Company, Thousand Oaks, California, USA and [2]Department of ECE, University of California, Santa
           Barbara, California, USA

(Corwin Pavilion)

Plenary Session

Monday AM, June 20th, 2011

Session Chairs: Patrick Fay and Miguel Urteaga

8:30 AM Welcoming Remarks
Presentations: IEEE Fellows, Best Student Paper Awards, and IEEE EDS
Celebrated Member Award for Prof. Herbert Kroemer, UCSB

8:50 AM I.-1 Plenary Paper
Soft, Curvilinear Semiconductor Devices for Bio-Integrated Electronics
J. Rogers, Department of Materials Science and Engineering University of Illinois
Urbana/Champaign, Champaign, Illinois, USA

9:50 AM Break

10:10 AM I.-2 Plenary Paper
Devices for high performance computing beyond 14nm node- Is there anything other than Si?
W. Haensch, IBM TJ Watson Research Center, Yorktown Heights, New York, USA

11:10 AM I.-3 Plenary Paper
The latest performance of GaN-based nonpolar and semipolar emitting devices
Shuji Nakamura, Materials Department, University of California Santa Barbara, California, USA

978-1-61284-243-1/11 $26.00 © 2011 IEEE

978-1-61284-243-1/11 $26.00 © 2011 IEEE

Soft, Curvilinear Semiconductor Devices for Bio-Integrated Electronics

J. Rogers, Department of Materials Science and Engineering University of Illinois Urbana/Champaign, Champaign, Illinois, USA

Biology is curved, soft and elastic; silicon wafers are not. Semiconductor technologies that can bridge this gap in form and mechanics will create new opportunities in devices that require intimate integration with the human body. This talk describes the development of ideas for electronics that offer the performance of state-of-the-art, wafer-based systems but with the mechanical properties of a rubber band. We explain the underlying materials science and mechanics of these approaches, and illustrate their use in bio-integrated, 'tissue-like' electronics with unique capabilities in electrocorticography and cardiac electrophysiology, in both endocardial and epicardial modes. In vivo demonstrations with animal models illustrate the functionality offered by these technologies, and suggest several clinically relevant applications.

978-1-61284-243-1/11 $26.00 © 2011 IEEE

Devices for high performance computing beyond 14nm node - is there anything other than Si?

Wilfried Haensch,
IBM T. J. Watson Research Center, Yorktown Heights, NY 10598

Introduction

Devices have to live with in the magic triangle: Power, performance and density. Power is maxed out and gives a constraint in this design space. Economy will drive continued density scaling, possibly at the cost of device performance. Decreasing contacted gate pitch towards sub 50 nm will put increasing pressure on gate length and contact scaling to keep capacitive and resistive parasites in check. At this point there is no clear path to continued dielectric and junction depth scaling at the horizon that is needed to push conventional doping controlled devices beyond the 20nm gate length mark. Fully depleted devices like, FinFETs and extremely thin SOI, will present an opportunity to extend gate length scaling without stressing dielectric or junction scaling. It is expected that these new devices will be introduced for the high end computing space as early as the 14nm node, but not later than 11nm node. Low power technologies might adapt these devices earlier. Alternate device concepts or channel materials are currently investigated. They are neither demonstrated at relevant dimensions nor are they proven to deliver what they promise at this point in time.

Power performance scaling for systems and devices

It is believed that conventional CMOS technology could be operated at around 0.5V without loss of functionality and design robustness [1]. It is however clear that with this reduction of supply voltage comes a clock frequency reduction that will significantly reduce single thread performance. On the system level we find that for the existing Si high performance technologies active power P_{active} and performance f are scaling according to:

$$f = \alpha\left(V - V_0\right)$$
(Eq 1)

$$P_{active} = \alpha C_{eff} V^2 \left(V - V_0\right) + I_{leak} V$$
(Eq 2)

where C_{eff} is the total effective load capacitance of a chip, V is the operating voltage V_0 is the voltage at which frequency approaches zero (~0.25 V for modern technologies) and α is a constant that depends on the circuit and technology, and I_{leak} is the total leakage current of the chip. This same relation also applies to circuits that are optimized at other supply voltages. For example low voltage technologies generally optimize to higher threshold voltages which result in a somewhat higher V_0 (~0.3–0.4 V). It should be noted that P_{active} is the power for an operating chip that has always active and quiescent components. The question at hand is: what are the knobs to continue performance increase with a power budget that needs to be kept in check. Eq's (1) and (2) give a rough idea of the cost in performance if power is reduced. From the device perspective performance scaling can be captured by

$$I_{on} = w C_{inv} v_{sat} \left(V - V_{th}\right)$$
(Eq 3)

978-1-61284-243-1/11 $26.00 © 2011 IEEE

$$I_{off} = wI_0 10^{-\frac{V_{th}}{S}} \qquad\qquad\qquad\text{(Eq 4)}$$

Where w is the average device width, C_{inv} the inversion capacitance, v_{sat} the saturation velocity in the device, V_{th} the threshold voltage ,S the sub threshold slope, and I_0 the target current at which the threshold voltage is defined. The threshold voltage and V_0 are closely correlated. Effective chip capacitance C_{eff} is related to the device width that will also determine wire length and pitch, and is therefore directly proportional to density of integration. C_{eff} will have a front end , or device related, and an a back-end, or wiring related, contribution. Since wire delay limitations can be mitigated by the insertion of repeaters, both parts are interrelated. It is notable that from the device perspective we have the material parameters C_{inv} (~ dielectric scaling) and v_{sat} (~ transport enhancement) and the device architecture parameters V_{th} and S to work with. To meet power and performance targets on the system level the industry has seen a slow down on voltage scaling to meet the trade-off between active switching power and leakage [2], shown in Figure 2. To maintain performance the dielectric scaling was improved with the introduction of hi-k materials and strain engineering was employed to boost carrier velocity. It is however noticeable that gate length scaling has slowed down due to the fact that the basic device architecture did not change. In Figure 3 we show the possible progression of device architecture for the coming technology nodes. The transition from doping controlled devices to fully depleted is expected to

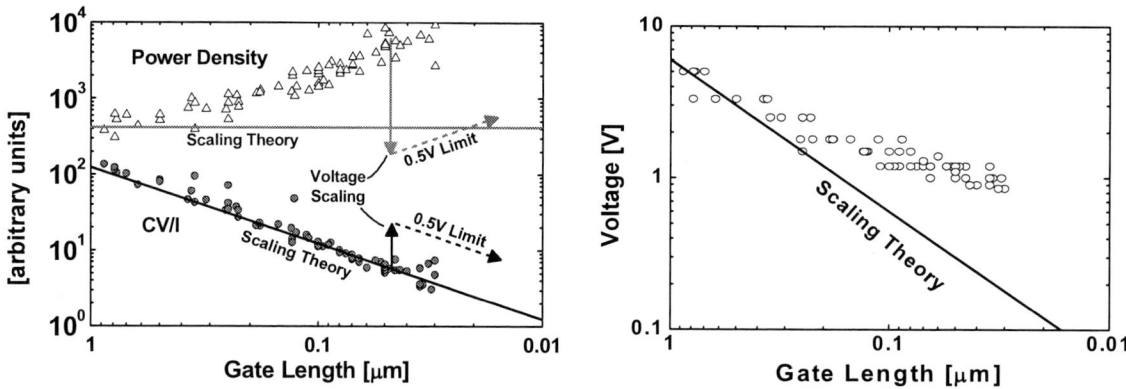

Figure 2: Voltage and performance scaling, adapted from [2]

take place earliest in the 14 nm and latest the 11 nm node technology [3] for high performance technologies. In the near future doping controlled devices will be replaced by the fully depleted body devices. In contrast to the doping controlled devices where threshold voltage and subthreshold slope are controlled by gate dielectric thickness, junction scaling, and channel doping, in the fully depleted devices they are controlled by the body thickness.

Typically the body thickness will determine the minimum supported gate length L_{min}. For the ETSOI device that is built on a very thin Si film on insulating buried oxide (BOX) we have $L_{min} \sim 4\ T_{si}$ and for the FinFET, which is a 3D device structure $L_{min} \sim 2\ D_{fin}$. The FinFET has a superior scaling behavior compared to the ETSOI device because of its double gate nature. However processing issues related to the 3D nature of the structure will make it more difficult to build. The FinFET would be able to support gate length of about 10nm, which is expected at the 8 nm technology node, at a Fin width of 5 nm. Reducing the Fin width further would result in a dramatic mobility loss due quantum confinement [4] and unacceptable threshold variation due to

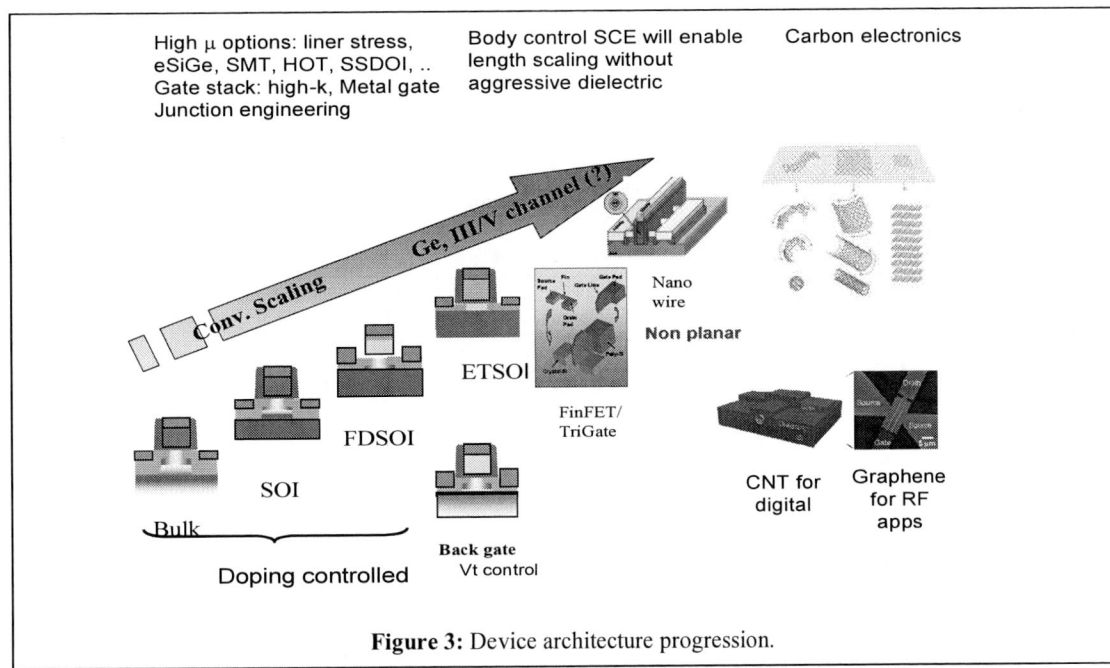

Figure 3: Device architecture progression.

Fin width tolerances in the manufacturing process. An evolution of the ETSOI device architecture on thick BOX (~ 120 nm) towards a device on thin BOX (~ Lg) is possible to enable gate length scaling at fixed body thickness Tsi. At a critical body thickness of Tsi=5 nm it can not, however, scale better than a FinFET in the extreme case of a very thin BOX. Therefore both FinFETs and ETSOI devices will not be able to be scaled below 10 nm gate length.

In the transition for 14 nm to 8 nm high mobility channel material like Ge and III/V compounds might emerge to boost device performance, similar to strain engineering in Si. However, the existing device architectures will prevent scaling beyond 10 nm gate length or even larger gate length. In particular for III/V materials quantum confinement effects are expected to dominate much earlier than in Si or Ge due to their small effective masses. Due to their light effective mass III/V channels will be more sensitive to direct S/D tunneling at shorter gate length. Figure 4 shows a material comparison of the materials [5] that are considered as possible candidates to replace or supplement Si to boost performance. Attention needs to be paid to the density of states that will provide the charge density and the two masses labeled m_{con} and m_{dens} that are indicators for mobility $\sim 1/m_{con}$ and quantum confinement effects $\sim 1/m_{dens}$, respectively. From this table we see the promise of III/V materials as having very high mobility, however, at the cost of a very low density of states. At this point of time the verdict is still out if these two opposing trends will result in a net gain for a properly scaled device.

The best possible scaled device that can be built with conventional semiconducting materials has wire geometry, as indicated in the evolution sequence in Figure 3. In principle this device would allow to push gate length beyond 10 nm. For wires a scaling behavior of $L_{min} \sim 1.5 * d_{wire}$ is expected. For the 5 nm technology node a gate length of 8 nm is required which would require a wire diameter of about 5 nm. III/V materials will not be feasible at this wire dimension because of the large quantum confinement effects due to their low effective mass. Variability impact on device parameters would be unacceptable. Si devices and circuits [6][7] have been demonstrated, however, it is not clear if they can deliver the needed performance at the desired integration densities. The net is that with the existing known materials and architectures there is no solution to build a high performance device beyond the 8 nm technology node.

978-1-61284-243-1/11 $26.00 © 2011 IEEE

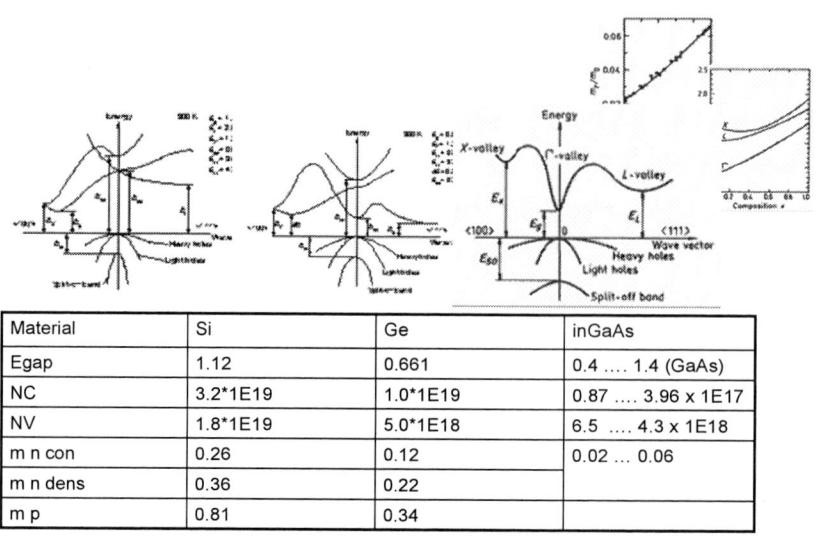

Material	Si	Ge	inGaAs
Egap	1.12	0.661	0.4 .... 1.4 (GaAs)
NC	3.2*1E19	1.0*1E19	0.87 .... 3.96 x 1E17
NV	1.8*1E19	5.0*1E18	6.5 .... 4.3 x 1E18
m n con	0.26	0.12	0.02 ... 0.06
m n dens	0.36	0.22	
m p	0.81	0.34	

Figure 4: Alternative channel materials

CNTs combine good transport properties and excellent electrostatic scaling [8]. Therefore they offer a device solution that needs to be ready for the 5 nm node time frame. Other more exotic device concepts are at most in early exploration [9] and not expected to be of commercial use in the next ten years.

Summary

A scaling path for Si based technology seems possible to the 8nm node. Power limitation will force to reduce the supply voltage at the expense of device performance and susceptibility to process variations. A lower limit of Vdd=0.5V seems feasible. Parallelism on system level will provide system through put which stresses architecture and software development. In particular legacy code will be a problem for a transition period that might require dual supply multi core architectures. In this scenario device technology has to cater to both high voltage and low voltage operation. Beyond the 8nm node new device concepts are needed. Considering the time frame of a 2019 manufacturing for this node a device has to be demonstrated now. At this point CNTs seem to be the only viable option for the post Si area.

[1] L. Chang, et al., "Practical Strategies for Power-Efficient Computing Technologies," submitted to Proc. IEEE, 2009.

[2] E. J. Nowak, "Maintaining the benefits of CMOS scaling when scaling bogs down," *IBM J. Res. Dev.*, vol. 46, pp. 169-180, Mar/May 2002.

[3] W. Haensch et al. ,"Silicon Scaling and Beyond", *IBM J. Res. Dev.*, vol. 50, pp. 339-361, July 2006.

[4] D. Esseni, M. Mastrapasqua, G. K. Celler, C. Fiegna L. Selmi, E. Sangiorgi, " Low field electron and hole mobility of SOI transistors fabricated on ultrathin silicon films for deep submicrometer technology application, " Transactions on Electron Devices 48 (12) pp 2842-2850 Dec 2001

[5] NSM Archive, Physical Properties of Semiconductors: http://www.ioffe.rssi.ru/SVA/ NSM/Semicond

[6] S. Bangsaruntip et al. , " Gate-all-around silicon nanowire 25-stage CMOS ring oscillators with diameter down to 3 nm", 2010 Symposium on VLSI Technology, Honolulu, June 15-17, 2010 pp 21-22

[7] H. Yan et al. ," Programmable nanowire circuits for nanoprocessors" , Nature 470 (7333), February 10, 2011

[8] [8]A.D. Franklin and Z. Chen, "Length scaling of carbon nanotube transistors," *Nature Nanotechnology*, vol. 5, Nov. 2010, pp. 858-862.

[9] K. Bernstein et al. , "Device and Architecture Outlook for Beyond CMOS Switches," Proceedings of the IEEE 98 (12) , pp 2169-2184 2010

The Latest Performance of GaN-Based Nonpolar and Semipolar Emitting Devices

Shuji Nakamura

Material Department

University of California Santa Barbara, CA

Phone: 805-893-5552, e-mail: shuji@engineering.ucsb.edu

All of conventional nitride-based light emitting diodes (LEDs) and laser diodes (LDs) available in a market have been grown on polar (C-plain) plain GaN. Nonpolar and semipolar plain GaN are getting popular recently to improve the performance of those devices further. In the case of LDs, the nonpolar and semipolar plains have a several advantages. Non-isotropic strain changes the density states of the valence band and, then increases the optical gain. There are no quantum confined Stark effects (QCSE) in nonpolar plains, which reduces the radiative recombination rate. We have recently confirmed that material gain of nonpolar plain is three times higher than that of polar plain. As a result, we succeeded in making nonpolar blue LDs and semipolar green LDs using those advantages.

Here, the latest performance of nonpolar/semipolar LEDs and LDs is described. We fabricated LEDs from violet to yellow color region. Conventionally there have been no efficient green LEDs and, as a result no green LDs, which are called as a green gap. The semipolar (20-21) plain was good in green region of LEDs and LDs to fill out the green gap. The performance of semipolar green LEDs was relatively good in comparison with commercially available polar green LEDs. Thus, we succeeded in fabricating green LDs with an emission wavelength of 516 nm at UCSB. Other groups such as Sumitomo Electric Inc. and Soraa Inc. also succeeded in fabricating green LDs with an emission wavelength of 520-530 nm with an output power of about 50 mW using non-polar plains. These promising results indicate a big advantage of nonpolar/semipolar plains to make higher efficient LDs from UV to red color range.

Next target is higher efficient nonpolr/semipolar LEDs. We have not confirmed the higher efficiency of nonpolar and semipolar LEDs from UV to red color yet except for green and yellow regions in comparison with that of polar LEDs. Due to a lack of QCSE, we could expect a higher efficiency of nonpolar/semipolar blue LEDs in comparison with that of blue polar LEDs. It would be a matter of optimization and time to achieve the higher efficiency.

One key issue of nonpolar/semipolar devices is a substrate. The size and price of the substrate is the biggest issue at this moment. We have worked for ammonothermal crystal growth of bulk GaN to overcome the problems of the substrate issue. The latest progress of the ammonothermal growth of bulk GaN is also described.

978-1-61284-243-1/11 $26.00 © 2011 IEEE

978-1-61284-243-1/11 $26.00 © 2011 IEEE

Session II.A (Corwin Pavilion East)

III-V FETs

Monday PM, June 20th, 2011

Session Chairs: Dae-Hyun Kim, Teledyne Scientific Company and Vijay Narayanan, IBM Corporation

1:30 PM II.A-1 Invited Paper
Ultra-thin Compound Semiconductors on Insulator (XOI) for MOSFETS and TFETS
R. Kapadia[1,2,3], K. Takei[1,2,3], A. C. Ford[1,2,3], H. Fang[1,2,3], S. Chuang[1,2,3], M. Madsen[1,2,3], S. Krishna[4,] and A. Javey[1,2,3], [1]Electrical Engineering and Computer Sciences, University of California, Berkeley, California, [2]Materials Sciences Division, Lawrence Berkeley National Laboratory, Berkeley, California, [3]Berkeley Sensor and Actuator Center, University of California, Berkeley, California, and [4]Center for High Technology Materials, University of New Mexico, Albuquerque, New Mexico, USA

2:10 PM II.A-2 Student Paper
Experimental Investigation of Scalability and Transport in $In_{0.7}Ga_{0.3}As$ Multi-Gate Quantum Well FET (MuQFET)
L. Liu, V. Saripalli, V. Narayanan and S. Datta, The Pennsylvania State University, University Park, Pennsylvania, USA

2:30 PM II.A-3 Student Paper
60 nm gate length Al_2O_3 / $In_{0.53}Ga_{0.47}As$ gate-first MOSFETs using InAs raised source-drain regrowth
A. D. Carter, J. J. M. Law, E. Lobisser, G. J. Burek, W. J. Mitchell, B. J. Thibeault, A. C. Gossard, and M. J. W. Rodwell, ECE Department, University of California, Santa Barbara, California, USA

2:50 PM II.A-4 Student Paper
15 nm diameter InAs nanowire MOSFETs
A. W. Dey[1], C. Thelander[2], M. Borgström[2], B. Mattias Borg[2], E. Lind[2], and L.-E. Wernersson[1], [1]Dept. of Electrical and Information Technology and [2]Dept. of Solid State Physics, Lund University, Lund, SWEDEN

3:10 PM Break

3:30 PM II.A-5 Invited Paper
The Nanoelectric Modeling Tool NEMO5: Capabilities, Validation and Applications to Sb-Heterostructures
S. Steiger[1], M. Povolotskyi[1], H.-H. Park[1], T. Kubis[1], G. Hegde[1], B. Haley[1], M. Rodwell[2] and G. Klimeck[1,3], [1]Network for Computational Nanotechnology, Purdue University, West Lafayette Indiana, USA and [2]University of California, Santa Barbara, California, USA

4:10 PM II.A-6 Student Paper
Experimental Determination of Dominant Scattering Mechanisms in Scaled InAsSb Quantum Well
A. Agrawal[1], A. Ali[1], R. Misra[2], P. E. Schiffer[2], B. R. Bennett[3], J. B. Boos[3] and S. Datta[1], [1]Department of Electrical Engineering, The Pennsylvania State University, University Park, Pennsylvania, USA [2]Department of Physics, The Pennsylvania State University, University Park, Pennsylvania, USA, and [3]Naval Research Laboratory, Washington, District of Columbia, USA

4:30 PM II.A-7
Late News

4:50 PM II.A-8
Late News

978-1-61284-243-1/11 $26.00 © 2011 IEEE

978-1-61284-243-1/11 $26.00 © 2011 IEEE   12

Ultra-thin Compound Semiconductor on Insulator (XOI) for MOSFETs and TFETs

Rehan Kapadia[1,2,3], Kuniharu Takei[1,2,3], Alexandra C. Ford[1,2,3], Hui Fang[1,2,3], Steven Chuang[1,2,3], Morten Madsen[1,2,3] Sanjay Krishna[4], Ali Javey[1,2,3]

[1]Electrical Engineering and Computer Sciences, University of California, Berkeley, CA, 94720.
[2]Materials Sciences Division, Lawrence Berkeley National Laboratory, Berkeley, CA 94720.
[3]Berkeley Sensor and Actuator Center, University of California, Berkeley, CA, 94720.
[4]Center for High Technology Materials, University of New Mexico, Albuquerque, NM 87106.

Due to their high electron mobility, III-V semiconductors are promising channel materials for future devices [1]. InAs is one such promising material; however, due to the small bandgap (E_g~0.36 eV) bulk devices are not feasible. In addition, heteroepitaxial growth of thin layers on Si is challenging due to the inherent lattice mismatch. Here, we present a platform developed for integration of single-crystalline ultra-thin compound semiconductor layers on insulator (XOI)[2], resembling the conventional SOI substrates.

Epitaxial lift-off and transfer has been widely explored in the past[3]. Here, we use a modified method for transfer of ultra-thin semiconductor layers (Fig. 1) to integrate InAs on an Si/SiO$_2$ substrate. Although InAs was chosen as the material system here, the method can be used for other materials as well. Briefly, a source substrate with the active layer and a sacrificial layer is patterned using nanoimprint lithography followed by a selective wet etch. Next, the sacrificial layer is partially removed with a second selective wet etch. The InAs is then detached from the source wafer and transferred to a Si/SiO$_2$ substrate. AFM characterization (Fig. 2) confirms the uniform transfer and smooth surface of the ribbons, essential for good device performance. The quality of the interface between the transferred InAs and the underlying SiO$_2$ is shown by TEM (Fig. 3), illustrating the near atomically flat interface between the InAs and the underlying substrate.

To characterize the electron transport in the transferred InAs layers, back-gated XOI MOSFETs (Fig 4) with L_G~5μm and varied InAs thickness were fabricated and measured. First, the peak field-effect mobility for 48 nm thick InAs is ~5500 cm^2/V-s. In addition, as the thickness is decreased, the mobility drops due to increased phonon scattering, caused by quantum confinement, and surface scattering. Furthermore, top gated MOSFETs were fabricated from an 18 nm InAs ribbon with nominal L_G~500 nm and a 7 nm ZrO2-Ni gate stack. As shown in Fig. 5, the device exhibited an on current of 1.4 mA/μm at $V_{GS} = V_D = 1$V, on/off ratio of ~10^4 and SS~146 mV/decade. In the future, scaling of the gate stack and channel length can be used to improve performance. The effects of various surface treatments prior to ZrO$_2$ ALD are demonstrated in Fig. 6. The best SS (~107 mV/decade) was obtained through a thermal oxidation of the InAs ribbons prior to the ALD.

Due to the lattice mismatch between the InAs and the underlying sacrificial layer, strain in the transferred InAs can be controlled by using a capping layer prior to InAs release[4]. Figure 7 shows the method for strain control as well as the range of strain possible by tuning the capping layer thickness. This presents an important degree of control in XOI and allows for the study of strained devices

978-1-61284-243-1/11 $26.00 © 2011 IEEE

This platform also enables fabrication of TFETs. Initial efforts to fabricate InAs homojunction TFETs using Zn gas phase doping[5] resulted in devices with characteristics as shown in Fig. 8 exhibited SS~190 mV/dec for $V_{DS} = 0.1$V and on current ~0.5 µA/µm at $V_{DS} = V_{GS} = 1$V.

In conclusion, the XOI platform presents a convenient route towards integration of high-performance III-V semiconductors on Si. Since it is possible to study the material properties without the constraints of the original growth substrates, elucidation of fundamental transport physics as well as fabrication of high performance devices are enabled.

[1] T. N. Theis et al., *Science*, vol. 327, p. 1600 (2010)
[2] H. Ko et al. *Nature*, vol. 468, p. 286 (2010)
[3] E. Yablonovitch et al., *Appl. Phys. Lett.*, vol. 56, p. 2419 (1990)
[4] H. Fang et al., *Appl. Phys. Lett.*, vol. 98, p. 012111 (2011)
[5] A. Ford et al., *Appl. Phys. Lett.*, vol. 98, 113105 (2011)

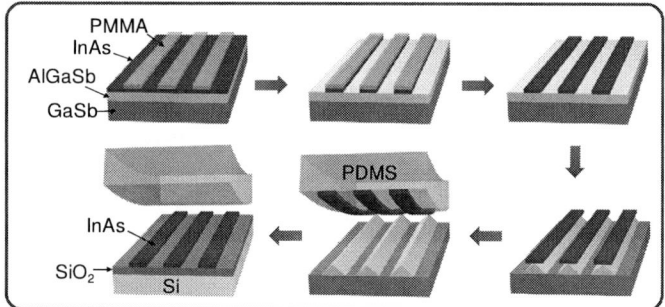

Fig. 1: Schematic of epitaxial transfer process Reprinted with permission from [2]

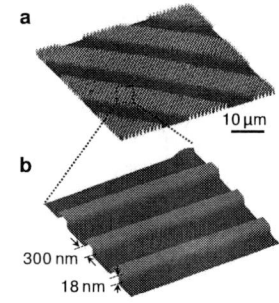

Fig. 2: AFM image of transferred InAs Reprinted with permission from [2]

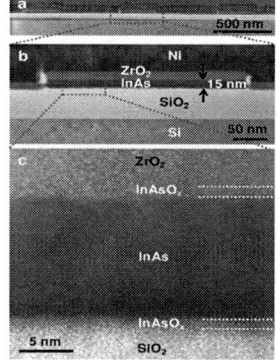

Fig 3: TEM image of InAs on Si/SiO2 with a ZrO2/Ni gate stack. Reprinted with permission from [2].

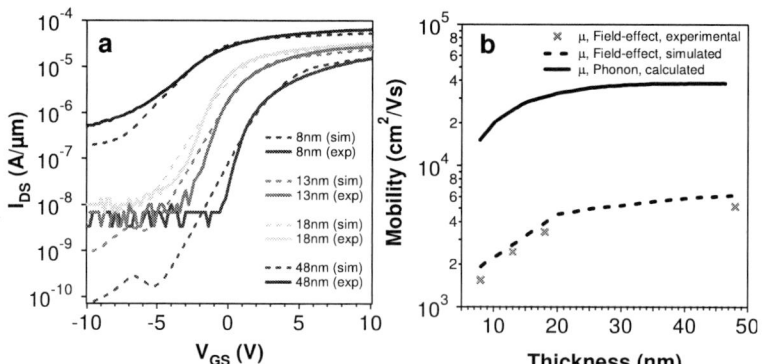

Fig. 4: I_D-V_G curves of back-gated InAs XOI MOSFETs and peak field effect mobility vs InAs thickness. Reprinted with permission from [2].

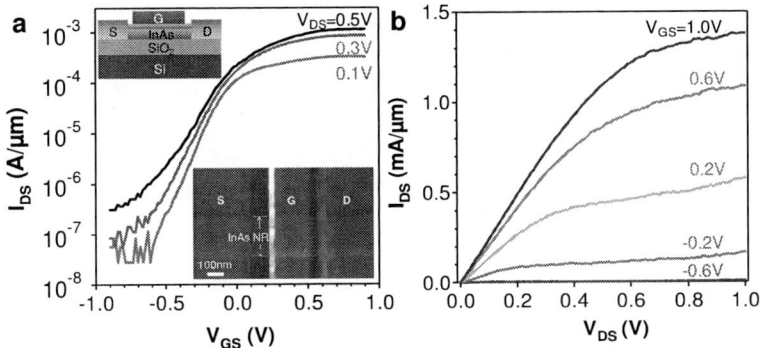
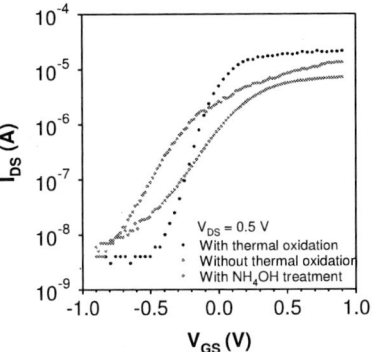

Fig. 5: I_D-V_D and I_D-V_G characteristics of a top-gated InAs MOSFETs. Reprinted with permission from [2]

Fig. 6: I_D-V_G curves for top-gated devices with three surface passivation methods. Reprinted with permission from [2]

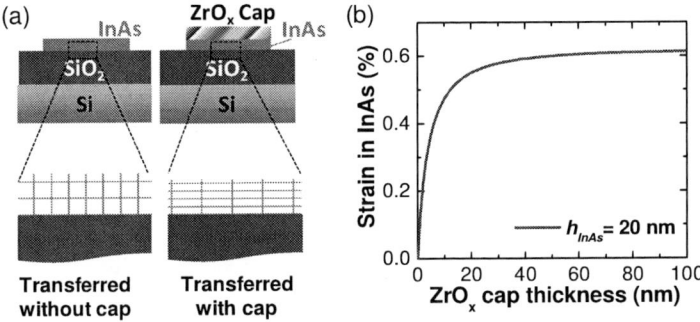
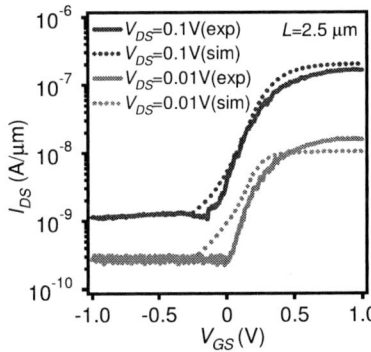

Fig. 7: Schematic of InAs transferred with and without a capping layer (ZrOx). The capping layer acts as a strain memory layer. The amount of strain relaxation reduces as the thickness of the capping layer decreases. Reprinted with permission from [4]

Fig. 8: IDS-VGS curves for Zn doped InAs homojunction TFET. Reprinted with with permission from [5]

978-1-61284-243-1/11 $26.00 © 2011 IEEE

978-1-61284-243-1/11 $26.00 © 2011 IEEE

Experimental Investigation of Scalability and Transport in In$_{0.7}$Ga$_{0.3}$As Multi-Gate Quantum Well <u>FET</u> (MuQFET)

L. Liu, V. Saripalli, V. Narayanan and S. Datta

The Pennsylvania State University, University Park, PA, 16802, USA; Ph: (814) 865-0519; Email: <u>lul137@psu.edu</u>

Compound semiconductors such as In$_{0.7}$Ga$_{0.3}$As and InSb are being actively researched as replacement for silicon channel materials for logic applications due to their superior transport properties [1,2]. Planar III-V quantum-well FETs have already demonstrated with superior performance than the state-of-the art Si MOSFETs for low supply voltage (Vcc) applications [1-3]. A key research challenge remains in addressing the scalability of III-V based quantum-well FETs to sub-14 nm node logic applications while still maintaining their excellent transport advantage. In this study, we demonstrate quasi-ballistic operation of non-planar, <u>m</u>ulti-gate, modulation doped, strained In$_{0.7}$Ga$_{0.3}$As <u>q</u>uantum well <u>FET</u> (MuQFET), combining the electrostatic robustness of multi-gate configuration with the excellent electron mobility of high mobility quantum well channel, In$_{0.7}$Ga$_{0.3}$As (Figure 1).

Device Fabrication: The In$_{0.7}$Ga$_{0.3}$As MuQFET devices were fabricated on MBE grown quantum well heterostructure as shown in Figure 1(a). The strained high mobility In$_{0.7}$Ga$_{0.3}$As channel is sandwiched between two adjacent high bandgap In$_{0.52}$Al$_{0.48}$As layers and is modulation doped by the delta doped layer placed in the bottom barrier and separated from the QW by a 3nm spacer. An assortment of fins with width varying between 40nm and 420nm were patterned using electron beam lithography with nested dummy fins mitigating the electron deflection issues. The source/drain contact was formed using Ni(10nm)/Ge(30nm)/Au(80nm) evaporation and lift-off. A 350 °C annealing in the nitrogen (N$_2$) ambient for 90s is used to form the alloyed ohmic contact to the quantum-well. A 10nm thick Al$_2$O$_3$ high-k dielectric is deposited using Atomic Layer Deposition (ALD) at 300°C. Ti(10nm)/Ni(10nm)/Au(10nm) based wrap-around gates are defined using a electron beam lithography and lift-off process. The fabrication flow is summarized in Table 1, and the titled view SEM image is shown in Figure 2.

Device characterization: The fabricated strained In$_{0.7}$Ga$_{0.3}$As MuQFETs were characterized using an HP4156A semiconductor parameter analyzer at room temperature. The typical transfer characteristics for an In$_{0.7}$Ga$_{0.3}$As MuQFET with Lg=150nm and W=40nm is shown in Figure 3(a). The Ion/Ioff ratio exceeds 2000, and Ion > 100uA/um at Vd=0.5V. The gate leakage is greatly suppressed due to the use of high-k gate stack instead of a Schottky gate. The output characteristic indicates that the channel is well pinched-off implying high self-gain at very low supply voltage (Figure 3(b)). In Figure 3(c), the transfer characteristics of devices with four different fin widths are compared. It is clear that, with decreasing fin width, the multi-gate configuration markedly improves the gate control over the charge in the channel. Thinner fin width also aids in depleting the channel and induces a positive threshold voltage shift implying the possibility of enhancement mode QWFET in a multi-gate configuration.

Short channel effect: The transfer characteristic of a 60nm Lg In$_{0.7}$Ga$_{0.3}$As MuQFET with 40nm fin width is shown in Figure 4(a) where the Ion/Ioff degrades to 300. The sub-threshold slope (SS) and drain induced barrier lowering (DIBL) as a function of gate length are plotted (Figure 4(b)-(c)) for various fin widths ranging from 420 nm to 10 nm. Using three-dimensional numerical simulation, we explore the potential profile in the channel of the fabricated MuQFET. We extract the electrostatic scaling length, Λ, from the electrostatic potential profile which drops off exponentially as a function of the distance into the channel from the source (Figure 5(a)). We also simulated the electrostatic potential profile for two other device configurations including a surface channel In$_{0.53}$Ga$_{0.47}$As FINFET [5] and a lattice matched In$_{0.53}$Ga$_{0.47}$As multi-gate III-V quantum-well FET [4]. It is evident that our fabricated MuQFETs with EOT of 58nm show superior scalability compared to the planar QWFET for the same EOT, which is consistent with the trend observed in Figure 4(a). If we scale the EOT of the dielectric to 2nm as reported in [4], the electrostatic scaling length is comparable (Table 5(b)). We also compare the SS and DIBL versus normalized gate length (Lg/Λ) for these devices in Figure 5(c). We expect further improvement in the sub-threshold slope of In$_{0.7}$Ga$_{0.3}$As MuQFET by proper surface pre-treatment and post-annealing before and after high-k deposition process.

Mobility extraction: 3D numerical simulation (self-consistent Schrodinger-Poisson included) is used to compare the electron density profile in the fin cross-section in MuQFET and classical FINFET (Figure 6(a)). For a given gate overdrive (Vg-Vt), the electron density is highest along the top and sidewall surfaces in FINFETs, whereas the MuQFET shows volume conduction. This suggests that transport in MuQFETs is less affected by surface scattering. Figure 6(a) shows the simulated transfer characteristics of the MuQFETs with and without access resistance. The MuQFET channel mobility is extracted as a function of Lg using the simulated channel carrier density (Figure 6(b)). It is evident that the effective mobility in In$_{0.7}$Ga$_{0.3}$As MuQFET is a combination of "ballistic mobility", μ_{ball}, (given by $2qL_g/2\pi mv_{th}$, v_{th} is thermal velocity) and the long channel mobility in the diffusive limit, $\mu_{ballistic}$ [6].

[1] S. Datta et al, *IEEE Elect. Dev. Lett.* vol.28, p. 685 (2007)   [2] M. K. Hudait et al, *IEDM Tech. Dig.*, p. 625 (2007)
[3] K. Majumdar et al, *IEEE Trans. on Nanotechnology*, vol. 9, p. 342 (2010)
[4] M. Radosavljevic et al, *IEDM Tech. Dig.*, p. 614 (2010)   [5] Y.Q. Wu et al, *IEDM Tech. Dig.*, p.309 (2009)
[6] M.S. Shur et al, *IEEE Elect. Dev. Lett.*, vol.23, p. 511 (2002)

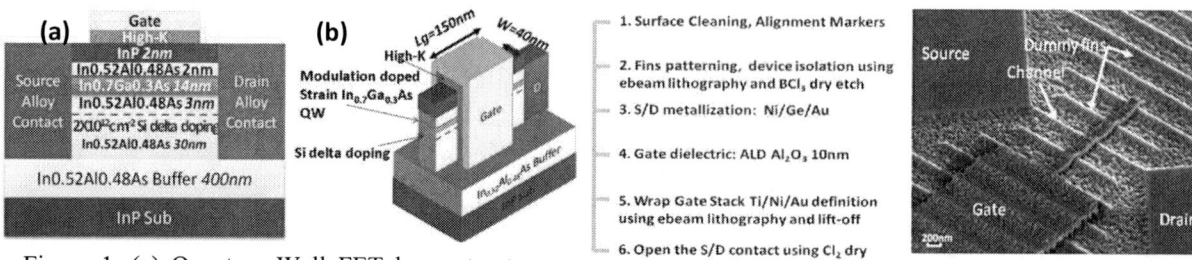

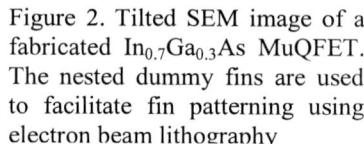

Table 1. Fabrication flow of In$_{0.7}$Ga$_{0.3}$As MuQFET

1. Surface Cleaning, Alignment Markers

2. Fins patterning, device isolation using ebeam lithography and BCl$_3$ dry etch

3. S/D metallization: Ni/Ge/Au

4. Gate dielectric: ALD Al$_2$O$_3$ 10nm

5. Wrap Gate Stack Ti/Ni/Au definition using ebeam lithography and lift-off

6. Open the S/D contact using Cl$_2$ dry etch

Figure 1. (a) Quantum Well FET layer structure; (b) 3D schematic of high In content In$_{0.7}$Ga$_{0.3}$As Multi-Gate Quantum Well FET (MuQFET)

Figure 2. Tilted SEM image of a fabricated In$_{0.7}$Ga$_{0.3}$As MuQFET. The nested dummy fins are used to facilitate fin patterning using electron beam lithography

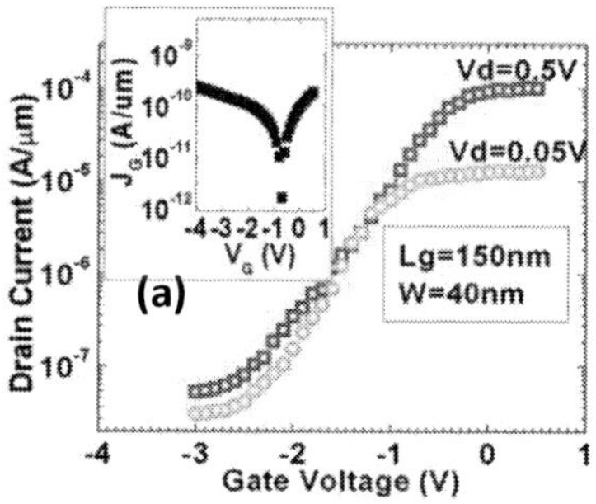

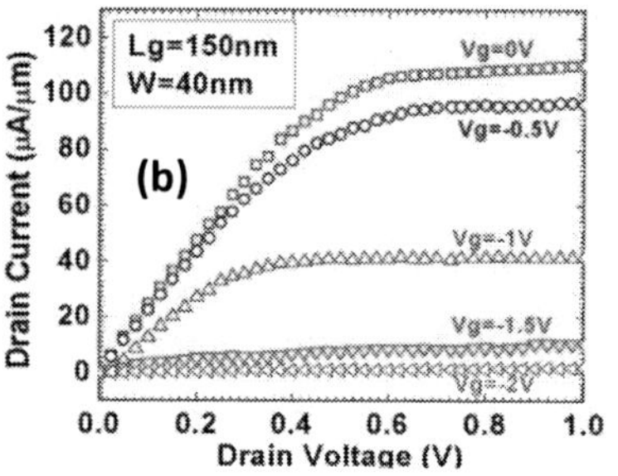

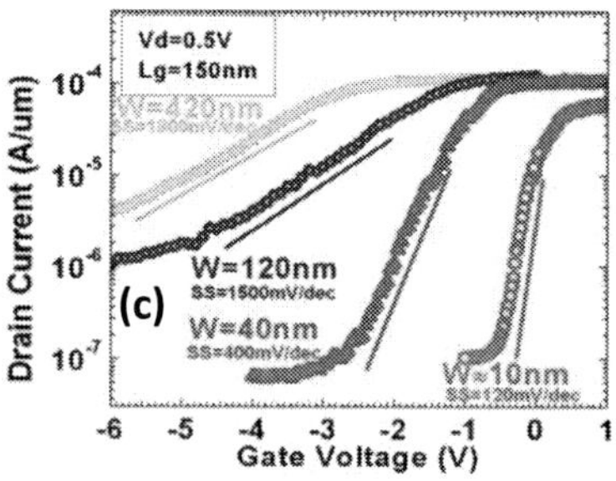

Figure 3. (a) Transfer (Id-Vg) characteristics of In$_{0.7}$Ga$_{0.3}$As MuQFET, Lg = 150 nm (Vd=0.05V, 0.5V), shows Ion/Ioff>2000, I$_{on}$>100uA/um and Ig<I0^{-9}uA/um; (b) Output characteristics (Id-Vd) shows low pinch-off voltage; (c) MuQFET transfer characteristics vs. fin width (W=10nm, 40nm, 120nm and 420nm); thinner Wsi improves electrostatics with positive Vt shift.

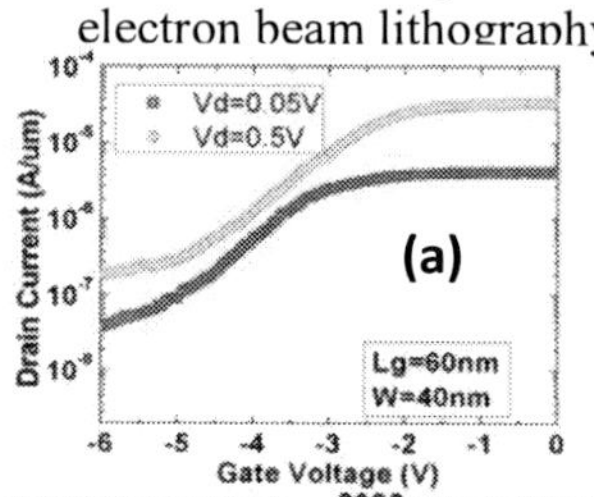

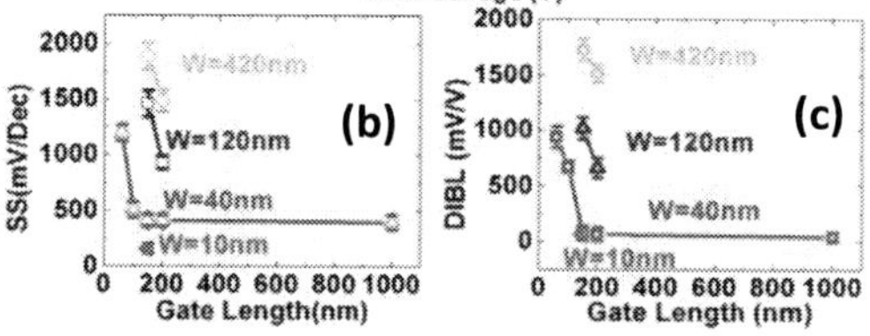

Figure 4. (a) Short Channel MuQFET (Lg = 60nm) characteristic; (b) SS versus Lg for W= 420nm, 120nm, 40nm, 10nm; (c) DIBL versus Lg for W= 420nm, 120nm, 40nm and 10nm

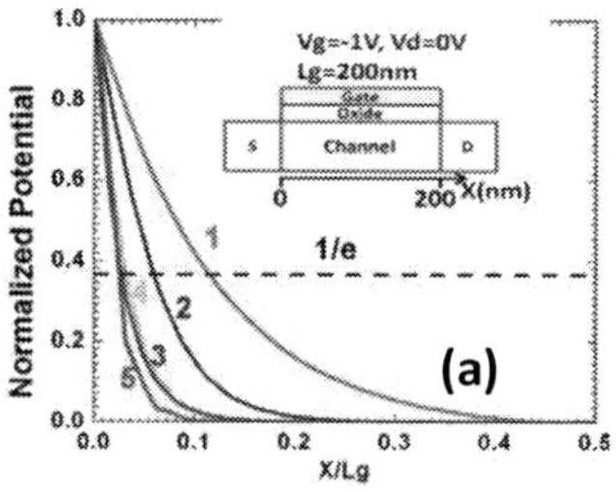

Device	Oxide	EOT(A)	Λ(nm)
1	Planar QWFET (this work)	10nm Al2O3	57.8	22.1
2	MuQFET (this work) W=40nm	10nm Al2O3	57.8	12
3	MuQFET (projected) W=40nm	2nm Al2O3	21.3	5.6
4	InGaAs FINFET W=40 [5]	5nm Al2O3	24.6	6.1
5	Multi-Gate QWFET W=35nm [4]	TaSiO	21.5	4.9

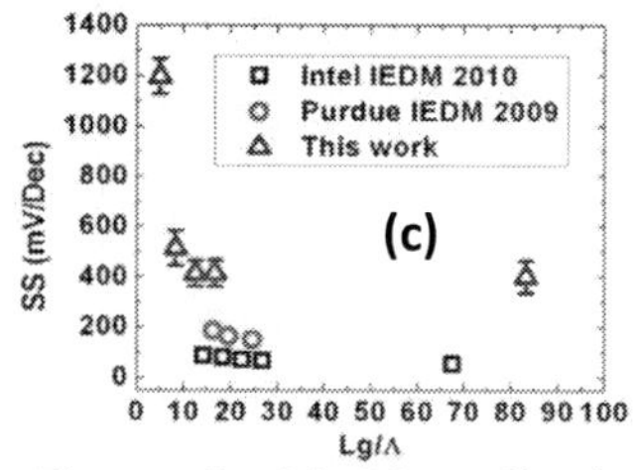

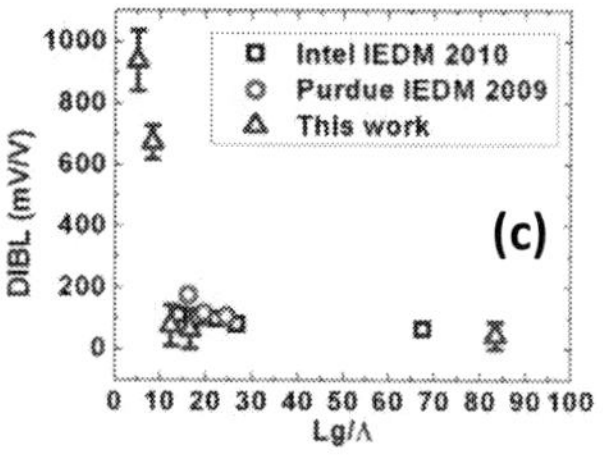

Figure 5. (a) Normalized electrostatic potential for planar, FINFET and MuQFET devices using numerical simulation, The scalng length , Λ, is extracted by fitting V(x)≈V(0)+(Vch-V(0))*(1-exp(-x/Λ)) to the exact potential profile; (b) Summary of scaling length in planar, MuQFET and FINFET devices (c) SS and (d) DIBL versus the normalized gate length, Lg/Λ comparing this work with previous results[4][5].

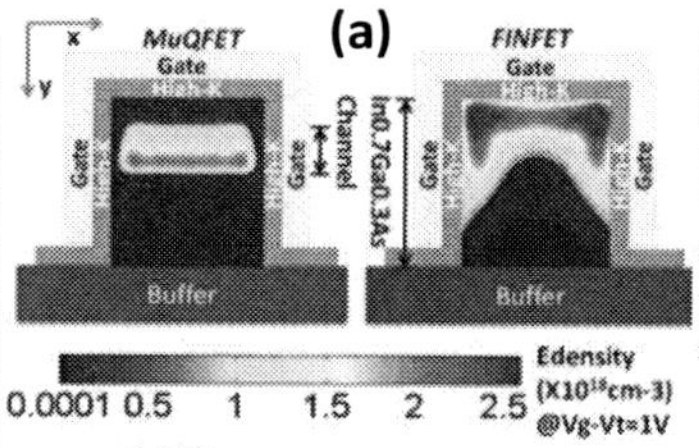

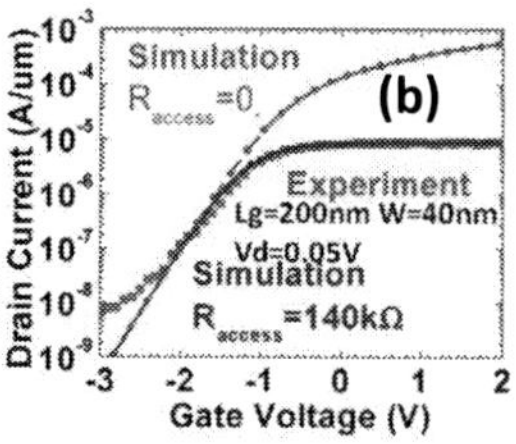

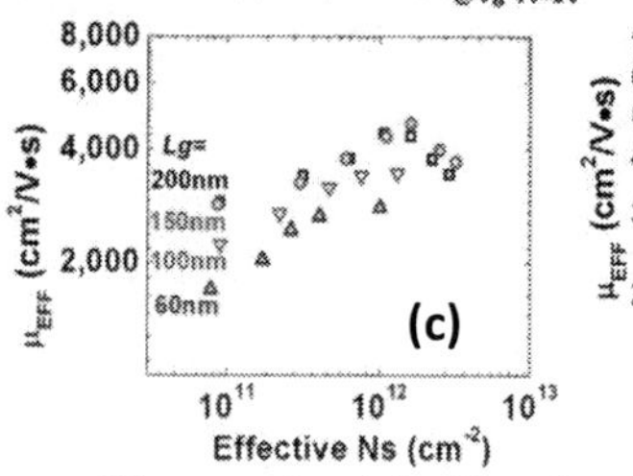

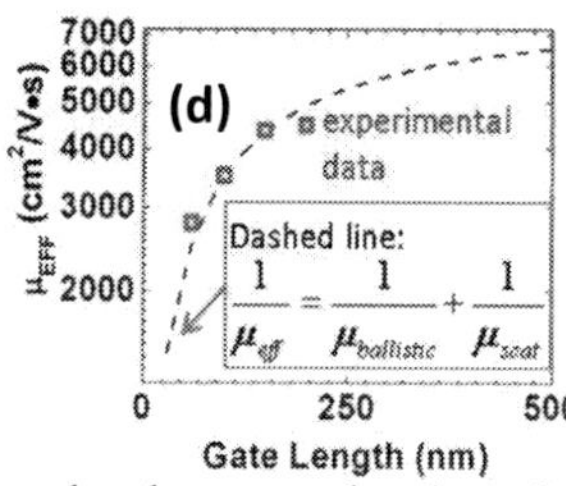

Figure 6. (a) Simulated electron density for MuQFET and FINFET at Vg-Vt=1V employing self consistent Schrodinger-Poisson in x-y plane; (b) Comparison of simulation results with experiment with and w/o access resistance; (c) Carrier density is extracted from simulation to estimate effective mobility at e-density of 10^{12} cm^{-2} vs. Lg (d) Effective mobility is affected by ballistic mobility (2qL$_g$/mv$_{th}$) [4] indicating quasi-ballistic transport in MuQFET.

978-1-61284-243-1/11 $26.00 © 2011 IEEE

60 nm gate length Al_2O_3 / $In_{0.53}Ga_{0.47}As$ gate-first MOSFETs using InAs raised source-drain regrowth

Andrew D. Carter, J. J. M. Law, E. Lobisser, G. J. Burek, W. J. Mitchell,
B. J. Thibeault, A. C. Gossard, and M. J. W. Rodwell

ECE Department, University of California, Santa Barbara, CA 93106-9560
Phone: 805-893-3273, Fax: 805-893-3262, Email: adc@ece.ucsb.edu

Given adequately low source/drain (S/D) access resistivity and dielectric interface trap density ($R_{access} < 50$ $\Omega–\mu m$,[1] and $D_{it} < 2\cdot10^{12} cm^{-2} eV^{-1}$,[2] respectively), InGaAs MOSFETs will provide greater on-state current than silicon MOSFETs at the same effective oxide thickness (EOT). The access resistance must be obtained in a self-aligned structure with a contacted gate pitch ~4 times the physical gate length (L_g), e.g. 116 nm at 32 nm L_g,[3] while control of short channel effects demands that the S/D region depth be only a fraction of gate length; low-resistance, ultra-shallow fully self-aligned III-V MOS processes must therefore be developed. Here we report a 60 nm L_g $In_{0.53}Ga_{0.47}As$ MOSFET fabricated in a gate-first process with self-aligned raised InAs S/D access regions formed by MBE regrowth. The devices have a peak drive current of 1.36 mA/μm at $V_{ds} = 1.25$ V and $V_{gs} = 3$ V and an $R_{on} =$ 341 ohm-μm. To our knowledge this is the lowest R_{on} and smallest L_g reported to date for $In_{0.53}Ga_{0.47}As$ surface channel MOSFETs.[4]

The epitaxial layer structure, grown by molecular beam epitaxy (MBE), has a semi-insulating Fe doped InP substrate, 300 nm not intentionally doped (NID) $In_{0.52}Al_{0.48}As$, 3 nm $In_{0.52}Al_{0.48}As$ n-type Si-doped at $3\cdot10^{19} cm^{-3}$, and 10nm NID $In_{0.53}Ga_{0.47}As$. Prior to atomic layer deposition (ALD) growth of the ~5 nm Al_2O_3 gate dielectric, the surface was treated by repeated hydrogen plasma / trimethylaluminum (TMA) cycles. A 60 nm sputtered W/15 nm electron beam evaporated Cr/400 nm PECVD SiO_2/15 nm electron beam evaporated Cr gate stack was blanket-deposited. Gate lengths between 60 nm to 1.3 μm were defined by patterning the upper Cr layer with a combination of electron beam and optical lithography. A high power inductively coupled (ICP) plasma SF_6/Ar etch defined vertical pillars in the SiO_2 layer. Cl_2/O_2 ICP etched the Cr, and a SF_6/Ar ICP etched the W gate. Etch undercuts in the W and Cr layers are less than 10 nm. 25nm SiN_x was deposited by PECVD and etched in a CF_4/O_2 ICP, defining gate sidewalls. After gate oxide removal in AZ400K, the semiconductor surface was oxidized by exposure to UV ozone and a subsequent removal of this oxide by a 10:1 DI H_2O:HCl etch prior to MBE loading. Regrowth of ~50 nm InAs n-type Si-doped at ~$1\cdot10^{20} cm^{-3}$ (~$5\cdot10^{19} cm^{-3}$ active doping) was done as outlined in ref.[5]. 20 nm Ti / 60 nm Pd/120 nm Au was lifted off for source-drain metallization, and devices were isolated in a 1:1:25 H_3PO_4:H_2O_2:DI H_2O solution.

Devices were characterized using an Agilent 4155C semiconductor parameter analyzer. A 60 nm L_g / 9 μm W_g device has a peak drain current density of 1.36 mA/μm at $V_{ds} = 1.25$ V and $V_{gs} = 3$ V and an $R_{on} = 341$ ohm-μm. Extrapolated source and drain resistances $R_S=R_D = 153$ ohm-μm for these devices. Short channel effects and large delta doping underneath the channel prevent complete channel depletion. Heavy pulse doping (~$1\cdot10^{13} cm^{-2}$) was used to overcome D_{it}-induced channel depletion underneath the ungated sidewall regions, which in turn increased device source to drain leakage. The ~60 nm L_g device has a DC transconductance g_m of 0.3 mS/μm at a $V_{GS} = 0.7$ V. Gate leakage current is < 20 nA/μm at all gate biases. The extracted $R_S=(6.5 mS/\mu m)^{-1}$ is too low to explain the measured $g_m =0.3$ mS/μm; suggesting that the transconductance is instead limited by the thick oxide and large D_{it}.. In conclusion, we have shown a 60 nm L_g Al_2O_3/InGaAs MOSFET with low R_{on} and low access resistance. Future work will include scaling the gate dielectric EOT, reducing process-damage-induced[6] D_{it} , and reduced width and increased carrier concentration in the sidewall regions surrounding the gate.

Acknowledgements: This research was supported by the SRC Non-classical CMOS Research Center (Task 1437.006). A portion of this work was done in the UCSB nanofabrication facility, part of NSF funded NNIN network and MRL Central Facilities supported by the MRSEC Program of the NSF under award No. MR05-20415.

---

[1] M. J. W. Rodwell, *et al*, 2010 Device Research Conference, Notre Dame, Indiana.

[2] A.D. Carter, *et al*, Jpn. J. Appl. Phys., to be submitted.

[3] S. Natarajan, *et al*, IEDM 2008.

[4] M. Radosavljevic, *et al*, IEDM 2009 and 2010, T. Kanazwa, *et al*, IPRM 2010, Y.Q. Wu, *et al*, IPRM 2010, and U. Singisetti, *et al*, IEEE Electron Device Lett. **30** (2009).

[5] M.A. Wistey, *et al*, North American MBE Conference 2009.

[6] G. Burek , *et al*, *submitted to* Applied Physics Letters.

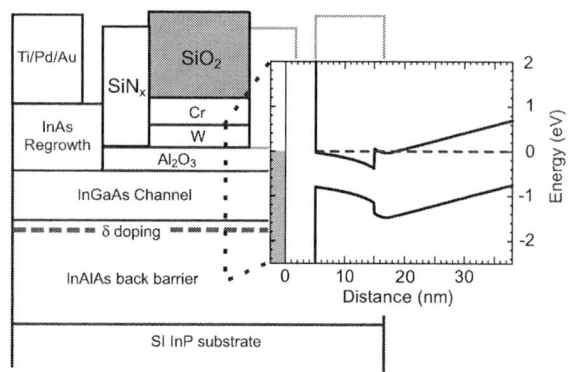
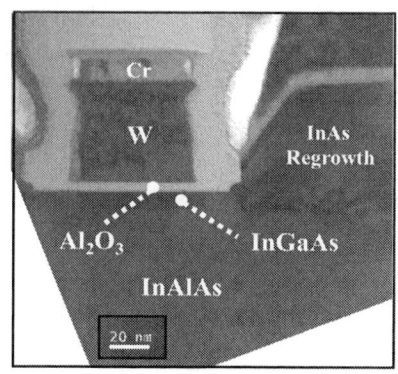

Figure 1: Schematic cross section of the gate first MOSFET. Inset: Energy band diagram across the gated channel region of the device.

Figure 2: TEM cross section of a gate first MOSFET.

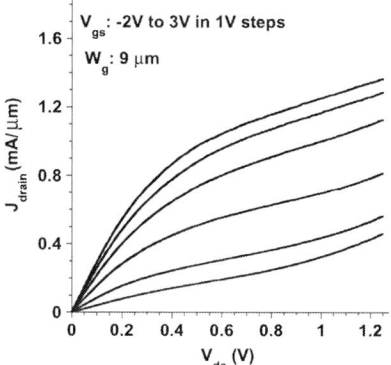
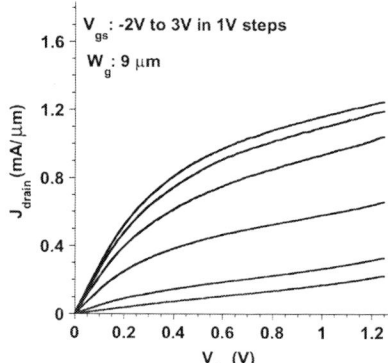
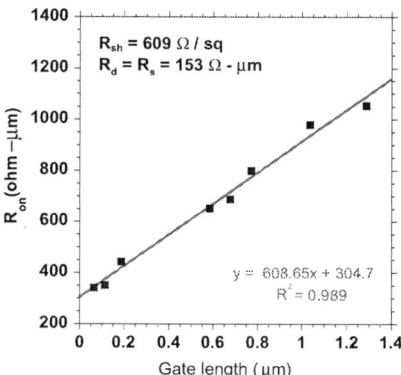

Figure 3: J_{drain} versus V_{ds} for 60 nm L_g depletion mode device.

Figure 4: J_{drain} versus V_{ds} for 115 nm L_g depletion mode device.

Figure 5: R_{on} as a function of gate length for 9 µm gate width devices, measured at $V_{gs} = 3$ V, $V_{ds} = 0$ to 0.1 V.

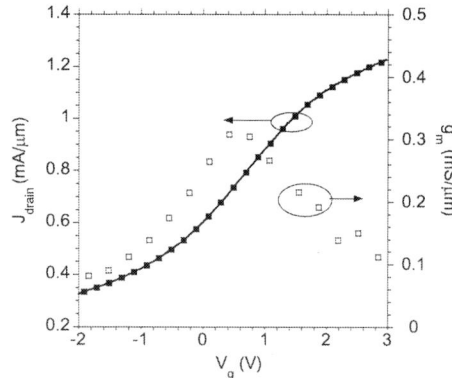
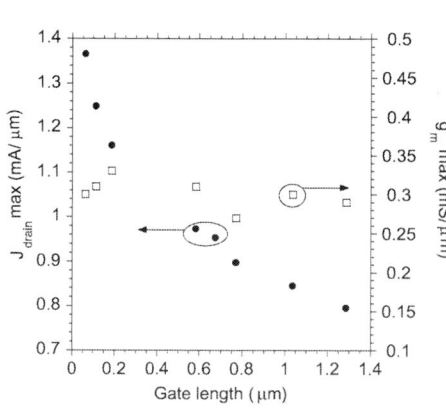
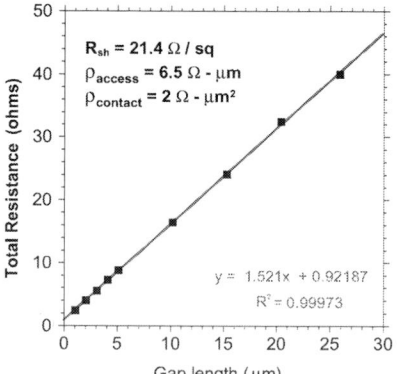

Figure 6: Measured J_{drain} versus V_{gs} ($V_{ds} = 1$ V) and DC transconductance for the 60 nm gate length device. Transconductance data is 5% weighting smoothed from raw data.

Figure 7: Maximum drain current density and DC transconductance versus gate length for 9µm gate width devices.

Figure 8: Transmission line measurement for 14 µm gap width of source-drain metalization and InAs regrowth.

978-1-61284-243-1/11 $26.00 © 2011 IEEE

15 nm diameter InAs nanowire MOSFETs

Anil W. Dey[†], Claes Thelander[‡], Magnus Borgström[‡], B. Mattias Borg[‡], Erik Lind[‡],
and Lars-Erik Wernersson[†]

[†]Dept. of Electrical and Information Technology and [‡]Dept. of Solid State Physics, Lund University, Box 118, S-221 00, Lund, Sweden.
E-mail: anil.dey@eit.lth.se. Phone: +46 46 2224560. Fax: +46 46 129948

Abstract. InAs is an attractive channel material for III-V nanowire MOSFETs and early prototype high performance nanowire transistors have been demonstrated[1]. As the gate length is reduced, the nanowire diameter must be scaled quite aggressively in order to suppress short-channel effects[2]. However, a reduction in transconductance (g_m) and drive current (I_{ON}) could be expected due to increased surface scattering for thin wires. We present data for the device properties of thin InAs nanowires, with diameters in the 15 nm range, and investigate possible improvements of the performance focusing on transistor applications. In order to boost I_{ON}, the source and drain resistance need to be reduced. Several doping sources were therefore evaluated in the study, among them selenium (Se), tin (Sn) and sulphur (S) to form n-i-n structures. We report very high current densities, up to 33 MA/cm^2, comparable to modern HEMTs[3], and a normalized transconductance of 1.8 S/mm for a nanowire with an intrinsic segment of nominally 150 nm and a diameter of 15 nm.

Device fabrication. The nanowires were grown from Au aerosols on an InAs substrate using metal organic vapor phase epitaxy (MOVPE)[4]. Different sets of nanowires were grown, using selenium (Se), tin (Sn) and sulphur (S) as dopant sources, forming i, n, i-n and n-i-n type of transistors. The precursors used were arsine (AsH$_3$) and trimethylindium (TMI). The average diameters for the different growths were; Sn-i-Se: 15 nm, undoped reference: 23 nm, Se: 20 nm, i-Se: 20 nm. The nanowires were transferred to a prepatterned chip where source and drain contacts were defined by means of electron beam lithography and metal evaporation, as shown in Fig 1b. An overlapping gate was formed over the source and drain, with a separating gate dielectric deposited by atomic layer deposition (ALD) of 80 cycles of HfO$_2$ (Picosun Sunale ALD), corresponding to approximately 7.2 nm thickness. The high-κ processing employed a lift-off technique to define the HfO$_2$ window, which limited the ALD deposition temperature to 100°C. Following the ALD step gate electrodes were defined by EBL, metallization and lift-off, Fig 1d.

Electrical characterization. DC electrical characterization was performed in vacuum. For the reference InAs nanowires, without any extrinsic doping, the current density was measured at 3.6 MA/cm^2. Several doping profiles were evaluated to increase the current density. We found a maximum current density for the Sn-i-Se nanowires at V_G = 1 V, V_{DS} = 0.75V, of 33 MA/cm^2 (I_{max} = 59 μA), where the intrinsic section of the Sn-i-Se nanowires had a length of nominally 150 nm. The Sn-i-Se nanowires showed a larger output conductance than the other doping profiles. R_{ON} was extracted to 12.8 kΩ for the Sn-i-Se device which corresponds to 3 x 10^{-4} Ω·cm. The source-drain separation of the devices were 750 nm, 750 nm, 300 nm and 450 nm for the Sn-i-Se, reference, Se and i-Se sample respectively. The maximum normalized transconductance (normalized to the diameter) for Sn-i-Se nanowires was measured to 1.8 S/mm (averaged over 500 mV), or 0.6 S/mm, normalized to the circumference. The maximum voltage gain, g_m/g_0, was extracted to 0.9, 10.4, 1.4, 5.7 for the Sn-i-Se, reference, Se and i-Se sample respectively (at V_D = 0.5 V). This figure of merit is mainly determined by the output resistance. The sulphur doped nanowires were strongly tapered due to parasitic radial growth and therefore not considered in the device analysis.

Conclusions. Very high drive currents with acceptable off-characteristics are reported for very thin nanowires. Although challenges remain in the material composition, specifically controlling the intrinsic segment of a n-i-n structure, we expect an improvement in off-state characteristics with wrap all around gates and high-T deposition of the HfO$_2$. We conclude that these prototype transistors may be a viable candidate for future low power, high-speed electronics.

[1] C. Thelander et al. Electron Device Letters, vol.29, pp.206–208, 2008
[2] E. Lind et al. Electron Devices, IEEE Transactions on, vol.56, no.2, pp.201-205, Feb. 2009.
[3] K. Shinohara et al. Electron Device Letters, vol.25, no.5, May 2004.
[4] C. Thelander et al. Nanotechnology 21, 205703, 2010.

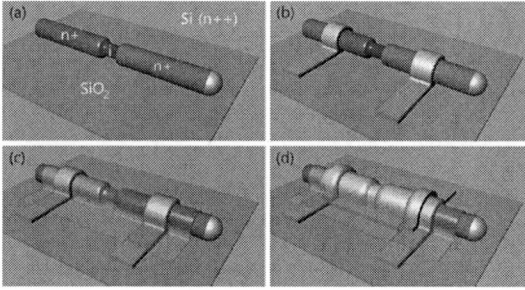

Fig 1. Schematic device fabrication process. (a) The nanowires are transferred to a prepatterned chip. (b) Source and drain contacts are formed by EBL patterning and lift-off. (c) A gate dielectric is deposited overlapping the source and drain contacts. (d) EBL is used to define a topgate.

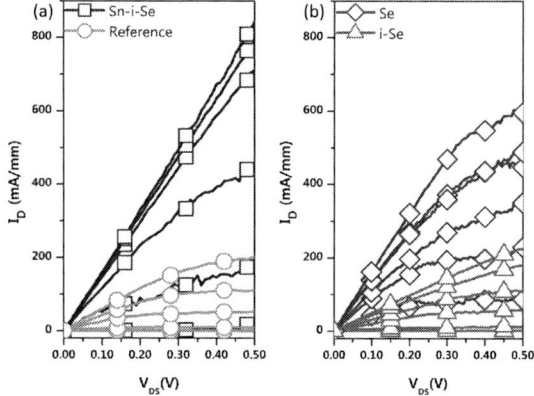

Fig 2. I_D-V_{DS} normalized to the wire circumference. Doping profiles compared. $V_{G,max}$ = 1 V and ΔV_G = 0.5 V. (a) Sn-i-Se (15 nm) and undoped reference device (23 nm). (b) Se (20 nm) and i-Se device (20 nm).

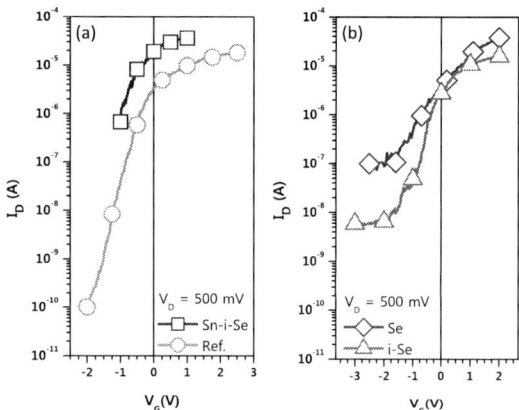

Fig 3. I_D-V_G sweeps for selected dopant profiles. Sweep direction: high to low. (a) Sn-i-Se (15 nm) and undoped reference wires (23 nm) (b) Se and i-Se doping (20 nm).

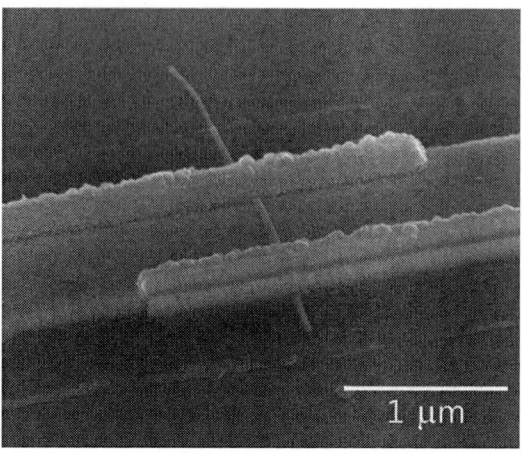

Fig 4. SEM image of a device based on an undoped InAs nanowire (52° tilt).

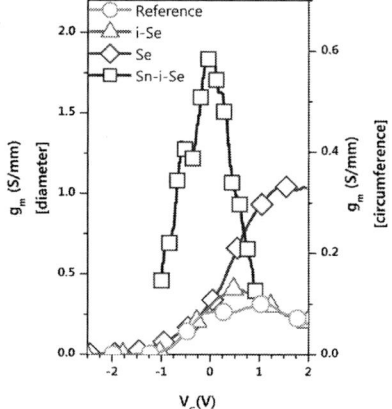

Fig 5. Transconductance for devices with different doping profiles normalized to the diameter/circumference of the nanowire. Source-drain distance approximately 300-750 nm. Data averaged over 500 mV.

The Nanoelectronic Modeling Tool NEMO 5: Capabilities, Validation, and Application to Sb-Heterostructures

Sebastian Steiger[1], Michael Povolotskyi[1], Hong-Hyun Park[1], Tillmann Kubis[1], Ganesh Hegde[1], Benjamin Haley[1], Mark Rodwell[2] and Gerhard Klimeck[1,3]

[1]Network for Computational Nanotechnology, Purdue University, West Lafayette IN 47906, USA
[2]University of California, Santa Barbara, CA 93106, USA
[3]Email: gekco@purdue.edu. Phone: (765) 494-9212

Introduction

Modeling and simulation take an important role in the exploration and design optimization of novel devices. As the downscaling of electronic devices continues, the description of interfaces, randomness, and disorder on an atomistic level gains importance and continuum descriptions lose their validity. Often a full-band description of the electronic structure is needed to model the interaction of different valleys and nonparabolicity effects. NEMO 5 [1] is a modeling tool that addresses these issues and is able to provide insight into a broad range of devices. It unifies the capabilities of prior projects: multiscale approaches to quantum transport in planar structures in NEMO-1D [2], multimillion-atom simulations of strain and electronic structure in NEMO-3D [3] and NEMO-3D-Peta [4], and quantum transport in nonplanar structures in OMEN [5]. NEMO 5 aims at becoming a community code whose structure, implementation, resource requirements and license allow experimental and theoretical researchers in academia and industry alike to use and extend the tool.

Capabilities and Validation

NEMO 5 currently handles pseudomorphic nanostructures composed of diamond, zincblende, simple-cubic, wurtzite, graphene and rhombohedral (trigonal) crystals. It is able to compute *strain* in large structures using an enhanced valence force field model [6] (Fig. 1). The same physical model can be used to find *phonon* spectra (Fig. 2). These capabilities have been validated against various literature results [7].

Electronic structure calculations are done using the empirical tight-binding method [8] using a variety of nearest-neighbor models ranging from an effective mass description to the 20-band $sp^3d^5s^*$ model. Bulk band structures for most IV and III-V materials were validated against literature results. Band diagrams of nanostructures were shown to coincide with independently developed codes. The influence of strain can be treated using the formulation of Ref. 9 or an enhanced version [10]. *Quantum Dot Lab* [11], an educational version of an effective mass solver with interactive 3D wavefunction and absorption visualization that uses NEMO 5 as an engine, is available online at *nanoHUB.org* without the need for any installation.

Selfconsistent *Schroedinger-Poisson* calculations can be performed using a multiscale approach where the density can be based on a mixture of quantum and semiclassical models. This enables the simulation of larger structures where the computationally intensive solution of the Schrödinger equation is restricted to the central parts. A 1D version of this capability can be accessed online [12] through *nanoHUB.org* (Fig. 3). This type of simulation was also validated with independently developed code.

Quantum transport through nanostructures can be computed using two approaches. For ballistic, coherent calculations an open-boundary wavefunction method is preferred [13] (Fig. 4). These calculations have been validated against literature [14] and prior developed code [5]. Electron-phonon scattering can be included in a deformation-potential description [15] using the NEGF formalism [14].

978-1-61284-243-1/11 $26.00 © 2011 IEEE

Application: Sb heterostructures

As an example application we present band structure calculations of two antimonide-based heterostructures. [111]-Sb heterostructures were recently proposed as high-current HEMT candidates [16]. The relative proximity of the Γ-, X- and L-valleys, their coupling through translational symmetry breaking as well as the large anisotropy of carrier masses make the validity of a standard effective mass approach questionable. The calculations in this section were performed using an $sp^3d^5s^*$ tight-binding model including spin-orbit coupling. The parameter sets were determined using genetic-algorithm fitting against literature values for conduction and valence band edges and masses. The fitting process of the parameters including strain effects is still ongoing and the final parameter sets will be published elsewhere [17]. The lattice mismatch of 0.6% (GaSb/AlSb) and 1.3% (InAs/AlSb) is neglected for this work.

In Fig. 5 a triple-QW GaSb-AlSb structure grown along [111] is displayed where the barriers between the wells consist of a single Al-atom. With a lattice constant of 0.61nm, a single [111]-AlSb monolayer has a thickness of about 0.35nm. Although the Γ-valley in bulk GaSb is slightly (30meV) lower than the L-valley, the enormous difference of the effective masses in the confinement direction (0.04 vs. 1.3) pushes the confined Γ-state far above the confined L-state. The high confinement masses also allow for ultrathin barrier layers, enabling exquisite electrostatic control and multiple conduction channels. Yet the in-plane effective masses of the 6 lowest bands remain small such that a fast device can be expected. A series of simulations with well and barrier thicknesses varying between 1 and 3 ML (Table 1) reveals that the in-plane effective masses remain at around 0.09-0.10, close to the bulk value. The masses were obtained by fitting the lowest QW subband within $0.15\pi/a$ of the minimum, as transport in high-current devices occurs above the band edge. A carrier density of $1e13cm^{-2}$ at a temperature of 300K positions the Fermilevel E_F about 70-90meV above the subband edge E_{0L} assuming six parabolic bands. Conversely, E_F–E_c~0.2eV is able to accommodate n_{2D}~2-3e13cm^{-2}. As Table 1 illustrates, the separation of higher bands with larger in-plane masses exceeds 0.2eV for most settings.

A second simulation example is the [100]-AlSb/InAs/GaSb/InAs/AlSb single quantum well depicted in Fig. 6a. This heterostructure features no-common-atom interfaces which are preferably of the type In-Sb [18]. The question of tight-binding parameter mixing for the interface bonds has not yet been fully answered. In this work InAs parameters were taken for the In-Sb interface. Especially the split-off valence subband depends on this choice as there are large differences in spin-orbit coupling. The resulting band structure (Fig. 6b) shows large band warping in the valence bands. As expected, the *k=0* eigenfunctions (Fig. 6c) of the conduction band are mostly located in the InAs layers whereas the valence band states are confined to the GaSb layer.

This paper introduces our latest NEMO 5 tool kit to the advanced research community and highlights capabilities relevant to very recent DRC-discussed results [16].

[1] S. Steiger et al., manuscript in preparation.
[2] R. Lake et al., J. Appl. Phys. 81, 7845 (1997)
[3] G. Klimeck et al., IEEE Trans. El. Dev. 54, 2090 (2007)
[4] S. Lee et al., Proc. IWCE 13, 1 (2009)
[5] G. Klimeck and M. Luisier, Proc. IEDM, 1 (2008)
[6] Z. Sui et al., Phys. Rev. B 48, 17938 (1993)
[7] A. Paul et al., J. Comp. Electron. 9, 160 (2010)
[8] G. Klimeck et al., CMES 3, 601 (2002)
[9] T. Boykin et al., Phys. Rev. B 66, 125207 (2002)
[10] T. Boykin et al., Phys. Rev. B 81, 125202 (2010)

[11] DOI: 10254/nanohub-r450.10
[12] DOI: 10254/nanohub-r5203.15
[13] M. Luisier et al., Phys. Rev. B 74, 205323 (2006)
[14] S. Datta, Superlatt. Microstruct. 28, 253 (2000)
[15] S. Jin et al., J. Appl. Phys. 99, 123719 (2006)
[16] M. Rodwell et al., Proc. DRC, 149 (2010)
[17] G. Hegde et al., manuscript in preparation
[18] H. Kroemer, Phys. E 20, 196 (2004)
[19] B. Dorner et al., J. Phys.:Cond. Mat. 2, 1475 (1990)

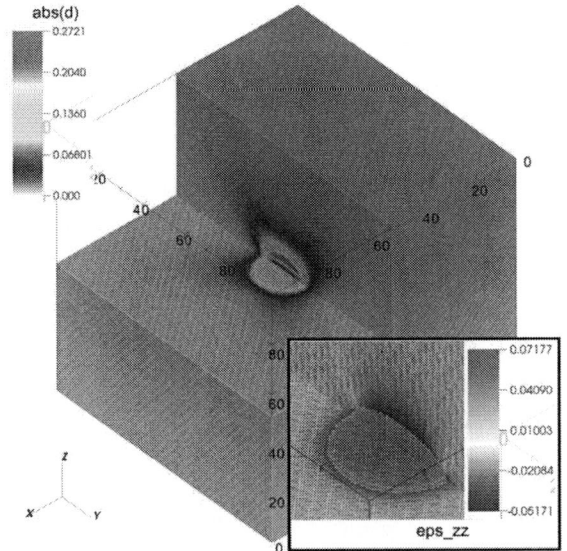

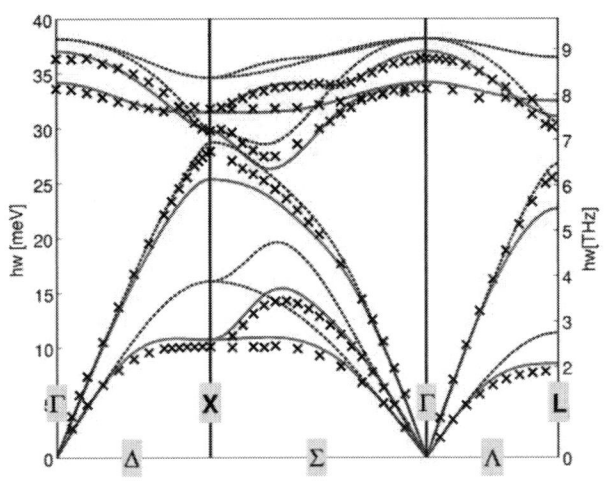

Fig. 1: Strain relaxation in a dome-shaped InAs quantum dot embedded in (100nm)^3 GaAs consisting of 44 million atoms. The dot is 20 nm wide and 5 nm high, consisting of about 38000 atoms. Shown is the magnitude of the displacement vector. *Inset:* Strain component ε_{zz}.

Fig. 2: Bulk phonon dispersions of GaAs computed using the Keating model *(blue – dashed)* and an extended VFF model with augmented Coulomb interaction *(red - solid)*. Also shown are experimental measurements [19] *(black crosses)*.

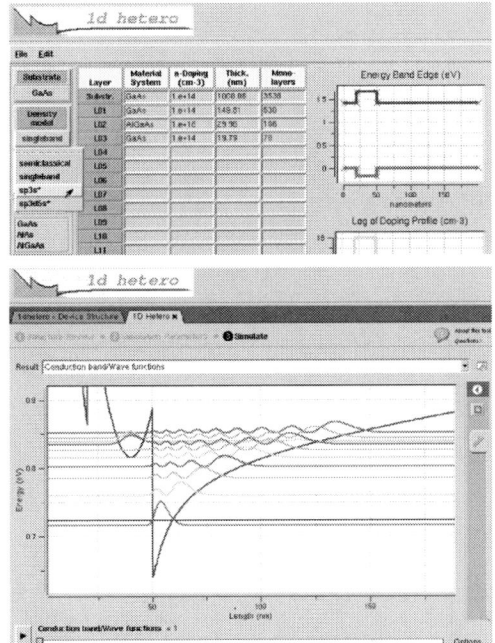

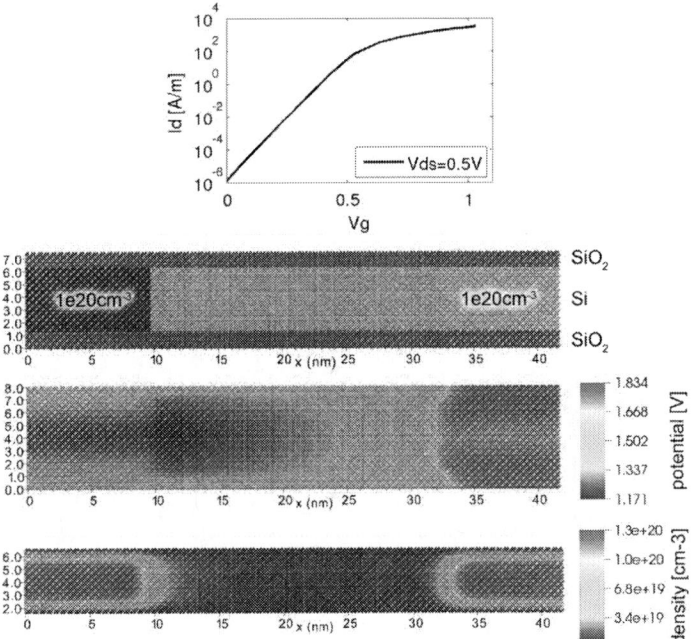

Fig. 3: *1D Heterostructure Design tool* on nanoHUB.org. *Top:* Structure definition screen. *Bottom:* Computed electron states and energies.

Fig. 4: Transport through an ultrathin-body transistor at Vds=0.5V. *From top to bottom:* Id-Vg curve, schematic of the structure, electrostatic potential and electron density at Vg=0.9V.

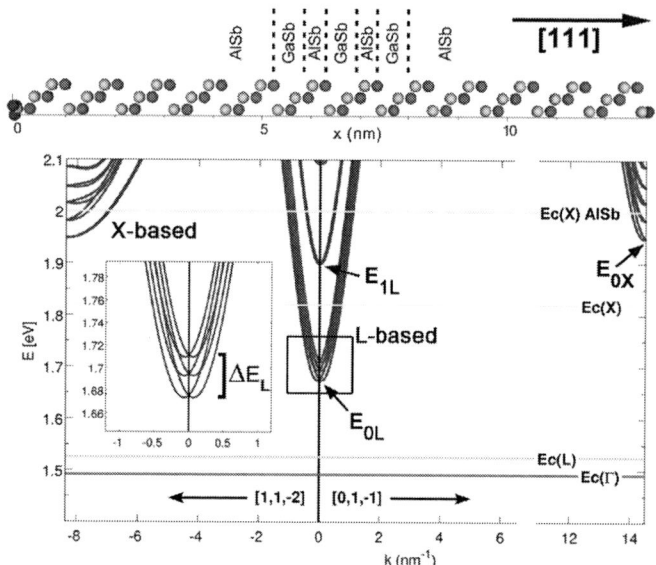

AlSb ╲ GaSb	1 ML (0.35nm)	2 ML (0.7nm)	3ML (1.05nm)
1 ML	m*=0.086 ΔE_L=248 ΔE=223	m*=0.088 ΔE_L=50 ΔE=177	m*=0.094 ΔE_L=48 ΔE=152
2 ML	m*=0.105 ΔE_L=35 ΔE=224	m*=0.102 ΔE_L=13 ΔE=270	m*=0.102 ΔE_L=37 ΔE=277
3 ML	m*=0.095 ΔE_L=58 ΔE=240	m*=0.098 ΔE_L=12 ΔE=263	m*=0.100 ΔE_L=9 ΔE=277

Fig. 5: [111]-GaSb/AlSb triple-QW structure (2 ML wells, 1ML barriers) and in-plane band diagram

Table 1: [111]-GaSb/AlSb triple-QW dependence of m*, the splitting ΔE_L and the distance $\Delta E=\min(E_{0X},E_{1L})-E_{0L}$ (meV) on well and barrier thicknesses

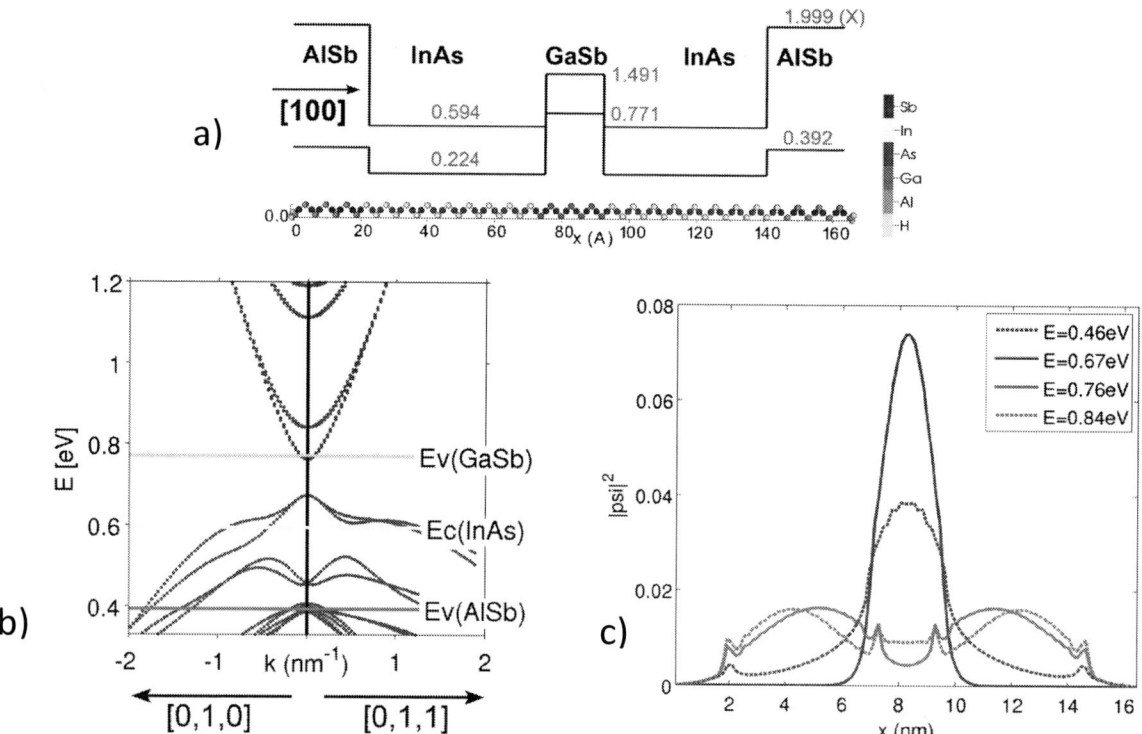

Fig. 6: AlSb/InAs/GaSb/InAs/AlSb double-QW. a) Atomistic structure. b) In-plane band dispersion. c) $k=0$ eigenstate wavefunctions for selected energies (average between anion and cation positions).

978-1-61284-243-1/11 $26.00 © 2011 IEEE

Experimental Determination of Dominant Scattering Mechanisms in Scaled InAsSb Quantum Well

A. Agrawal[1], A. Ali[1], R. Misra[2], P. E. Schiffer[2], B. R. Bennett[3], J. B. Boos[3] and S. Datta[1]

[1]Department of Electrical Engineering, The Pennsylvania State University, University Park, PA 16802, USA

[2]Department of Physics, The Pennsylvania State University, University Park, PA 16802, USA

[3]Naval Research Laboratory, Washington, DC 20375, USA

Antimonide based compound semiconductors have gained considerable interest in recent years due to their superior electron and hole transport properties [1]. A Mixed anion $InAs_ySb_{1-y}$ quantum well heterostructure with high electron mobility of 13,300 cm^2/Vs has already been demonstrated at a sheet carrier density of $2x10^{12}$ /cm^2, albeit for a thick EOT quantum well (QW) structure [2]. A thin EOT structure is desired for improving short channel effects while maintaining the high electron mobility in the QW. In this paper, we study the low field electron transport properties in the high mobility $InAs_{0.8}Sb_{0.2}$ quantum well as we scale the QW heterostructure. Fig. 1(a),(b) show the schematic of the thick (T_{QW}=12nm) and scaled (T_{QW}=7.5nm) quantum well FET structure using $InAs_{0.8}Sb_{0.2}$ as channel material, $In_{0.2}Al_{0.8}Sb$ barrier layer and an ultra-thin GaSb surface layer for avoiding surface oxidation of Al in the barrier [2]. Fig. 2(a),(b) show the simulated energy band diagram of the two structures using self-consistent Schrodinger-Poisson simulation, indicating strong electron confinement in the QW. The effect of nonparabolicity on thick QW with T_{QW}=12nm has already been studied and an effective mass (m^*) of $0.043m_0$ has been extracted experimentally [3]. For scaled QW the subband spacing was adjusted in order to achieve electron sheet charge density as a function of temperature, and the extracted density of states m^*=$0.05m_0$ was correlated to the transport effective mass. Experimental work to verify the obtained effective mass for scaled QW is underway.

Hall measurements were performed on the device layers by varying the temperature from 4K-300K. We extract the dominant scattering mechanisms that are responsible for limiting the mobility at low and high temperatures. The relaxation time approximation (RTA) applied to the Boltzmann transport equation is the theoretical framework employed to estimate the scattering rates. The scattering of the electrons confined in the quantum-well consists of a variety of mechanisms having unique temperature dependence which explain the overall mobility characteristics of the fabricated heterostructure. The inverse of the total relaxation time, τ_{tot} , can be calculated from the sum of the scattering rates for the individual scattering processes. There are six scattering mechanisms limiting the electron mobility: acoustic deformation potential scattering, polar optical phonon scattering, remote ionized impurity scattering, alloy disorder scattering, interface roughness scattering, and coulomb scattering due to charge trapped at the barrier and channel interface (Tamm States) [4]. The scattering rates are finally derived using the Fermi's golden rule.

Fig. 3 shows experimental and modeled sheet carrier concentration as a function of temperature in the channel. The significant contribution from second subband mandates the inclusion of intersubband scattering along with intrasubband scattering. Fig. 4 shows the experimental and modeled hall mobility vs temperature for electrons in the InAsSb channel, indicating the dominant scattering mechanisms at various temperatures. For 12nm thick quantum well, interface charge scattering from Tamm States is the primary source for limiting mobility at both room temperature and low temperature. In the case of scaled quantum well of 7.5nm thickness and 5nm thick barrier, the remote ionized impurity scattering increases by 3x due to reduction in spacer layer thickness, and interface roughness scattering increases by 75x because electrons are closer to the interface and the perturbation in potential is much stronger than that in thick quantum well. Table I lists the parameter values used for scattering calculation for mobility analysis using RTA.

Fig. 5 and 6 show transfer and output characteristics for the thick and scaled MOS-QWFET at room temperature depicting enhancement mode operation for the scaled device. Excellent performance with an I_{ON}/I_{OFF} = 700 over 1V V_{GS} swing is obtained for QW with T_{QW}=7.5nm. Fig. 7 shows the pareto plot indicating % contribution of different scattering components to total mobility at 300K for thick and scaled QW. This provides a clear picture regarding the detrimental scattering mechanisms in each device.

In conclusion, high electron Hall mobility of 13,300 cm^2/Vs at Ns=$2x10^{12}$ /cm^2 is achieved on the fabricated $InAs_{0.8}Sb_{0.2}$ QW FET device with 12nm thick QW and 5,500 cm^2/Vs at Ns=$1.7x10^{12}$ /cm^2 is obtained on scaled QWFET with 7.5nm QW thickness. Excellent agreement between experimental and modeled results for a wide range of temperature has been obtained.

[1] B. R. Bennett, et. al., *Journal of Crystal Growth*, Vol. 312, pp.37 (2009), [2] A. Ali et. al., *Electron Devices Meeting (IEDM)*, pp.6.3.1-6.3.4, Dec. 2010 , [3] A. Ali et. al., *IEEE TED*, doi: 10.1109/TED.2011.2110652 [4] H. Kroemer et.al., *J. Vac. Sci. Technol. B*, 10(4), pp.1769, Jul/Aug 1992

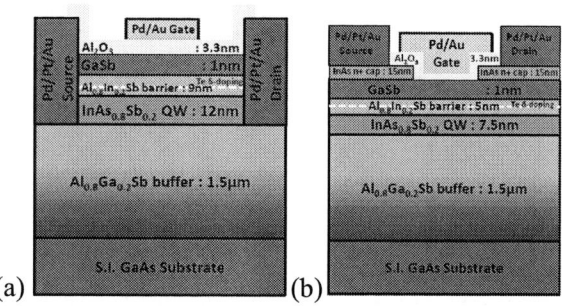

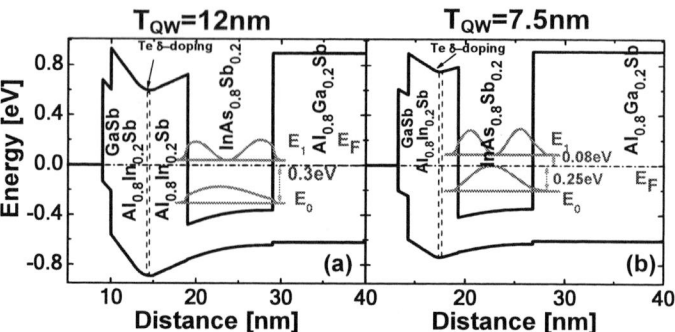

Fig. 1(a) Schematic of the InAs$_{0.8}$Sb$_{0.2}$ MOS-QWFET on GaAs substrate with 12nm QW thickness and 9nm barrier layer (b) Schematic of scaled InAs$_{0.8}$Sb$_{0.2}$ MOS QWFET with 7.5nm QW thickness and 5nm barrier layer

Fig. 2 Band diagram of InAs$_{0.8}$Sb$_{0.2}$ QW heterostructure with 12nm and 7.5nm QW thickness and 1nm GaSb layer from Schrodinger-Poisson simulation indicating strong electron confinement.

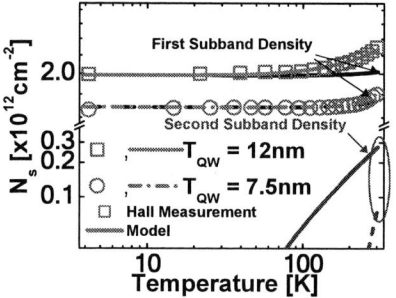

Fig. 3 Experimental and modeled sheet charge density vs temperature in InAs$_{0.8}$Sb$_{0.2}$ channel for 12nm and 7.5nm QW thickness. The contribution from first and second subband is indicated.

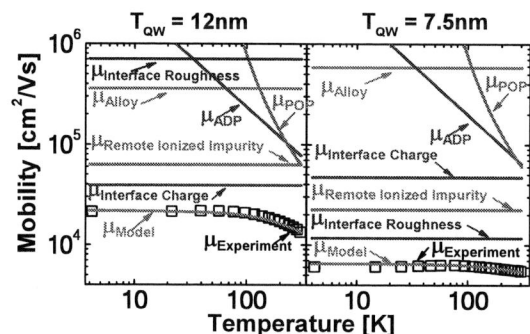

Fig. 4 Experimental and modeled electron mobility in InAs$_{0.8}$Sb$_{0.2}$ QW channel of 12nm and 7.5nm thickness, depicting dominant scattering mechanisms at low and room temperature.

Parameter	Value
Acoustic Deformation Potential	4.8 eV
Polar Optical Phonon Energy	27.8 meV
Alloy Disorder Potential	0.3eV
Interface Charge	6x10^{11} /cm^2
Mean height of Roughness	6.2Å(T$_{QW}$=12nm), 6.8Å(T$_{QW}$=7.5nm)
Correlation Length	20nm
Remote Ionized Impurity	1.8x10^{12} /cm^2

Table I. Values of different parameters used for scattering rate calculation using Relaxation Time Approximation.

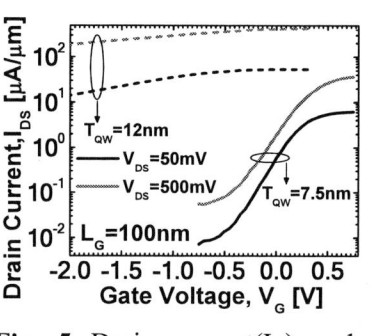

Fig. 5 Drain current(I$_D$) and gate current(I$_G$) vs. gate voltage(V$_G$) of the two InAs$_{0.8}$Sb$_{0.2}$ MOS-QWFET with QW thickness=7.5nm and 12nm. Scaled QWFET shows superior enhancement mode characteristics than thick QWFET.

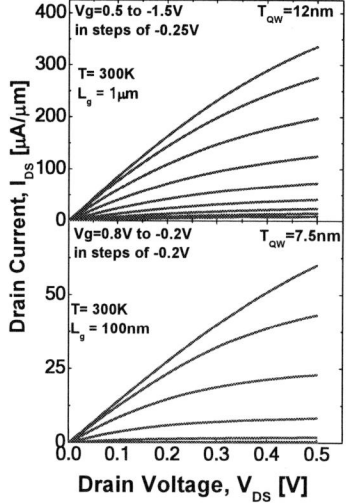

Fig. 6 I$_D$ vs. V$_{DS}$ of InAs$_{0.8}$Sb$_{0.2}$ MOS-QWFET with QW thickness=7.5nm and 12nm and Al$_2$O$_3$-GaSb composite stack at 300K.

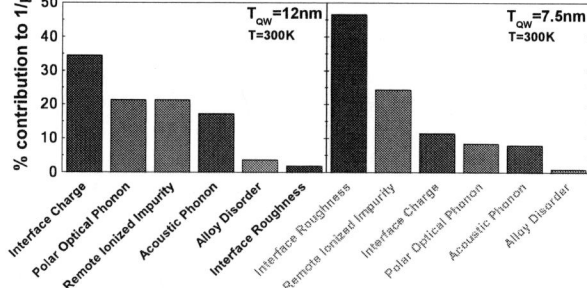

Fig. 7 Pareto plot showing %contribution of different scattering mechanisms to total mobility at 300K for thick and scaled QW. Interface charge scattering dominates for T$_{QW}$=12nm and interface roughness scattering dominates for T$_{QW}$=7.5nm.

978-1-61284-243-1/11 $26.00 © 2011 IEEE

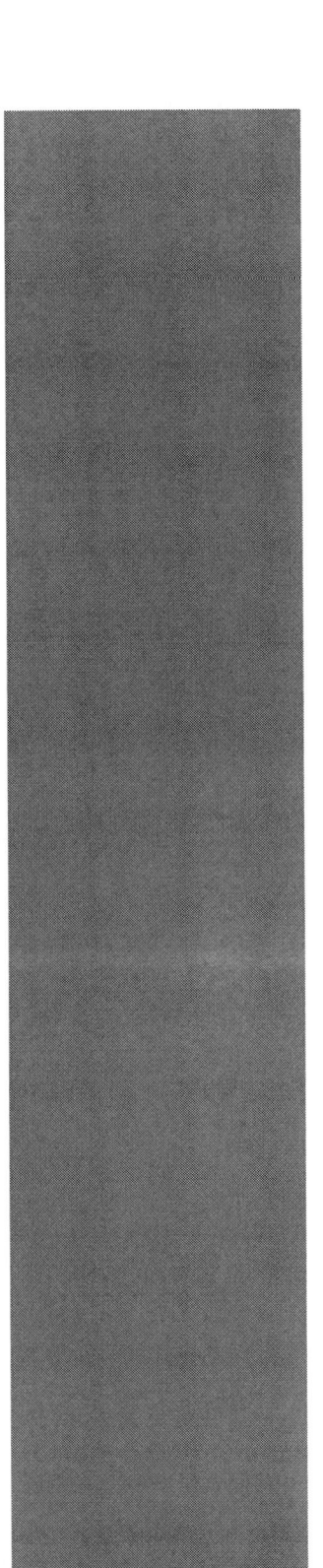

Session II.B (Corwin Pavilion West)

Graphene

Monday PM, June 20th, 2011

Session Chairs: Kaustav Banerjee, University of California, Santa Barbara and Grace Xing, University of Notre Dame

1:30 PM II.B-1
Integration of High Quality Top-Gated Graphene Field Effect Devices on 150 mm Substrate
J. Heo, H.-J. Chung, S.-H. Lee, H. Yang, J. Shin, U-I. Chung, S. Seo, Semiconductor Device Lab, Samsung Advanced Institute of Technology, Yongin-si, KOREA

1:50 PM II.B-2 Student Paper
Complementary-Type Graphene Inverters Operating at Room-Temperature
H.-Y. Chen and J. Appenzeller, ECE Department and Birck Nanotechnology Center, Purdue University, West Lafayette, Indiana, USA

2:10 PM II.B-3 Student Paper
Gate capacitance scaling and graphene field-effect transistors with ultra-thin top-gate dielectrics
B. Fallahazad, K. Lee, S. Kim, C. Corbet, E. Tutuc, Microelectronics Research Center, University of Texas at Austin, Austin, Texas, USA

2:30 PM II.B-4 I Invited Paper
Synthesis and Applications of Graphene for Flexible Electronics
B. H. Hong, Department of Chemistry & SKKU Advanced Institute of Nanotechnology, Sungkyungkwan University, Suwon, KOREA

3:10 PM Break

3:30 PM II.B-5 Student Paper
Sub-10 nm Epitaxial Graphene Nanoribbon FETs
K. Tahy[1], W. S. Hwang[1], J. L. Tedesco[2], R .L. Myers-Ward[2], P.M. Campbell[2], C. R. Eddy[2] Jr., D. K. Gaskill[2], H. Xing[1], A. Seabaugh[1], and D. Jena[1], [1]Department of Electrical Engineering, University of Notre Dame, Notre Dame, Indiana, USA and [2]U.S. Naval Research Laboratory, Washington, District of Columbia, USA

3:50 PM II.B-6 Student Paper
Effect of Oxide Thickness Scaling on Self-Heating in Graphene Transistors
S. Islam, M.-H. Bae, V. Dorgan, and E. Pop, Dept. of Electrical and Computer Engineering, University of Illinois, Urbana-Champaign, Illinois, USA and Micro and Nanotechnology Lab, Urbana, Illinois, USA

4:10 PM II.B-7
Graphene Quantum Capacitance Varactors for Wireless Sensing Applications
S. J. Koester, University of Minnesota-Twin Cities, Minneapolis, Minnesota, USA

4:30 PM II.B-8
Late News

4:50 PM II.B-9
Late News

978-1-61284-243-1/11 $26.00 © 2011 IEEE

978-1-61284-243-1/11 $26.00 © 2011 IEEE 30

Integration of High Quality Top-Gated Graphene Field Effect Devices on 150 mm Substrate

Jinseong Heo, Hyun-Jong Chung, Sung-Hoon Lee, Heejun Yang,

Jaikwang Shin, U-In Chung, Sunae Seo

Semiconductor Device Lab, Samsung Advanced Institute of Technology, Yongin-si, 446-712 Korea

phone: +82 31 280 9345, fax: +82 31 280 9158, email: jinseong.heo@samsung.com

Recent success of inexpensive and high-throughput chemical vapor deposition (CVD) growth[1] of graphene on Ni or Cu substrates has shown promises for potential industrial applications such as transparent electrodes[2] and field effect transistors (FET).[3] However, high-coverage uniform growth of monolayer graphene on a wafer scale is still a major obstruction, which impedes high yield integration of high performance field effect devices. Here, we report the first demonstration of high quality top-gated graphene field effect devices on 150 mm substrates exploiting unprecedented homogeneous CVD growth of monolayer graphene.

Graphene in this study was synthesized by inductively coupled plasma enhanced chemical vapor deposition (ICP-CVD) where we adopted a low-temperature-plasma-supported growth. Growth substrate prepared by electron beam evaporation of Cu on 300 nm SiO_2. During growth, the substrate was heated to 650℃ within 10min under a base pressure of ~10^{-7} torr, then treated with H_2 plasma. After purging with Ar for a couple of minutes, the C_2H_2 gas mixture was added (C_2H_2: Ar =1: 40) for graphene growth at the same temperature. Monolayer nature of our grown film was determined from scanning tunneling microscope (STM) image of characteristic hexagonal networks of carbon atoms in Fig. 1(a), Raman spectrum with single Lorentzian fit of 2D peak at 2780 cm^{-1} in Fig. 1(c) and optical transmittance of 97.7% at 550nm in Fig. 1(c) inset. Furthermore, uniformity of the film was confirmed with mapping of Raman spectrum on various points of 150mm wafer in Fig. 1(d)-(f).

Topgated graphene field effect devices presented in Fig. 2 were fabricated through patterning source/drain electrodes, etching graphene channel by O_2 plasma, depositing 35nm of Al_2O_3 as gate dielectric, and finally positioning gate electrode between source and drain. 4-probe measurement using Hall bar geometry of an individual device in Fig. 2(b) enables us to assess channel resistance without contact resistance and also simultaneously extract carrier density by Hall resistance. Statistics of field effect characteristics on arrays of devices across the entire 6" wafer is summarized in Fig. 2(d) and (e). Maximum channel resistance occurred at Dirac point is mapped in Fig. 2(d) with color contrast of different values. Around 60% of designed devices exhibit field effect by top-gate. Fig. 2(e) shows cumulative plots of R_{dirac} (maximum channel resistance), V_{dirac} (topgate voltage at R_{dirac}) and mobility. Majority of samples falls in the range of R_{dirac} = 20-80kohm, V_{dirac} = 2-4V and more importantly 5% of devices has mobility over 3,000 $cm^2/V \cdot s$. Note that sample distribution shown here is not coming from non-uniform growth or variable quality of graphene, but rather caused by extrinsic effects like charged defects or impurities introduced during graphene transfer or substrate interaction via polar phonons. Influence of charged impurities and surface phonons on graphene devices are investigated through temperature dependent resistivity(Fig. 3(c)).[4,5]

Channel conductivity versus gate voltage (G_{ds}-V_g) curves of different samples (w=6μm, l=4μm) are plotted in Fig. 3(a). Stable operation of our devices in ambient condition results from passivation by the Al_2O_3 gate dielectric, but observed asymmetry between hole and electron transport (always poor electron mobility) is possibly due to electron trap sites induced at the interface between graphene and gate dielectric. 4probe I_{ds}-V_{ds} characteristics are presented at gate voltages from -5 to 0 (left) and from +5V to 0 (right) with a step of 0.5V in Fig. 3(b). Channel current continues to decrease approaching V_{dirac} from both positive and negative. In P-channel mode (left) the curve changes from linear to sublinear approaching V_{dirac}. On the other hand, in the N-channel region (right) slight saturation behavior is observed at all gate voltages in low current regimes. In addition, there is pronounced negative differential resistance in electron channel. The emergence of negative differential resistance moves to lower voltage when gate voltage is closer to V_{dirac}. Our results show significant progress closer to the realization of development of graphene field effect devices.

[1] K. S. Kim et al., *Nature* **457**, 706 (2009).

[2] S. Bae et al., *Nature Nanotechnology* **5**, 574 (2010).

[3] J. Kedzierski et al., *IEEE* **30**, 745 (2009).

[4] J.-H. Chen et al., *Nature Nanotechnology* **3**, 206 (2008).

[5] J. Heo et al., arXiv:1005.2506.

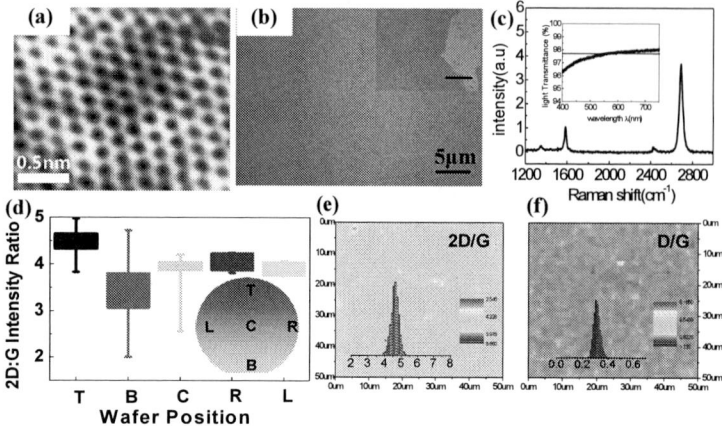

Fig. 1. (a) Scanning Tunneling Microscope (STM) image of graphene. (b) Optical microscope image. (c) Raman spectroscopy after transferred on SiO2(100nm)/Si. Inset: light transmittance on Graphene/Glass. (d) Distribution of 2D over G ratio on 5 points of 150mm wafer. (e), (f) False color images of 2D over G ratio and D over G ratio over 50X50µm, respectively. Insets: histograms of the ratio.

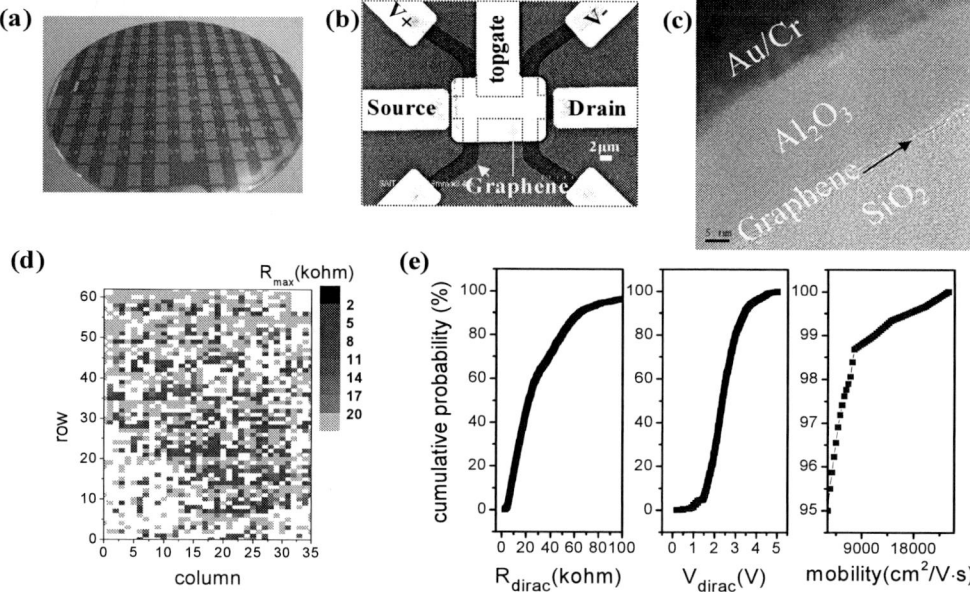

Fig. 2. (a) 6" wafer with integrated top-gated graphene devices. (b) Scanning electron microscope image of fabricated top-gated devices. (c) Transmission electron microscope image of cross-section of Au/Al2O3/Graphene/SiO2/Si. (d) Maximum resistance maping of the entire wafer. (e) Cumulative probability of channel resistance at Dirac voltage (left), Dirac voltage (center) and mobility (right).

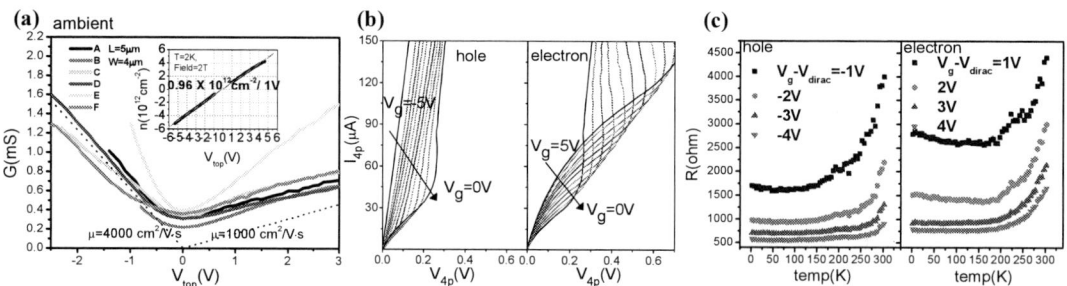

Fig. 3. (a) Conductivity vs. top gate voltage for various samples. Inset: carrier density vs. top gate voltage from Hall measurement. (b) 4probe I-V characteristics upon varying top gate voltage. (c) Temperature dependent resistivity for each carrier at different top gate voltages.

978-1-61284-243-1/11 $26.00 © 2011 IEEE   32

Complementary-Type Graphene Inverters Operating at Room-Temperature

Hong-Yan Chen, Joerg Appenzeller

ECE Department and Birck Nanotechnology Center, Purdue University, West Lafayette, IN 47906, USA

Email: chen200@purdue.edu, appenzeller@purdue.edu, Phone: +1-765-494-0215

Graphene has recently emerged as a promising candidate for a number of electronic applications. However, the fact that graphene is a zero band gap material by nature has raised many questions in terms of graphene's usefulness for digital applications. Several recent experimental studies have demonstrated graphene based inverters, but issues remain, such as, low inverter gain (0.044[1], 0.02[2]) and mismatch between input/output voltage levels[1,2]. Li et al.[3,4] reported top-gated complementary-like graphene inverters exhibiting a gain larger than 1. However, all data were obtained at 77K, and the implementation of a p-type and n-type FET was accomplished by relying on the intrinsic dependence of graphene's transfer characteristics on the supply voltage, an effect that is hardly controllable and that poses major problems for further device optimization. In this paper, focusing on inverter characteristics without attempting to build a highly scaled device, we report the first room-temperature, electrostatic doping controlled complementary graphene inverter with a gain larger than one.

A typical device structure as fabricated by us is shown in Fig 1. Graphene flakes from highly oriented pyrolytic graphite (HOPG) were mechanically exfoliated onto 90nm SiO_2 on a doped Si substrate, and through their reflectivity of light initially characterized in terms of the number of graphene layers. All metal contacts, defined by e-beam lithography, consisted of a stack of e-beam evaporated Ti/Pd/Au (5nm/20nm/30nm). Metal contacted graphene flakes were then further patterned using e-beam lithography and an O_2 plasma dry etching approach to define two parallel channels of width ~ 60nm, as well as two side gates located ~40nm apart from each channel. The channel length was chosen to be 3.5µm for optimization purposes as explained below. Devices were inspected for shorts between the channel and the side gate by both, imaging with an SEM and electrically testing in a probe station set-up. All measurements were conducted with an Agilent 4156C Precision Semiconductor Parameter Analyzer at room-temperature in a gas N_2 ambient.

To enable the desired n-type and p-type FET operation from the two channels in figure 1, we employed an electrostatic doping approach in which an electric field is created through a voltage difference (ΔV_{sg}) between the two graphene side gates. An increasing ΔV_{sg} shifts the threshold voltages of the p- and n-FETs away from each other with little degradation of the transconductance (g_m) (Fig 2). (Here V_{th} is defined as the back gate voltage at which the current is minimum.) We find that $\Delta V_{th} = V_{th,p} - V_{th,n}$ depends linearly on ΔV_{sg}. Clear inverter characteristics (Fig 3) are obtained with our approach in the voltage range between $V_{th,p}$ and $V_{th,n}$. Figure 4 analyzes the impact of V_{th} shift on the output voltage swing, as well as on the inverter gain. The larger ΔV_{th}, the larger the inverter-like region, and hence the output voltage swing (Fig 4b). The dependence of maximum gain on ΔV_{sg} is more complicated. Assuming two completely identical graphene channels, maximum inverter gain is expected to occur at $V_{in} = V_{in,max\ gain} = (V_{th,p} + V_{th,n})/2$. An analytical expression for the maximum gain is given in Fig 4, where $g_m(V_{in,max\ gain})$ and the current $I(V_{in,max\ gain})$ are functions of ΔV_{sg}. Fig 4c illustrates the various component trends and shows the overall experimental result consistent with the analytical expression. Figure 5 analyzes the dependence of the inverter gain on the supply voltage V_{SS}. The device channel was designed to be 3.5µm long to enable characterizing the impact of a large range of V_{SS} values. (Note that in addition scattering in our long channel is resulting in some current saturation at large V_{SS} which is believed to be beneficial for the achievable output voltage swing and noise margin.) While in fact the V_{out} swing only shows a rather mild dependence on V_{SS}, we observed a very strong dependence of the maximum gain on V_{SS}, allowing us to operate under optimized voltage conditions. The inset of figure 5c explains our findings qualitatively. It is in particular the fact that g_m at $V_{in,max\ gain}$ increases first with increasing V_{SS} but then decreases due to some "rounding" of the current transfer characteristics close to the Dirac point at high V_{SS} that is responsible for the allover maximum gain dependence. One of the issues that our complementary graphene inverter, as well as all other graphene based inverters reported so far, has not addressed is the issue of operating with an appreciable noise margin. In other words, it is desirable to obtain a stable output signal even when the input signal is affected by fluctuations due to noise. In order to address this issue, we propose and have fabricated an upgraded version of a complementary graphene inverter structure (Fig 6a), in which two channels are connected in series and side-gated to create a 'flat-band' in the off-current for the "new" p-FET and the "new" n-FET respectively (Fig 6b). These 'flat-bands', after shifted properly, are expected to translate into flat output voltages as highlighted in figure 6c.

In summary, we have reported the first demonstration of a room-temperature complementary-like graphene inverter with an inverter gain larger than 1. Electrostatic doping allowed to study for the first time the detailed impact of n-type and p-type graphene FETs' threshold voltages on inverter gain and output voltage swing, which, together with the study on supply voltage (V_{SS}), enabled optimization of the inverter gain in our devices.

Acknowledgements: We greatly appreciate Dr. James Cooper's inspiration of this work and insightful discussions with Saptarshi Das.

[1] Traversi et al, *APL* 94, 223312, 2009
[2] Harada et al, *APL* 96, 012102, 2010
[3] Li et al, *Nano Lett.*, 10, p2357, 2010
[4] Li et al, *ACS Nano*, v.5, no.1, p.500, 2011

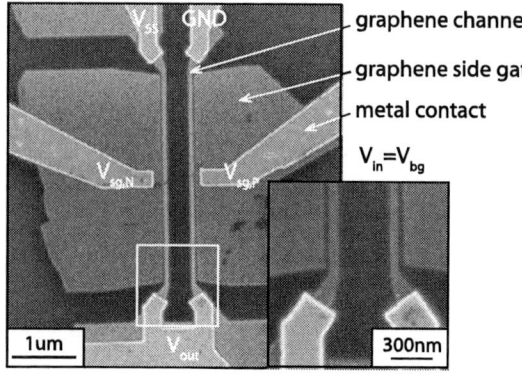

Fig 1. SEM images of a graphene inverter. Channels (3.5µm long) are separated by air gaps to side gates.

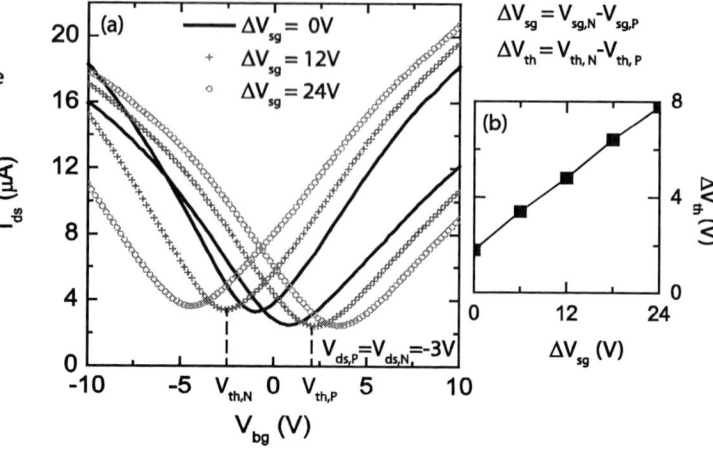

Fig 2. (a) Transfer characteristics of p-type (right) and n-type (left) graphene FETs for various ΔV_{sg}-values. (b) Extracted ΔV_{th} vs. ΔV_{sg}.

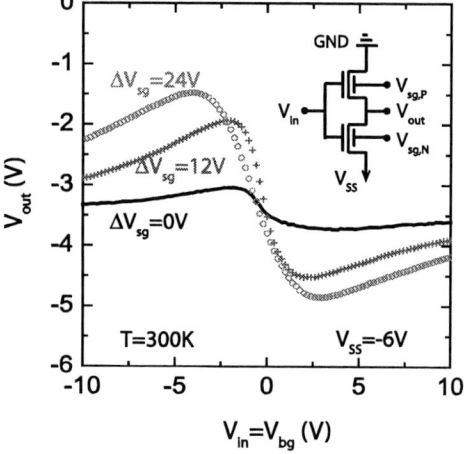

Fig 3. Graphene inverter characteristics measured according to the schematics shown in the inset.

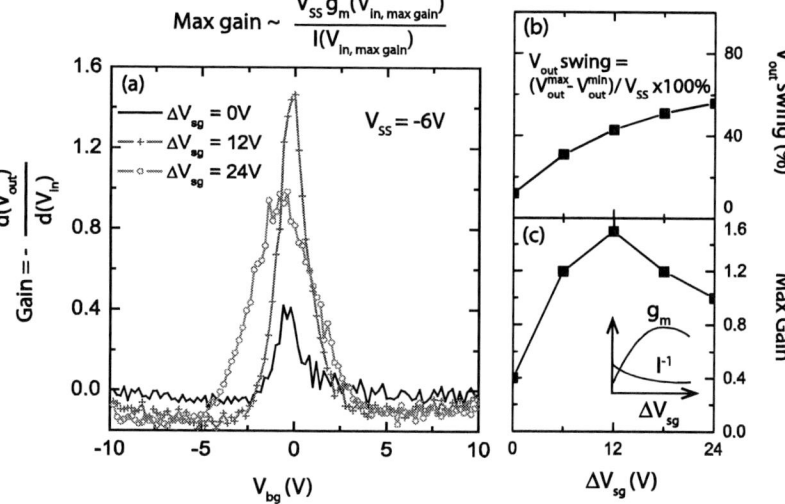

Fig 4. (a) Graphene inverter gain for different ΔV_{sg}. (b) Extracted output voltage swing and (c) maximum gain vs. ΔV_{sg}.

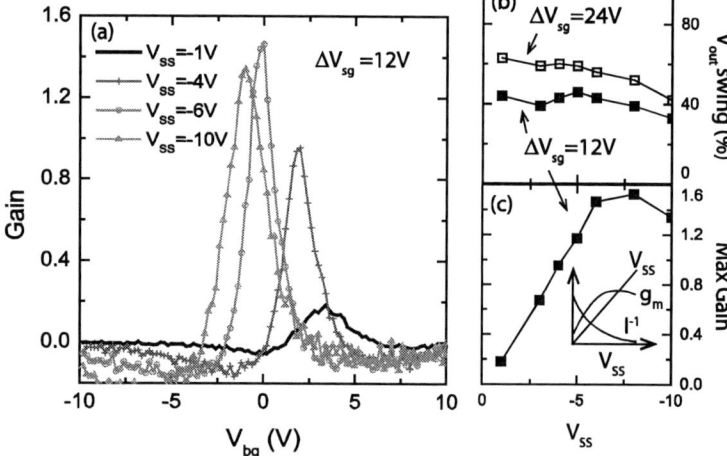

Fig 5. (a) Graphene inverter gain for different supply voltages. (b) Extracted output voltage swing and (c) maximum gain vs. V_{SS}.

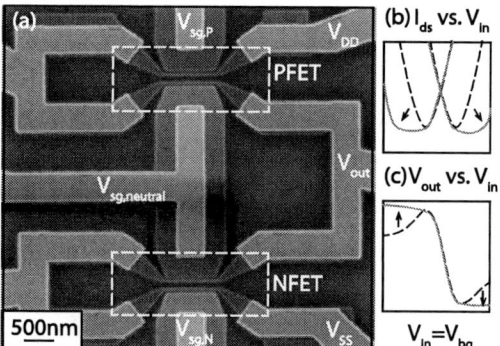

Fig 6. SEM image of a "next generation" structure of a graphene inverter and projected characteristics (solid red lines).

978-1-61284-243-1/11 $26.00 © 2011 IEEE

Gate capacitance scaling and graphene field-effect transistors with ultra-thin top-gate dielectrics

Babak Fallahazad, Kayoung Lee, Seyoung Kim, Chris Corbet, Emanuel Tutuc

Microelectronics Research Center, University of Texas at Austin, 10100 Burnet Rd., Austin, TX 78758
Phone: (512) 471-4960, Fax: (512) 471-8575, email: babak@mail.utexas.edu

Graphene has emerged recently as an attractive channel material for high frequency analog device applications. High carrier mobility and large gate capacitance are both desirable attributes for such devices. A main obstacle however in depositing thin dielectrics on graphene, with high dielectric constant is its chemical inertness. This obstacle can be overcome by either directly depositing the dielectric, e.g. using sputtering or e-beam evaporation, or by using a seed layer which provides nucleation sites for atomic layer deposition (ALD). The interfacial layer however reduces the gate capacitance and can also impact the quality of the ALD dielectric subsequently grown. Here we provide a systematic study of gate capacitance scaling of graphene field effect transistors with Al_2O_3 gate dielectric with two seed layers, oxidized aluminum and oxidized titanium. Our results show the oxidized Ti film on graphene provides a smooth surface, which allows us to use a Ti nucleation layer as thin as 6Å, and achieve uniform coverage required for the subsequent ALD. The k-value of the ALD Al_2O_3 grown on graphene using oxidized Ti as nucleation layer is 12.7, a value 2.5 times larger than the ALD Al_2O_3 grown using oxidized Al. We demonstrate graphene devices with ultra-thin top gate dielectrics, with EOT values as low as 3.5 nm.

Single-layer graphene flakes are identified using optical contrast and Raman spectroscopy, after mechanical exfoliation on 285nm thermally grown SiO_2 on highly doped Si substrate. The active areas of the devices are then isolated by e-beam lithography and oxygen plasma etching. After defining metal contacts (Ni) through a second e-beam lithography and lift-off process, a layer of Ti or Al is thermally evaporated on graphene. The Ti (Al) thin film becomes oxidized once exposed to the air and provides the nucleation sites for the ALD. The thickness of the deposited Ti and Al films are 6Å and 15Å respectively. The surface roughness of as deposited films on several monolayer graphene devices was measured using atomic force microscopy. Figure 1 data show that the surface roughness of the oxidized Ti layer is 0.244 nm, a value twice lower than that of the oxidized Al layer, with the average value of 0.520 nm. The lower surface roughness of the Ti layer, a consequence of a lower surface diffusion, allows that oxidized Ti to fully cover the graphene surface at lower thickness by comparison to Al, and facilitates the deposition of ultra-thin gate dielectrics. After the nucleation layer deposition, the sample is transferred to the ALD chamber, where an Al_2O_3 film is grown using Trimethyl Aluminum (TMA) and H_2O as precursors. A top gate is defined with final e-beam lithography, metal deposition, and lift-off. The schematic view of a dual-gated device is illustrated in Fig. 2(a), and an optical micrograph of a sample device is shown in Fig. 2(b).

Figure 3 shows an example of the sample resistance (R) vs. top-gate voltage (V_{TG}), measured at different back-gate voltages (V_{BG}); the device gate stack consists of 6 Å of oxidized Ti and 20 Å of Al_2O_3. The ratio of the top-gate capacitance (C_{TG}) to back-gate capacitance (C_{BG}) is calculated from the slope of Dirac voltage ($V_{Dirac,TG}$) shift as a function of V_{BG} (Fig. 3 inset). The typical C_{TG}/C_{BG} value for devices with this top dielectric stack is 65, corresponding to a top gate capacitance of C_{TG}=910 nF/cm^2; the value of C_{BG} =14 nF/cm^2 is determined using C-V measurements.

Figure 4(a) shows the measured C_{TG} vs. the Al_2O_3 film thickness (t_{Al2O3}), for the two interfacial layers, namely oxidized Ti and oxidized Al. The devices with Ti interface have a higher C_{TG} at the same Al_2O_3 thickness. The top-gate capacitance consists of interface metal oxide capacitance (C_{Int}) in series with the C_{Al2O3} (Fig. 4-b). To decouple the effect of interfacial layer from Al_2O_3 capacitance, in Fig. 4(c) we plot C_{TG}^{-1} vs. t_{Al2O3} graph for both interfaces. The C_{TG}^{-1} shows a linear dependence on t_{Al2O3}, with slopes inversely proportional to Al_2O_3 dielectric constant. Remarkably, the Al_2O_3 dielectric constants are very different for the two interfaces, with extracted k-values of 5 and 12.7 for the films grown on oxidized Al, and oxidized Ti respectively. While the origin of this finding remains to be clarified, Fig. 4 data clearly show that the interfacial layer critically impacts the quality of ALD dielectric.

Figure 5 shows the mobility (μ), extracted from four-point, gate-dependent measurements, vs. t_{Al2O3} for devices using oxidized Ti as interfacial layer. These data show that the first ~4 nm of Al_2O_3 lead to a mobility decrease, with the μ values remaining constant for thicker Al_2O_3 layers; a similar finding has been reported for HfO_2 deposited on graphene and was attributed to charged point defects due to oxygen vacancies [2]. Combining the C_{TG} vs. t_{Al2O3} of Fig. 4, with the μ vs. t_{Al2O3} data of Fig. 5, in Fig. 6 we show the $\mu \cdot C_{TG}$ product as a function of t_{Al2O3}.

The top gate leakage current density (J) versus the electric field (E) is shown in Fig. 7, for different stack thicknesses. The measured gate resistance of devices with the thinnest top-gate dielectric, 6 Å oxidized Ti followed by 2 nm Al_2O_3 is ~1 MΩ. The leakage current density for devices with stack thicknesses of 6 nm or more are less than 10^{-5} A·cm^{-2}.

[1] S. Kim *et al., Appl. Phys. Lett.* **94**, 062107 (2009). [2] B. Fallahazad *et al., Appl. Phys. Lett.* **97**, 123105 (2010).

978-1-61284-243-1/11 $26.00 © 2011 IEEE

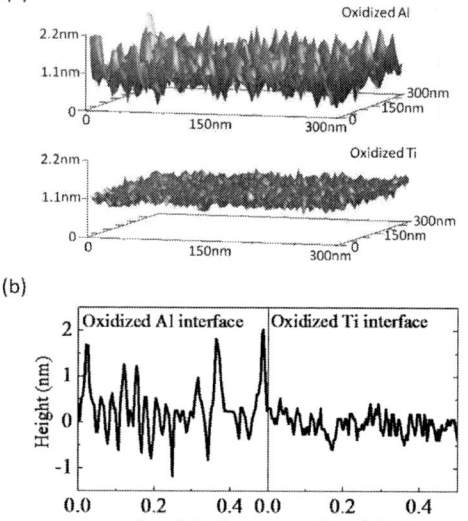

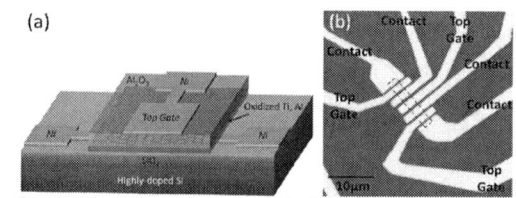

Fig. 2: (a) Schematic of a graphene device with a top dielectric stack consisting of Ti oxidized nucleation layer, and ALD Al_2O_3 (b) Optical micrograph of a dual-gated graphene FET with Ni contacts. The dashed contour marks the graphene area.

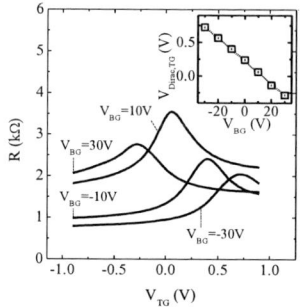

Fig. 1: (a) Topography comparison of 15 Å oxidized Al and 6 Å oxidized Ti deposited on of graphene, measured with AFM. (b) Scan along a 500 nm line. The oxidized Ti surface is smoother (R_a=0.240 nm) than with of oxidized Al (R_a=0.520 nm).

Fig. 3: Resistance (R) of the dual gated GFET measured as a function of the top-gate bias (V_{TG}), for different back-gate voltages. Inset: Dirac voltage ($V_{Dirac,TG}$) shift as a function of the back-gate voltage (slope: C_{TG}/C_{BG}=65).

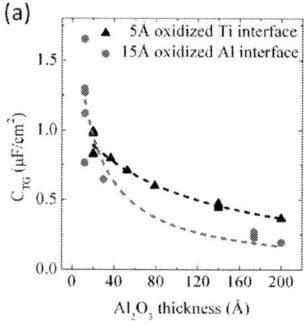

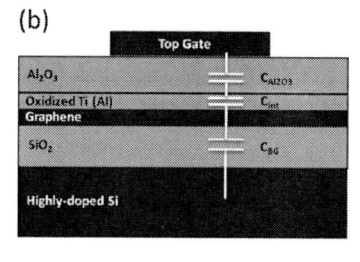

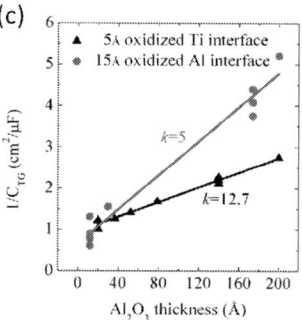

Fig. 4: (a) The top gate capacitance (C_{TG}) of dual-gated GFETs as a function of Al_2O_3 thickness for devices with oxidized Al interface and oxidized Ti interface used as nucleation layers. (b) Schematic view of the different capacitor components in a dual-gated graphene device. (c) $1/C_{TG}$ vs. Al_2O_3 thickness; the data are used to decouple the interface and the ALD Al_2O_3 capacitances.

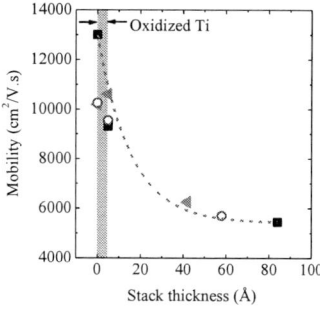

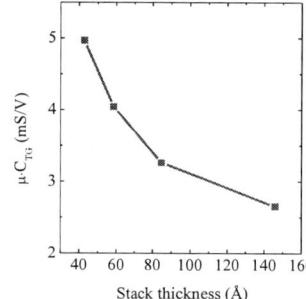

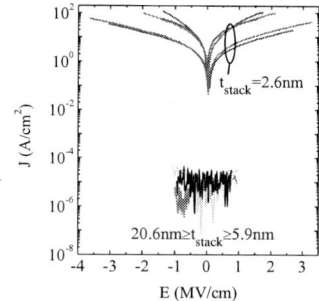

Fig. 5: μ vs. top dielectric stack thickness for three mono-layer graphene devices. The stack consists of 6 Å oxidized Ti followed Al_2O_3 grown by ALD.

Fig. 6: $\mu \cdot C_{TG}$ as a function of top gate stack (consists of oxidized Ti (6 Å) and Al_2O_3) thickness. The $\mu \cdot C_{TG}$ product determines the transconductance (g_m). $g_m = \left(\frac{W}{L}\right) \cdot \mu \cdot C_{TG} \cdot V_D$, where W and L are the device width and length, V_D is the drain bias.

Fig. 7: Leakage current density (J) as a function of vertical electric field (E) for different top dielectric thicknesses. The gate resistance for devices with 2.6 nm top dielectric stack is ~1 MΩ.

978-1-61284-243-1/11 $26.00 © 2011 IEEE

SYNTHESIS AND APPLICATIONS OF GRAPHENE FOR FLEXIBLE ELECTRONICS

Byung Hee Hong

Department of Chemistry & SKKU Advanced Institute of Nanotechnology, Sungkyungkwan University, Suwon 440-746, Korea

As the paradigm of electronic devices changes toward flexible electronics, the development of new materials that can stand high strains becomes more and more important. In particular, flexible transparent electrodes are essential to develop a new type of displays and solar cells that are flexible, foldable or stretchable. However, the current material for transparent electrodes such as indium tin oxides (ITO) is not suitable as flexible electrodes due to its fragility. Graphene, an atom thick carbon materials, is not only highly transparent and conducting but also extremely flexible, which is expected to replace the use of ITO both for flexible and non-flexible electronics in the future. Recently, a method to produce graphene films in large scale using roll-to-roll process has been developed, and the ITO replacement of touch screen panels was successfully demonstrated. In this talk, the overview of recent progresses in the macroscopic applications of graphene for various macroscopic electronics including flexible light emitting diodes, solar cells, batteries, and thin-film transistors will be presented, and the future directions of graphene-based macroelectronics will be discussed.

978-1-61284-243-1/11 $26.00 © 2011 IEEE   38

Sub-10 nm Epitaxial Graphene Nanoribbon FETs

K. Tahy[1], W. S. Hwang[1], J.L. Tedesco[2], R.L. Myers-Ward[2], P.M. Campbell[2], C.R. Eddy[2] Jr., D.K. Gaskill[2], H. Xing[1], A. Seabaugh[1], and D. Jena[1]

[1]Department of Electrical Engineering, University of Notre Dame, Notre Dame, IN 46556, USA

phone: (574) 631-8835, fax: (574)631-4393, email: djena@nd.edu

[2]U.S. Naval Research Laboratory, Washington, DC 20375, USA

Graphene is being investigated as a promising candidate for electronic devices. For digital electronic devices, a substantial bandgap is necessary. It is possible to open a bandgap in graphene by quantum confinement of the carriers in patterned graphene nanoribbons (GNRs); GNRs with width W nm have a bandgap $E_g\sim1.3/W$ eV [1]. This implies that sub-10 nm wide ribbons can enable room-temperature operation of GNRs as traditional semiconductors, but with ultimate vertical scaling, and still take advantage of high current drives. To date, GNRs have been fabricated from exfoliated graphene [2] and operated by back gates, or nanometer scale ribbons produced by 'explosive' methods [3] that are neither controlled nor reproducible. These methods are not suitable for large-area device fabrication. In this work, we report lithographically patterned GNRs on epitaxial graphene on SiC substrates. Specifically, we show the first top-gated GNR field-effect transistors (FETs) on epi-graphene substrates that exhibit the opening of a substantial energy bandgap (exceeding ~0.15 eV at a ribbon width of 10 nm), respectable carrier mobility (700 - 800 cm^2/Vs), high current modulation (10:1 at 300 K), and high current carrying capacity (0.3 mA/μm at V_{DS} = 1 V) at the same time. Both single GNR and GNR array devices are reported.

Epitaxial graphene was grown on Si-face 6H-SiC [4]. Raman measurements indicate predominantly single layer graphene coverage. The graphene was patterned by e-beam lithography and etched in an O$_2$ plasma. Hydrogen silsesquioxane (HSQ), a negative-tone electron-beam resist, was used to form sub-10 nm GNRs. A 15 nm thick ALD Al$_2$O$_3$ film on spin-coated HSQ was used as the top-gate insulator [Fig. 1]. The HSQ was used to seed the ALD deposition. The gate stack was found to have a small hysteresis in the I-V characteristics. E-beam evaporated Cr/Au source/drain contacts were then deposited to form FETs.

The device transfer characteristics of a top-gated 10 nm wide GNR device are shown in Fig. 2(a). Room temperature modulation is ~10, but the 4 K modulation increases to ~10^6, clearly indicating the opening of a bandgap. The opening of a 0.15 eV bandgap was confirmed by temperature dependent measurements of the off-state current. Measurement of the differential conductance as a function of V_{DS} and V_{GS} at 4 K shows a transport gap exceeding 0.15 eV [Fig. 2(b)]; similar gaps were measured on multiple devices and are plotted in Fig. 1(e). The family I-V curves shown in Fig. 2(c) indicate that the GNR FETs a) switch off, b) carry ~0.3 mA/μm current density at V_{DS} = 1 V, and c) are yet to saturate. One way to further increase the net current drive is to use a parallel array of GNRs between the source and the drain. In Fig. 3, the characteristics of such a GNR array FET with 30 parallel 13 nm wide GNRs with a 30 nm pitch is shown. That the drain current scales with the number of ribbons is expected, but the net transfer characteristics of individual GNRs is also preserved in the array and reproducible over multiple devices. This implies that possible non-uniformities in GNR widths and edge roughnesses have a minimal effect in the measured devices. The maximum high-field current drive measured before breakdown was extremely high – approaching ~7.5 mA/μm if scaled to the active ribbon width (390 nm). Scaling by the total channel width (30 x 13 nm + 30 x 17 nm gap) would lead to 3.25 mA/μm, which is still remarkable. Such high current drives have never been reported in graphene before. We attribute this high current carrying capability to the high electrical and thermal conductivity of the 1D graphene channels (due to absence of lateral scattering), coupled with the excellent thermal conductivity of the underlying SiC substrate.

The above results are the first report of top-gated GNR FETs on large-area epitaxial graphene exhibiting exceptionally high drive currents, the opening of a substantial bandgap, and linear scaling of properties with the number of GNRs in parallel arrays. In this light, the disadvantages typically attributed to GNRs (such as edge-scattering and resultant degradation of device performance) need to be carefully re-examined. By scaling the widths down further, substantial modulation at room temperature is expected. This work was supported by the Semiconductor Research Corporation Nanoelectronics Research Initiative and the National Institute of Standards and Technology through the Midwest Institute for Nanoelectronics Discovery (MIND) and, in part, by the Office of Naval Research (ONR).

[1] T. Fang, A. Konar, H. Xing, and D. Jena, Phys. Rev. B, vol. 78, p. 205403 (2008).

[2] X. Li, X. Wang, L. Zhang, S. Lee, and H. Dai, Science, vol. 319, p. 1229 (2008).

[3] P. Kim, M. Y. Han, A. F. Young, I. Meric, and K. L. Shepard, IEDM, p. 241 (2009).

[4] J. L. Tedesco et al., ECS Trans. Vol. 19, pp. 137-150 (2009).

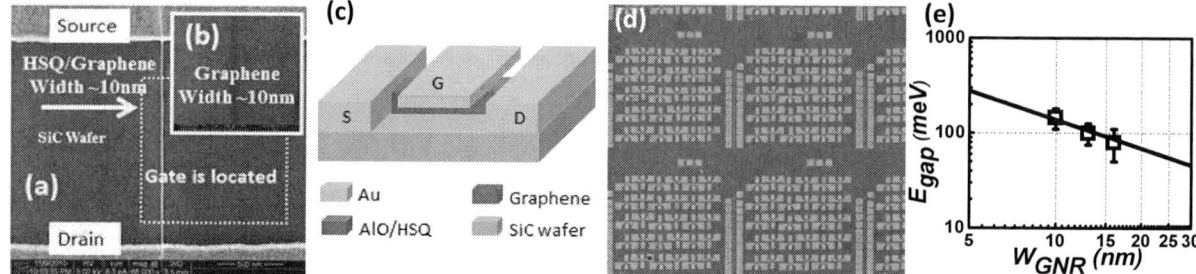

Fig. 1. (a) SEM micrograph of a GNR FET showing the 10 nm HSQ mask and the source/drain electrodes. **(b)** The 10 nm graphene channel after the removal of the HSQ mask. **(c)** Schematic of GNR FET with Au/Al$_2$O$_3$/HSQ gate stack. **(d)** Wafer-scale optical image of the GNR FETs. **(e)** The scaling of the band gap E_{gap} with respect to decreasing GNR width follows the expected $1.3/W_{GNR}$ trend.

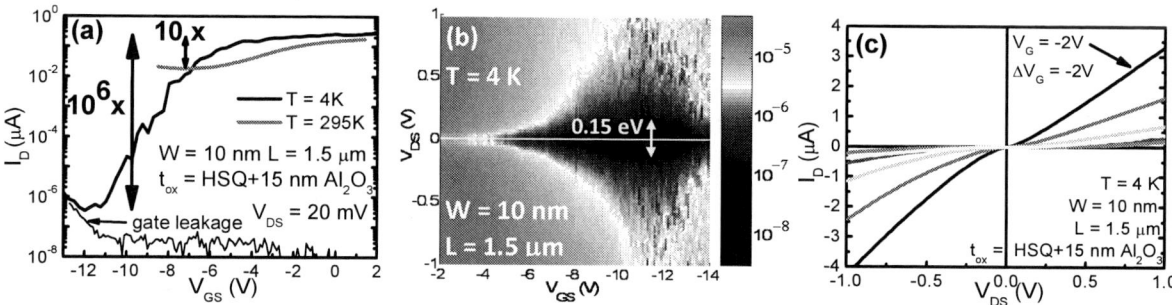

Fig. 2. (a) Transfer characteristic of a 10 nm GNR FET on room temperature and at 4 K. **(b)** Differential conductance dI_D/dV_{DS} as a function of V_{DS} and V_{GS}. The color bar is the exponent, $\log_{10}(dI_{DS}/dV_{DS})$ in Ω^{-1}. Transport gap of ~ 0.15 eV is observed. **(c)** Low temperature family $I-V$ of the same device showing good on/off ratio.

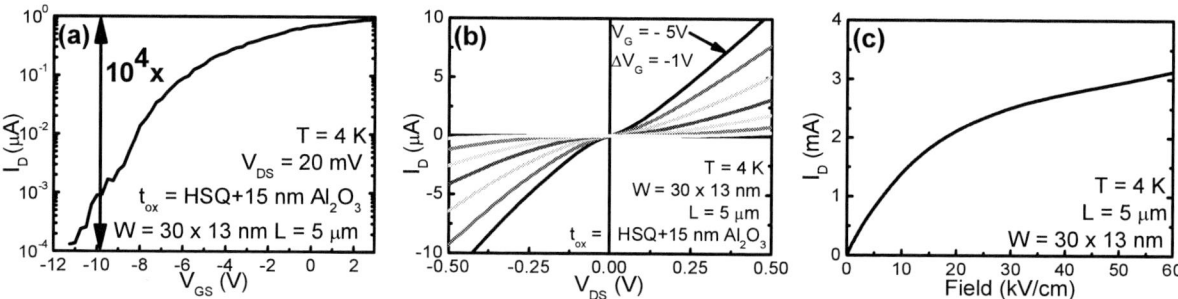

Fig. 3. (a) Gate dependence of I_{DS} in a FET with 30 parallel 13 nm GNR channel with a 30 nm pitch. Despite the possible width variations, an on/off ratio of 10^4 is achieved at low temperature. **(b)** Output characteristics of the 30 GNR array-FET at 4 K. **(c)** The current drive of the 30 GNR array-FET is approximately 30 times of the current of a single GNR FET. The very high measured maximum current density of ~7.5 mA/μm is attributed to the excellent thermal conductivity of the SiC substrate. Current density is scaled for the active ribbon area (390 nm). Scaling by the total channel width (30 x 13 nm + 30 x 17 nm gap) would lead to 3.25 mA/μm, which is still remarkable.

978-1-61284-243-1/11 $26.00 © 2011 IEEE   40

Effect of Oxide Thickness Scaling on Self-Heating in Graphene Transistors

Sharnali Islam, Myung-Ho Bae, Vincent Dorgan, and Eric Pop

Dept. of Electrical and Computer Engineering, University of Illinois, Urbana-Champaign

Micro and Nanotechnology Lab, 208 N Wright St, Urbana IL 61801. E-mail: epop@illinois.edu

Recent studies using infrared (IR) imaging of graphene transistors [1,2] have revealed substantial Joule heating under realistic operating conditions for graphene-on-insulator (GOI) devices. Here we use simulations calibrated against experimental data to examine the trends of performance degradation caused by self-heating as a function of insulator (SiO_2) thickness. We also examine both unipolar and ambipolar operating conditions, and find that peak channel temperatures are proportional to oxide thickness for the unipolar case (as would be expected), but for ambipolar operation an optimum oxide substrate thickness exists (~80 nm) which minimizes peak temperature, due to competing electrostatic and thermal effects.

To calibrate our simulations, we performed infrared imaging of a functioning GOI device on $t_{ox} = 100$ nm SiO_2 (Fig. 1), which was combined with our previous observations on 300 nm SiO_2 [1]. The measured temperature of the graphene channel is shown in Fig. 1(b), at $V_{SD} = -12$ V and various gate voltages V_{GD} (Dirac voltage $V_0 = 5.2$ V). As V_{GD} is varied from -5 V to 5 V, the channel is n-type near the source and p-type near the drain, operating in the ambipolar regime. The peak temperature (hot spot) is the location of maximum electric field and minimum charge density [1], i.e. the charge neutrality point (CNP) where the n- and p-regions meet. Changing V_{GD} alters the n- and p- concentrations and moves the hot spot within the channel. Fig. 2 shows excellent agreement between our measurements (symbols) and simulations (lines).

The simulations self-consistently calculate the charge densities, field, potential, and temperature along the graphene channel [1,3]. Mobility and saturation velocity are modeled as functions of both charge density and temperature [3]. The carrier densities and temperature profiles at various V_{GD} are shown in Figs. 2(b-c) respectively. To understand the role of oxide thickness scaling, in Fig. 3 we examine the temperature profiles during ambipolar transport for $t_{ox} = 20$–300 nm, at the same power density. Fig. 3(b) plots the *average* and *peak* temperature of the graphene channel vs. oxide thickness. The average temperature decreases with scaled oxide, but the peak temperature begins to increase when t_{ox} is reduced below ~80 nm. This occurs because the Joule heating effect induced by the high electric field at the CNP overcomes the cooling ability of the thinner oxide at $t_{ox} \leq 80$ nm. Thus, thinner oxides give a lower average temperature, but can cause the formation of a highly localized hot spot in the ambipolar regime.

We now extend our simulations to devices smaller than the IR wavelength (Figs. 4-5), where IR thermal imaging cannot be directly applied. We consider high-bias conditions under unipolar transport in Fig. 4, which shows I_D-V_{SD} curves with (solid, red) and without self-heating (dashed, blue) for different gate voltages. Thermal effects at high bias can lead to current degradation up to ~10% when current densities approach 1 mA/μm (Fig. 4a) and nearly 20% at higher carrier density (Fig. 5b, 2 MV/cm vertical field). Interestingly, since graphene is a gapless material, we find that self-heating during device operation can alter the majority carrier concentration through thermal generation (Fig. 4b). Changes in carrier density lead to changes in the field distribution and electrostatics along the channel (Fig. 4c), all effects that must be carefully taken into account in such graphene devices. Finally, in Fig. 5 we examine the peak temperature and the degradation of saturation current as a function of oxide thickness. In unipolar operation, both the peak temperature and the current degradation scale proportionally with oxide thickness, as expected, due to scaling of the thermal resistance under the graphene channel.

To summarize, we have studied the effects of self-heating on graphene-on-insulator transistors. Physically, we find that ambipolar and unipolar operation lead to different scaling of peak channel temperature. For realistic devices, although the current degradation can be kept <10% by careful choice of oxide thickness (~80 nm), sharply peaked temperatures can nevertheless have an impact on long-term device reliability [4] and must be carefully considered in future device designs.

[1] M.-H. Bae et al., *Nano Lett.* 10, 4787 (2010). [2] M. Freitag et al, *Nature Nano* 5, 497 (2010). [3] V. Dorgan et al., *Appl. Phys. Lett.* 97, 082112 (2010). [4] D. K. Schroder et al., *J. Appl. Phys.* 94, 1-18 (2003).

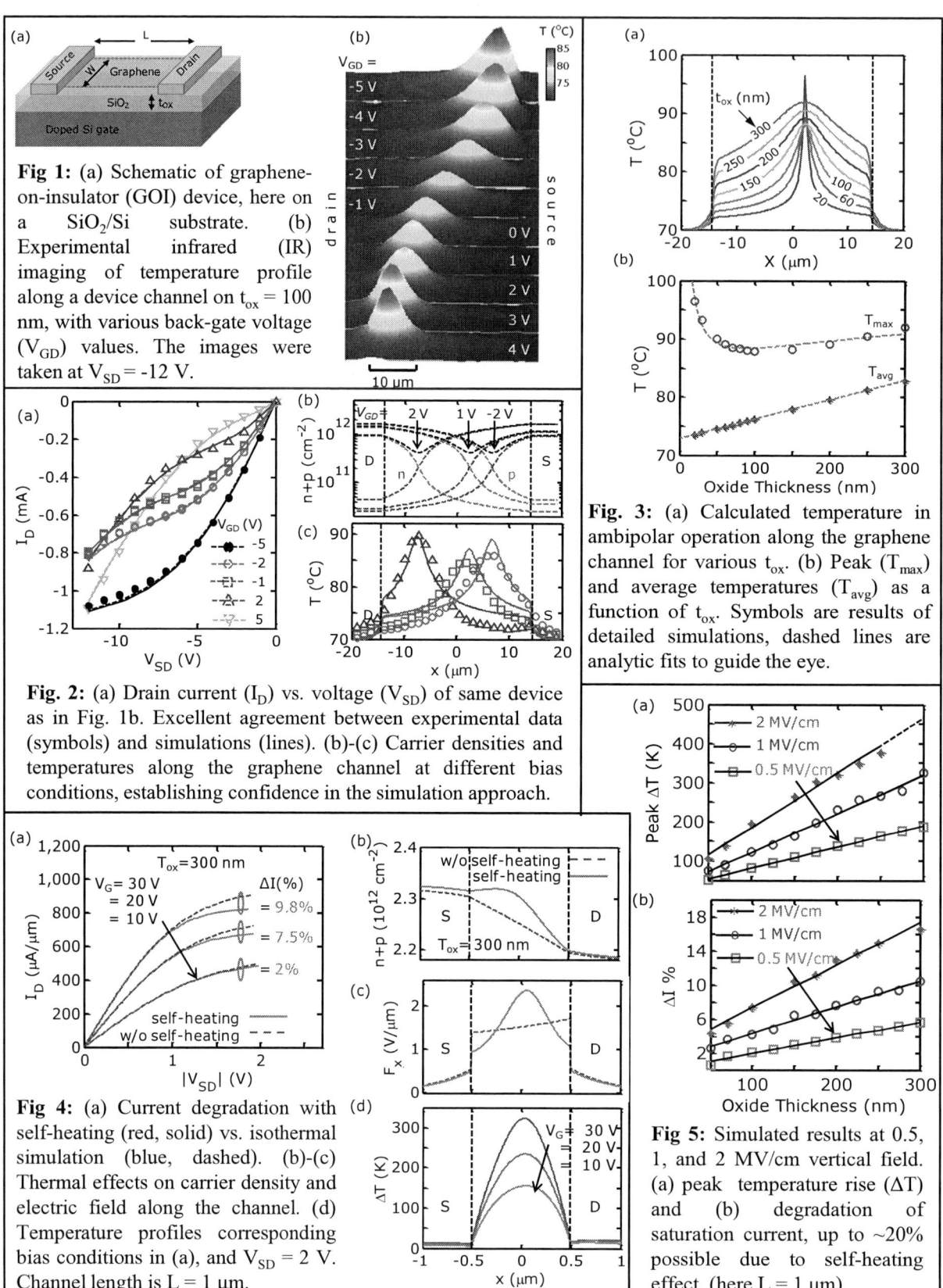

Fig 1: (a) Schematic of graphene-on-insulator (GOI) device, here on a SiO$_2$/Si substrate. (b) Experimental infrared (IR) imaging of temperature profile along a device channel on t$_{ox}$ = 100 nm, with various back-gate voltage (V$_{GD}$) values. The images were taken at V$_{SD}$ = -12 V.

Fig. 2: (a) Drain current (I$_D$) vs. voltage (V$_{SD}$) of same device as in Fig. 1b. Excellent agreement between experimental data (symbols) and simulations (lines). (b)-(c) Carrier densities and temperatures along the graphene channel at different bias conditions, establishing confidence in the simulation approach.

Fig. 3: (a) Calculated temperature in ambipolar operation along the graphene channel for various t$_{ox}$. (b) Peak (T$_{max}$) and average temperatures (T$_{avg}$) as a function of t$_{ox}$. Symbols are results of detailed simulations, dashed lines are analytic fits to guide the eye.

Fig 4: (a) Current degradation with self-heating (red, solid) vs. isothermal simulation (blue, dashed). (b)-(c) Thermal effects on carrier density and electric field along the channel. (d) Temperature profiles corresponding bias conditions in (a), and V$_{SD}$ = 2 V. Channel length is L = 1 µm.

Fig 5: Simulated results at 0.5, 1, and 2 MV/cm vertical field. (a) peak temperature rise (ΔT) and (b) degradation of saturation current, up to ~20% possible due to self-heating effect. (here L = 1 µm).

978-1-61284-243-1/11 $26.00 © 2011 IEEE

Graphene Quantum Capacitance Varactors for Wireless Sensing Applications

S. J. Koester

University of Minnesota-Twin Cities, 200 Union St. SE, Minneapolis, MN 55455
Ph: (612) 625-1316, FAX: (612) 625-4583, Email: skoester@umn.edu

Introduction: The low density of states in graphene makes it possible for the quantum capacitance to be of the same order of magnitude as the oxide capacitance for experimentally achievable gate dielectric thicknesses [1]. This property, combined with the fact that the density of states varies as a function of energy, means that the capacitance in a metal-oxide-graphene capacitor can be tuned by varying the carrier concentration [2]. The very high mobility and zero band gap in graphene also allow it to remain conductive throughout the entire tuning range, making graphene an idea material to realize a high quality factor (Q) variable capacitor (varactor). If combined with an on-chip inductor to form an LC oscillator circuit, graphene varactors could enable a new class of ultra-compact sensors with wireless readout capability. Compared to MEMS-based varactors [3], the extremely-large capacitance per unit area of graphene varactors should allow orders-of-magnitude improvement in scalability, a vital feature for numerous applications including *in vivo* sensing where small size is critical. In this abstract, the device concept is described and simulated performance projections are provided. The main findings in this study are that wide frequency tuning ratios (> 50%) and high Q (> 40 at 1 GHz) are possible using realistic assumptions for the graphene properties, device dimensions and parasitic resistances.

Device Description and Simulation Assumptions: The physical structure of the varactor is shown in Fig. 1. The cross-sectional geometry could vary depending upon the quantity to be sensed, and therefore may either utilize a standard top-gate geometry (Fig. 1(a), e.g., for sensing radiation-induced trapped charges in the buried oxide [4]) or an inverted geometry (Fig. 1(b) e.g., for sensing absorbed molecules, pH, etc. [5]). In either case, a multi-finger layout such as the one shown in Fig. 1(c) is likely to be needed to enable the capacitance to be increased while minimizing series resistance. A simple equivalent circuit model as shown in Fig. 2 has been utilized to analyze the potential performance of these devices. In this model, the relations in [2] were utilized for the quantum capacitance and the electron and hole concentrations, n and p, respectively. The channel resistance, R_{ch}, was determined using the relation: $R_{ch} = 0.25 \cdot (L_g + 2L_{ext})/(q\mu(n + p))/W_g/N_{fingers}$. The electron and hole mobilities, μ, were assumed to be equal, and μ was further assumed to be invariant with carrier concentration. Additional series resistance components associated with the contact and metallization (Au) resistances were also included in the model. Parallel conductance associated with gate leakage was further modeling assuming typical values for high-κ oxides on Si [6].

Simulation Results and Discussion: Simulations were performed at room temperature using the net carrier concentration, $n_{net} = n - p$ as the independent variable. Results at fixed frequency showing capacitance and quality factor plotted vs. n_{net} are shown in Fig. 3. These simulations utilized an equivalent oxide thickness (EOT) of 0.7 nm, $\mu = 5000$ cm^2/Vs and contact resistance (R_c) of 500 Ω-μm. Additional parameters are shown in Fig. 3. The frequency tuning range was calculated assuming an ideal inductance, L, of 8.8 nH in series with the varactor circuit shown in Fig. 2. The results in Fig. 4 show that the frequency can be tuned by roughly 50% between $n_{net} = 5 \times 10^{11}$ cm^{-2} and 1×10^{13} cm^{-2}, while maintaining $Q \geq 88$ over the entire tuning range. Again using the parameters in Fig. 3, the dependence of the varactor Q on various device parameters was calculated and the results are shown in Fig. 5. While the Q degrades with increasing L_g and reduced R_c and μ, reasonable values can be achieved using values already demonstrated in the literature [7]-[8]. Reducing EOT improves the tuning range (Fig. 6(a), and at extremely-small EOT values, tuning ranges approaching 100% can be achieved (Fig. 6(b)). Using high-κ-on-Si gate leakage values no degradation of Q was found even for gate leakage conductance values above 10^6 μS/cm^2, demonstrating that scaling to very small EOT values is feasible. The form factor benefit of the graphene varactors is shown in Fig. 7. The plot shows that the high capacitance density of the graphene varactor allows frequencies comparable to MEMS based LC circuits, but with much smaller capacitor area. This capability is critical for wireless sensor scaling, since it allows the inductor to shrink while still allowing the tuning frequency range to be below the inductor self-resonance frequency. Finally, despite the fact that large voltages cannot be sustained across the gate oxide, with proper varactor/inductor matching, relatively high ac current levels can be sustained, as shown in Fig. 8.

[1] Z. Chen and J. Appenzeller, IEDM, 2008; [2] T. Fang, et al., *Appl. Phys. Lett.*, 2007; [3] C. Son and B. Ziaie, *IEEE Trans. Biomed. Eng.*, 2008; [4] J.-B. Yau, et al., VLSI-TSA, 2011; [5] C. Shan, et al., *Anal. Chem.*, 2009; [6] J. F. Kang, et al., *Electrochem. & Solid-State Lett.*, 2005; [7] B. Fallahazad, et al., *Appl. Phys. Lett.*, 2010; [8] K. Nagashio, et al., *Appl. Phys. Lett.*, 2010.

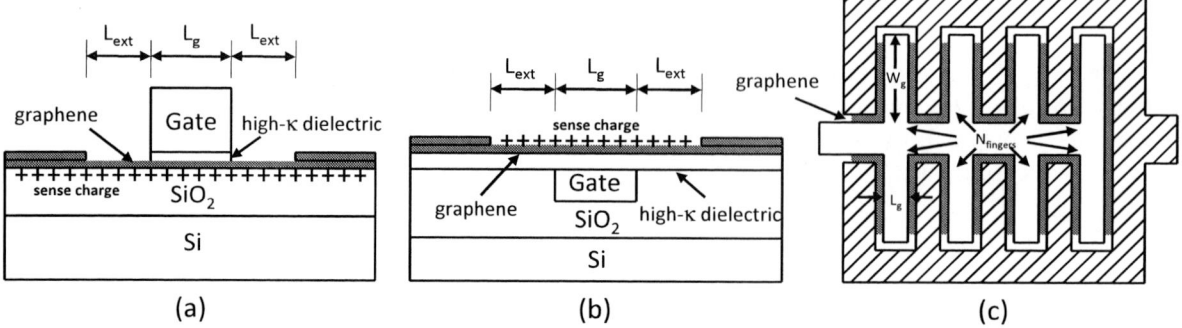

Fig. 1 Schematic layout of graphene varactor sensor geometries. (a) Buried charge-trapping sensor geometry, (b) Surface sensor design with buried gate electrode, (c) Top-view layout of multi-finger varactor geometry.

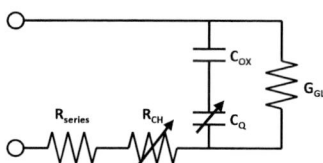

Fig. 2. Equivalent circuit model of varactor. Tuning occurs as the quantum capacitance changes with charge density in the graphene.

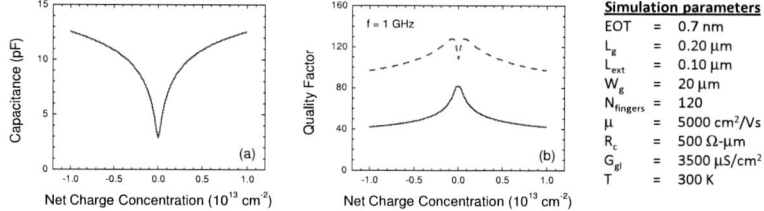

Fig. 3. Plot of (a) capacitance and (b) quality factor vs. net carrier concentration using the parameters shown at right. Solid: with metal resistances, dashed: without metal resistances.

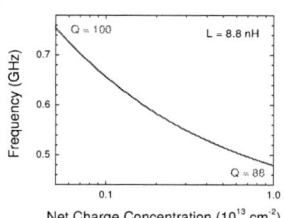

Fig. 4. Resonant frequency vs. net carrier concentration assuming an ideal series inductor ($L = 8.8$ nH). $Q > 88$ over the entire tuning range.

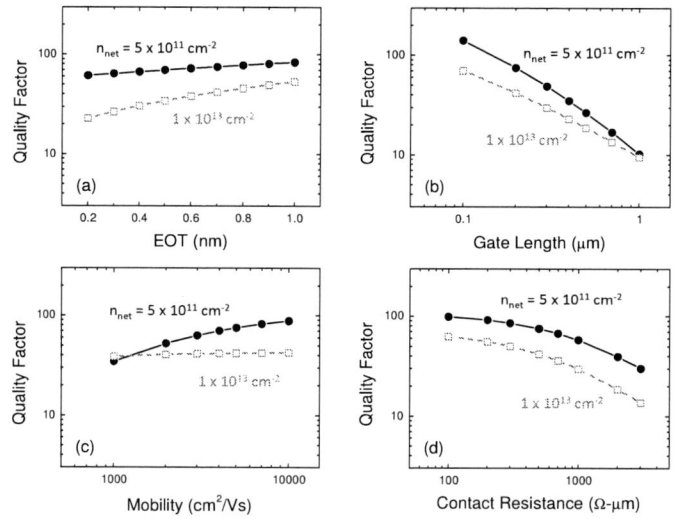

Fig. 5. Quality factor at $f = 1$ GHz for two carrier concentrations plotted vs. (a) equivalent oxide thickness, (b) gate length, (c) mobility, and (d) contact resistance.

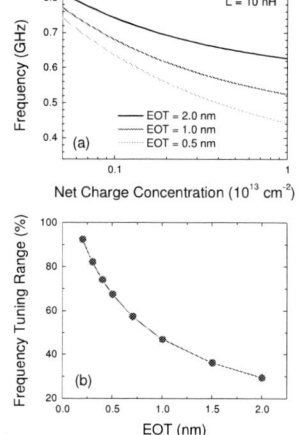

Fig. 6. (b) Plot of frequency vs. net carrier concentration for 3 different values of EOT. (b) Tuning range plotted vs. EOT.

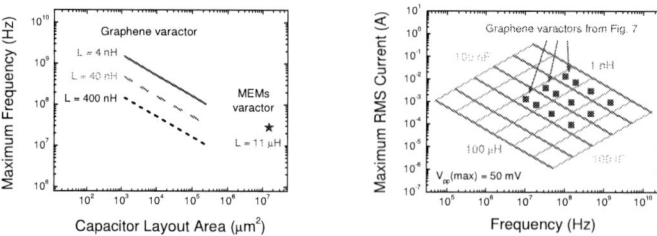

Fig. 7. Max. frequency vs. varactor layout area for 3 L values. Values from [3] are also shown.

Fig. 8. Max. current vs. f for LC resonators where $V_{max-pp} = 50$ mV. Graphene resonator values from Fig. 7 are shown in blue.

978-1-61284-243-1/11 $26.00 © 2011 IEEE

Session III (Lagoon Plaza)

Poster Session

Monday PM, June 20th, 2011

III-1 Student Paper
Effects of Heavily Doped Source on the Subthreshold Characteristics of Nanowire Tunneling Transistors
M. A. Khayer and R. K. Lake, Laboratory for Terascale and Terahertz Electronics (LATTE), Department of Electrical Engineering, University of California, Riverside, California, USA

III-2
Spatially resolved photovoltaic performance of axial GaAs nanowire pn-diodes
A. Lysov, C. Gutsche, M. Offer, I. Regolin, W. Prost, F.-J. Tegude, Center for Nanointegration Duisburg-Essen, University of Duisburg-Essen, Duisburg, GERMANY

III-3
Low loss AlInN/GaN Monolithic Microwave Integrated Circuit Switch
A. Sattu[1], D. Billingsley[1], J. Deng[1], J.Yang[1], R. Gaska[1], M. Shur[2], G. Simin[3], [1]Sensor Electronic Technology, Inc., Columbia, South Carolina, USA, [2]Electrical and Computer Science Engineering, Rensselaer Polytechnic Institute, Troy, New York, USA, and [3]Electrical Engineering, University of South Carolina, Columbia, South Carolina, USA

III-4
High-Mobility Organic Thin-Film Transistors with Photolithographically Patterned Top Contacts
U. Zschieschang[1], N. H. Hansen[2], J. Pflaum[2], T. Yamamoto[3], K. Takimiya[3], H. Kuwabara[4], M. Ikeda[4], T. Sekitani[5], T. Someya[5], and H. Klauk[1], [1]Max Planck Institute for Solid State Research, Stuttgart, GERMANY, [2]University Würzburg and ZAE Bayer e.V., GERMANY, [3]Hiroshima University, Higashi-Hiroshima, JAPAN, [4]Nippon Kayaku Co., Ltd., Kita-ku, Tokyo, JAPAN, and [5]University of Tokyo, Tokyo, JAPAN

III-5
High Breakdown Voltage ZnMgO/In-Ga-Zn-O Heterostructure Transistors
J. Yamaguchi, I. Soga, and T. Iwai, Fujitsu Laboratories Ltd., Atsugi, Kanagawa, JAPAN

III-6
High-Resolution Temperature Sensing with Source-Gated Transistors
R. A. Sporea, J. M. Shannon, and S. R. P. Silva, Advanced Technology Institute, FEPS, University of Surrey, Guildford, Surrey, UNITED KINGDOM

III-7 Student Paper
Voltage-Controlled Spin-Wave-Based Logic Gate
T. Liu and G. Vignale, Department of Physics and Astronomy, University of Missouri, Columbia, Missouri, USA

III-8 Student Paper
Effect of Disorder on Superfluidity in Double Layer Graphene
B. Dellabetta and M. J. Gilbert, Department of Electrical and Computer Engineering and Micro and Nanotechnology Laboratory, University of Illinois, Urbana, Illinois, USA

III-9 Student Paper
Dual Pillar Spin Transfer Torque MRAM with tilted magnetic anisotropy for fast and error-free switching and near-disturb-free read operations
N. N. Mojumder, S. K. Gupta and K. Roy, School of Electrical and Computer Engineering, Purdue University, West Lafayette, Indiana, USA

III-10 Student Paper
Giant Magnetoelectric Effect in Nanofabricated $Pb(Zr_{0.52}Ti_{0.48})O_3$-$Fe_{85}B_5Si_{10}$ Cantilevers and Resonant Gate Transistors
F. Li[1], Z. Fang[1], R. Misra[3], S. Tadigadapa1, Q. Zhang[1,2] and Suman Datta[1], [1]Electrical Engineering, [2]Materials Research Institute, and [3]Physics Department, The Pennsylvania State University, University Park, Pennsylvania, USA

III-11 Student Paper
Observation of Trap-Assisted Steep Sub-threshold Swing in Schottky Source/Drain Al_2O_3/InAlN/GaN MISHEMT
Q. Zhou[1], H. Chen[1], C. Zhou[1], Z. Feng[2], S. Cai[2], K. J. Chen[1], [1]Department of Electronic and Computer Engineering, Hong Kong University of Science and Technology, Kowloon, Hong Kong and [2]National Key Laboratory of Application Specific Integrated Circuit, Hebei Semiconductor Research Institute, Shijiazhuan, CHINA

III-12 Student Paper
Towards Electronics at 1000 °C
D. Maier[1], M. Alomari[1], N. Grandjean[2], J.-F. Carlin[2], M.-A. Diforte-Poisson[3],C. Dua[3], S. L. Delage[3], and E. Kohn[1], [1]Institute of Electron Devices and Circuits, University of Ulm, Ulm, GERMANY, [2]École polytechnique fédérale de Lausanne, Lausanne, SWITZERLAND, and [3]III/V Lab, Marcoussis, FRANCE

III-13 Student Paper
Bias Temperature Stress Analysis of ZnO Thin Film Transistors with HfO_2 Gate Dielectrics
J. J. Siddiqui[1], J. D. Phillips[1], K. Leedy[2], and B. Bayraktaroglu[2], [1]EECS Department, University of Michigan, Ann Arbor, Michigan, USA and [2]Air Force Research Laboratory, Sensors Directorate, Wright-Patterson AFB, Ohio, USA

III-14 Student Paper
Carbon Nanotube Purified Ink-Based Printed Thin Film Transistors: Novel Approach in Controlling the Electrical Performance
N. Rouhi, D. Jain, and P. J. Burke, Electrical Engineering and Computer Science Department, University of California-Irvine, Irvine, California, USA

III-15
Monolayer MoS_2 Transistors – Ballistic Performance Limit Analysis
K. Ganapathi, Y. Yoon, and S. Salahuddin, Department of Electrical Engineering and Computer Sciences, University of California, Berkeley, California, USA

III-16 Student Paper
RF performance projections for 2D Graphene Transistors: Role of Parasitics at the Ballistic transport limit
P. Zhao[1], D. Jena[1], and, S. O. Koswatta[2], [1]Department of Electrical Engineering, University of Notre Dame, Notre Dame, Indiana, USA and [2]IBM Research Division, T. J. Watson Research Center, Yorktown Heights, New York, USA

III-17
High Performance N- and P-Type Gate-All-Around Nanowire MOSFETs Fabricated on Bulk Si by CMOS-Compatible Process
Y. Song, H. Zhou, Q. Xu, J. Luo, C. Zhao and Q, Liang, Key Laboratory of Microelectronics Devices & Integrated Technology, Institute of Microelectronics, Chinese Academy of Sciences, Beijing, CHINA

III-18 Student Paper
The effect of field effect device channel dimensions on the effective mobility of graphene
A. Venugopal[1], J. Chan[1], W. P. Kirk[1], L. Colombo[2] and E. M. Vogel[1], [1]University of Texas at Dallas, and [2]Texas Instruments Incorporated, Dallas, Texas, USA

III-19 Student Paper
Top-gated single-electron transistor in germanium nanowires
S.-K. Shin, S. Huang, N. Fukata, and K. Ishibashi, Advanced Device Laboratory, RIKEN, Wako, Saitama, JAPAN

978-1-61284-243-1/11 $26.00 © 2011 IEEE

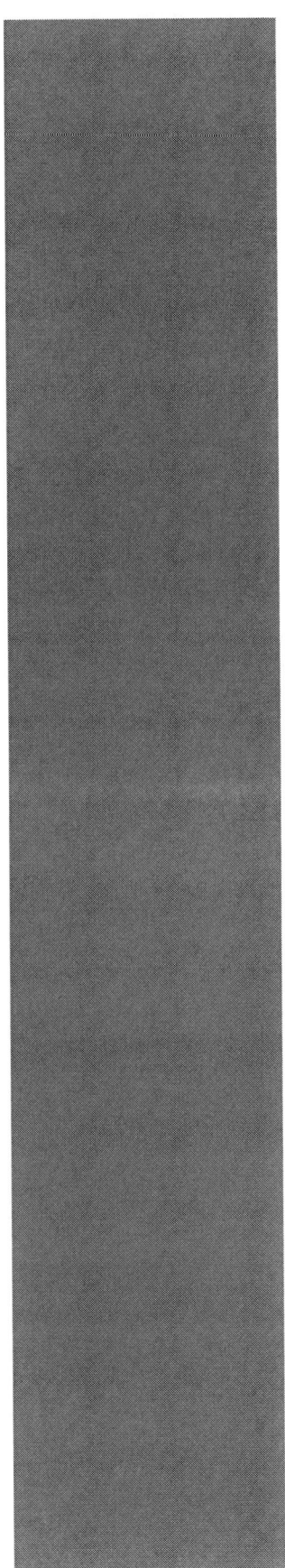

III-20 Student Paper
Reliability of Ambipolar Switching Poly-Si Diodes for Cross-Point Memory Applications
M. H. Lee[1], C.-Y. Kao[1], C.-L. Yang[1], Y.-S. Chen[2], H. Y. Lee[2], F. Chen[2], and M.-J. Tsai[2]
[1]Institute of Electro-Optical Science and Technology, National Taiwan Normal University, Taipei, TAIWAN and [2]Electronics and Optoelectronics Research Laboratories, Industrial Technology Research Institute, Hsinchu, TAIWAN

III-21 Student Paper
Electrochemical supercapacitor based on flexible pillar graphene nanostructures
J. Lin[1], J. Zhong[1], D. Bao[2], J. Reiber-kyle[3], W. Wang[4], V. Vullev[2], M. Ozkan[3], C. S. Ozkan[1,4], [1]Department of Mechanical Engineering, [2]Department of Bioengineering, [3]Department of Electrical Engineering, and [4]Materials Science and Engineering Program, University of California, Riverside, California, USA

III-22 Student Paper
"Zero" Drain-Current Drift of Inversion-Mode NMOSFET on InP (111)A Surface
C. Wang, M. Xu, R. Colby, E. A. Stach and P. D. Ye, School of Electrical and Computer Engineering and Birck Nanotechnology Center,Purdue University, West Lafayette, Indiana, USA

III-23 Student Paper
Improvement of efficiency in inverted bottom-emission white OLEDs by doping the hole transport layer
H. Lee[1], J. Kwak[2], J. Lim[3], K. Char[3], S. Lee[4] and C. Lee[1], [1]School of Electrical Engineering and Computer Science, Inter-University Semiconductor Research Center, Seoul National University, Seoul, KOREA, [2]Department of Electronics Engineering, Dong-A University, Busan, KOREA, [3]School of Chemical and Biological Engineering, Intelligent Hybrids Research Center, Seoul National University, Seoul, KOREA, and [4]Department of Chemistry, Seoul National University, Seoul, KOREA

III-24 Student Paper
Vertical Organic Field-Effect Transistor Array Fabrication Based on Laser Holography Lithography Process
D. Kim and Y. Hong, Department of Electrical Engineering and Computer Science, Seoul National University, Seoul, KOREA

III-25 Student Paper
Ambipolar Nano-crystalline-silicon TFTs with Submicron Dimensions and Reduced Threshold Voltage Shift
A. Subramaniam, K. D. Cantley, R. A. Chapman, B. Chakrabarti, and E. M. Vogel, Department of Electrical Engineering, The University of Texas at Dallas, Richardson, Texas, USA

III-26
Monolithically Grown $In_xGa_{1-x}As$ Nanowire on Silicon Tandem Solar Cells with High Efficiency
J. C. Shin[1], K. H. Kim[2], H. Hu[2], K. J. Yu[1], J. A. Rogers[2,1], J.-M. Zuo[2], and X. Li[1,2], [1]Department of Electrical and Computer Engineering and [2]Department of Materials Science and Engineering, University of Illinois, Urbana, Illinois, USA

III-27 Student Paper
3D Simulation of Electrical Characteristic Fluctuation Induced by Interface Traps at Si/high-κ Oxide Interface and Random Dopants in 16-nm-Gate CMOS Devices
H.-W. Cheng, Y.-Y. Chiu, F.-H. Li, and Y. Li, Department of Electrical Engineering, National Chiao Tung University, Hsinchu, TAIWAN

III-28 Student Paper
Creating dynamic nanowire devices using wrapped gates
K. Storm, G. Nylund, M. Borgström, J. Wallentin, C. Fasth, C. Thelander and L. Samuelson, Solid State Physics, the Nanometer Structure Consortium, Lund University, Lund, SWEDEN

978-1-61284-243-1/11 $26.00 © 2011 IEEE

III-29 Student Paper
InAlAs/InGaAs Metamorphic HEMT and MOS-HEMT with Regrown Source/Drain by MOCVD
X. Zhou, Q. Li, and K. M. Lau, Department of Electronic and Computer Engineering, Hong Kong University of Science and Technology, Clear Water Bay, Kowloon, HONG KONG

III-30 Student Paper
Numerical Study of Electronic Transport Through Bilayer Graphene Nanoribbons
K. M. M. Habib and R. K. Lake, Department of Electrical Engineering, University of California, Riverside, California, USA

III-31 Student Paper
Tunnel-FET Architecture with Improved Performance due to Enhanced Gate Modulation of the Tunneling Barrier
L. De Michielis[1], L. Lattanzio[1], P. Palestri[2], L. Selmi[2], and A. M. Ionescu[1], [1]Nanoelectronic Devices Laboratory (Nanolab), Ecole Polytechnique Fédérale de Lausanne, SWITZERLAND and [2]Department of Electrical, Managerial and Mechanical Engineering, University of Udine, Udine, ITALY

III-32 Student Paper
Orientation dependent complex bandstructure of $Si_{1-x}Ge_x$ alloys
A. Ajoy[1], K. V. R. M. Murali[2], S. Karmalkar[1], and S.E. Laux[3], [1]Indian Institute of Technology Madras, INDIA, [2]IBM SRDC, Bangalore, INDIA, and [3]IBM SRDC, T. J. Watson Center, Yorktown Heights, New York, USA

III-33 Student Paper
Towards Planar GaAs Nanowire Array High Electron Mobility Transistor
X. Miao, and X. Li, Department of Electrical and Computer Engineering, Micro and Nanotechnology Laboratory, University of Illinois at Urbana-Champaign, Urbana, Illinois, USA

III-34 Student Paper
Interface States at high-κ/InGaAs interface: H_2O vs. O_3 based ALD Dielectric
H. Madan[1,2], D. Veksler[1], Y.T. Chen[1,3], J. Huang[1], N. Goel[1], G. Bersuker[1] and S. Datta[2], [1]SEMATECH, Albany, New York, USA, [2]The Pennsylvania State University, University Park, Pennsylvania, USA, and [3]University of Texas at Austin, Texas, USA

III-35
C-V Measurements of Single Vertical Nanowire Capacitors
P. Mensch, K. E. Moselund, S. Karg, E. Lörtscher, M. T. Björk, H. Schmid and H. Riel, IBM Research – Zurich, Rüschlikon, SWITZERLAND

III-36 Student Paper
Barrier Height, Interface Charge & Tunneling Effective Mass in ALD Al_2O_3/AlN/GaN HEMTs;
S. Ganguly, J. Verma, G. Li, T. Zimmermann, H. Xing, D. Jena, University of Notre Dame, Department of Electrical Engineering, Notre Dame, Indiana, USA

III-37 Student Paper
Graphene Field-Effect Transistors Using Large-Area Monolayer Graphene Grown by Chemical Vapor Deposition on Co Thin Films
M. E. Ramón[1], A. Gupta[1], C. Corbet[1], D. A. Ferrer[1], H.C.P. Movva[1], G. Carpenter[2], L. Colombo[3], G. Bourianoff[4], M. Doczy[4], D. Akinwande[1], E. Tutuc[1], and S.K. Banerjee[1], [1]Microelectronics Research Center, The University of Texas at Austin, Austin, Texas, USA, [2]IBM Research, Austin, Texas, USA, [3]Texas Instruments Incorporated, Dallas, Texas, USA, and [4]Intel Corporation, Austin, Texas, USA

III-38 Student Paper
Modeling of Dielectric Breakdown-Induced Time-Dependent STT-MRAM Performance Degradation
G. Panagopoulos, C. Augustine and K. Roy, School of ECE, Purdue University, West Lafayette, Indiana, USA

978-1-61284-243-1/11 $26.00 © 2011 IEEE

III-39 Student Paper
Introduction of ALD Beryllium Oxide Gate Dielectric for III-V MOS Devices
T. Akyol[1], J. H. Yum[1,5], D. A. Ferrer[1], M. Lei[2], M. Downer[2], C. W. Bielawski[3], T. W. Hudnall[4], G. Bersuker[5], J.C. Lee[1], S. K. Banerjee[1], [1]Microelectronics Research Center – Dept. of Electrical and Computer Eng., [2]Dept. of Physics, [3]Dept. of Chemistry, University of Texas at Austin, Texas, USA, [4]Texas State University at San Marcos, San Marcos, Texas, USA and [5]Sematech Austin, Texas, USA

III-40 Student Paper
Transport Properties of CVD-Grown Graphene Nanoribbon Field-Effect Transistors
A. S. Lyons, A. Behnam, E. K. Chow, and E. Pop, Dept. of Electrical and Computer Engineering, Micro and Nanotechnology Lab, University of Illinois, Urbana-Champagne, Illinois, USA

III-41 Student Paper
Protein Nanopore-gated Bio-transistor for Membrane Ionic Current Recording
T.-S. Lim, D. Jain, and P. J. Burke, Integrated Nanosystems Research Facility, Department of Electrical Engineering and Computer Science, University of California Irvine, Irvine, California, USA

III-42 Student Paper
Tunnel FET-Based Pass-Transistor Logic for Ultra-Low-Power Applications
S. H. Kim, Z. A. Jacobson, P. Patel, C. Hu, and T.-J. K. Liu, EECS Department, University of California, Berkeley, California, USA

III-43 Student Paper
Metal/III-V Effective Barrier Height Tuning using ALD High-κ Dipoles
J. Hu, K. Saraswat, and H.-S. P. Wong, Department of Electrical Engineering, Stanford University, Stanford, California, USA

III-44
Intrinsic DC Operation and Performance Potential of 50nm Gate Length Hydrogen-terminated Diamond Field Effect Transistors
D. A. J. Moran[1], O. J. L. Fox[2], H. McLelland[1], S. Russell[1], P. W. May[2], [1]The School of Engineering, The University of Glasgow, Glasgow, United Kingdom and [2]The School of Chemistry, The University of Bristol, Bristol, United Kingdom

III-45 Student Paper
Improvement of f_T in InAl(Ga)N barrier HEMTs by Plasma Treatments
R. Wang[1], G. Li[1], T. Fang[1], O. Laboutin[2], Y. Cao[2], J. W. Johnson[2], G. Snider[1], P. Fay[1], D. Jena[1], and H. Xing[1], [1]Department of Electrical Engineering, University of Notre Dame, Notre Dame, Indiana, USA and [2]Kopin Corporation, Tauton, Massachusetts, USA

III-46 Student Paper
Scaling behavior and velocity enhancement in Self-aligned N-polar GaN/AlGaN HEMTs with maximum f_T of 163 GHz
N. Nidhi[1], S. Dasgupta[1], D. F. Brown[2], J. S. Speck[1] and U. K. Mishra[1], [1]ECE Department, University of California Santa Barbara, Santa Barbara, California, USA and [2]Hughes Research Laboratories, Malibu, California, USA

III-47 Student Paper
Fermi-level Pinning at Metal/Antimonides Interface and Demonstration of Antimonides-based Metal S/D Schottky pMOSFETs
Z. Yuan[1], A. Nainani[1], J.-Y. Lin[1], B. R. Bennett[2], J. B. Boos[2], M. G. Ancona[2] and K. C. Saraswat[1], [1]Dept. of Electrical Engineering, Stanford University, California, USA and [2]Naval Research Laboratory, Washington, District of Columbia, USA

III-48 Student Paper
Uniaxially Tensile Strained Accumulation-Mode Gate-All-Around Si Nanowire nMOSFETs
M. Najmzadeh, D. Bouvet, W. Grabinski, and A. M. Ionescu, Nanoelectronic devices lab., Swiss Federal Institute of Technology (EPFL), Lausanne, SWITZERLAND

978-1-61284-243-1/11 $26.00 © 2011 IEEE

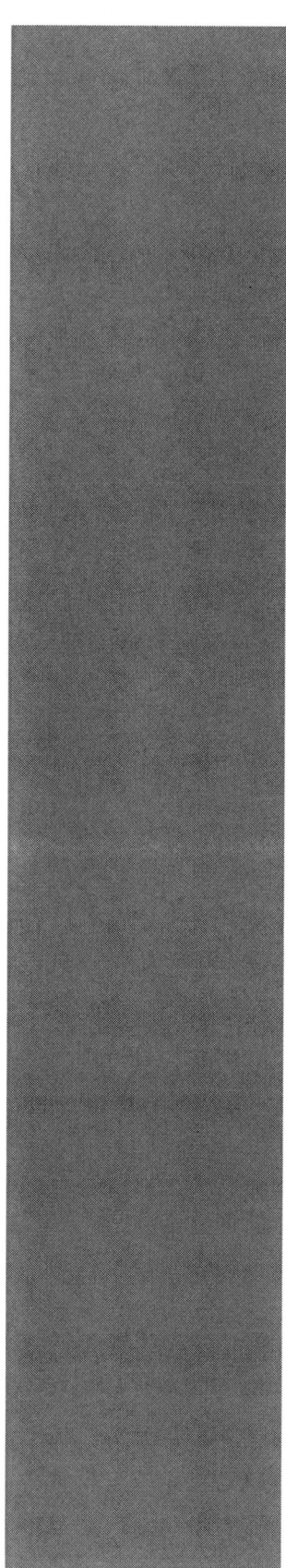

III-49

Spintronics Search Engines

H. Dery[1], H. Wu[1], B. Ciftcioglu[1], M. Huang[1], Y. Song[2], R. Kawakami[3], J. Shi[3], I. Krivorotov[4], I. Zutic[5], and L. J. Sham[6], [1]Department of Electrical and Computer Engineering, University of Rochester, Rochester, New York, USA, [2]Department of Physics, University of Rochester, Rochester, New York, USA, [3]Department of Physics, University of California Riverside, Riverside, California, USA, [4]Department of Physics, University of California Irvine, Irvine, California, USA, [5]Department of Physics, State University of New-York at Buffalo, Buffalo, New York, USA, and [6]Department of Physics, University of California San Diego, San Diego, California, USA

III-50 Student Paper

Low Frequency Transconductance and Output Resistance Dispersion of Epitaxial Graphene Nanoribbon-based Field Effect Transistors

G. Aroshvili[1], N. Meng[2], D. Vignaud[2], D. Pavlidis[1] and H. Happy[2], [1]Department of High Frequency Electronics, Darmstadt University of Technology, Darmstadt, GERMANY and [2]Institute of Electronics, Microelectronics and Nanotechnology, CNRS and Univ.p Lille 1, Villeneuve d'Ascq, FRANCE

III-51 Student Paper

InAs/SiGe on Si Nanowire Tunneling Field Effect Transistors

C. Kshirsagar and S. J. Koester, University of Minnesota-Twin Cities, Minneapolis, Minnesota, USA

978-1-61284-243-1/11 $26.00 © 2011 IEEE

Effects of Heavily Doped Source on the Subthreshold Characteristics of Nanowire Tunneling Transistors

M. Abul Khayer, and Roger K. Lake

LAboratory for Terascale and Terahertz Electronics (LATTE), Department of Electrical Engineering, University of California, Riverside, CA 92521, USA. Phone: (951) 827 4515, email: mkhayer@ee.ucr.edu

Band-to-band tunneling field-effect transistors (TFETs) have recently gained interest due to their operation in the sub-60 mV/decade limit which makes them ideal for reducing power dissipation in integrated circuit. III-V nanowire (NW) such as InSb NW TFETs show promise for ultra-low power and high-speed devices [1] due to its narrow direct bandgap.

To support the high electric fields required to achieve large interband tunnel current [1], heavy doping is required in the source and/or drain. When the doping is so high, the impurity bands merge with the conduction and valence bands giving rise to a large band-tail in the density-of-states which decays exponentially [2] in the bandgap as $\exp\{-|E-E_{C,V}|/\alpha\}$, with $E_{C,V}$ being the conduction(valence) band-edge and α is the decay factor. The off-current and the inverse subthreshold slope are limited by such density-of-states tails in the bandgap which results in mid-gap trap assisted band-to-band tunneling [3]. A larger value of α means a larger band-tail and a wider trap distribution and vice versa. It is important to investigate the effects of such impurity states in the bandgap to determine the super cut-off nature of TFETs in order to ensure a steep S ($\ll$ 60 mV/decade) at a higher on-current.

In this abstract, we report on the effects of heavily doped source (impurity states) on the off-current and the inverse subthreshold slope of NW TFETs. The investigation is done for InSb NW TFETs. We show that even with α comparable with or greater than kT, there is still a significant reduction of the inverse subthreshold slope in TFETs from its ideal thermally-limited value in FETs. To investigate the effects of heavily doped source on the off-current and the inverse subthreshold slope of NW TFETs, we have developed a generalized full band quantum mechanical numerical model which is based on the non-equilibrium Green's function (NEGF) method within recursive Green's function (RGF) algorithm. The tunneling current in the presence of incoherent scattering is calculated and compared with the coherent scattering.

A square [100] InSb/InP core-shell NW with a core cross-section of 2 nm is considered. The device utilizes a p^+-i-n^+ structure. Two gate lengths, 10 nm and 20 nm, are considered. A source Femi level of 0.1 eV is used which corresponds to a doping density of $N_A=3.8\times10^{19}$ /cm^3. Two tunneling processes in the subthreshold conduction mode are considered: (1) promoted by a mid-gap impurity state in the source, an electron can directly tunnel from the source to the gate region and subsequently collected by the drain, and (2) an electron can hop by the impurity states through the bandgap in the gate region and tunnel to the drain. These tunneling mechanisms affect the subthreshold conduction of TFET devices which have been investigated in this work.

It is found that current is dominated by tunneling through the bandgap for all devices. For wider impurity distribution in the source, the number of available states within the energy bandgap increases, which in turn, promotes larger tunneling current in the subthreshold mode of conduction. For gate length of 20 nm, current is significantly enhanced by the impurity states in the gate region. As the doping level in the source increases, the value of the inverse subthreshold slope (S) increases. However, with reasonably heavy source doping, S can still be minimized ($<$ 60 mV/decade) for both 10 nm and 20 nm gate length devices. The off-current for coherent transport is significantly reduced than those with incoherent scattering.

In conclusion, we have developed a generalized full band quantum mechanical model to investigate the effects of heavily doped source on the subthreshold characteristics of NW TFETs. The method is applied to InSb NW TFETs. It is found that necessary heavy doping of the source is not a show-stopper for TFETs.

This work is supported by the Focus Center Research Program (FCRP) on Functional Engineered Nano-Architectonics (FENA).

[1] M. A. Khayer, and R. K. Lake, *IEEE Electron Devices Lett.,* vol. 30, no. 12, pp. 1257, 2009.

[2] T. Walter, et. al., *J. Appl. Phys.* vol. 80, pp. 4411, 1996.

[3] N. N. Mojumder and K. Roy, *IEEE Trans. Electron Dev.*, vol. 56, pp. 2193, 2009.

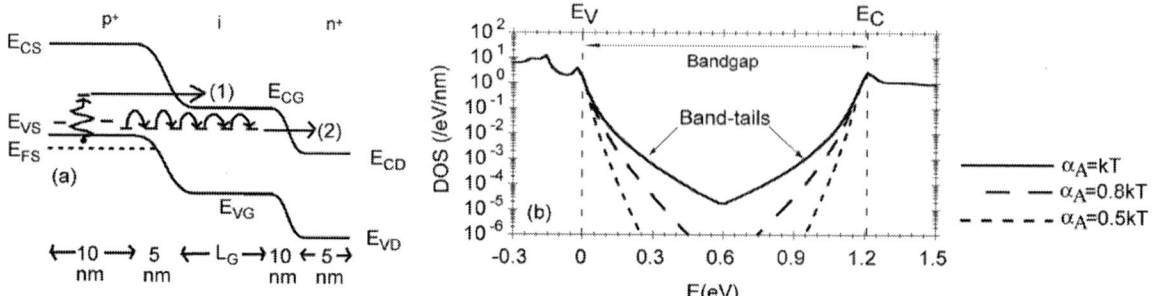

Fig. 1: (a) The band profile (not to scale) for a p$^+$-i-n$^+$ TFET in the off state. Impurity states resulting from heavy source doping promotes carrier tunneling from the source to the gate region. Two tunneling processes in the subthreshold conduction mode are indicated: (1) mid-gap trap assisted tunneling, and (2) tunneling by carrier hopping. (b) Density of states as a function of energy of 2 nm NW diameter InSb NW showing the band-edge and the band-tails. Away from the one-dimensional band-edge singularity, the band-tails decay in the bandgap as exp$\{-|E-E_{C,V}|/\alpha_A\}$, with E_C and E_V being the conduction and valence band-edges, respectively. Plots for a range of α_A are investigated. α_A is extracted from the initial decay part of the density-of-states plots.

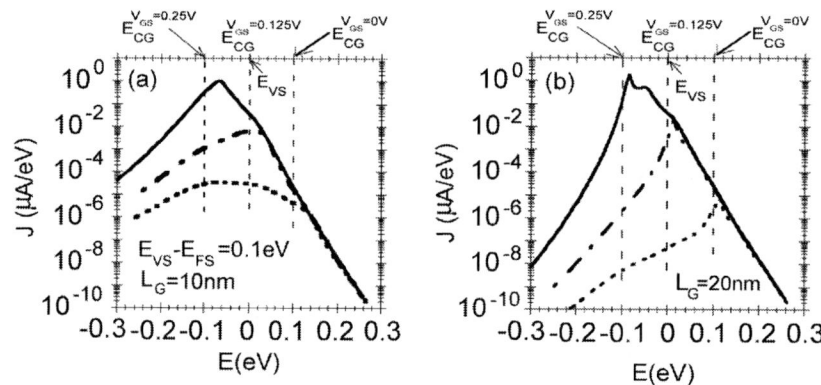

Fig. 2: Current density as a function of energy of InSb NW TFETs with heavily doped source region for two gate lengths (a) 10 nm and (b) 20 nm. For all plots, E_{VS}=0 eV is the valence band-edge in the source which is taken as reference. α_A =1 kT and source Fermi level of 0.1 eV is considered. Three gate biases are used, ie., V_{GS}=0 V (dotted lines), V_{GS}=0.125 V (dash dotted lines), and V_{GS}=0.25 V (solid lines). On top of each plot,

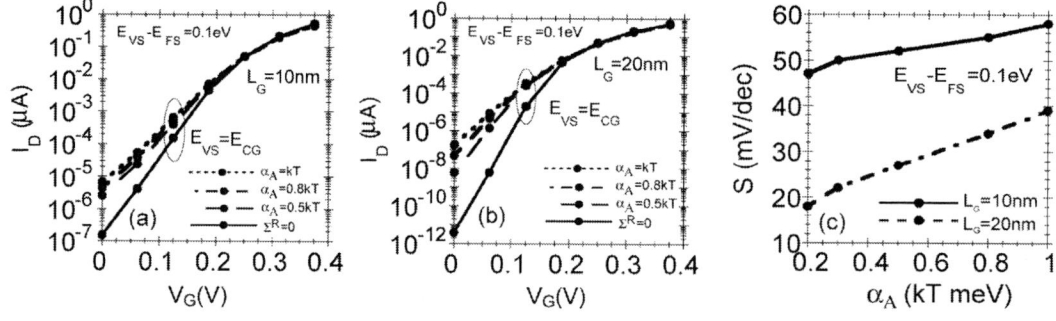

Fig. 3: I_D-V_G characteristics on a log scale of InSb NW TFETs for a range of values of α_A for two gate lengths (a) 10 nm and (b) 20 nm. For comparison, I_D-V_G curves for coherent scattering case is also presented. For all plots, a source Fermi level of 0.1 eV is used. Fig. (c) shows the inverse subthreshold swing (S) as a function of α_A for the heavily doped source for both 10 nm and 20 nm gate length devices. S can be reduced and maintained below the fundamental limit 60 mV/decade with the heavy source doping considered.

978-1-61284-243-1/11 $26.00 © 2011 IEEE

Spatially resolved photovoltaic performance of axial GaAs nanowire pn-diodes

Andrey Lysov, Christoph Gutsche, Matthias Offer, Ingo Regolin, Werner Prost, Franz-Josef Tegude

Center for Nanointegration Duisburg-Essen, University of Duisburg-Essen, D-47048 Duisburg, Germany
Email: andrey.lysov@uni-due.de; phone: +49/(0) 203 379 3880, fax: +49/(0) 203 379 3400;

Nanowires allow for a combination of highly mismatched materials needed for optimized broad spectrum absorption [1], and the use of appropriate substrates. III-V nanowires offer in addition a high absorption coefficient which is indispensable if a high efficiency energy transformation is aimed at the nanoscale. Currently, both radial [2] as well as axial III-V nanowire pn-junctions are under discussion for photovoltaic applications. The axial device enables a high forward current, ultra-low leakage, and a possible staggered integration of multiple cells. Moreover, the axial device gives full access for a high resolution photovoltaic analysis that is especially important at the current status of evaluation since a detailed device understanding is pending and the published yield data of nanowire devices are still low [1-2].

In this work we have investigated the photovoltaic properties of GaAs nanowires with an axial pn-junction with high spatial resolution in order to quantify the volume of absorption and the efficiency obtained with the available growth technology [3]. Spatially resolved photocurrent spectroscopy was used to investigate mechanism of carrier photo generation. The *I-V* characteristics of nanowire pn-diode were measured, while nanowires were locally illuminated by focused *CW* laser *(λ = 532 nm)* at different positions (Figure 1a). The short circuit current is maximal when the diode is illuminated at the position of the pn-junction (position 5 in the Figure 1a). No photocurrent was detected either in the vicinity of contacts or in the p- and n-diode parts. This demonstrates that charge separation by an electric field takes place only in the vicinity of the depletion region, while p- and n- nanowire regions are field free and contacts are well ohmic. The charge separation efficiency at the pn-junction was estimated to $\eta_{cc} = 42\,\%$ and was limited by the recombination of carriers at the nanowire surface traps.

To determine solar conversion efficiency of a fabricated nanowire photovoltaic device with 100 nm diameter the current voltage measurement under standard 1-sun conditions was carried out (Figure 2a). A short circuit current of 88 pA and an open circuit voltage of 0.56 V were obtained yielding a fill factor of 69 % under 1-sun conditions without any antireflection coating. The obtained open circuit voltage is approximately three times larger, than achieved with single axial Si-nanowire pin-diodes [1]. The achieved fill factor of 69 % is similar to the 65 % value, reported for the single GaAs coaxial nanowire pn-diodes [2]. The power conversion efficiency of the nanowire photovoltaic device under standard illumination was estimated with the formula $\eta = \frac{V_{oc} \cdot I_{sc} \cdot FF}{P_{in}} = 9\,\%$.

Power dependent photocurrent measurements (Figure 5b) made under monochromatic homogeneous laser illumination $(\lambda = 532\,nm)$ demonstrate linear scaling of photocurrent with illumination intensity (Figure 2b and 2c). The photosensitivity of the nanowire diode directly follows the increasing photo excitation with 0.24 A/W with respect to the absorbed illumination power up to an illumination density of at least 90 W/cm². Open circuit voltage of 0.85 V remains almost constant in the entire intensity range (Figure 2d), representing the maximal achievable photo voltage. The fill factor is independent of the illumination power and stays at the level of 72 % (Figure 2d).

The first investigation of photovoltaic performance of axial GaAs nanowire pn-diodes underlines the potential of axial photo cells with high absorption coefficient for solar cell application. The excellent transport properties of the highly doped device enable at least 100-sun condensed illumination without any loss at the cell level. These results will guide the development of nanowire based photovoltaic and photonic devices.

[1] T. J. Kempa, B. Tian, D. R. Kim, J Hu, X. Zheng, and C. M. Lieber, "Single and tandem axial p-i-n nanowire photovoltaic devices ," Nano Lett., vol. 8, pp. 3456-3460, 2008.

[2] C. Colombo, M. Heiß, M. Grätzel, and A. Fontcuberta i Morral, "Gallium arsenide p-i-n radial structures for photovoltaic applications," *Appl. Phys. Lett.*, vol. 94, no. 17, p. 173108, 2009.

[3] I. Regolin, C. Gutsche, A. Lysov, K. Blekker, Z.-A. Li, M. Spasova, W. Prost, and F.-J. Tegude, "Axial pn-junctions formed by MOVPE using DEZn and TESn in vapour-liquid-solid grown GaAs nanowires," *J. Cryst. Growth*, vol. 315, pp. 143-147, 2011.

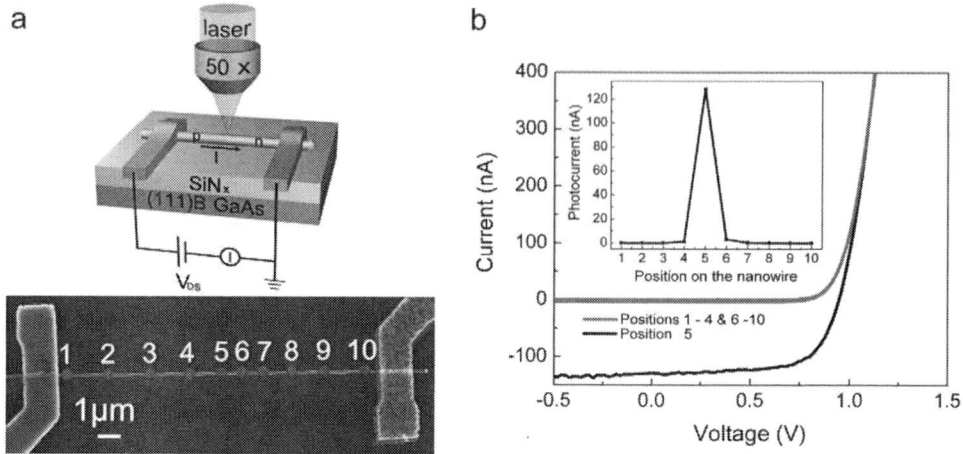

Figure 1. (a) Schematic of a measurement setup for photocurrent microscopy. The lower inset shows a SEM micrograph of the investigated nanowire-diode. (b) I(V) characteristics of the GaAs nanowire pn-diode illuminated by focused laser spot at different positions. I(V) curves from all positions except for position 5 are superimposed on each other. Upper inset shows photocurrent as a function of a laser spot position. The corresponding positions are denoted by numbers on the SEM image in Figure 2a.

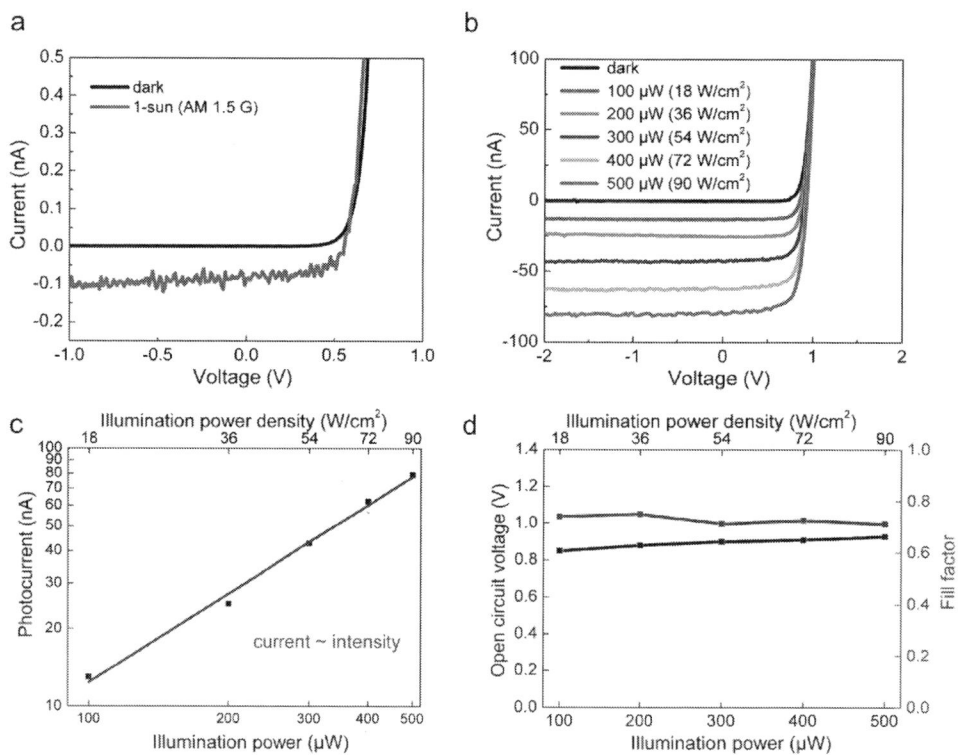

Figure 2. a) Dark and 1-sun illuminated I-V characteristics of nanowire pn-diode. b) I-V characteristics of the nanowire pn-diode measured under monochromatic homogeneous laser illumination (λ = 532 nm) with various illumination powers. c) Photocurrent as a function of illumination power. The line is a linear fit to the data. d) Power dependence of the open circuit voltage and fill factor measured under monochromatic homogeneous laser illumination (λ = 532 nm).

978-1-61284-243-1/11 $26.00 © 2011 IEEE

Low loss AlInN/GaN Monolithic Microwave Integrated Circuit Switch

A. Sattu[1], D. Billingsley[1], J. Deng[1], J.Yang[1], R. Gaska[1], M. Shur[2], G. Simin[3]

[1]Sensor Electronic Technology, Inc., 1195 Atlas Road, Columbia, SC 29209
[2]Electrical and Computer Science Engineering, Rensselaer Polytechnic Institute, Troy, NY
[3]Electrical Engineering, University of South Carolina, Columbia, SC 29209

We report on the first AlInN/GaN Heterojunction Field Effect Transistor (HFET) based Monolithic Microwave Integrated Circuit (MMIC) switch. Lattice-matched AlInN/GaN heterostructures with indium contents of ~17% exhibit a very large conduction band discontinuity, ΔE_C, of 1.7 eV. This large discontinuity results in 2DEG densities as high as 4.7×10^{13} cm^{-2} [1] and electron mobilities as high as 1617 cm^2/V-s [2]. As a result these heterostructures can achieve record low sheet resistances, making them very attractive candidates for ultra-low loss microwave and other switching devices.

The AlInN/GaN and AlGaN/GaN heterostructures in this comparative study were grown using a proprietary MEMOCVD® deposition technique. MMIC switches were fabricated using $Al_{0.83}In_{0.17}$N/GaN and $Al_{0.25}Ga_{0.75}$N/GaN structures on sapphire substrates. Both heterostructures have a 20 nm AlN nucleation layer followed by a 2 µm thick unintentionally doped GaN layer and 1 nm thick AlN spacer. The barrier layers for two device types were 7.5 nm $Al_{0.83}In_{0.17}$N and 25 nm $Al_{0.25}Ga_{0.75}$N. As grown, the measured sheet resistance was ~215 and 280 Ω/□ in AlInN/GaN and AlGaN/GaN structures, respectively.

Test devices with source-drain spacing L_{SD} = 5 µm and gate length L_G = 1.5 µm have been fabricated to study the I-V characteristics, ON-resistance and OFF-capacitance. The device width was 100 µm; the gate was centered in the source-drain region. The DC I-V characteristics of AlInN/GaN and AlGaN/GaN HFET are shown in Fig. 1. The drain saturation current of AlInN/GaN HFET was ~1.05 A/mm compared to 0.64 A/mm in the AlGaN/GaN HFET. The gate leakage currents at -10V gate bias were ~10-90 and 10-20 µA/mm for AlInN/GaN and AlGaN/GaN devices.

The HFET 'ON' resistance and "OFF" capacitance were extracted from the S-parameters of the test devices connected in series into 50Ω CPW signal line. The small signal analysis of the device in the ON state (V_G =0) shows frequency-independent $|S_{21}|$ = 1.3 and 2 dB in AlInN and AlGaN devices respectively (Fig. 2). The corresponding ON-resistances extracted from the S-parameters were 1.5 and 2.3 Ω×mm. The OFF-state capacitances C_{OFF} extracted from the $|S_{21}|$-frequency dependence at the gate bias V_G = - 10 V (Fig. 2) were ~0.196 and 0.15 pF/mm for AlInN/GaN and AlGaN/GaN structures respectively. A higher C_{OFF} in AlInN devices is attributed to the lower barrier thickness.

A single-pole-double-throw (SPDT) MMIC switch was also fabricated using these two types of the heterostructures. The layout (shown in the inset to Fig. 3) consisted of series and shunt HFETs with the widths of 0.6 mm and 0.4 mm correspondingly. The dimensions of the HFETs used in the MMIC were L_{GS} = L_{GD} = 2.25 µm, L_{SD} = 6 µm and L_G = 1.5 µm.

The measured insertion losses and isolation of SPDT switches are shown in the Fig. 3 and Fig. 4 respectively. As seen from Fig. 3, the AlInN-based switch exhibits a lower insertion loss as a result of the lower ON-resistance. The isolation of AlInN SPDT switch (Fig. 4) is higher than that of the AlGaN switch, in spite of a higher OFF-capacitance. This is due to a very low 'ON' resistance of the shunt element comprising the MMIC layout. The isolation of both types of SPDT switches shows the resonance behavior in the frequency range of 6-8 GHz resulting from the interaction between the OFF capacitance and CPW non uniformities.

The novel AlInN SPDT switch shows superior performance compared to recently published data on III-Nitride switches. At 1 GHz, the developed SPDT switch has ~20 dB higher isolation compared to that of [3] with nearly the same insertion loss of 0.3 dB. Compared to the results of [4], the AlInN-based SPDT switch shows 12 -25 dB higher isolation in the frequency range of 6 – 8 GHz with only 0.2 dB higher insertion loss (due to a larger source-drain spacing used in out device layout).

In summary, the first low-loss AlInN/GaN switch MMIC demonstrated superior performance compared to AlGaN/GaN switches with a similar layout and to the published data. A combination of low loss ≈ 0.3dB (that can be further decreased using smaller source-drain spacing) and excellent isolation of 40-45 dB makes AlInN/GaN HFETs extremely promising devices for microwave switching applications.

[1] J. Yang; X. Hu; J. Deng; Gaska, R.; Shur, M.; Simin, G,. "AlInN/ GaN heterostructure field-effect transistors", International Semiconductor Device Research Symposium, 2009. 9-11 Dec. 2009 Page(s): 1 – 2, Digital Object Identifier 10.1109/ISDRS.2009.5378095

[2] S. Guo, X. Gao, D. Gorka, J. W. Chung, H. Wang, T. Palacios, A. Crespo, J. K. Gillespie, K. Chabak, M. Trejo, V. Miller, M. Bellot, G. Via, M. Kossler, H. Smith, and David Tomich, "AlInN HEMT grown on SiC by metalorganic vapor phase epitaxy for Millimeter-wave applications", Phys. Status Solidi A 207, No. 6, 1348–1352 (2010) / DOI 10.1002/pssa.200983621

[3] H. Ishida, Y. Hirose, Y. Ikeda, T. Matsuno, K. Inoue, Y. Uemoto, T. Tanaka, T. Egawa, D. Ueda , "A High-Power RF Switch IC Using AlGaN/GaN HFETs with Single-Stage Configuration", IEEE transactions on Electron devices, Vol. 52, No. 8, pp 1893-1899, Aug 2005

[4] C. Cambell, D. Dumka, "Wideband High Power GaN on SiC SPDT Switch MMICs", IMS 2010, pp-145-148, 2010

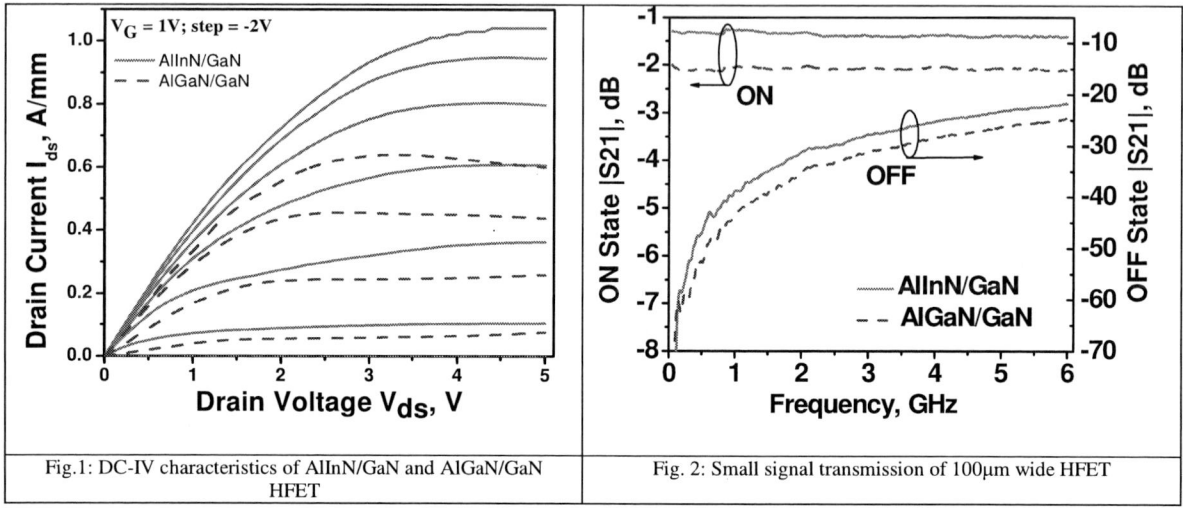

| Fig.1: DC-IV characteristics of AlInN/GaN and AlGaN/GaN HFET | Fig. 2: Small signal transmission of 100µm wide HFET |

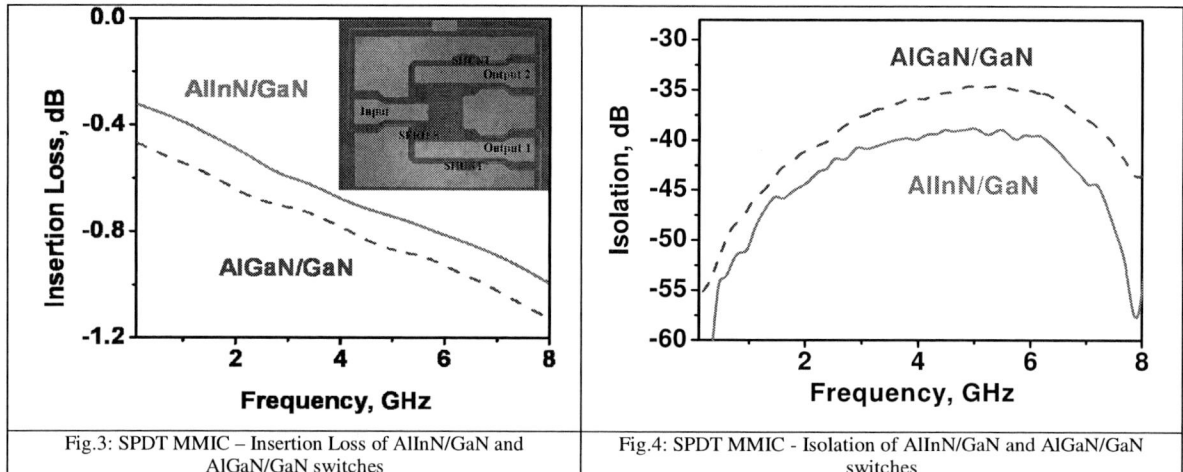

| Fig.3: SPDT MMIC – Insertion Loss of AlInN/GaN and AlGaN/GaN switches | Fig.4: SPDT MMIC - Isolation of AlInN/GaN and AlGaN/GaN switches |

978-1-61284-243-1/11 $26.00 © 2011 IEEE

High-Mobility Organic Thin-Film Transistors with Photolithographically Patterned Top Contacts

Ute Zschieschang,[a] Nis Hauke Hansen,[b] Jens Pflaum,[b] Tatsuya Yamamoto,[c] Kazuo Takimiya,[c] Hirokazu Kuwabara,[d] Masaaki Ikeda,[d] Tsuyoshi Sekitani,[e] Takao Someya,[e] Hagen Klauk[a]

a) Max Planck Institute for Solid State Research, Heisenbergstr. 1, 70569 Stuttgart, Germany
(phone: +49 711 689-1401; email: U.Zschieschang@fkf.mpg.de)
(b) University Würzburg and ZAE Bayer e.V., Germany; (c) Hiroshima University, Higashi-Hiroshima, Japan;
(d) Nippon Kayaku Co., Ltd., Kita-ku, Tokyo, Japan; (e) University of Tokyo, Tokyo, Japan

Due to its large-area capability and high resolution, photolithography is the preferred patterning method for pentacene thin-film transistors (TFTs) for display and circuit applications [1,2]. Since the morphology of thin pentacene films is very sensitive to solvents and heat [3,4], the photolithographic patterning of the source/drain contacts is ideally performed prior to the pentacene deposition, which explains the general preference for the bottom-contact (coplanar) TFT structure. However, as experiments [5] and simulations [6,7] have shown, the bottom-contact TFT structure is associated with substantially larger contact resistance than the top-contact (staggered) structure, which means that for the same channel length, top-contact TFTs are expected to provide larger transconductance and higher cutoff frequency than bottom-contact TFTs. Here we report on organic TFTs with Au top contacts patterned by ordinary photolithography and wet etching (using common solvents, photoresists, and etchants) having field-effect mobilities (0.4 cm^2/Vs) and on/off current ratios (10^7) similar to those of optimized bottom-contact pentacene TFTs [1,2,5].

The key to realizing high-mobility organic TFTs with photolithographically patterned top contacts is a conjugated semiconductor which in the pristine state has a mobility similar to (or greater than) that of pentacene, but is less sensitive to solvents and heat than pentacene. An example is dinaphtho-thieno-thiophene (DNTT), which provides mobilities similar to pentacene in shadow-mask-patterned top-contact TFTs, but unlike pentacene crystallizes in a single polymorph [8]. Fig. 1 shows the electrical characteristics of a DNTT TFT fabricated on a silicon substrate with a gate dielectric of 100 nm thermal SiO_2 + 8 nm Al_2O_3 (deposited by atomic-layer deposition and covered with an alkylphosphonic acid self-assembled monolayer (SAM)) and Au top contacts patterned by vacuum evaporation through a shadow mask. These shadow-mask-patterned DNTT TFTs have on/off current ratios of 10^7 and a carrier mobility of 2.2 cm^2/Vs, which is slightly better than the mobility of pentacene TFTs with shadow-mask-patterned Au top contacts (1.5 cm^2/Vs; see Table 1).

DNTT and pentacene TFTs with photolithographically patterned top contacts were fabricated as follows: (1) a 30 nm thick layer of Au was uniformly deposited onto the vacuum-deposited semiconductor by evaporation; (2) a 1.3 µm thick layer of Shipley 1813 photoresist (dissolved in PGMEA) was deposited onto the Au by spin-coating, baked for 1 min at a temperature of 90 °C on a hotplate, exposed to UV light through a photomask, and developed in an aqueous tetramethylammonium hydroxide (TMAH) solution; (3) the Au regions no longer protected by photoresist were etched in an aqueous I_2/KI solution; (4) the photoresist was stripped in acetone; (5) the substrate was rinsed with DI water and dried with nitrogen.

The results are summarized in Table 1. Photolithographic patterning of top contacts on pentacene causes the mobility in the pentacene to drop by a factor of 10 (from 1.5 to 0.16 cm^2/Vs), consistent with the results by Gundlach et al. [3]. DNTT appears much less affected by the photolithographic process, providing a mobility of 0.46 cm^2/Vs and an on/off ratio of 10^7 in photolithographically patterned top-contact TFTs (see Figure 2).

X-ray diffraction measurements performed on 25 nm thick pentacene and DNTT films before and after the films were subjected to a complete photolithography process reveal a significant structural phase change in the pentacene film, associated with a distinct evolution of the pentacene bulk phase at d = 14.4 Å (see Fig. 3). This bulk-phase evolution, which was also reported by Gundlach et al. [3]. and which is presumably initiated by the solvent exposure, is correlated with the pronounced drop in mobility in the case of pentacene. The DNTT films, in contrast, are characterized by a single phase, both before and after the photolithography process.

[1] I. Yagi et al., *J. Soc. Inf. Display* **2008**, *16*, 15
[2] K. Myny et al., *Org. Electronics* **2010**, *11*, 1176
[3] D. J. Gundlach et al., *Appl. Phys. Lett.* **1999**, *74*, 3302
[4] T. Ji et al., *Org. Electronics* **2008**, *9*, 895

[5] D. J. Gundlach et al., *J. Appl. Phys.* **2006**, *100*, 024509
[6] I. G. Hill, *Appl. Phys. Lett.* **2005**, *87*, 163505
[7] C. H. Shim et al., *IEEE Trans. Electr. Dev.* **2010**, *57*, 195
[8] T. Yamamoto et al., *J. Am. Chem. Soc.* **2007**, *129*, 2224

Fig 1. Schematic cross-section and electrical characteristics of a DNTT TFT with top contacts patterned with a **shadow mask**. The field-effect mobility is 2.2 cm^2/Vs, the on/off current ratio is 10^7, and the subthreshold swing is 0.6 V/decade.

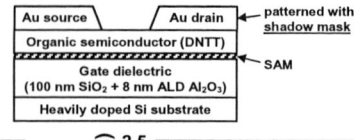

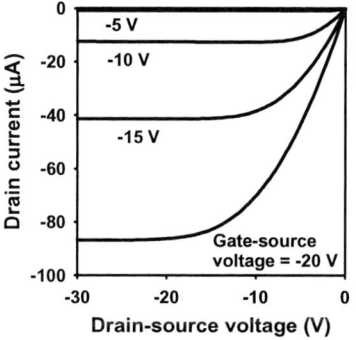 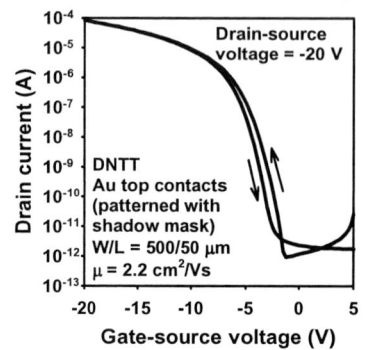 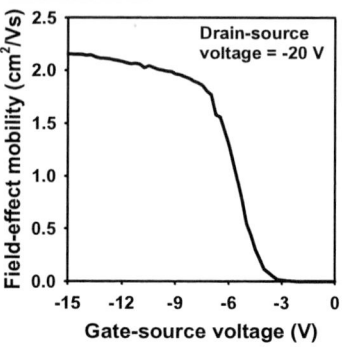

Fig. 2. Schematic cross-section and electrical characteristics of a DNTT TFT with top contacts patterned by **photolithography**. The field-effect mobility is 0.46 cm^2/Vs, the on/off current ratio is 10^7, and the subthreshold swing is 0.6 V/decade.

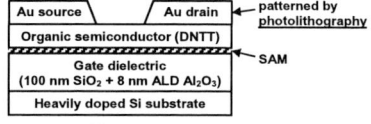

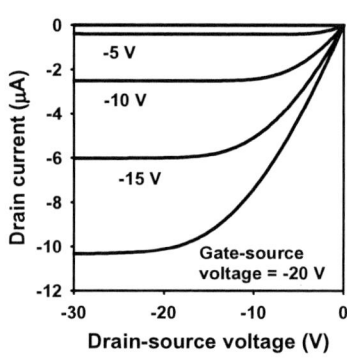 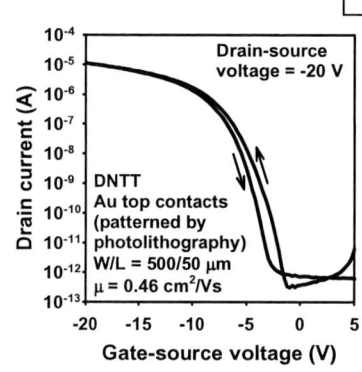 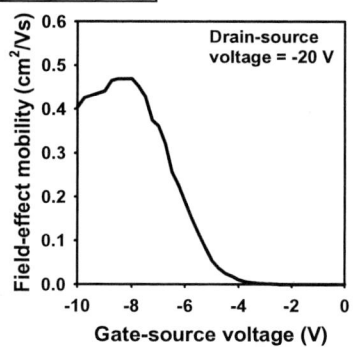

Table 1. Summary of the field-effect mobility (μ) and on/off current ratio measured for pentacene and DNTT TFTs with either <u>Au bottom contacts</u>, or with <u>Au top contacts</u> patterned by evaporation through a <u>shadow mask</u>, or with <u>Au top contacts</u> patterned by <u>photolithography</u>, wet etching, and resist strip. All TFTs have a channel length of 50 μm and a channel width of 500 μm.

	pentacene	DNTT
bottom contacts	μ = 0.13 cm^2/Vs on/off ratio = 10^7	μ = 0.002 cm^2/Vs on/off ratio = 10^4
top contacts patterned with a shadow mask	μ = 1.5 cm^2/Vs on/off ratio = 10^7	μ = 2.2 cm^2/Vs on/off ratio = 10^7
top contacts patterned by photolithography	μ = 0.16 cm^2/Vs on/off ratio = 10^2	**μ = 0.46 cm^2/Vs** on/off ratio = 10^7

Fig. 3. X-ray diffraction spectra of a 25 nm thick pentacene film deposited by vacuum evaporation onto an SiO$_2$ / Al$_2$O$_3$ / SAM dielectric before and after subjecting the pentacene film to photolithography. The evolution of the pentacene bulk phase at q_z = 0.435 Å^{-1} (d = 14.4 Å) upon photolithography is indicated. In contrast to pentacene, DNTT films are characterized by a complete absence of a second structural phase.

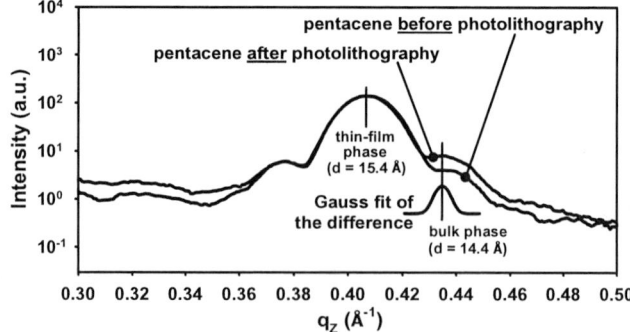

978-1-61284-243-1/11 $26.00 © 2011 IEEE

High Breakdown Voltage ZnMgO/In-Ga-Zn-O Heterostructure Transistors

Junichi Yamaguchi, Ikuo Soga, and Taisuke Iwai

Fujitsu Laboratories Ltd., 10-1 Morinosato-Wakamiya, Atsugi, Kanagawa 243-0197, Japan

Email: yamaguchi.j@jp.fujitsu.com, Phone: +81-46-250-8234, Fax: +81-46-250-8844

ZnO-based semiconductors with the wide-band gap have attracted great interest for electronic and optical applications [1]. Among them, an amorphous In-Ga-Zn-O (a-IGZO) has been intensively studied [2,3]. The thin-film transistors using of a-IGZO as an active n-channel layer exhibit good performances such as the high field-effect mobility [$\mu_{FE} \sim 10$ cm^2 (Vs)$^{-1}$], I_{on}/I_{off} ratio of $\sim 10^8$, and excellent process stability, even when the channel layer was deposited at room temperature (RT). The band gap of a-IGZO was estimated to $E_g = 3.1-3.4$ eV from several spectroscopic techniques [4], which is a promising candidate for high voltage applications such as switching devices for power supplies. Moreover, a-IGZO films have the high uniformity in large area owing to the amorphous character, which results in low-cost fabrication.

In order to develop high breakdown voltage (BV) a-IGZO transistors, the carrier concentration (N_e) of the channel layer should be suppressed to less than $N_e \sim 10^{18}$ cm^{-3}. It is, however, reported that the high-N_e layer exists in the surface of a-IGZO films due to the O defects and structural disorder [4]. In addition, a-IGZO films indicate the reduction of the Hall mobility (μ_{Hall}) with decreasing N_e [2] which results in the decrease of an operating speed of the device. The combination of the low-N_e and high-μ_{FE} (and μ_{Hall}), are strongly required to achieve the high performance transistor under high operating voltages.

In this work, we have first reported on the fabrication of high BV a-IGZO transistors. To suppress the O defects in the surface of a-IGZO films and/or the interface between the a-IGZO films and substrates, the heterostructures sandwiched a-IGZO between $Zn_{0.6}Mg_{0.4}O$ (ZMO: $E_g \sim 4$ eV) were used as a channel. The ZMO(4)/a-IGZO(25)/ZMO(4) (in nm) heterostructures were deposited on oxidized Si substrates by rf magnetron sputtering at RT, and then annealed in O$_2$ atmosphere at 400 °C for 1 h.

Figure 1 shows μ_{Hall} as a function of N_e for the ZMO/a-IGZO/ZMO heterostructure, as a comparison, that for the ZMO/a-IGZO and a-IGZO films is also shown. For the ZMO/a-IGZO/ZMO heterostructure, the highest μ_{Hall} [26.4 cm^2 (Vs)$^{-1}$] is obtained at the lowest N_e (1.8×10^{17} cm^{-3}). This result indicates that the ZMO films are effective to suppress the O defects for the a-IGZO film. The schematic of the top-gate type transistor with the ZMO/a-IGZO/ZMO heterostructure is shown in Fig. 2(a). A high-κ dielectric MgO(35 nm) film was deposited as a gate insulator by rf magnetron sputtering at RT. The source, drain, and gate electrodes were formed with the electron-beam evaporated Pt(100 nm)/Ti(10 nm) by lift-off processes. The channel width (W), length (L) and the length between gate and drain (L_{gd}) were designed to be $W/L = 100/1.5$ μm and $L_{gd} = 20$ μm as shown in Fig. 2(b). The transfer (I_d–V_{gs}) characteristics are shown in Fig. 3. The transistor parameters extracted form the results are $\mu_{FE} = 27.5$ cm^2 (Vs)$^{-1}$, the transconductance, $g_m = 8.2$ mS/mm at $V_{gs} = 0.5$ V, and the I_{on}/I_{off} ratio of $\sim 10^7$. The obtained μ_{FE} is consistent with μ_{Hall}. However, the transistor indicates the normally-on behavior (the threshold voltage: $V_{th} = -2.1$ V) (see the inset of Fig. 3). The BV at the off-state (BV_{off}) is at most ~ 50 V due to the increase of the gate leakage current (data not shown). To improve BV_{off}, the gate insulator was replaced with the thicker MgO(60 nm)/ZMO(20 nm) film. The I_d–V_{gs} and output (I_d–V_{ds}) characteristics are shown in Figs. 4(a) and 4(b). The transistor operated at $V_{ds} = 100$ V and the characteristic parameters are as follows: $\mu_{FE} = 6.3$ cm^2 (Vs)$^{-1}$, $g_m = 2.1$ mS/mm at $V_{gs} = -1.2$ V, I_{on}/I_{off} ratio of $\sim 10^{10}$, and $V_{th} = -6.1$ V. As shown in Fig. 4(c), the transistor exhibited high BV_{off} of ~ 240 V.

In summary, we have demonstrated the fabrication process and the characterization for the high BV ZMO/a-IGZO/ZMO heterostructure transistors. The transistor exhibited the high μ_{FE} of 27.5 cm^2 (Vs)$^{-1}$ and g_m of 8.2 mS/mm, which indicates that the passivation using the ZMO(4 nm) films is effective. The BV_{off} exhibited ~ 240 V by optimizing the thickness of the gate insulator.

978-1-61284-243-1/11 $26.00 © 2011 IEEE

[1] J. Nishii *et al.*, Jpn. J. Appl. Phys. **42**, L347 (2003).
[2] K. Nomura *et al.*, Nature (London) **432**, 488 (2004).
[3] H. Yabuta *et al.*, Appl. Phys. Lett. **89**, 112123 (2006).
[4] K. Nomura *et al.*, Appl. Phys. Lett. **92**, 202117 (2008).

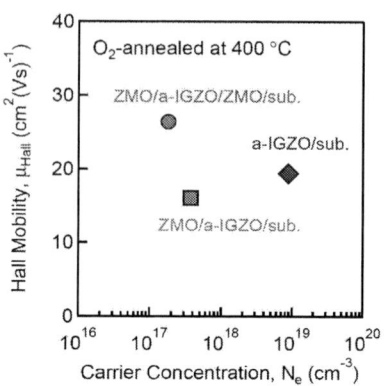

Fig. 1. Relationship between Hall mobility and carrier concentration for the ZMO/a-IGZO/ZMO, ZMO/a-IGZO, and a-IGZO films on an oxidized Si substrate. Each film thickness was 25 nm for a-IGZO and 4 nm for ZMO.

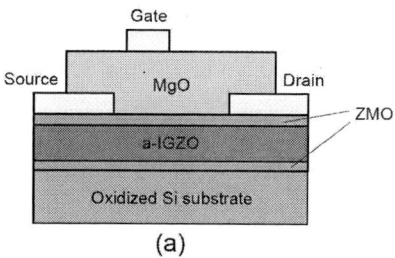

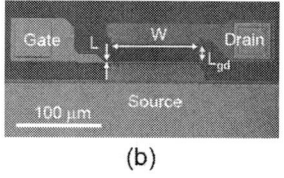

Fig. 2. (a) Schematic cross-section and (b) top view of the ZMO/a-IGZO/ZMO heterostructure transistor. In (b), W and L indicate the channel width and length. L_{gd} is the length between the gate and drain.

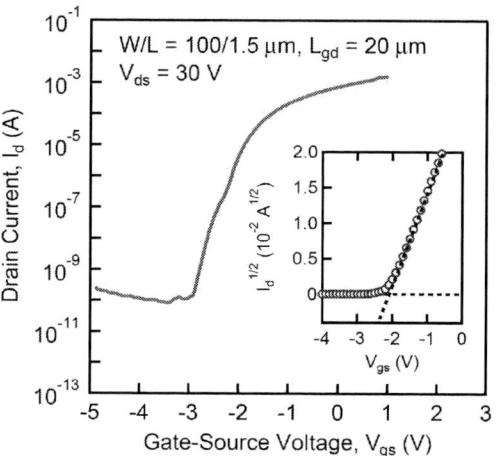

Fig. 3. I_d–V_{gs} characteristics for the transistor with the MgO(35 nm) gate insulator. The inset shows the square root of I_d as a function of V_{gs}.

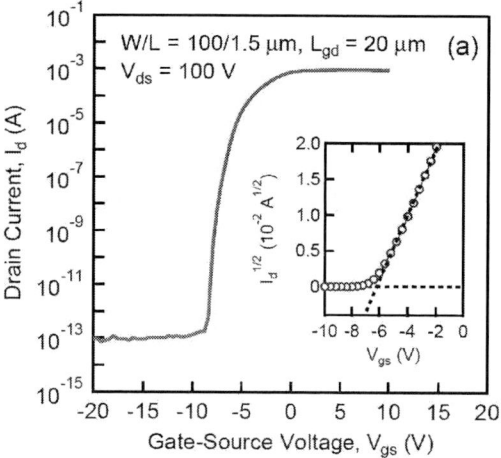

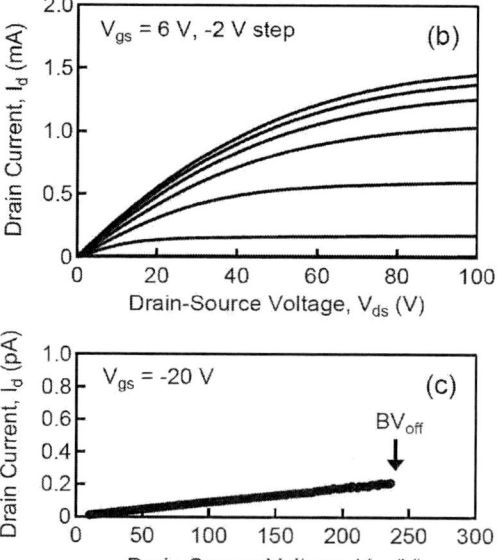

Fig.4. Operation results for the transistor with the MgO(60 nm)/ZMO(20 nm) gate insulator. (a) I_d–V_{gs} and (b) I_d–V_{ds} characteristics. The inset in (a) is the $I_d^{1/2}$–V_{gs} curve. (c) I_d–V_{ds} characteristics at the off-state (V_{gs} = -20 V). The arrow shows the breakdown voltage.

978-1-61284-243-1/11 $26.00 © 2011 IEEE   60

High-Resolution Temperature Sensing with Source-Gated Transistors
R. A. Sporea, J. M. Shannon, S. R. P. Silva
Advanced Technology Institute, FEPS, University of Surrey, Guildford, Surrey, GU2 7XH, U.K.
email: r.a.sporea@surrey.ac.uk; phone: +44 1483 686083

Source-gated transistors (SGTs) [1] are three-terminal devices in which the current is controlled by a potential barrier at the source. The gate voltage is used primarily to modulate the effective height of the source barrier. These devices have a number of operational advantages over conventional field-effect transistors, including a potentially much smaller saturation voltage and very low output conductance in saturation, which lead to low power operation and high intrinsic gain [2].

Thin-film SGTs with Schottky source barriers have been fabricated on glass, in a self-aligned, back-gate polysilicon process. A microphotograph of a typical SGT is shown in Figure 1, while Figure 2 shows schematically the cross-sectional structure of a device (processing described in [2]). Electrical measurements (Figure 3) reveal on/off ratios in excess of 10^5 and typical SGT behaviour: small change in saturation voltage with gate potential and flat output characteristics in the saturation regime up to relatively high drain voltages. These devices seem to be well suited for signal amplification applications and as active loads in large area electronic circuits, and represent the experimental basis of the current study.

A feature of devices containing Schottky barriers in general and of the Schottky SGTs in particular is the temperature dependence of the drain current; high Schottky barriers produce a large variation of drain current against temperature, which is undesirable in most applications. In order to keep the current variation with temperature small, low barrier heights have to be used. However, the very large activation energy which can be obtained by designing the SGT with a high Schottky barrier can be exploited for highly sensitive temperature measurements.

In order to act as a temperature sensor, the SGT can be operated at a constant gate voltage, to produce an exponentially temperature-dependent drain current, or at constant drain current in a feedback loop, leading to a change in gate voltage with temperature given by: $\dfrac{\partial V_G}{\partial T} \approx \dfrac{1}{T}\left(V_G - V_T - \dfrac{t_s \phi_B \left(C_i + C_S\right)}{\alpha C_i}\right)$, where t_S is the semiconductor thickness, α is the barrier lowering constant, ϕ_B is the height of the source barrier at zero applied bias, V_T is the threshold voltage, C_i and C_S are the capacitances per unit area of the gate insulator and semiconductor, respectively. The constant V_G operation can be applied to a sensing scheme such as [3] and can make use of the high intrinsic gain of the SGT [2], while a (linearly) temperature dependent gate voltage scheme can be integrated into a sensor of the type described in [4]. Figure 4 shows the change of V_G with temperature for two devices (low and high barrier) operated at constant current. A simulation of a similar structure with a high barrier (Figure 5) reveals a *1/T* dependence of V_G. From both the measurements and the simulation it can be concluded that devices with high barriers exhibit a very strong sensitivity of V_G to temperature (several 100mV/^{O}C), which is far more than that of typical p-n junctions used in integrated temperature sensors (~2mV/^{O}C) and could translate into greatly improved resolution when measuring small changes in temperature. Figure 6 describes the measured linear dependence between sensitivity and gate voltage which is in accordance with the equation above. As the threshold of the SGT can be set during the design phase [2], highly sensitive SGTs can be made which have a very negative V_T and operate at low V_G to meet the requirements of low voltage systems.

Source-gated transistors are inherently very stable under electrical stress, as has been described in [5]. Figure 7 shows the excellent stability of the drain current over a number of days in a fabricated polysilicon SGT, together with the very small threshold shift due to stress. This behaviour, together with the low saturation voltage and high output conductance is very desirable for low power, sensor matrices where power supply ripple rejection, uniformity of signals across a wide area, sensitivity and power are important design considerations. The versatility of the SGT concept is extended by the possibility of realizing, in principle, of an SGT structure in any semiconductor or fabrication process as long as a reliable potential barrier can be fabricated at the source.

The authors would like to acknowledge the contributions to the design and fabrication of the polysilicon SGTs by Dr. Nigel Young and Dr. Michael Trainor from Philips Research. The work of R. A. Sporea is supported by EPSRC, U.K.

[1] J. M. Shannon and E. G. Gerstner, *IEEE Electron Dev. Lett.*, **24**, no. 6, pp. 405-407, 2003; [2] R. A. Sporea et al., *IEEE Trans. Electron. Devices*, **57**, iss. 6, Oct. 2010; [3] A. Nakashima, Y. Sagawa and M. Kimura, *IEEE Sensors Journal*, **11**, iss. 4, pp. 995-99, 2011; [4] A. Bakker, J. H. Huijsing, *IEEE J. Solid-State Circuits*, **31**, iss.7, pp. 933-937, 1996; [5] J. M. Shannon, *Appl. Phys. Lett.*, **85**, no. 2, pp. 326-328, 2004.

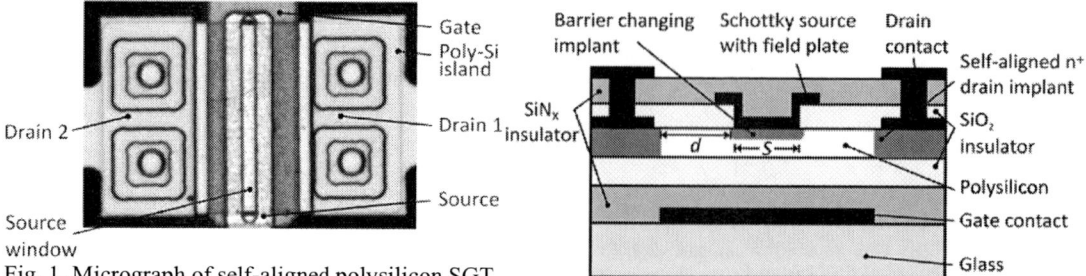

Fig. 1. Micrograph of self-aligned polysilicon SGT. Width W=50μm, source length S=2μm, source-drain separation d=10μm.

Fig. 2. Cross-section of the self-aligned structure.

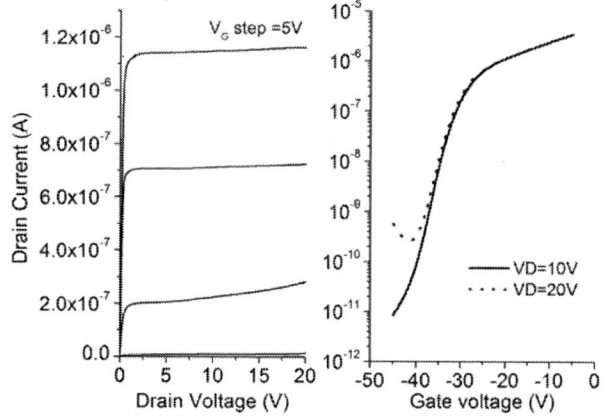

Fig. 3. Output characteristics (left) and corresponding transfer curve (right) for an SGT with high source barrier and W=50μm, S=2μm, d=6μm.

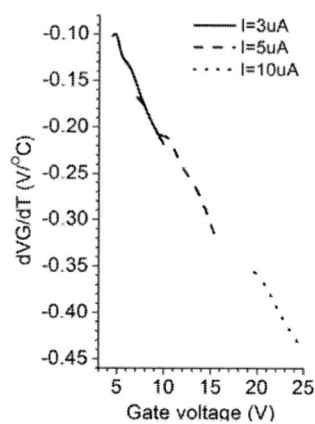

Fig. 6. Measured dependence of dV_G/dT on V_G as temperature changes.

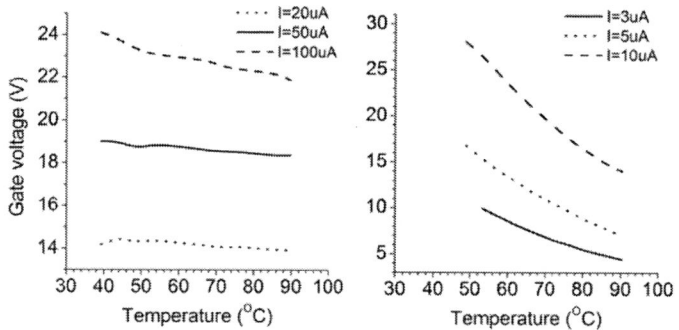

Fig. 4. Measured variation of the gate voltage with temperature for SGTs with low (left) and high (right) barriers operated at constant current.

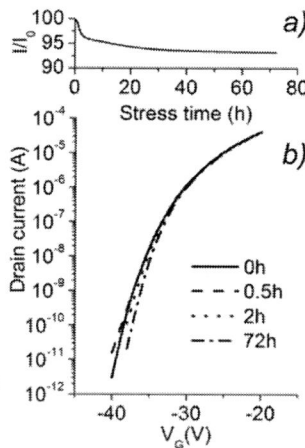

Fig. 7. a) Stability of drain current to prolonged electrical stress for polysilicon SGTs; b) Change of transfer characteristic during stressing period.

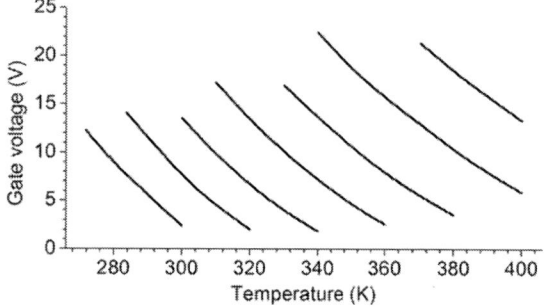

Fig. 5. Simulation of change in gate voltage with temperature for a Schottky barrier SGT operated at constant current. Left to right: I_D=3nA, 10nA, 30nA, 100nA, 300nA, 1μA, 3μA.

978-1-61284-243-1/11 $26.00 © 2011 IEEE

Voltage-Controlled Spin-Wave-Based Logic Gate

Tianyu Liu and Giovanni Vignale

Department of Physics and Astronomy, University of Missouri, Columbia, Missoiuri 65211,
USA/ Email: tl5w8@mail.missouri.edu/ Phone: 573-882-7051/ Fax: 573-882-4195

Spin wave spintronics (also known as magnonics) processes information by propagating spin waves with no charge displaced. Because dissipation is thus minimized this is rapidly becoming an important subject of research within the larger area of spintronics. The logic states in magnonic circuitry can be defined either by the phase or by the amplitude of the spin wave. In both cases, a π-phase shifter plays a crucial role in performing logical operations. The first spin wave logic gate was experimentally demonstrated by Kostylev *et al* [1]. They utilized an inhomogeneous magnetic field to control the phase difference between spin waves propagating in different arms of a Mach-Zehnder interferometer – and thus the amplitude of the output spin wave. Later, Schneider *et al* [2] and Lee *et al* [3] developed a complete set of logic gates such as NOR, XOR and AND, based on spin wave interferometry. However, all of theses gates are controlled by a current-induced magnetic field. As the devices shrink down, π-phase shift requires a larger electric current to induce stronger magnetic field, which inevitably increases the power-loss. Therefore, voltage-controlled spin wave electronics becomes an attractive alternative avenue towards nano-scale magnonics, where exchange spin waves are of primary interest.

In this study, we propose a prototype voltage-controlled spin wave NOT gate based on a ring-shaped Mach-Zehnder interferometer made of yttrium-iron-garnet (YIG), as shown in Fig. 1. The phase of the spin wave is manipulated by a radial electric field through the spin-orbit (SO) coupling. For the spin wave with a given energy, its wave vector will be shifted by $A = \frac{eES\cos\theta_0}{\hbar^2/(2m\lambda^2)}$ where λ is a characteristic length that controls the strength of SO coupling, S is the magnitude of saturated magnetization, m is the the mass of an electron and θ_0 denotes the direction of equilibrium magnetization as shown in Fig. 1. To understand the basic principle of our device, one can go back to the concept of magnon – the particle-like counterpart of spin waves. The idea is to exploit the Aharonov-Casher [4] phase induced by an electric field on such a particle [5]. However, due to the extremely small value of the characteristic length $\lambda = \frac{\hbar}{mc}$ for electrons in vacuum, it seems impossible to obtain a π-phase shift within the coherence length of spin waves. Fortunately, recent theoretical research [6] shows that the SO length can be enhanced by three orders of magnitude in magnetic materials due to the existence of intrinsically spin-orbit entangled molecular orbitals. For example, we find $\lambda \simeq 1.5\,\text{Å}$ in YIG, which is sufficiently large to allow the realization of nanometer-scale spin-wave interferometer working as a NOT gate. By using scattering matrix methods, we have obtained an analytical expression for the transmission probability of spin waves propagating in such a ring interferometer,

$$|T|^2 = \frac{4\sin^2(\pi kR)\cos^2\alpha}{\cos^4\alpha - \cos(2\alpha)\cos(2\pi kR) + 2\sin^2(\pi kR)}, \tag{1}$$

where $\alpha = \pi\cos\theta_0(1 + \frac{eV_{in}}{\ln(r_0/R)}\frac{2m\lambda^2}{\hbar^2})$, with R being the radius of the ring, r_0 the radius of cylindrical conductor in the center of the ring, and V_{in} the voltage applied to the conductor. Fig. 2 shows clearly how to accomplish a NOT logic functionality by adjusting the input voltage.

Work supported by ARO Grant No. W911NF-08-1-0317

[1] M.P. Kostylev, *et al.*, Appl. Phys. Lett. **87**, 153501 (2005).
[2] Schneider, *et al.*, Appl. Phys. Lett. **92**, 022505 (2008).
[3] K. Lee and S. Kim, J. Appl. Phys. **104**, 053909 (2008).
[4] Y. Aharonov and A. Casher, Phys. Rev. Lett. **53**, 319 (1984).
[5] Z. Cao, X. Yu, and R. Han, Phys. Rev. B **56**, 5077 (1997).
[6] H. Katsura, N. Nagaosa, and A.V. Balatsky, Phys. Rev. Lett. **95**, 057205 (2005).

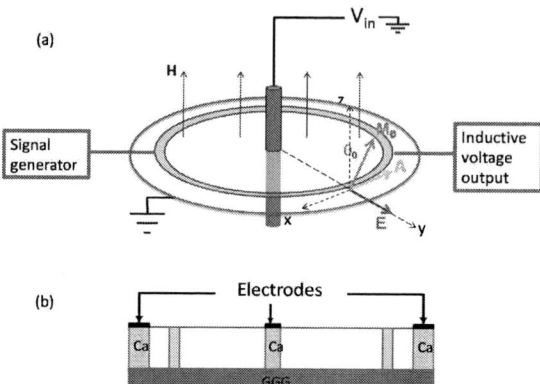

Figure 1: Schematic view of the voltage-controlled NOT gate: (a) a ferromagnetic ring in the presence of a radial **E** field. A weak magnetic field is applied perpendicular to the ring plane in order to tilt the equilibrium magnetization away from the ring but still in the plane tangent to the ring. In the absence of the magnetic field, the electric field has no influence on the phase shift of spin waves; (b) Possible fabrication of the ring (cross section view): a single-crystal YIG film is grown on $Gd_3Ga_5O_{12}$ single-crystal substrate by liquid phase epitaxy; a ring pattern is produced by ultraviolet lithography and chemical wet etching; implanting calcium ions into the regions marked by "Ca" changes those regions from insulators to conductors so that a radial electric field appears as applying an input voltage on the center cylindrical conductor.

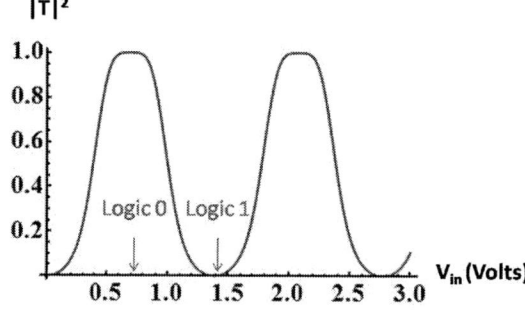

Figure 2: Transmission probability as a function of input voltage. A high voltage input yields a destructive interference and thus the zero induced voltage output, and a low voltage input yields a non-zero voltage output. In other words, this ring interferometer forms a voltage-controlled NOT gate where high and low voltage represents logic states 1 and 0, respectively.

Effect of Disorder on Superfluidity in Double Layer Graphene

B. Dellabetta and M. J. Gilbert

Department of Electrical and Computer Engineering and Micro and Nanotechnology Laboratory, University of Illinois
208 N. Wright Street, Urbana, IL 61801
Phone: 217-333-3064 Fax: 217-244-6375 E-mail: dellabe1@illinois.edu

Abstract

Post-CMOS logic in bilayer graphene is very promising due to the possibility of observing room temperature collective states. An excitonic superfluid is predicted to form in double layer graphene systems at room temperature if the two individual monolayers of graphene are separated by an oxide no more than a few nanometers thick [1]. Recent experiments have shown evidence of interaction enhanced transport in double layer graphene [2], but there is a significant discrepancy in the quality of the two graphene layers which may be occluding the phase transition. We present and compare the performance characteristics of ideal and disordered double layer graphene systems at room temperature in the purported regime of superfluidity.

We perform quantum transport calculations on double layer graphene using the Non-Equilibrium Green's Function (NEGF) formalism in an effort to elucidate the evolution of a BEC under non-equilibrium conditions in the presence of lattice defects. We find that lattice defects spread throughout the channel can degrade interlayer current by 30%, but disorder concentrated near the contacts causes a much more significant reduction of 80% in interlayer current. We also find that steady-state spontaneous coherence is lost for defect concentrations greater than 4%; a very clean system is therefore necessary for potential post-CMOS logic applications.

Introduction

One of the most striking features of the formation of a Bose-Einstein Condensate (BEC) is an enhanced interlayer tunneling current. Initially discovered in cryogenic GaAs heterostructures in the Quantum Hall regime [3], this type of behavior may be viewed as a spontaneous coherence, or "pseudospin ferromagnetism", where a particular superposition of states in the two layers is selected for the entire system.

Pseudospin ferromagnetism must survive at room temperature in order to be useful in the context of a device. Room temperature operation is possible if the carrier densities in the two layers are essentially equal but with opposite polarity – electrons in one layer, holes in the other. This is uniquely possible in graphene at sufficiently high, yet still experimentally obtainable, carrier densities in the respective layers due to the symmetric band dispersion relationship and holes and atomically two-dimensional nature [4].

Device Structure and Simulation Method

Fig. 1 depicts the device configuration considered in this work. Each layer is 30 nm long by 10 nm wide. They are separated by a SiO_2 dielectric 1 nm thick. Top and bottom gates (V_{TG} and V_{BG}) manipulate individual concentrations in each layer, and are set to be $V_{TG} = -V_{BG} = 0.4$ V which results in a quasiparticle density of 10^{13} cm^{-2} in each layer. Contact

biases are set in the drag-counterflow geometry to maximize condensate current driven through the system [1].

To simulate this system with NEGF [5], we expand the tight-binding description of the top and bottom graphene monolayers (H_{TL}, H_{BL}) to the following paired Hamiltonian,

$$H_{sys} = \begin{bmatrix} H_{TL} & \Delta \\ \Delta & H_{BL} \end{bmatrix}. \quad (1)$$

The interlayer coupling matrix Δ is defined as

$$\Delta = \Delta_{sas}\hat{x} + Um_{ex}\left(\cos(\phi_{ex}), \sin(\phi_{ex})\right). \quad (2)$$

In eq. (2), Δ_{sas} is the single particle tunnel splitting, which acts to destroy excitons. The interlayer Coulomb term, U, is the screened pairing potential between an electron and a hole between the SiO_2 dielectric. In eq. (2), m_{ex} is the magnitude of the pseudospin polarization and ϕ_{ex} is its planar orientation. We iterate over the mean-field equations and couple to a 3D Poisson solver to obtain a self-consistent solution with compatible particle densities and potential profiles.

The Hamiltonian may be modified in a number of ways to model disorder arising as artifacts of fabrication. In this work, we model vacancies by setting all intralayer and interlayer coupling with the vacant site to zero. We uniformly remove carbon atoms from the top graphene layer, so as to better model the experiments in [2], and then iterate the disordered Hamiltonian to self-consistency.

Device Results

Fig. 2 shows a plot of the magnitude of the order parameter, $|\Delta|$, with 2% vacancies randomly placed near the contacts. The order parameter magnitude clearly decays near vacancies, with the largest drop occurring at sites close to multiple vacancies. Interlayer quasiparticle current density, which is directly proportional to $|\Delta|$, therefore can drop nonlinearly with disorder strength based on the randomly selected configuration. Quasiparticle current is only nonzero near the contacts, thus channel disorder does not show the same relation. This qualitative difference can be seen in Figs. 3 and 4, which plot the interlayer current achieved for various percentages of randomly configured channel and contact vacancies relative to the ideal case.

Figs. 3 and 4 display a significant amount of variance, particularly in the case of contact disorder. Fig. 5 expands on this to show how maximum interlayer current varies from device to device based on vacancy location. Channel disorder does not prevent the quasiparticle tunneling action. Critical current therefore does not vary significantly with configuration, and performance decreases up to 30% before self-consistency is lost. Contact disorder, however, varies appreciably relative to configuration. The average critical current follows a linear correlation to disorder strength, with a much more significant drop of 80% in performance before self-consistency is lost.

978-1-61284-243-1/11 $26.00 © 2011 IEEE

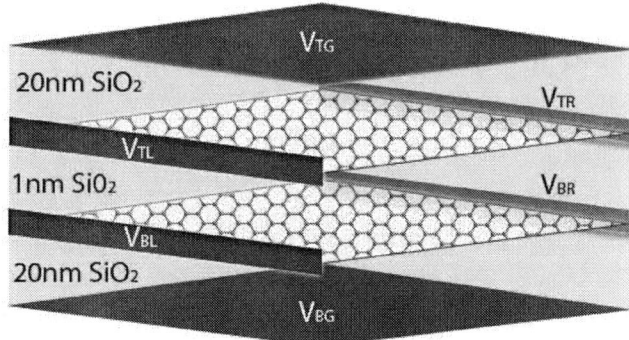

Fig. 1: Schematic of the device under consideration. Each monolayer is 10 nm wide by 30 nm long. The two graphene monolayers are separated by a tunnel oxide of SiO_2 1 nm thick. An oxide 20 nm thick separates the monolayers from the top and bottom gates.

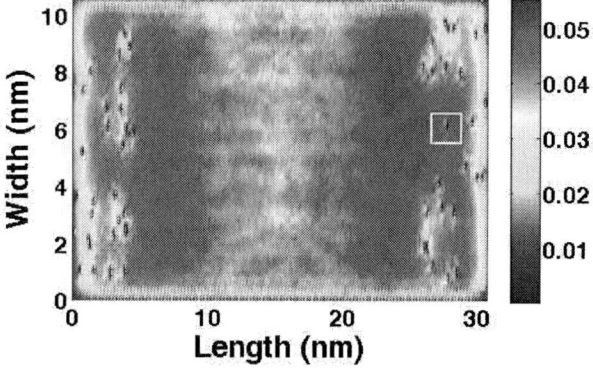

Fig. 2: Plot of order parameter magnitude, $|\Delta|$, for a system in the condensed state with 2% vacancies in the contact. Isolated vacancies (white box) have a little long range effect, but bunched disorder has a nonlinear effect on $|\Delta|$ at longer distances. Interlayer current is therefore very dependent on how closely vacancies are bundled to one another.

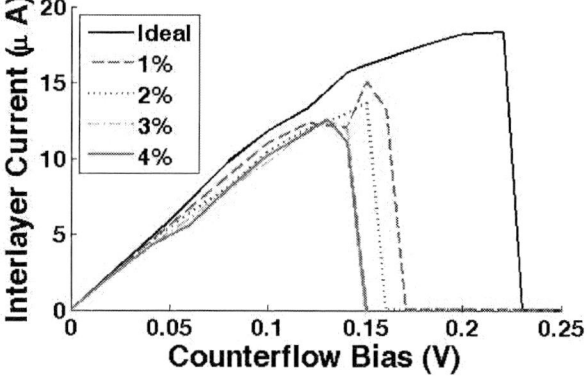

Fig. 3: Interlayer current vs. counterflow bias for a range of vacancies randomly inserted within the channel of the top graphene monolayer. A localized density of states forms near the vacancies which provides a small change in interlayer conductivity, but decreases the voltage at which there is a phase transition back to the normal state.

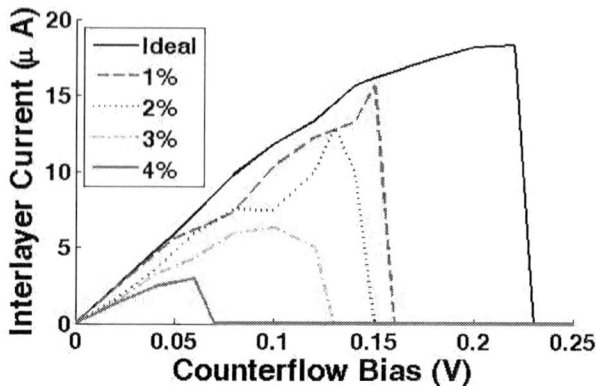

Fig. 4: Similar I-V curve for a range of vacancies randomly inserted within 5nm of the contacts of the top graphene monolayer. Disorder induces backscattering states that prevent injected electrons from triggering a condensate current through the channel.

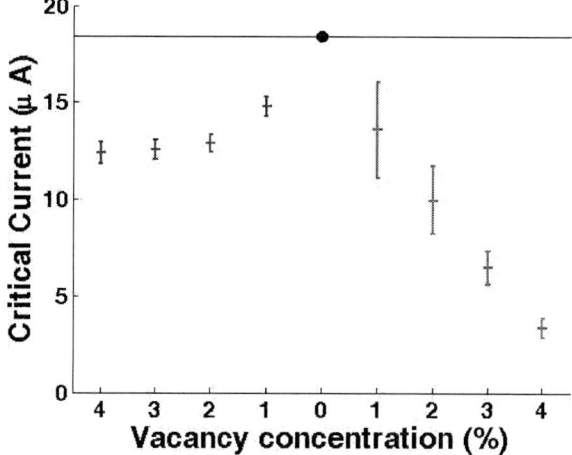

Fig. 5: Statistical distribution of maximum interlayer current that can be driven through the condensate before the phase transition to a non-interacting state. Channel (contact) vacancy concentrations are plotted on the left (right) of the ideal maximum interlayer current.

References

[1] M. J. Gilbert, "Finite temperature pseudospin torque effect in bilayer graphene," *Physical Review B* **82**, 165408 (2010).

[2] S. Kim, I. Jo, J. Nah, Z. Yao, S. K. Banerjee and E. Tutuc, "Coulomb drag of massless fermions in graphene," *arXiv:1010.2113* (2010).

[3] B. Spielman, J. P. Eisenstein, L. N. Pfeiffer and K. W. West, "Resonantly enhanced tunneling in a double layer quantum Hall ferromagnet," *Physical Review Letters* **84**, 5808 (2000).

[4] H. Min, R. Bistritzer, J.-J. Su and A. H. MacDonald, "Room-temperature superfluidity in graphene bilayers" *Physical Review B* **78**, 121401(R) (2008).

[5] A. Svizhenko, M. P. Anantram, T. R. Govindan, B. Biegel, and R. Venugopal, "Two-dimensional quantum mechanical modeling of nanotransistors," *J. Appl. Phys.* **91**, 2343 (2002).

Dual Pillar Spin Transfer Torque MRAM with tilted magnetic anisotropy for fast and error-free switching and near-disturb-free read operations

Niladri N. Mojumder, Sumeet K. Gupta and Kaushik Roy

School of Electrical and Computer Engineering, Purdue University, West Lafayette, Indiana 47907, USA

niladri@ecn.purdue.edu, guptask@purdue.edu, kaushik@purdue.edu

We propose a three terminal, dual pillar magnetic tunnel junction (MTJ) with tilted magnetic anisotropy for fast and error-free precessional magnetic switching with near-disturb-free magneto-resistive data sensing. Marginal tilting of magnetic anisotropy of the pinned layer in the write-in port enables fast (~2ns) and error-free magnetic switching, subject to an electric current density of almost 70% lower than that required in a conventional STT-MRAM with perpendicular magnetic anisotropy (PMA). A thicker tunnel barrier is incorporated in the spatially and electrically isolated read-out port for higher tunneling magneto-resistance (TMR) and near-disturb-free read operations. Dual bit line memory architecture with just one access transistor per bit-cell is also proposed. The technology-circuit co-optimization of the proposed one transistor Dual Pillar Spin Transfer Torque (DPSTT) MRAM cell is carried out using effective mass-based spin transport [1] and finite temperature macro-magnetic simulations involving Landau-Lifshitz-Gilbert-Slonczewski (LLGS) equation [2-4]. The proposed DPSTT-MRAM bit-cell outperforms the state-of-the-art 1T-1MTJ STT-MRAM cell in terms of higher cell TMR, single supply voltage for read/write, near-disturb-free data access under parametric process variations with comparable or even lower critical switching current.

In the proposed dual pillar MTJ (Fig. 1), the magnetic anisotropy of the reference layer in the write-in port (pinned layer 1) is tilted by an angle θ_{Tilt} relative to that of the free layer with PMA. To achieve uncompromised cell TMR and low disturb failure probabilities during read, the magnetic anisotropy of the reference layer in the read-out port (pinned layer 2) is kept collinear to the free layer anisotropy. The access transistor is connected between the bit line BLL and port 1. Port 2 and 3 are connected to the source line SL and bit line BLR respectively (Fig. 7). To write into the cell, word line is asserted after appropriate polarity of voltage is applied between BLL and SL. During write, as most of the applied voltage drops across oxide 1 with higher resistance than the ferromagnetic metals, the current is injected through the entire free layer area ($4\lambda \times 10\lambda$), but not just the area underneath the nonmagnetic contact ($4\lambda \times 4\lambda$). To sense the data previously written, both BLL and SL are pre-charged to V_{DD} and BLR is pre-discharged. The word line is asserted to send a small sensing current between the bit lines BLL and BLR.

The advantage of a relative angular tilt between the free layer and pinned layer easy-axes on precessional STT switching is estimated analytically in Fig. 2. With collinear magnetic anisotropies ($\theta_{Tilt}=0^0$), electric current induced magnetization switching is initiated by thermally induced initial angular precession of the free-layer moment around its easy-axis (z-direction). The single domain magnetic simulation framework used to capture the magnetization dynamics of the free layer is shown in Fig. 4. A faster switching (τ_{sw}~2ns) with sufficiently low switching failure probability ($P_{F,SW}=10^{-9}$) requires a current density (J) close to $6.5MA/cm^2$ at T=300K, sufficient to cause the tunnel oxide breakdown. However, a marginal tilt of the pinned layer 1 anisotropy (θ_{Tilt}~5^0-20^0) in dual pillar MTJ offers almost 3X reduction in J for τ_{sw}=2ns with $P_{F,SW}$ lower than 10^{-9} (Fig. 3). With sufficient anisotropy tilting (>5^0), the current induced magnetization reversal becomes essentially independent of the initial thermal fluctuation that makes the free layer switching almost error-free. As estimated in Fig. 3, a 3X lower switching current density in the proposed DPSTT-MRAM reduces the critical switching current (I_C) by almost 25% as compared to the conventional STT-MRAM with 2.5X smaller free layer cross-section. Fig. 5 shows the write margin enhancement in the proposed DPSTT-MRAM bit cell as compared to the conventional 1T-1MTJ bit-cell.

In DPSTT MTJ, incorporation of a thicker tunnel barrier ($T_{OXIDE,2}$~1.8nm) in the read port improves the bit-cell TMR (Fig. 6) and reduces the number of disturb failures under parametric process variations (Fig. 7). During read, as the current is injected through a smaller cross-section of the free layer underneath the pinned layer 2, magnetic domain-wall dragging increases the cell disturb current ($I_{DISTURB}$) significantly. The presence of a thicker tunnel barrier in the read port combined with the effect of magnetic domain-wall dragging offer a higher read disturb margin ($I_{READ}/I_{DISTURB}$) in DPSTT-MRAM, subject to the same bit line voltage as used during write. This eliminates the necessity of a second lower supply voltage during read (~$0.25V_{DD}$), as mandatory in conventional 1T-1MTJ bit cell. Single ended voltage sensing is used for reading the DPSTT-MRAM bit-cell with a read speed of close to 1GHz (Fig. 8).

The schematic of the dual bit line memory array architecture for the three-terminal DPSTT-MRAM bit cell is shown in Fig. 9. Unlike conventional 1T-1MTJ architecture, the proposed architecture includes an additional bit-line connected to the third terminal (port 3 in Fig. 1) of all the DPSTT devices placed in a column. The data from individual bit cells in a column are read using single ended voltage sensing. The word lines (WL's) are shared among all the bit cells placed in a row. In this new matrix-architecture, multiple words can be placed in a given row and accessed independently as desired. Left bit lines (BLL's), right bit lines (BLR's) and source lines (SL's) can all be shared among bit cells placed in individual columns. The programming and data sensing schemes in the DPSTT-MRAM array with dual bit line architecture have been illustrated in Fig. 10. In Fig. 11, we schematically show a possible 3D lay-out scheme suitable for integrating the DPSTT-MRAM in a planar CMOS technology. Dual finger transistor lay-out in Fig. 11 achieves higher density with S/D diffusions of consecutive access transistors in a column being shared. The provision for an additional metal track for the third terminal of the DPSTT device is kept in the layout without any area overhead.

To summarize, we show a thorough comparison of the proposed DPSTT-MRAM bit cell with the conventional 1T-1MTJ case in terms of various memory design attributes in Fig. 12. Lower energy consumption during read and write, error-free switching, higher cell TMR and near-disturb-free access make DPSTT-MRAMs a competitive choice for high performance, ultra-dense non-volatile memories for future stand-alone and embedded applications.

Acknowledgement: This research was funded in part by Qualcomm Inc., Intel, and Nano Research Initiative (NRI).

Reference:
[1] N. N. Mojumder et.al. J. Appl. Phys. 108, 104306 (2010); [2] J. C. Slonczewski, J. Magn. Mater. 159, L1-L7, 1996;
[3] L. Berger, Phys. Rev. B, vol. 54, pp. 9353-9358, 1996;   [4] W. Brown, Phys. Rev. 130, 1677 (1963).

978-1-61284-243-1/11 $26.00 © 2011 IEEE

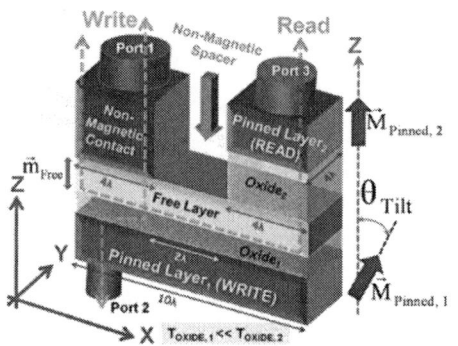

Fig. 1 Dual Pillar MTJ with pinned layer-1 anisotropy in the write port tilted by angle θ_{Tilt} relative to PMA

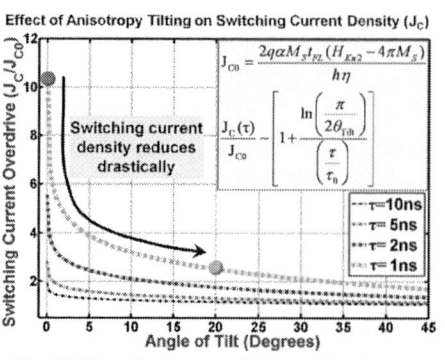

Fig. 2 Effect of anisotropy tilting on electric current density (J_C) for precessional magnetic switching

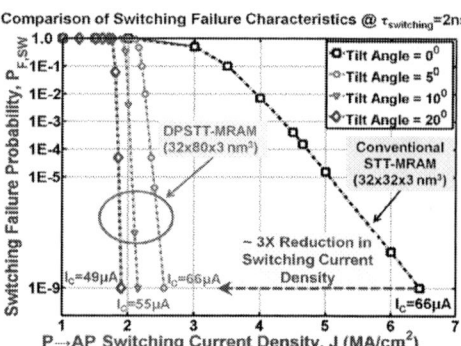

Fig. 3 Anisotropy tilting reduces the J_C in DPSTT-MRAM by 3X relative to the conventional 1T-1R case

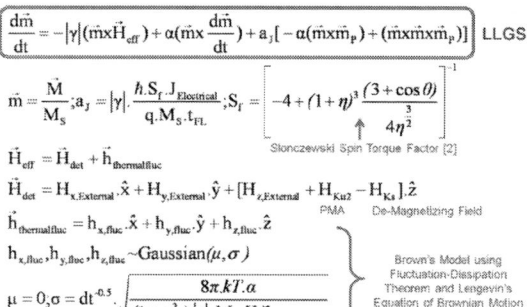

Fig. 4 Single domain magnetic simulation framework for the free layer in presence of stochastic thermal noise [2-4]

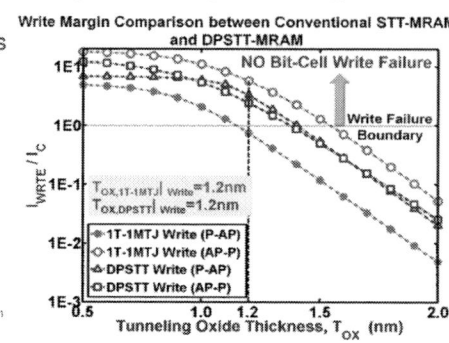

Fig. 5 Electrical write margin comparison between DPSTT-MRAM and 1T-1MTJ bit cell @ τ_{sw}=2ns

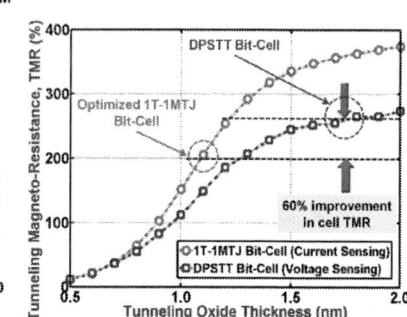

Fig. 6 Higher cell TMR (~265%) in DPSTT-MRAM reduces the number of cell decision failures

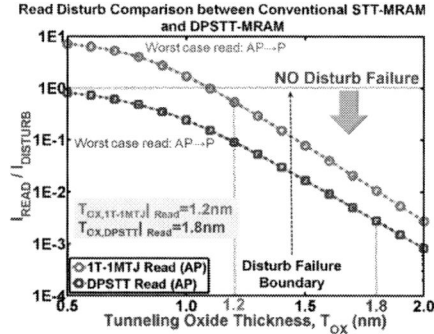

Fig. 6 Higher disturb current for DPSTT bit cell drastically reduces the number of disturb failures

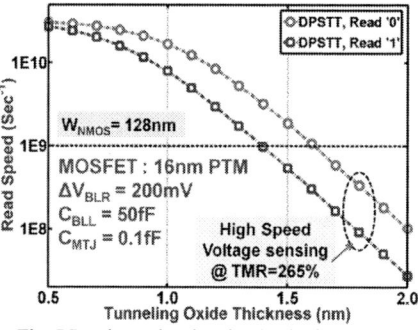

Fig. 5 Read speed estimation in single ended voltage sensing for the DPSTT-MRAM bit cell

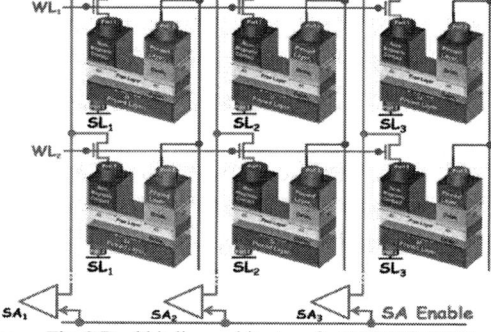

Fig. 9 Dual bit-line architecture for DPSTT-MRAM array with just one access transistor per bit cell.

For a memory array of size M x N (M rows and N columns), we assume t consecutive bit cells in a row forming a word (N=t*n, where n is natural number). The steps to write a word in the pth row with its initial bit on $(q+1)^{th}$ column:

(i) For each j=q+1,q+2,..,q+t; BLL_j=BLR_j=$\overline{SL_j}$;
    For writing '1', BLL_j=0 and
    For writing '0', BLL_j=V_{DD}
    For all other j, BLL_j=BLR_j=SL_j=0
(ii) SA remains turned OFF
(iii) WL_i=V_{DD} for i=p and WL_i=0 for all other i

Steps to read a word in the pth row with its initial bit on $(q+1)^{th}$ column:

(i) For each j=q+1,q+2,...,q+t: BLL_j=SL_j=V_{DD} and
    BLR_j=0. For all other j, BLL_j=BLR_j=SL_j=0
(ii) SA is turned ON
(iii) WL_i=V_{DD} for i=p and WL_i=0 for all other i
(iv) The voltage drop across BLL_j's are sensed through corresponding SA's

Fig. 10 Programming and data read-out scheme in DPSTT-MRAM array with dual bit-line architecture.

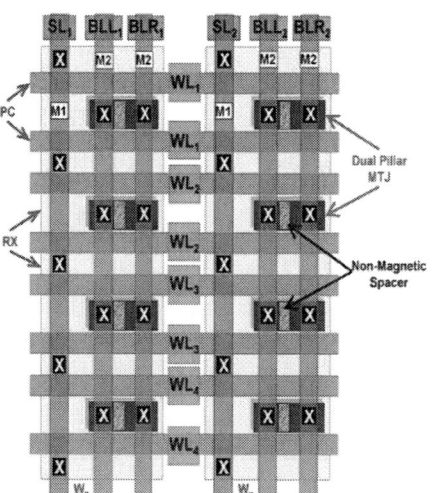

Fig. 11 Lay-out of the DPSTT-MRAM memory array. Insertion of an additional metal track is necessary, unlike the conventional 1T-1R case.

Memory Design Attributes @T=300K	DPSTT-MRAM Bit-cell: (T_{OX1}= 1.2nm,T_{OX2}=1.8nm, θ_{Tilt}=20° wrt PMA)	1T-1MTJ Bit-cell: (T_{ox}=1.2nm, PMA)
MTJ Size (Feature Size, F=16nm)	$32 \times 80 \times 3\ nm^3$ (PMA)	$32 \times 32 \times 3\ nm^3$ (PMA)
$J_{CRITICAL}$ (P→AP) @P_{sw}=1E-9, EA=80kT	1.9MA/cm² @2ns	6.5MA/cm² @2ns
I_C(P→AP) @τ_{sw}=1E-9	49µA @τ_{sw}=2ns	66µA @τ_{sw}=2ns
Read/Write Voltage	1.0V/1.0V	0.25V/1.0V
Worst ($I_{READ}/I_{DISTURB}$): Read '1 (AP-state)'	0.20%	54%
Read Frequency	1.0 GHz	1.0 GHz
Full bit cell TMR including transistor	265% (voltage sensing)	205% (current sensing)
Read Frequency	1.0 GHz	1.0 GHz
Access Transistor Width (L_{ch}=16nm)	128nm (Iso-area bit cell)	128nm (Iso-area bit cell)
Write Energy/Bit @ P_{sw}=1E-9, τ_{sw}=2ns	0.10 Pico-Joule	0.13 Pico-Joule
Read Energy/Bit @ τ_{READ} = 1ns	0.9Femto-Joule	9.0Femto-Joule

Fig. 12 Comparison of the proposed DPSTT-MRAM bit cell with the conventional 1T-1R case in terms of various memory design attributes

978-1-61284-243-1/11 $26.00 © 2011 IEEE

Giant Magnetoelectric Effect in Nanofabricated $Pb(Zr_{0.52}Ti_{0.48})O_3$-$Fe_{85}B_5Si_{10}$ Cantilevers and Resonant Gate Transistors

Feng Li[1], Zhao Fang[1], Rajiv Misra[3], Srinivas Tadigadapa[1], Qiming Zhang[1,2] and Suman Datta[1]

[1]Electrical Engineering, [2]Materials Research Institute, [3]Physics Department, The Pennsylvania State University, University Park, PA 16802 Email: fxl135@psu.edu, Phone: (814)321-6158

Magnetoelectric (ME) laminates show higher ME coefficients than that of natural multiferroics (e.g. Cr_2O_3, BiTiO) by up to several orders of magnitude. Recent studies on bulk ME sensors using $Fe_{85}B_5Si_{10}$ (Metglas) /polyvinylidene fluoride composite show a high ME voltage coefficient of 21V/cm·Oe at 20 Hz [1]. However, bulk sensors suffer from poor epoxy bonding, aging and difficulty of integration with CMOS electronics. Here, we report, for the first time, the monolithic nanofabrication of $Pb(Zr_{0.52}Ti_{0.48})O_3$ (PZT)-$Fe_{85}B_5Si_{10}$ ME cantilevers (Fig.1(a)) on silicon substrate which achieve 0.46 V/cm·Oe at 20 Hz and 1.8 V/cm·Oe at a resonance frequency of 8.4 KHz. Also, ME cantilever based resonant gate transistors (RGT) (Fig.1 (b)) has been designed and analyzed in comparison with ME cantilever. A 10X signal to noise ratio improvement can be reached by ME RGT. This shows the compatibility of the nanofabricated cantilever ME sensors with the Si process technology and paves the way for the future integration of MEMS based ultra-sensitive magnetic sensors with advanced Si nanoelectronics.

ME cantilever fabrication and characterization: Fig.2 shows the SEM picture and process flow of a fully released ME cantilever. First, a 0.93um thick PZT was deposited on Pt (100nm) coated SOI substrate using a sol-gel technique, followed by 50nm Pt adhesion layer (sputtered at 5mTorr, 200W dc), 260nm Metglas (5mTorr, 150W rf) and a 60nm Cr/Au top electrode deposition. The oxide etcher, Tegal 6500, was used to etching through the PZT layer and accesses the bottom electrodes. The cantilever beams (area ~ 200um*40um) were then patterned and etched down to the Si substrate. Finally, XeF_2 based dry etch was used to release the cantilevers. We individually characterize the sputtered Metglas and the PZT thin films. The Metglas/silicon cantilever was made to characterize the magnetostriction using laser vibrametry. A 0.4ppm/Oe magnetostrictive coefficient ($d_{33,m}$) was achieved without annealing, as shown in Fig. 3(a). The PZT films showed an effective piezoelectric coefficient $e_{31,f}$ of 7 C/m^2 (Fig. 3(b)) [2]. Fig. 3(c) gives the ME coefficient (α_{ME}=0.46V/cmOe at H_{dc}=25Oe) of the composite PZT/Metglas cantilever. The measured α_{ME} is samller than the expected value because of the lower Fe composition (~70%) in the Metglas layer. Table I compares the low frequency (at 20Hz) α_{ME} in this work with the one reported in recent literature [3, 4]. This shows the good ME response for the scaled PZT/Metglas cantilever sensor. And fig. 3(d) gives the frequency response where α_{ME} (1.8 V/cm·Oe) increases further by a factor of 4 at the resonance frequency. The noise performance of the sensor has also been investigated in order to evaluate the signal to noise ratio (SNR) of the cantilever sensor. The equivalent circuit with noise source of the whole measurement system and charge amplifier output with 0.7 Oe ac input magnetic field are shown in Fig. 4 (a) and (b). We measure the sensor capacitance as well as the dielectric loss (Fig. 4(c)) in order to calculate the sensor noise. Fig. 4(d) compares the calculated noise versus the measured noise of both sensor and amplifier. It shows that the total noise is dominated by the dielectric loss noise of the cantilever sensor. From the above results, we have the SNR of the whole system of 790 ($Hz^{1/2}$/Oe) and the minimum detectable magnetic field of 100nT/$Hz^{1/2}$.

ME RGT design and analysis: In Fig. 5(a), the equivalent circuit of ME RGT is demonstrated. The PZT/Metglas resonant gate is sensing the input ac magnetic field and results in the capacitance (air gap) change between the resonant gate and transistor gate. And the corresponding oxide charge change can be detected by the transistor and is reflected by the drain current. The voltage mode readout circuit [3] will be used to detect the response of the ME RGT (Fig.5 (b)) and Fig. 5(c) shows the layout of the readout circuit on which the resonant gate will be built. The noise performance of the voltage mode readout circuit is shown in Fig. 5 (d). This 1/f noise from the circuit is considered to be the dominated system noise since the resonance gate has the lowest dielectric loss (air) noise. Finally, we compare the SNR of ME cantilever and ME RGT at resonance in Table II. The output noise PSD at resonance frequency is reduced by 100X because of the less sensor noise, so the total SNR of ME RGT will be enhanced 10X in comparison to ME cantilever and a minimum detectable field of 10nT/$Hz^{1/2}$ can be achieved. The ME RGT is still under fabrication.

[1] Z. Fang, S. G. Lu, F. Li, S. Datta, Q. M. Zhang, and M. El Tahchi, Appl. Phys. Lett. 95, 112903 (2009)

[2] S. Trolier-McKinstry and P. Muralt, J. Electroceram., 12, 7 (2004)

[3] Peng Zhao, et al., Appl. Phys. Lett. 94, 243507 (2009)

[4] Zhiguang Wang, et al., J. Appl. Phys., **109**, 034102 (2011)

[5] F. Li, F. Zhao, Q.M. Zhang and S. Datta, Electronics Letters, Vol. 46 No. 16 (2010)

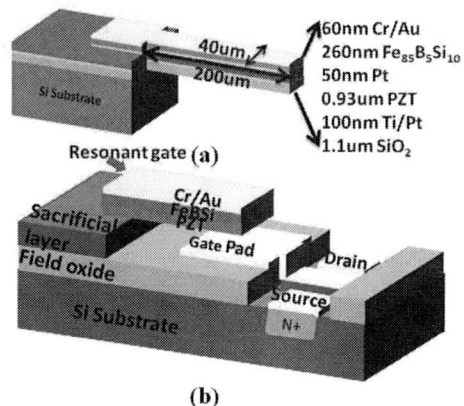

Fig.1 Schematic for Magnetoelectric(ME) (a) cantilever and (b) resonant gate transistor(RGT).

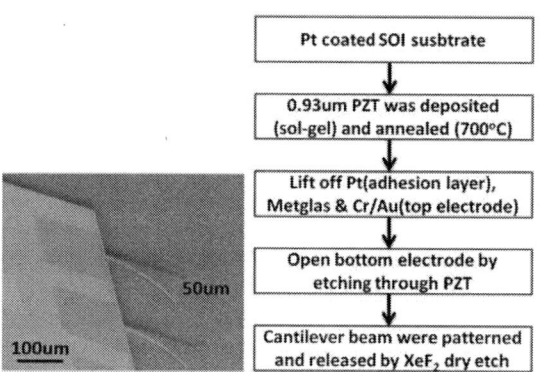

Fig. 2 SEM and process flow for ME cantilevers.

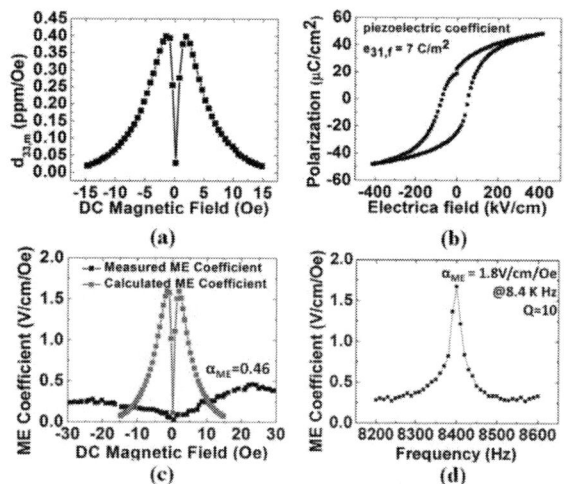

Fig.3 ME cantilever characterization: (a) Magnetostriction of Metglas (b) PE loop & piezo coefficient of PZT (c) Measurement vs. calculated ME coefficient (d) ME coefficient at resonance(8.4K Hz)

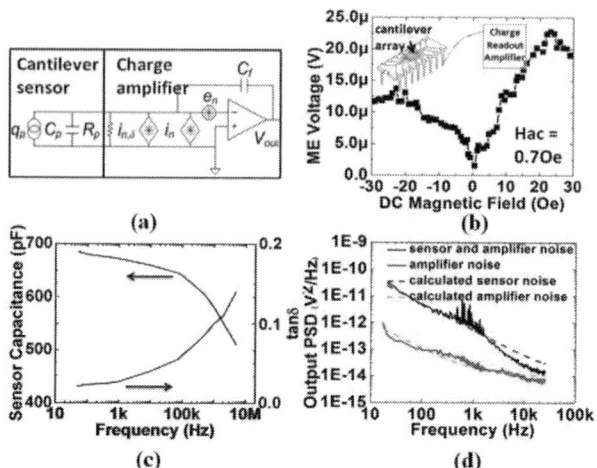

Fig. 4 (a) Equivalent circuit with noise source (b) ME cantilever output voltage versus DC magnetic field(input ac magnetic field equals 0.7Oe) (c) Sensor capacitance and dielectric loss vs. frequency (d) noise power spectrum density(PSD) of amplifier and sensor

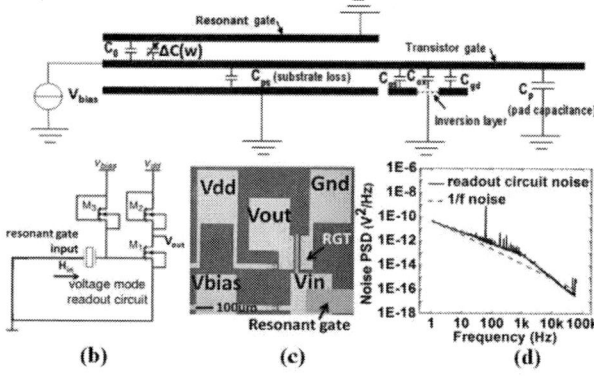

Fig. 5 (a) Equivalent circuit of ME RGT (b) Voltage mode readout circuit diagram (c) ME RGT readout circuit layout (c) Noise PSD for the readout circuit

Table I ME coefficient comparison

At 20 Hz	PZT/FeBSi (this work)	PZT/FeGa [3]	BaTiO/FeB Si [4]
α_{ME} (V/cmOe)	0.46	0.1	0.055

Table II SNR comparison

Sensor at resonance	α_{ME} (V/cmOe)	Noise (V^2/Hz)	SNR ($Hz^{1/2}$/Oe)
Cantilever	1.7	1e-13	790
RGT	1.6	1e-15	79,00

978-1-61284-243-1/11 $26.00 © 2011 IEEE

Observation of Trap-Assisted Steep Sub-threshold Swing in Schottky Source/Drain Al₂O₃/InAlN/GaN MISHEMT

Qi Zhou[1], Hongwei Chen[1], Chunhua Zhou[1], Zhihong Feng[2], Shujun Cai[2], Kevin J. Chen[1]

1. *Department of Electronic and Computer Engineering, Hong Kong University of Science and Technology, Kowloon, Hong Kong*
2. *National Key Laboratory of Application Specific Integrated Circuit, Hebei Semiconductor Research Institute, Shijiazhuan, China*
Phone: (852) 23588530, Fax: (852) 23581485, Email: zhouqi@ust.hk

Devices with steep subthreshold swing (*SS*) are of great interest and significance in view of increasing subthreshold leakage current with the continuous MOSFET scaling. The standby power dissipation has grown due to the nonscalability of the SS to below 60 mV/dec at room temperature (RT). To circumvent this obstacle, novel devices that employ various turn-on mechanisms have been proposed[1-4]. In this work, we report the observation of steep SS~20 mV/dec in Schottky source/drain (SSD) Al₂O₃/InAlN/GaN MIS-HEMTs over a drain bias range of 0.1 to 5 V. The devices also feature high I_{ON}/I_{OFF} ratio (~10^9) and appreciable current drive of I_{Dmax}=230 mA/mm at room temperature. The devices are also characterized at elevated temperature (*T*) up to 155 °C. Steep SS lower than the theoretical diffusion limit is consistently observed over the testing temperature range. It is suggested that the steep switching behavior is obtained through the means of a dynamic de-trapping process at the Al₂O₃/InAlN interface. The dynamic de-trapping enables a dynamic negative shift in the threshold voltage during the gate upswing and effectively facilitates the formation of a sub-threshold swing as steep as 18 mV/dec.

The InAlN/AlN/GaN sample (Fig. 1) used in this study consists of a 7-nm In₀.₁₇Al₀.₈₃N barrier, a 1-nm AlN spacer and a 1.8 μm GaN buffer grown on sapphire substrate by MOCVD. The device fabrication commenced with mesa etching, followed by the formation of Ti/Au source/drain electrodes. A 12-nm Al₂O₃ dielectric layer was then deposited on the sample using atomic layer deposition (ALD), after which Ni/Au gate electrode was formed. Since larger off-state leakage could degrade *SS*[5], Schottky metal Ti/Au was used in the source/drain junction to further suppress the leakage current, in addition to the MIS structure that suppresses the gate leakage. The off-state leakage is two orders of magnitude lower in the Schottky source/drain devices than the control devices with Ohmic source (not shown). The I-V characteristics between two Schottky contacts with a variety of distances are shown in Fig. 2. In addition, the high-temperature annealing required for Ohmic contact formation is eliminated from the device process so that excellent metal morphology is obtained for further down-scaling of the device. For simple proof-of-concept, the devices designed in this work have short gate-to-source distance of 0.25~0.5 μm.

The representative I-V and transfer characteristics of the SSD MIS-HEMTs are shown in Fig. 3. The device exhibits substantial transistor behavior, with a maximum drain current I_{Dmax} of ~230 mA/mm and an off-set voltage of 0.12 V (at a drain current of 1 mA/mm). The switching behavior during the gate turn-on (forward) and turn-off (reverse) sweeps are depicted in Fig. 4. Hysteresis is observed with the threshold voltage V_T in the down-sweep more negative than that in the up-sweep. The SS during the up-sweep is less than 22 mV/dec. The root cause for this hysteresis is proposed to be the dynamic charging and discharging behavior of the deep acceptor-like trap states at the Al₂O₃/InAlN interface (Fig. 5). These acceptor-like traps are charged with electrons initially. During the forward-sweep starting from a large negative gate bias, these traps release electrons to the 2DEG channel and dynamically shift V_T negatively, accelerating the turn-on process. During the reverse-sweep, however, the same discharging process and corresponding negative shift in V_T would decelerate the turn-off process, leading to the SS larger than 60 mV/dec. The SS measured at a variety of V_{DS} are shown in Fig. 4c. The steepest SS is measured to be 18 mV/dec at a drain bias voltage of 3 V at room temperature.

Figure 6(a) shows the temperature dependence of I_{DS}-V_{GS} characteristics in a device with a 1 μm gate length. The SS measured in the up-sweep is 40 mV/dec at 155 °C, also considerably lower than the theoretical diffusion limit of 91 mV/dec. For the same device, the *T*-dependence of SS is plotted in Fig. 6b. The SS measured during the up-sweep exhibits moderate increase with temperature, but consistently lower than the theoretical limit *kT/q*. The results suggest that such MIS-HEMT at least shows steep SS during the switch-on.

In conclusion, the Schottky source/drain Al₂O₃/InAlN/GaN MISHEMTs feature steep switch-on behavior assisted by a dynamic charge-detrapping process is experimentally demonstrated. The SSD MISHEMTs exhibit steep SS (~20 mV/dec), high On/Off ratio (~10^9) and appreciable drive current (230mA/mm). The low SS is maintained over a wide range of drain bias voltage and temperature.

Acknowledgment: This work is supported by Hong Kong Research Grant Council under GRF 611610.

[1] K. Gopalakrishnan, et al., *IEEE IEDM* 02-289, (2002); [2] J. Appenzeller, et al., *Phys. Rev. Lett.*, Vol 93, 196805 (2004); [3] A. Padilla, et al., *IEEE IEDM*, (2008); [4] T. Krishnamohan, et al., *IEEE IEDM*, (2008); [5] J. Chung, et al., *IEEE Electron Device Lett.*, Vol. 29, p. 1196, 2008.

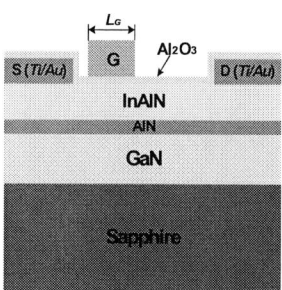

Fig. 1. Schematic cross section of Schottky source/drain MISHEMT.

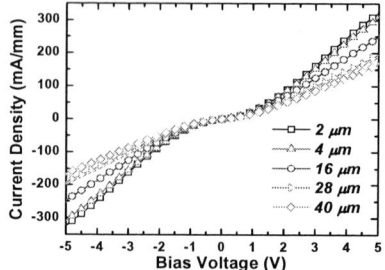

Fig. 2. Current-Voltage characteristics of adjacent Schottky contacts with various distance on InAlN/AlN/GaN sample.

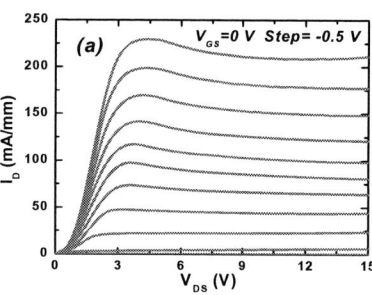

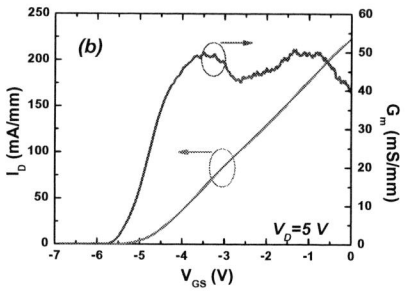

Fig. 3. Experimentally measured I_D-V_{DS} (a) and transfer curves (b) at room temperature. The device dimensions are: L_G=1 μm, L_{GS}=0.25 μm, L_{GD}=0.5 μm, W=10 μm.

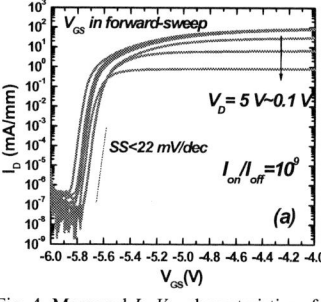

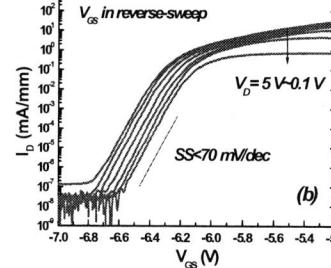

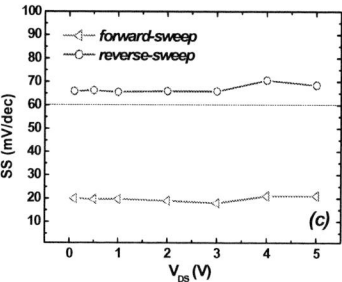

Fig. 4. Measured I_D-V_{GS} characteristics of a SSD MISHEMT at RT: (a) Forward sweep; (b) Reverse sweep and (c) SS extracted from I_D-V_{GS} against V_{DS}. Step size of V_{GS}=2 mV. The sweeping rate is 85 mV/Sec in the transfer measurement.

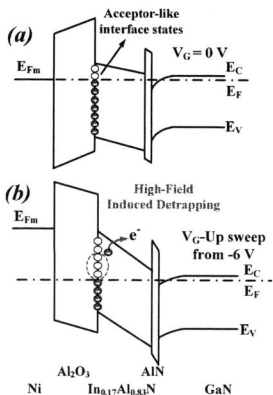

Fig. 5. Schematic band diagram of the MISHEMT structure showing (a) initial state (V_G=0 V) and (b) up-sweep of gate bias voltage. Discharging of deep level acceptor-like traps occurs while gate bias voltage in forward sweep.

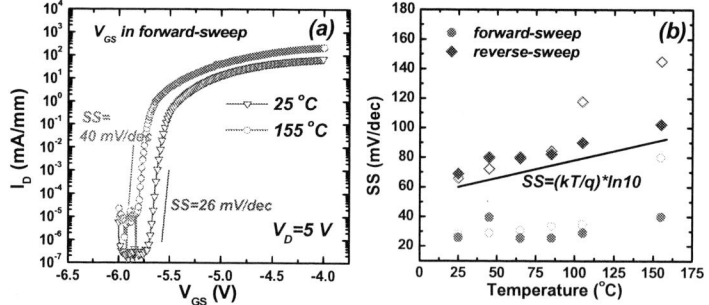

Fig. 6. Temperature dependence of (a) the transfer characteristics and (b) the subthreshold swing for forward-sweep (circle) and reverse-sweep (diamond) at drain voltage of 0.1 V (open symbol) and 5 V (solid symbol).

978-1-61284-243-1/11 $26.00 © 2011 IEEE

72

Towards Electronics at 1000 °C

D. Maier[1],[*], M. Alomari[1], N. Grandjean[2], J-F Carlin[2],
M-A Diforte-Poisson[3], C. Dua[3], S.L. Delage[3], E. Kohn[1]

[1]*Institute of Electron Devices and Circuits, University of Ulm, 89069 Ulm, Germany*
[2]*École polytechnique fédérale de Lausanne, 1015 Lausanne, Switzerland*
[3]*III/V Lab, 91461 Marcoussis, France*
[*] *E-mail: david.maier@uni-ulm.de , Phone: +49-731-50-26187*

High temperature electronics is up to now essentially limited to approx. 500 °C by the high temperature properties of the active semiconductor elements mostly based on SiC [1]. Sensing at even higher temperature relies therefore mostly on non-semiconductor components essentially limiting the systems complexities. However in recent years III-Nitride heterostructures, namely lattice matched InAlN/GaN heterostructures, have become an alternative. In an initial proof-of-concept experiment in 2006 [2] 1000 °C operation could be demonstrated for a short period of time.

Triggered by this experiment, commonly used HEMT processing steps have been refined to stabilize the high temperature device performance. This has resulted in 1 MHz large-signal operation for 50 hrs at 900 °C [3]. It was felt, that final failure was caused by metal electromigration and cracks formed in the passivation layer. In the first generation of experiments Au had been used as conductive overlayer material, limiting operation to approx. 700 °C. In the second generation technology the Au overlayer was replaced by Cu, allowing to extend testing to 900 °C, however already approaching its melting temperature of 1084 °C. Refining the contact technologies further has now resulted in a first large signal 1 MHz operation at 1000 °C (in vacuum) for 25 hrs. No significant change in pinch-off voltage could be observed, indicating no change in the heterostructure polarization properties and thus its structural integrity. It appears therefore that this heterostructure possesses indeed ceramic-like stability and device properties are essentially linked to the contact technologies, overlay metallization materials and passivation dielectrics used.

The devices were fabricated on MOCVD grown InAlN/GaN on sapphire with a barrier layer thickness of 12 nm. Device insulation was achieved by dry mesa etching in Argon plasma. Ohmic contact stack is Ti/Al/Ni/Pt annealed at 800 °C; the gate metal used was Mo; the device passivation was 30 nm Si_3N_4 deposited by PECVD at 340 °C. The devices have 2 parallel gate fingers without field-plate and air-bridge. Device geometry is L_G = 0.25 μm and $w_{G'}$ = 2 x 50 μm. The initial device pinch-off voltage is V_P = -2.1 V. The maximum drain current of the semi-enhancement mode device was 400 mA/mm at V_{GS} = +2 V, the maximum transconductance was 150 mS/mm.

In the experiment the temperature has been ramped up to 1000 °C within 48 hrs. Precision in temperature reading at 1000 °C is thought to be ΔT = 30 °C. The devices were tested in vacuum in 1 MHz large-signal class-A operation between V_g = +1 V and V_g = -2 V. The drain-source voltage was V_{DS} = 10 V, the external drain load resistor was R_l = 620 Ω. During testing, the mean DC output current (Mean I_D) has been recorded every 60 sec (see fig 1).

During the ramping period the channel sheet resistance increased from 700 Ω (RT) to 3.9 kΩ (1000 °C). The open channel output current decreased from 400 mA/mm to 200 mA/mm, dominated by the change in 2DEG mobility. The pinch-off voltage remained essentially unchanged at V_p = -2.1 V up to 1000 °C. During 1000 °C operation, buffer layer and gate leakage became noticeable. Fig. 2 shows the development of the mean output current (Mean I_D) with testing time, where the steady increase could be correlated with increase in buffer and gate leakage. This may be illustrated with the inserts of DC output characteristics measured initially at 1000 °C and after 1st hr, 3rd hr and 25 hrs respectively when the test was terminated (inserts fig.2). Despite of the degradation in electrical performance, no major effect of metal accumulation or depletion due to electromigration could be noticed (see fig 3).

To our knowledge his test represents the first semiconductor large signal operation of any semiconductor transistor at a temperature of 1000 °C. Certainly life time is still limited and especially contact technologies need further refinement. Nevertheless the experiment may serve as feasibility study towards the development of electronics at very high temperatures.

This work was in part supported by the European Union Project MORGAN (FP7 contract 214610).

[1] Neudeck, P.G.; Spry, D.J.; Liang-Yu Chen; Beheim, G.M.; Okojie, R.S.; Chang, C.W.; Meredith, R.D.; Ferrier, T.L.; Evans, L.J.; Krasowski, M.J.; Prokop, N.F.; , "Stable Electrical Operation of 6H–SiC JFETs and ICs for Thousands of Hours at 500 °C," *Electron Device Letters, IEEE* , vol.29, no.5, pp.456-459, May 2008. doi: 10.1109/LED.2008.919787

[2] Medjdoub, F.; Carlin, J.-F.; Gonschorek, M.; Feltin, E.; Py, M.A.; Ducatteau, D.; Gaquiere, C.; Grandjean, N.; Kohn, E., Electron Devices Meeting, 2006. IEDM '06. International , vol., no., pp.1-4, 11-13 Dec. 2006

[3] Maier, D.; Alomari, M.; Grandjean, N.; Carlin, J.-F.; Diforte-Poisson, M.-A.; Dua, C.; Chuvilin, A.; Troadec, D.; Gaquière, C.; Kaiser, U.; Delage, S.L.; Kohn, E.; IEEE TDMR, vol.10, no.4, pp.427-436, Dec. 2010. doi: 10.1109/TDMR.2010.2072507

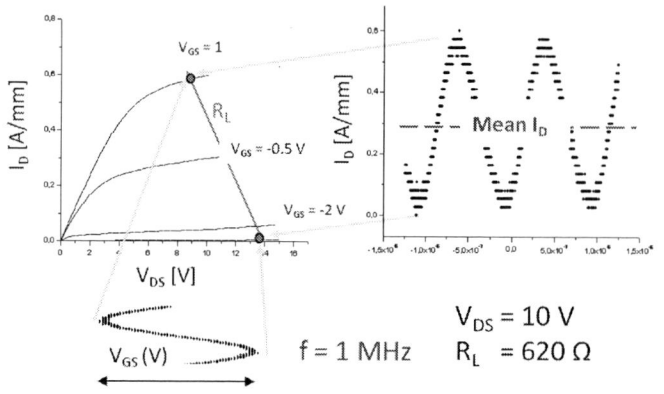

Fig. 1: Device large signal test configuration as described in text.

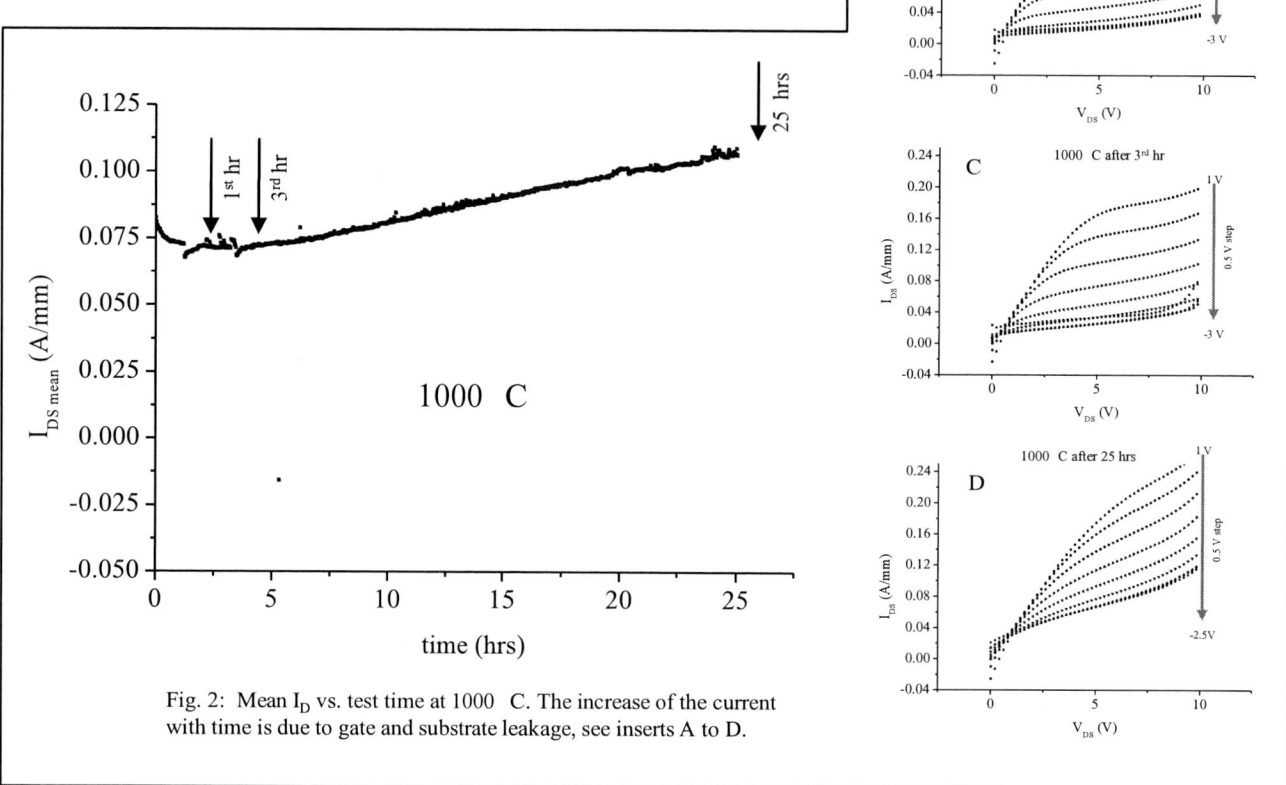

Fig. 2: Mean I_D vs. test time at 1000 C. The increase of the current with time is due to gate and substrate leakage, see inserts A to D.

Fig. 3: Passivated device after 25 hrs testing at 1000 C. No visible damage to the device structure can be seen.

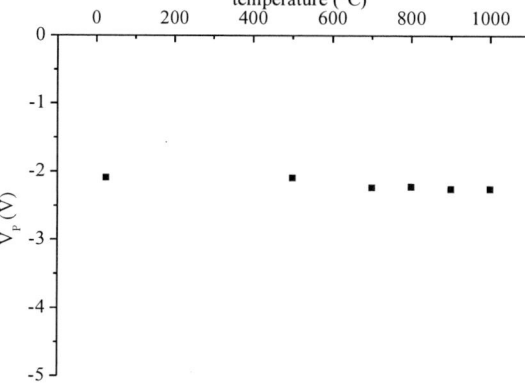

Fig. 4: Device pinch-off voltage (V_P) stability with temperature from $\sqrt{(I_D)} = f(V_G)$ extrapolation and after correction of residual leakage.

978-1-61284-243-1/11 $26.00 © 2011 IEEE

Bias Temperature Stress Analysis of ZnO Thin Film Transistors with HfO$_2$ Gate Dielectrics

J.J. Siddiqui[a][c], J.D. Phillips[a], K. Leedy[b], and B. Bayraktaroglu[b]

(a) EECS Department, University of Michigan, Ann Arbor, MI 48109-2122
(b) Air Force Research Laboratory, Sensors Directorate, Wright-Patterson AFB, OH
(c) Corresponding Author: Email: jjameel@umich.edu, Phone: 734-763-6132, Fax: 734-763-9324

ZnO thin film electronics have received much attention due to the relatively high electron mobility of ZnO thin films in comparison to amorphous silicon (a-Si) and organic thin films. There is significant interest in using ZnO thin film transistors (TFTs), or similar oxides such as InGaZnO and zinc tin oxide, to replace a-Si TFTs in large area display technologies such as active matrix liquid crystal display devices and active matrix organic light-emitting displays where transparency in the visible range and high carrier mobilities are significant advantages. In addition, the integration of high dielectric constant (high-k) dielectrics in ZnO TFTs has demonstrated performance advantages including reduced operating voltage, increased I_{on}/I_{off} ratios, and larger transconductance. HfO$_2$ has emerged as a high-k dielectric of choice for both silicon microelectronics and thin film electronics due to the high dielectric constant ($\varepsilon_r \sim 25\varepsilon_0$), low leakage current, and low synthesis temperature. Voltage stability is an important figure of merit for many TFT applications and much work has been done to characterize the voltage stability of a-Si and poly-crystalline silicon (p-Si) TFTs. Extensive Bias-Temperature-Stress (BTS) studies have been carried out on a-Si and p-Si TFTs to track the threshold voltage (V_{TH}), subthreshold slope (S), mobility (μ), and grain boundary trap creation (N_{TG}) over time and to correlate TFT parameter instabilities with physical mechanisms that include charge trapping in the gate oxide and charge state creation in the oxide, interface, and p-Si grain boundaries. Prior studies on the stability of ZnO TFTs have indicated threshold voltage shifts (ΔV_{TH}) with the same polarity as the stress voltage (V_{STR}) that increase with time and that S remains unchanged below a certain V_{STR}, but will degrade with time above this value[1-3]. Ability to recover pre-stress characteristics with and without post-stress treatments has also been reported. Further investigation is desired to both understand the device instability behavior dependence on temperature and gate-bias and to determine the physical origins governing the instabilities in this important material system. In this work, the instabilities of HfO$_2$/ZnO TFTs are studied by BTS investigation.

Thin-films of ZnO and HfO$_2$ were deposited by pulsed laser deposition and atomic layer deposition, respectively, and device structures were fabricated using standard photolithography and etching techniques. BTS and the TFT I-V parametric testing were carried out using a Keithley 4200 SCS with a temperature-controlled probe station. Individual TFTs from the same sample were diced apart and used for each bias-temperature experimental point in the BTS study. Each die was brought to the desired temperature, pre-tested, and then stress biased up to 10^4 seconds. Stress bias was periodically removed in order to obtain TFT parameters. V_{TH} and μ were extracted from a linear fit of $\sqrt{I_{DS}}$ vs V_{GS} in the TFT saturation region. Initial device performance indicates typical values of V_{TH} = 1.5 V, μ = 8.8 cm^2/Vs, S = 144 mV/decade, and I_{on}/I_{off} = 6x10^9. Positive-BTS results in a clear trend of increasing V_{TH} over time, while no clear trend is observed for μ, S, or I_{on}/I_{off}. ΔV_{TH} is calculated as $\Delta V_{TH} = V_{TH}(t)-V_{TH}(t=0sec)$. ΔV_{TH} over time demonstrates a logarithmic dependence that is consistent with electron tunneling into fixed charges in the gate dielectric according to $\Delta V_{TH} \approx r_d (N_T,\lambda)ln(1+t/t_0)$ where r_d is temperature independent coefficient related to the density of traps in the oxide (N_T) and a tunneling parameter (λ). ΔV_{TH} under BTS also shows weak temperature dependence and an exponential dependence with the inverse of the voltage stress bias, $1/V_{STR}$, further indicating behavior related to charge trapping in the oxide. No clear evidence is observed to suggest that charge trapping at grain boundaries in the polycrystalline ZnO channel plays any significant role in the observed ΔV_{TH}, where stronger temperature dependence would be expected, along with power law dependence on time and exponential dependence with V_{STR}. Work is currently underway on complementary negative-BTS and on the recovery characteristics of both the positive-BTS and negative-BTS cases and will be presented. Further understanding of charge trapping in the HfO$_2$/ZnO system is an important next step to understanding reliability of associated TFT devices.

[1] R.B.M Cross and M.M. De Souza, Appl. Phys. Lett. 89, 263513 (2006)
[2] R. Navamathavan et al, J. Electrochem. Soc. 153, 5 (2006)
[3] T.C. Fung et al, J. Disp. Tech. 5, 12 (2009)

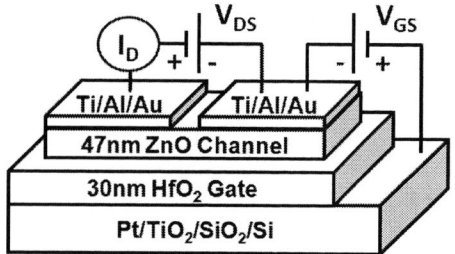

Figure 1: Schematic of HfO$_2$/ZnO TFT and biasing scheme for applied bias stress V$_{GS}$=V$_{STR}$ and V$_{DS}$=0.

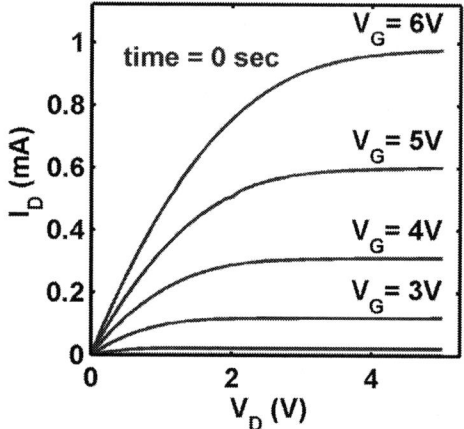

Figure 2: Representative Output (I$_D$-V$_{DS}$) TFT characteristics measured prior to BTS.

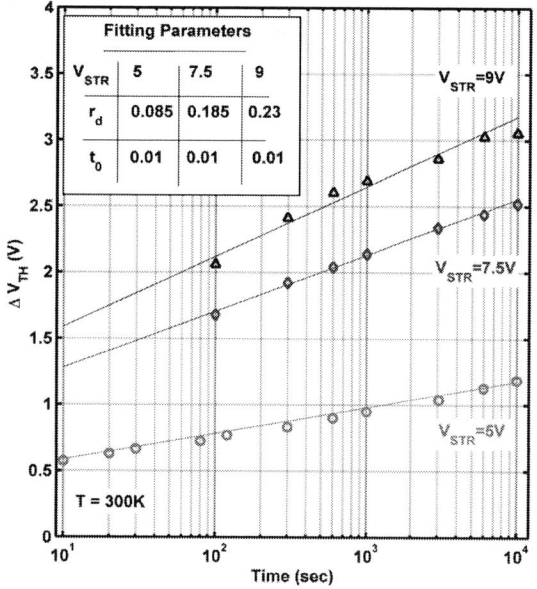

Figure 4: Threshold voltage shift with time at T=300K and variable V$_{STR}$ (data points). Fit to model representative of charge trapping in the HfO$_2$ dielectric (solid lines).

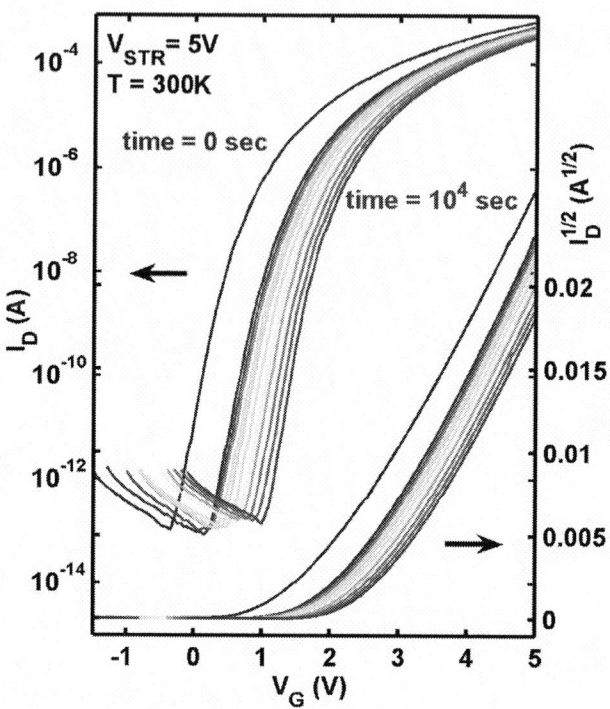

Figure 3: Series of transfer (I$_D$-V$_{GS}$) characteristics under V$_{STR}$=5V at T=300K.

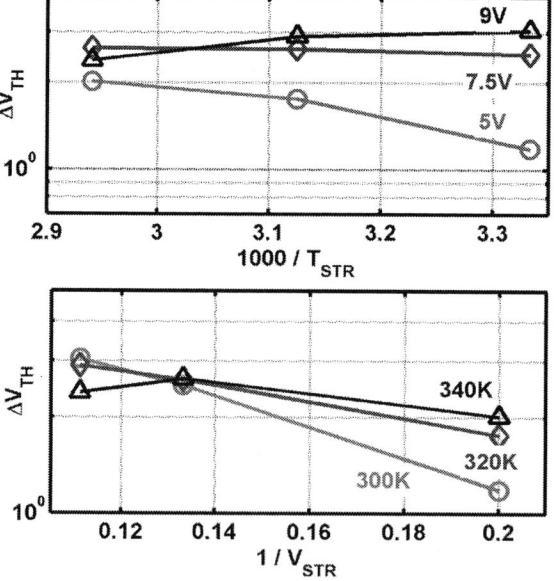

Figure 5: Threshold voltage shift with varying temperature and stress bias.

978-1-61284-243-1/11 $26.00 © 2011 IEEE   76

Carbon Nanotube Purified Ink-Based Printed Thin Film Transistors: Novel Approach in Controlling the Electrical Performance

Nima Rouhi, Dheeraj Jain, Peter J. Burke

Electrical Engineering and Computer Science Department, University of California-Irvine, Irvine, CA 92697
Tel: 949-824-9326, Fax: 949-824-3732, email: pburke@uci.edu

In this paper we present a comprehensive study of the solution-based printed carbon nanotube purified-ink devices while introducing a new idea of controlling the electronic performance of these devices. One of the most important concerns in nanoelectronics is whether the nanotube-based devices will ever enter the reality world of circuit designs? What are the fundamental and critical issues to be resolved? Which parameters affect the device performance most? A comprehensive study of the relationship between mobility, on/off ratio, and nanotube network density is presented for the first time in detail. This study reveals a clear road map towards experimental control over the performance of solution-based nanotube thin film transistors for a wide range of state-of-the-art applications.

The devices are fabricated on Si wafer (as a back-gate) with 300 nm oxide cap as the dielectric. The surface of the oxide is chemically treated with the self-assembled monolayer (SAM) of 3-aminopropyltriethoxysilane (APTES) as described in [1, 2]. Then about 20-40 µL of 99% semiconducting (determined by spectroscopic techniques) nanotube ink is put on top of the modified wafer and left for drying in ambient temperature. The source drain electrodes (Pd/Au) are deposited using e-beam evaporation. Different gate lengths from 10 to 100 µm are designed to study the effect of channel length while the channel width is fixed at 100 µm. The schematic of the process and fabricated thin film transistor is shown in figure 1 followed by the I-V characteristics of a sample device in figure 2.

Mobility and on/off ratio are the most important figures of merit, which have been explored in detail in this work. In addition, for the first time the network density and its critical impact on the device characteristics has been studied. Here we established a new idea to control the nanotube network density for modifying the performance of the devices. A fundamental relationship between mobility and on/off ratio was found, which has not been illustrated in any previous works (demonstrated in figure 5). More importantly, we discovered a tight correlation between the density of the network versus the mobility and on/off ratio that has also been overlooked in all printed nanoelectronic research studies so far. It has been presented here that the impact of network density is very crucial therefore, no comparison in any research work can be validated without taking this parameter into account. In fact, we offer that by controlling the network density we are able to adjust the device's characteristics in the most efficient way. The network density variation is obtained through managing both the surface treatment and the nanotube solution density prior to deposition. Mobility of more than 90 cm^2/V-s is obtained which is the highest reported so far. Comparing to state-of-the-art literature [2, 3], we are able to improve the mobility by at least a factor of 2X (more than 10X comparing to some other works) while reasonably maintaining the on/off ratio [2]. Besides, the on/off ratio of more than 100,000 shows 10X improvement comparing to similar works in the field [3]. Figure 3 to 5 demonstrate the relationships between mobility, on/off ratio, network density, and channel length. Figure 3 shows that by increasing the channel length, mobility will also increase. On the other hand, at a fixed channel length, mobility can be changed by altering the network density (as described in figure 4) which is the most critical key point for designers. Indeed, the effect of density on on/off ratio can be explained as well (more detail in [1]).

Thin film transistors (TFTs) were fabricated using solution-processed purified carbon nanotubes. Electrical measurements show mobility of more than 90 cm^2/V-s (2~10X improvement) and on/off ratio of more than 100,000 (10X improvement), compared to previous works. For the first time, the density dependence of the mobility and on/off ratio has been investigated showing a strong relationship between the density of the nanotube network and the devices' performance. Higher network density will increase the mobility while decreasing the on/off ratio. A comprehensive range of density from 10 tubes/µm^2 up to high end of 100 tubes/µm^2 was presented. This control method on the network density results in obtaining a wide range of mobilities and on/off ratios, which is critical for the broad market in printed electronics as well as CMOS and RF technologies. This work introduces a clear road map for a comprehensive range of nanotube-based transistor applications using semiconducting nanotube network. The presented parameters and techniques play a critical role in performance of carbon nanotube network transistors, which indeed is valuable for circuit designers. Based on these results, practitioners can now modify their circuits for a wide range of applications.

---

[1]   N. Rouhi, *et al.*, "Fundamental Limits on the Mobility of Nanotube-Based Semiconducting Inks," *Advanced Materials,* vol. 23, pp. 94-99, Oct 26 2010.

[2]   M. C. LeMieux, *et al.*, "Self-sorted, aligned nanotube networks for thin-film transistors," *Science,* vol. 321, pp. 101-104, Jul 4 2008.

[3]   D.-m. Sun, *et al.*, "Flexible high-performance carbon nanotube integrated circuits," *Nature Nanotechnology,* February 2011.

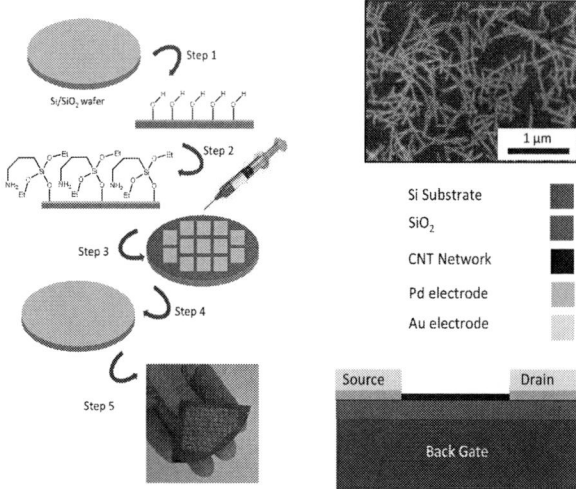

Fig. 1. Fabrication process and SEM image of nanotube network

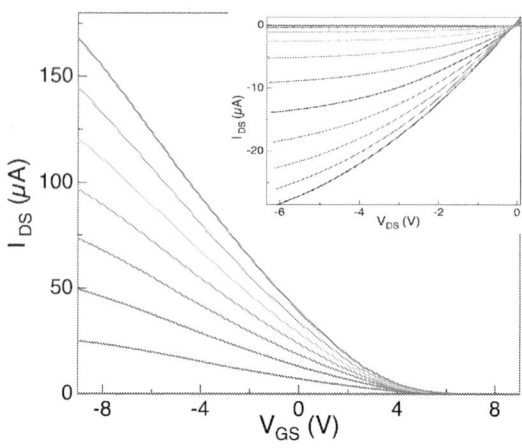

Fig. 2. Depletion curve (V_G : +10 ~ -10 V, with 2 V increment, V_{DS} : 0~7 V with 1 V increment). I_{DS}-V_{DS} in the inset

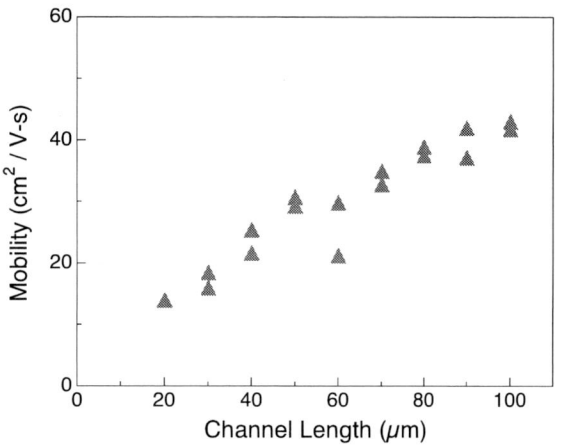

Fig. 3. Mobility vs. Channel length (moderate network density)

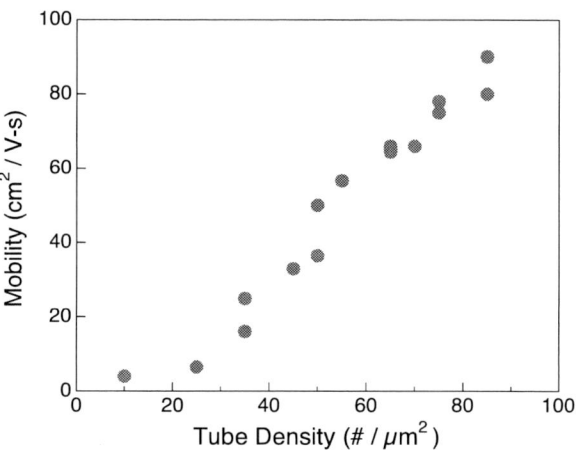

Fig. 4. Mobility vs. Network Density

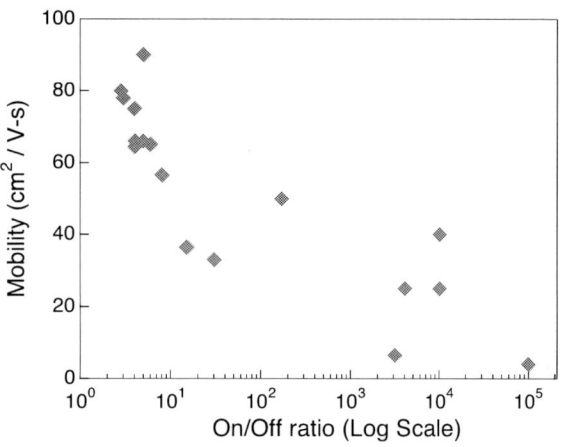

Fig. 5. Mobility vs. On/Off ratio

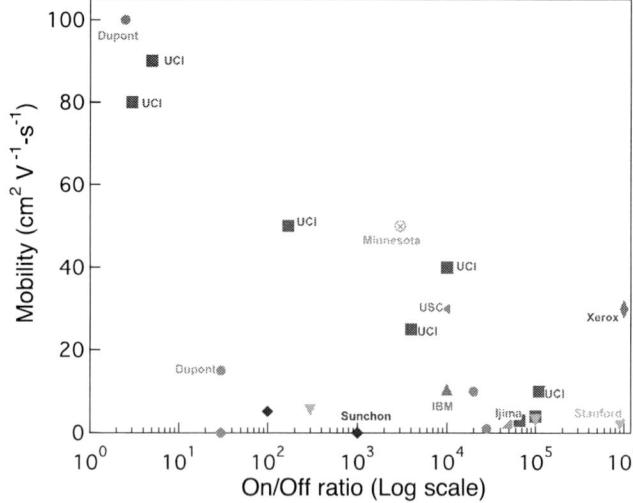

Fig. 6. Mobility vs. On/Off ratio (literature review)

978-1-61284-243-1/11 $26.00 © 2011 IEEE   78

Monolayer MoS₂ Transistors – Ballistic Performance Limit Analysis

Kartik Ganapathi*, Youngki Yoon*, and Sayeef Salahuddin

Department of Electrical Engineering and Computer Sciences, University of California, Berkeley, CA 94720, USA.
Phone: +1 510-501-9369. Email: kartik@eecs.berkeley.edu

INTRODUCTION: Monolayer MoS₂, a 2-D material like graphene, has gained significant interest recently since field-effect transistors (FETs) fabricated with it show better performance than most of the novel materials investigated in recent times [1]. With the electrostatic integrity of a 2-D material and a large bandgap (E_g = 1.8 eV) [2, 3], monolayer MoS₂ promises to fulfill what graphene has not. Rapid advancement in fabrication techniques of monolayer MoS₂ transistors, leveraging the knowledge in graphene, can be expected. Of significant technological relevance, before an aggressive pursuit of scaling down such devices, is the evaluation of maximum performance extractable of them. In this paper, we report the ultimate performance limits of MoS₂ transistors using ballistic quantum transport simulations within the Non-Equilibrium Green's Function (NEGF) formalism. Our simulations show that monolayer MoS₂ transistors can, in principle, provide (i) ON-OFF current ratio as high as 10^{11}, (ii) ON current of ~1.5 mA/μm, and (iii) a transconductance as high as about 3 mS/μm. We also analyze the effect of some key device parameters like gate underlap, Schottky barrier height, and contact resistance on the device performance, thereby outlining directions for optimization.

SIMULATION METHOD: Figures 1(a) and 1(b) show the schematic of an MoS₂ monolayer and that of the simulated MoS₂ transistor, respectively. The device dimensions and other parameters used in this study are mentioned therein. The reported mobility of MoS₂ monolayers with HfO₂ dielectric on top is around 200 cm²/V-s [1], which corresponds to a mean-free path of ~20 nm. We choose similar lengths for our simulations to ensure that the projections are realistic. The effective mass at the bottom of conduction band, calculated from dispersion relations in Ref. [4], is used to construct a single-band effective mass Hamiltonian, which is sufficient to describe carrier transport through bottom of conduction band. Metallic source and drain contacts are modeled differently than the channel in order to treat Schottky contacts and metal-induced-gap-states properly. Ballistic transport NEGF equations are solved with the Poisson's equation until a self-consistency between charge and electrostatic potential is reached.

RESULTS AND DISCUSSION: Figure 2(a) shows the transfer characteristics for the nominal device at a drain voltage V_D of 0.4 V. Owing to its large bandgap and excellent electrostatics, a very large ON-OFF ratio (~10^{11} ignoring gate leakage) along with an excellent subthreshold swing of 60 mV/decade can be obtained. The DIBL for the device is also expectedly low. The output characteristics for two different gate voltages are shown in Fig. 2(b). Figure 3 illustrates the conduction band profile along the device for several gate voltages during turn-on. The plots of ON current with vs. ON-OFF ratio with a constant V_D window, shown in Fig. 4(a), demonstrates that MoS₂ devices can provide large ON current (0.5 mA/μm) together with high I_{ON}/I_{OFF} (>10^4) for V_D = 0.5 V. The variation of transconductance (g_m) with V_G, plotted in Fig. 4(b), shows a g_m of about 3 mS/μm. The high value of g_m indicates that MoS₂ devices far away from the oxide capacitance regime, which is also reaffirmed by the potential profiles along the channel that show a large $\partial\psi_s/\partial V_G$, ψ_s being the channel electrostatic potential. The choice of metal for source and drain contacts invariably affects the device performance. Figure 5 shows the effect of variation of Schottky barrier height on the device characteristics. The current in ON state, which is determined mainly by the transmission probability through the metal contact and MoS₂ interface, expectedly degrades with increase in Schottky barrier height. Contact resistance, which arises due to different dispersion relations in contact and channel materials, is an important parameter affecting device performance. $I_D - V_G$ characteristics plotted for two different values of contact resistance indicate that it can severely affect ON current (Fig. 6). The underlap between gate and contact is an important parameter in optimizing for performance. Figure 7(a) shows $I_D - V_G$ characteristics for an underlap of 2 and 6 nm on each side. While a large underlap reduces gate-source capacitance, it increases the resistance in ON-state since the effective Schottky barrier is increased [Fig. 7(b)].

CONCLUSIONS: To summarize, using ballistic NEGF-based transport simulations, we project the maximum performance achievable with monolayer MoS₂ transistors. Our simulations show that these devices can provide (i) excellent switching behavior with very high ON current, (ii) a g_m of about 3 mS/μm, and (iii) immunity to short channel effects thanks to the electrostatistically efficient 2-D geometry. We have also investigated the effect of underlap, barrier height and contact resistance on the device performance. We note that while these numbers are representative of the best performance MoS₂ transistors can offer, the fact that they are significantly better than those for either state-of-the-art silicon, III-V or graphene makes MoS₂ devices promising for future electronic applications.

REFERENCES: [1] B. Radisavljevic et al., *Nature Nanotechnology*, vol. 6, pp. 147-151 (2011). [2] A. Splendiani et al., *Nano Lett.*, vol. 10, no. 4, pp. 1271-75 (2010). [3] K. F. Mak et al., *Phys. Rev. Lett.*, vol. 105, no. 13, pp. 136805-08 (2010). [4] S. Lebègue and O. Eriksson, *Phys. Rev. B*, vol. 79, no. 11, pp. 115409-12 (2009).

* These authors contributed equally to this work.

978-1-61284-243-1/11 $26.00 © 2011 IEEE

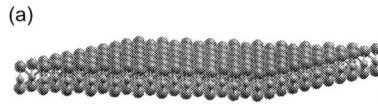

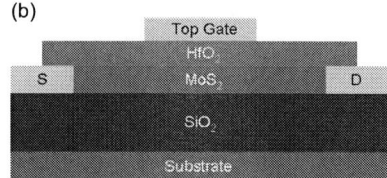

Figure 1. (a) Atomistic configuration of monolayer MoS$_2$. (b) Schematic structure of MoS$_2$ transistor. Material and nominal device parameters used in this study are as follows: $m^* = 0.45m_e$, $E_g = 1.8$ eV, $L_{ch} = 15$ nm, gate underlap is 2 nm on each side, 2.5 nm HfO$_2$ top gate, Schottky barrier height is 0.1 eV. $V_D = 0.4$ V. A grid spacing of 0.1 nm is used along the direction of transport.

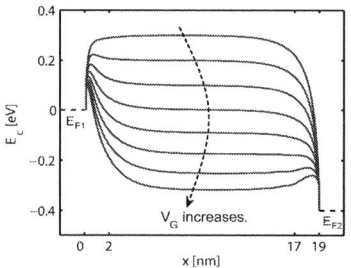

Figure 3. Conduction band profile (E_c) along the channel at various gate voltages from $V_G = -0.2$ to 0.5 V, in steps of 0.1V, are shown. Dashed lines indicate Fermi levels at the source and the drain. A very good electrostatic integrity, characteristic of an ideal 2-D material, is evident.

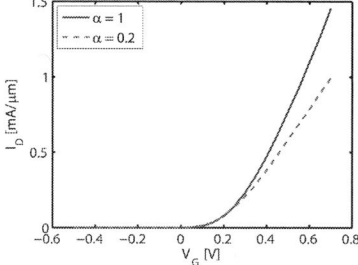

Figure 6. $I_D - V_G$ plots for different channel-contact coupling strengths. The coupling parameter used at the metal-semiconductor interface is $t_i = \alpha(t_m + t_s)/2$, where t_m and t_s are the coupling parameters within the metal contact and the semiconductor MoS$_2$, respectively. The contact resistance is modeled by varying α. A larger α indicates lesser reflections at the contact-channel interface and vice versa.

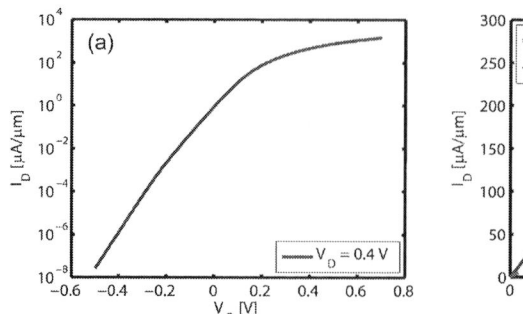

Figure 2. (a) $I_D - V_G$ characteristics at $V_D = 0.4$ V. (b) $I_D - V_D$ characteristics at $V_G = 0.2$ and 0.3 V. The maximum ON current achieved within the voltage range in our simulation is 1.5 mA/μm. The minimum subthreshold swing is 60 mV/dec. Output conductance is 126 μS/μm at $V_D = 0.4$ V and $V_G = 0.3$ V.

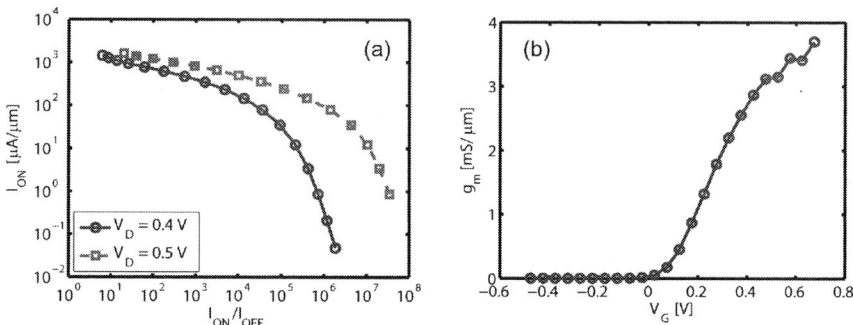

Figure 4. (a) I_{ON} vs. I_{ON}/I_{OFF}. (b) Transconductance ($g_m = \partial I_D/\partial V_G$) as a function of gate voltage at $V_D = 0.4$ V. The maximum ON-OFF ratio with $V_D = 0.4$ V is more than 10^6. The ON current can be as high as 0.5 mA/μm with 4 orders of magnitude in ON-OFF ratio for $V_D = 0.5$ V. The transconductance is about 3 mS/μm. The obtained value of g_m indicates that the device operates far away from the oxide capacitance regime, also evidenced by the conduction band profiles in Fig. 3 showing large $\partial \psi_s/\partial V_G$.

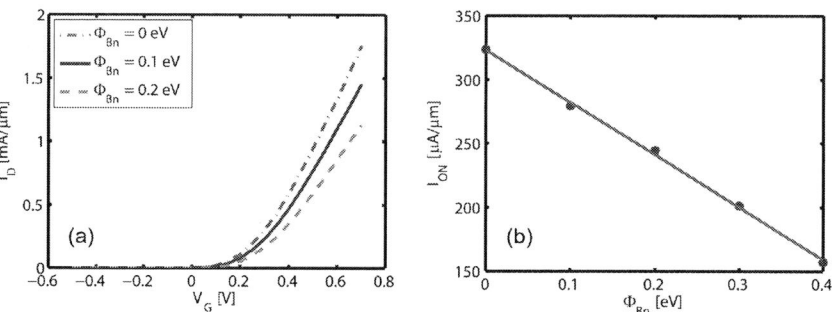

Figure 5. (a) $I_D - V_G$ plots for various Schottky barrier heights. Schottky barrier height is varied from 0 to 0.2 eV. (b) Plot of ON current, calculated at identical gate overdrive, as a function of barrier height. The current in ON state is significantly affected by the tunneling probability at the metal–MoS$_2$ interface, which reduces with increased tunneling barrier.

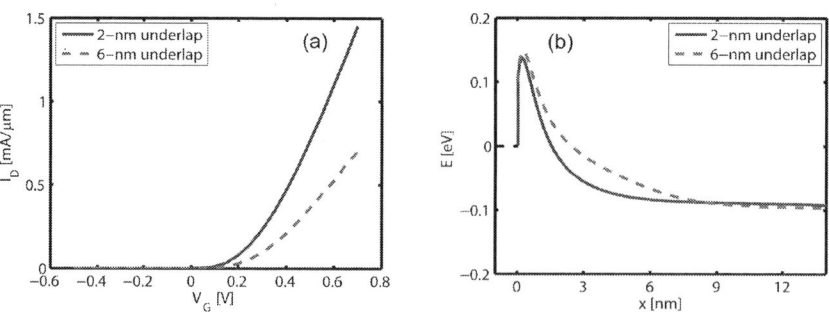

Figure 7. (a) Transfer characteristics and (b) conduction band profile near the source for devices with underlap of 2 and 6 nm at $V_G = 0.2$ V. A larger underlap results in a smaller ON current due to the increased effective Schottky barrier at the source-channel interface.

978-1-61284-243-1/11 $26.00 © 2011 IEEE

RF performance projections for 2D Graphene Transistors: Role of Parasitics at the Ballistic transport limit

Pei Zhao[1*], Debdeep Jena[1], and, Siyuranga O. Koswatta[2]

[1] Department of Electrical Engineering, University of Notre Dame, Notre Dame, IN 46556, USA

2 IBM Research Division, T. J. Watson Research Center, Yorktown Heights, NY 10598, USA

Phone: (574)631-1290 Email: *pzhao@nd.edu

Background: Graphene is a two-dimensional zero bandgap material with carbon atoms arranged in a honeycomb lattice [1, 2]. Although 2D monolayer graphene lacks a bandgap, it still shows promising potential for applications in high frequency analog devices that do not require a high on/off ratio as demanded by digital logic [3-5]. A cutoff frequency, f_T, as high as 170GHz is achieved in a 90 nm channel length Graphene Field-Effect-Transistor (GFET) with back-gated structure [4]. Projected f_T = 300GHz is reported for channel length L_{ch} = 140nm top-gated structure [5]. The motivation of this work is to explore the potential of high frequency performance of GFETs, and to elucidate the major factors that limit their performance. Our model captures the degradation of intrinsic performance due to parasitics, and the effect of metal-graphene (M-G) contacts as: (1) contact doping effect due to M-G work function difference [6], and (2) DOS broadening by M-G coupling and metal-induced states in the channel [7, 8].

Model: The simulations in this work utilize a mode-space based non-equilibrium Green's function formalism for ballistic transport with self-consistent electrostatics. The 2D electrostatics is solved by a finite-difference method. A channel width of 150nm is assumed, and the 2D graphene Hamiltonian matrix is decoupled into 1D modes. The qth mode retarded Green's function is $G_q(E)=[(E+i0^+)I - H_q - \Sigma_{Sq} - \Sigma_{Dq}]^{-1}$, and $\Sigma_{Sq}^{1,1} = (t_0)^2 g_{Sq}$, t_0 = 3eV. The surface Green's function is $g_{Sq}(E)= [(E+i\Delta)I-H_{contact}]^{-1}$, where Δ is the M-G coupling strength [7,8]. Inside the M-G contact, the Fermi level of the graphene layer under the metal electrode shifts by $\Delta E_{contact} = E_F-E_{Dirac}$ [6] which is captured in the electrostatic solution.

Results and discussions: The modeled device structures are shown in Figure 1, (a) top-gated structure ε_{ox} = 20 and t_{ox} = 1.5nm, and (b) back-gated structure with 90nm thick SiO$_2$. The contact resistance and parasitic capacitance have also been taken into consideration in the simulations. Figure 2 shows the effect of M-G contacts on the transfer characteristics and transconductance, g_m for the top-gated structure. The on-current, I_{on}, can be increased with strong M-G coupling strength Δ or heavy contact induced doping (i.e. larger $\Delta E_{contact}$). The off-current, I_{off}, does not increase. Large $\Delta E_{contact}$ increases g_m, but the maximum g_m does not show a strong dependence on Δ. We use Δ=50meV and ΔE_{contac} = -0.4eV for rest of the simulations. Figure 3 shows the I_{DS} vs. V_{GS} and g_m vs. V_{GS} at different V_{DS} for the top-gated structure. Large V_{DS} yields a higher maximum g_m, but low V_{DS} shows better linearity with a broader f_T peak. Transfer characteristics with different channel lengths are shown in Figure 4. For the top-gated structure excellent gate electrostatics helps avoid short channel effects (SCE). I_{on} remains the same for all channel lengths. I_{off} increases about 1.5 times when L_{ch} decreases from 100nm to 15nm due to direct source to drain tunneling. The rise in I_{off} leads to g_m degradation at L_{ch}=15nm. In the back-gated structure, the on/off ratio is degraded at shorter channel lengths, and the minimum conduction point shifts. Figure 5 shows the effect of contact resistance on I_D - V_{GS} characteristics at L_{ch} = 100nm. At V_{DS} = 0.3V, compared with the intrinsic case, when $R_{S/D}$ = 0.5Ωmm, on/off ratio decreases 3x for the top-gated structure and 1.2x for the back-gated structure. I_{on} reduces 22x for the top-gated structure and 6x for the back-gated structure. Figure 6 shows the comparison of f_T -V_{GS} with different channel lengths at V_{DS} = 0.3V. The cutoff frequency is calculated as $f_T = 1/2\pi\tau_{tot}$, where $\tau_{tot} = L_{ch}C_{gs}/g_m + C_{gd}/g_m + C_{gd}(R_S+R_D)$, $C_{gs} = \partial Q_{ch}/\partial V_{GS}$, $R_{S/D}$ = 0.5Ωmm, and C_{gd} = 2pF/cm and 0.5pF/cm for the top-gated and the back-gated structures, respectively. Charging/discharging process is faster at shorter channel lengths, thus the peak f_T increases. In the back-gated structure, SCE is strong, thus the on/off ratio decreases and g_m drops dramatically at short L_{ch}. When L_{ch} is shorter than 30nm, even the peak f_T drops. Figure 7 summarizes the f_T vs. L_{ch} at V_{DS} = 0.3V. The intrinsic f_T = $<v>/2\pi L_{ch}$ is added as a reference with the average ballistic velocity $<v>$ = $2v_F/\pi$ in 2D graphene. With $R_{S/D}$ = 0.5Ωmm and C_{gd} = 2pF/cm, f_T drops 2x at L_{ch}=100nm and 8x and L_{ch}=15nm for the top-gated structure. For the back-gated structure, with $R_{S/D}$ = 0.5Ωmm and C_{gd} = 0.5pF/cm, when L_{ch} is below 70nm, f_T does not increase, and it even decreases when the channel length is below 30nm. Thus, parasitics currently dominate the performance, and major gains are expected with their reduction. This work is supported by the Semiconductor Research Corporation Nanoelectronics Research Initiative and the National Institute of Standards and Technology through the Midwest Institute for Nanoelectronics Discovery (MIND).

978-1-61284-243-1/11 $26.00 © 2011 IEEE

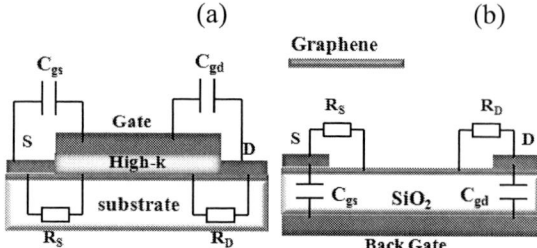

Figure 1. Modeled device structure with contact resistances and parasitic capacitances. (a) top-gated structure, $\varepsilon_{ox} = 20$ and $t_{ox} = 1.5$nm and (b) back-gated structure with 90nm thick SiO_2.

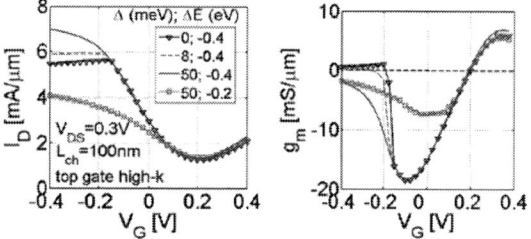

Figure 2. Effect of contact induced doping and metal/graphene coupling Δ. I_{on} can be increased with strong coupling Δ and high contact doping $\Delta E_{contact}$. I_{off} will keep the same with different Δ. Maximum g_m shows strong dependence on $\Delta E_{contact}$.

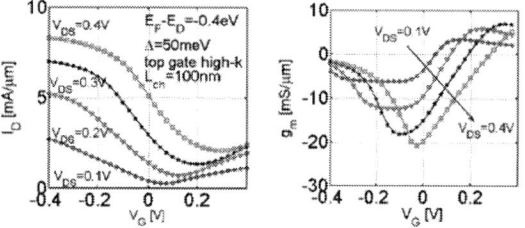

Figure 3. I_{DS} vs. V_G and g_m vs. V_G for the top-gated structure. Large V_{DS} gives higher maximum g_m, but low V_{DS} shows better linearity with broader f_T peak.

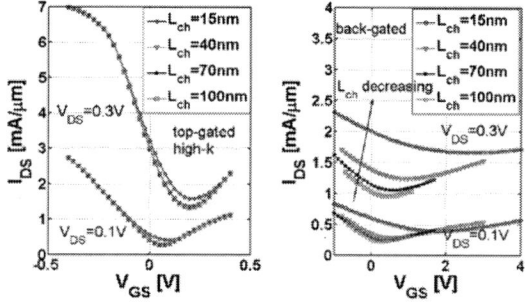

Figure 4. Transfer characteristics at different channel length for both device structures. For the top-gated structure with excellent gate electrostatics, I_{on} remains the same, and I_{off} increases about 1.5 times. For the back-gated structure, short channel effect is strong and the minimum conduction point shifts with different channel lengths.

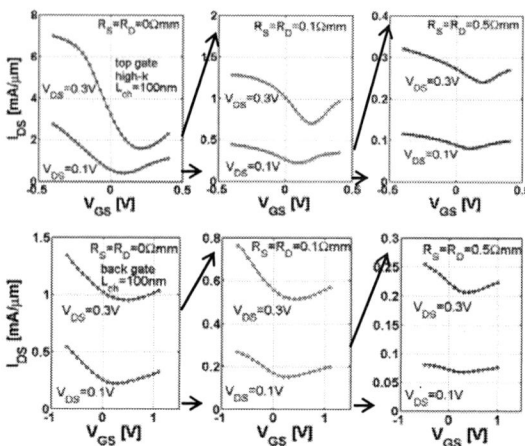

Figure 5. Effect of contact resistances on the transfer characteristics. The increase of contact resistance, dramatically reduces the on/off ratio and I_{on}.

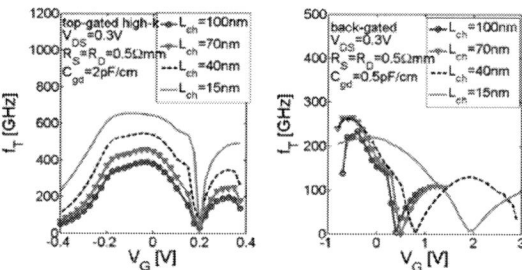

Figure 6. Effect of $R_{S/D}$, and C_{gd} on cut-off frequency f_T vs. V_{GS} with different channel lengths. For the top-gated structure, peak f_T increases while channel lengths is shrinking, but for the back-gated structure, peak f_T drops when $L_{ch} < 30$nm.

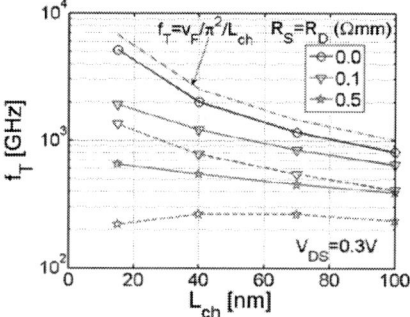

Figure 7. Prediction of peak f_T with the impact of contact resistances and parasitic capacitances (solid lines for the top-gated structure, $C_{gd} = 2$pf/cm and dashed lines for the back-gated structure, $C_{gd} = 0.5$pf/cm). With $R_{S/D} = 0.5\Omega$mm and $C_{gd} = 2$pF/cm, f_T drops 2x at $L_{ch} = 100$nm and 8x and at $L_{ch} = 15$nm for the top-gated structure. For the back-gated structure, with $R_{S/D} = 0.5\Omega$mm and $C_{gd} = 0.5$pF/cm, when $L_{ch} < 70$nm, f_T does not increase, in addition, f_T even drops for L_{ch} below 30nm.

High Performance N- and P-Type Gate-All-Around Nanowire MOSFETs Fabricated on Bulk Si by CMOS-Compatible Process

Yi Song, Huajie Zhou, Qiuxia Xu, Jun Luo, Chao Zhao and Qingqing Liang

Key Laboratory of Microelectronics Devices & Integrated Technology, Institute of Microelectronics, Chinese Academy of Sciences, Beijing 100029, China
Phone: +86 10 82995723; Fax: +86 10 82995684; Email: songyi@ime.ac.cn

Abstract

We demonstrate high performance silicon nanowire gate-all-around MOSFETs (SNWFETs) fabricated on bulk Si by a novel top-down CMOS-compatible method. The fabricated N- and P-type SNWFETs of sub-50 nm gate length and of ~5 nm in diameter show excellent short channel effects (SCEs) immunity with subthreshold slope (SS) of 90/69 mV/dec, DIBL of 47/10 mV/V, and high driving current of $2 \times 10^3 / 5.4 \times 10^3$ μA/μm at 0.1 nA/μm off-current.

Keywords: nanowire, gate-all-around, CMOS, bulk substrate

Introduction

Silicon nanowire gate-all-around MOSFETs (SNWFETs) have drawn intensive attention for the scaling of MOSFETs to the end of the roadmap. Most of the reported SNWFETs have been fabricated on SOI substrate due to its relatively simple realization [1-2]. The pervious reported SNWFETs on bulk substrate were mainly made using SiGe epitaxy dummy-gate process [3]. However, the usage of SiGe epitaxy increases the cost and the processing complexity. We have developed a novel scheme for fabricating SNWNFETs on bulk substrate by CMOS-compatible process [4]. In this paper, the fabrication of both N- and P-type SNWFETs using this new scheme is presented. High performance SNWFETs with superior short effects (SCEs) immunity have been demonstrated.

Fabrication

The fabrication procedure of N- and P-type SNWFETs is summarized in Fig. 1 and detailed in [4]. The key of the process is the formation of localized SOI structure and stress limited oxidation [5]. We conducted a bottom trench implantation followed by oxidation to grow a thick SiO_2. 3-nm-thick SiO_2/poly-silicon was used as the gate stack for the SNWFETs. The Si nanowire of 5 nm in diameter and 16-nm-thick SiO_2 underneath the Si nanowire were confirmed by SEM as shown in Fig. 2. By skipping the second low temperature stress-limited-oxidation in the SNWFETs flow, bulk FinFETs were fabricated as the control devices.

Electrical characteristics of N- and P-type SNWFETs

As shown in Fig. 3a, the I_d-V_g curves indicate excellent sub-threshold characteristic, SS=90 mV/dec and DIBL=47 mV/V for N-type SNWFETs, and SS=69 mV/dec and DIBL=10 mV/V for P-type SNWFETs. High I_{on}/I_{off} ratio of 10^7 has been achieved for both N- and P-type SNWFETs. In Fig. 3b, the I_d-V_d curves of N- and P-type SNWFETs show I_{on} of $2 \times 10^3 / 5.4 \times 10^3$ μA/μm at $|V_d| = |V_g| = 1.5$ V. Such high I_{on} should be attributed to the volume inversion and quasi-ballistic transportation nature. The small SS and DIBL show short channel immunity clearly. Table 1 summarizes the main electrical characteristics of N- and P-type SNWFETs along with reference data cited from other publications. It is obvious that our SNWFETs outperform the previously reported ones fabricated using various methods.

Discussions

A. The suppression of parasitic conduction

As shown in Fig. 1, a parasitic bottom trench MOSFET exists below the SNWFETs. In order to suppress the parasitic conduction, a bottom trench implantation has been conducted. Fig. 4 compares I_d-V_g curves for SNWFETs with/without bottom trench implantation. It is clearly shown that the implantation can effectively suppress the parasitic conduction via the channel of the parasitic MOSFETs. Though on current is sacrificed a little bit by adopting the implantation, the off current is greatly reduced because of the better SCEs immunity. As a result, the overall performance of SNWFETs is significantly improved.

B. Optimization of the key parameters

The extension of SNWFETs is a big problem due to the parasitic resistance, and it could not be shortened infinitely because of the fabrication difficulties such as 3D oxidation nature and lithography limitations. Fig. 5 shows correlations of I_{on}-I_{off} of SNWFETs with different nanowire extension lengths L_{ext}. Fig. 6 shows the extracted R_{on} vs. L_{ext}. R_{on} is decreased drastically as the reduced L_{ext}, which leads to the significant increase of I_{on}. L_{ext} has little impact on I_{off} because the Si nanowire is not the main leakage current path. Compared to the leakage via bottom trench MOSFETs, the off current through the Si nanowire is much less important. As indicated in Fig. 7, the I_{on}/I_{off} of SNWFETs can be further enhanced by shrinking the diameter of silicon nanowire (D_{nw}).

C. Comparison with the control bulk FinFETs

Fig. 8 shows the electrical characteristics of the SNWPFETs and the control bulk FinFETs. It is found that the bulk FinFETs has slightly higher driving currents with a comparable off current. The extracted SS/DIBL for SNWFETs (71 mV/dec/20 mV/V) and for bulk FinFETs (127 mV/ dec/120 mV/V) indicates that SNWFETs have superior electrostatic potential controllability than bulk FinFETs.

Summary

High performance N- and P-type gate-all-around SNWFETs fabricated on bulk Si by CMOS-compatible process were demonstrated. The subthreshold characteristics and SCEs immunity are significantly improved by adopting the bottom trench implantation. For N-type SNWFETs, on-current is 2×10^3 μA /μm, and the SS/DIBL are 90 mV/dec/47 mV/V. For P-type SNWFETs, on-current is 5.4×10^3 μA/μm, and the SS/DIBL are 69 mV/dec/10 mV/V. The demonstrated excellent device performance in this work suggests the method to fabricating SNWFETs on bulk is promising and worthy of future investigation.

978-1-61284-243-1/11 $26.00 © 2011 IEEE

Reference

[1]Hoong-Shing Wong et al., *VLSI*, p. 92, (2009). [2]A.M. S. Bangsaruntip et al., *VLSI* p. 21 (2010).[3]K. H. Y. Ming Li et al., *VLSI* p. 94 (2009).[4]Y. Song et al., *EDL* 31 p. 13 (2010). [5]J. Kedzierski et al., *JVST B*, 15 p. 2825 (1997). [6] Yu Tian et al., *IEDM*, p. 895, (2007). [7] V. Pott et al., *TNANO*, p.733, (2008). [8] K. H. Yeo et al., *IEDM*, p.286, (2006). [9] S. D. Suk et al., *TNANO*, p.733, (2008).

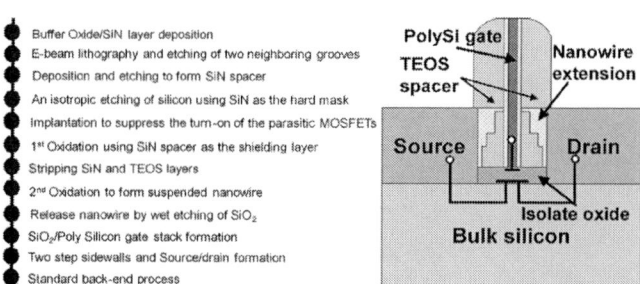

- Buffer Oxide/SiN layer deposition
- E-beam lithography and etching of two neighboring grooves
- Deposition and etching to form SiN spacer
- An isotropic etching of silicon using SiN as the hard mask
- Implantation to suppress the turn-on of the parasitic MOSFETs
- 1st Oxidation using SiN spacer as the shielding layer
- Stripping SiN and TEOS layers
- 2nd Oxidation to form suspended nanowire
- Release nanowire by wet etching of SiO₂
- SiO₂/Poly Silicon gate stack formation
- Two step sidewalls and Source/drain formation
- Standard back-end process

Fig.1 Schematic of SNWFETs fabrication process and device structure

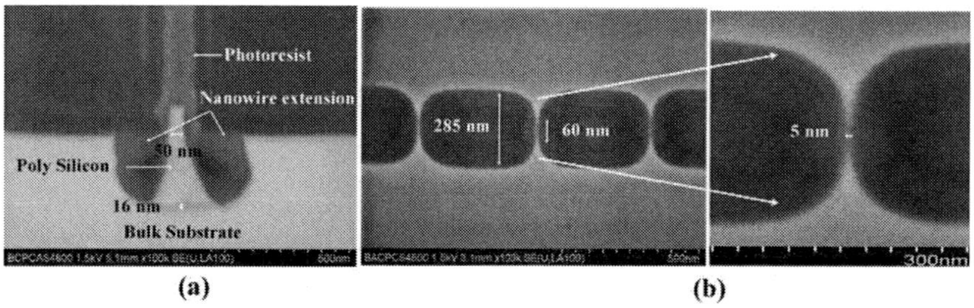

Fig.2 SEM image of (a) SNEFETs after gate formation and (b) 5 nm diameter of nanowires with high uniformity

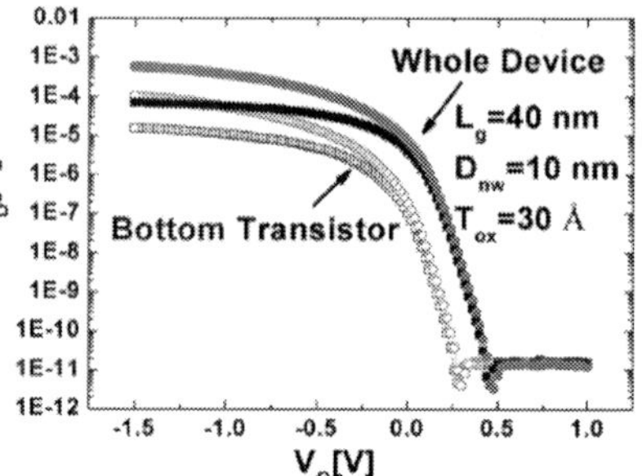

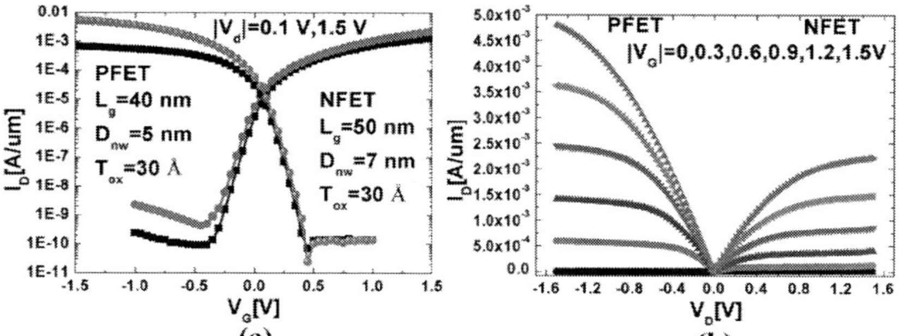

Fig.3 (a) I_D-V_G and (b) I_D-V_D curves of N/P SNWMOSFETs

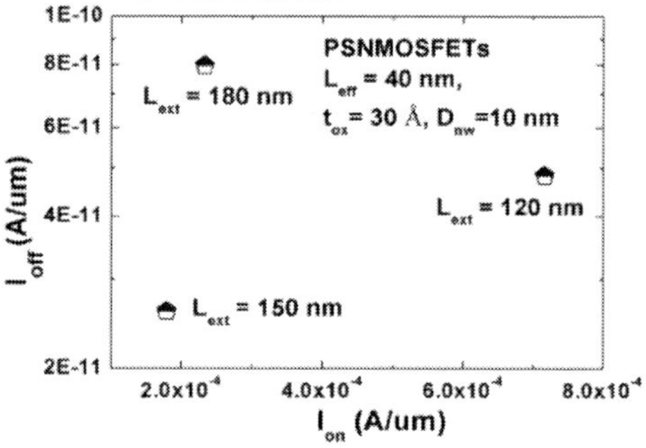

Fig.4 I_D-V_G comparison of bottom and whole device

Fig.5 Correlations of I_{on}-I_{off} with different lengths of nanowire extension L_{ext}

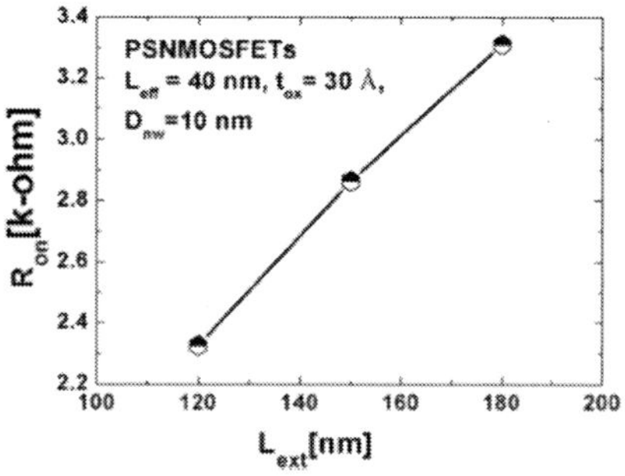

Fig.6 The extracted R_{on} with different L_{ext}

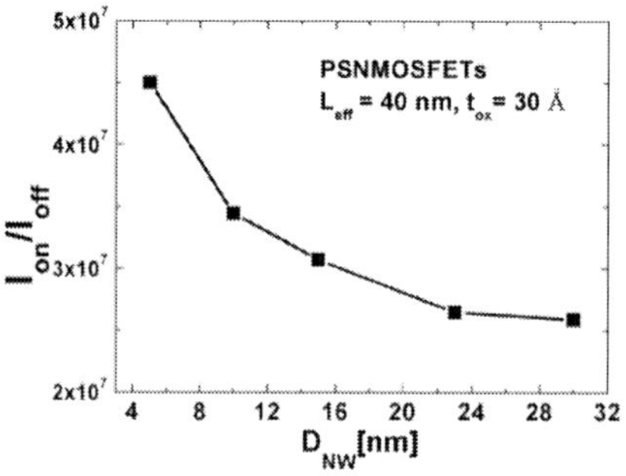

Fig.7 D_{NW} dependence of I_{on}/I_{off} at $|V_G|$=1.5 V of L_{eff}=40 nm

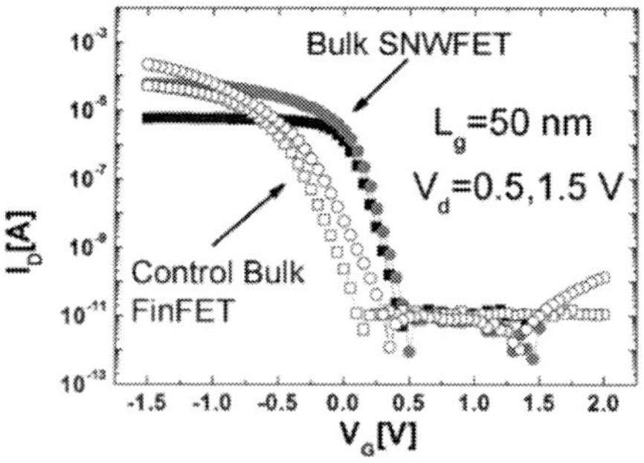

Fig.8 Electrical characteristics comparison between bulk GAA SNWPFETs and control bulk PFinFETs

Table. I Device performance of this work compared to other published results for bulk GAA SNWFETs

	V_{dd} (V)	L_g (nm)	D_{nw} (nm)	Nmos, I_{on}/I_{off} (μA/μm)	Pmos, I_{on}/I_{off} (μA/μm)	I_{on}/I_{off} (N/P)	SS (mV/dec)	DIBL (mV/V)
[6]	1.5	130	10	1e3/1e-5	-/-	10^8/-	73/-	8/-
[7]	2/-5	10^4	5	65.6/1e-4	5.43/1.7e-4	10^5/10^4	65/99	-/-
[8]	1/-1	15	8	1.5e3/2e-3	1.9e3/1e-3	10^6/10^6	72/71	50/43
[9]	1/-1	30	10	1.5e3/4e-3	1.2e3/9e-6	10^6/10^8	69/67	24/14
[4]	1	35	7	2.5e3/1e-2	-/-	10^5/-	140/-	167/-
This work	1.5/-1.5	50/40	7/5	2e3/2.4e-4	5.4e3/1e-4	10^7/10^7	90/69	47/10

978-1-61284-243-1/11 $26.00 © 2011 IEEE

The effect of field effect device channel dimensions on the effective mobility of graphene

Archana Venugopal, [1]Jack Chan,[1] Wiley P. Kirk,[1] Luigi Colombo[2] and Eric M. Vogel[1]

[1] University of Texas at Dallas,[2] Texas Instruments Incorporated, Dallas, Texas

Phone: 972-883-6637, fax: 972 -883 -5725, email : archana.venugopal@student.utdallas.edu

Graphene is a possible candidate for post CMOS applications and mobility is a material characteristic that has been utilized to gauge the quality of the material[1]. Mobility of exfoliated graphene transferred on SiO_2 has been reported to range from 2,000 to 25,000 $cm^2/V\cdot s$ [1, 2]. The large variation is typically attributed to factors such as scattering by defects in the underlying substrate, residue from processing, charged impurity scattering and phonon scattering [3]. In most previous studies one of the primary assumptions made is that the mobility is independent of channel dimensions. In this study, we performed room temperature effective mobility measurements as a function of channel dimensions. The mobility exhibits clear channel length (L_{ch}) and width (W_{ch}) dependence and varies from less than 1,000 $cm^2/V\cdot s$ to 7,000 $cm^2/V\cdot s$. Theoretical analysis of the conductivity (σ) in graphene devices as a function of W_{ch} performed by Vasko *et al*[4]. is in agreement with our experimental results. Mobility values for back gated devices with well defined channel dimensions in literature [5] are seen to be consistent with the trend that we report here.

Graphene was formed by mechanical exfoliation of natural graphite and transferred onto thermal SiO_2 grown on a *p*-type Si wafer (doping $\sim 10^{17}/cm^3$) with various thicknesses d (15, 90, and 300 nm). The monolayers were identified using optical microscopy, Scanning Electron Microscopy and Raman spectroscopy. A two step electron beam lithography process was used to etch the graphene flakes to desired dimensions and to define transfer length method (TLM) structures (Figure 1). Electrical measurements were performed on back-gated devices at room temperature in air using a HP 4155 Semiconductor Parameter Analyzer and Cascade Probe Station.

A comparison of three mobility models was performed (Figure 2) to confirm that the trend was not a result of the extraction technique. Figure 3 shows the effective mobility, extracted using the constant mobility model [6] as a function of L_{ch} for given W_{ch} on exfoliated graphene. The mobility increases with increasing L_{ch} and eventually saturates to a constant value at channel lengths of several micrometers. The observed trend is attributed to the fact that the device is operating in both a quasi ballistic regime and diffusive regime, depending on L_{ch} [7]. Figure 4 compares the effective mobility (μ_{eff}) as a function of different widths for specific channel lengths. The mobility and conductivity (Figure 3(a)) first increase and then decrease as a function of channel width W_{ch}. For W_{ch} in the range of a few hundred nm, where $d/W_{ch} \geq 1$, we observed that the measured conductivity and mobility extracted by fitting the data decrease with channel width. The effective mobility, μ_{eff}, and σ have a strong inverse dependence on W_{ch}, especially for $W_{ch} > d$ and are seen to saturate for large widths. This behavior is in agreement with the theoretical prediction that the d/W_{ch} ratio is dependent on the enhanced charge accumulation at the edges of the graphene strip [4].

Figure 5 compares mobility values for exfoliated graphene devices with similar L_{ch} and W_{ch} on thermal oxides with varying d, namely 15 nm, 90 nm and 300 nm. For a given L_{ch} and W_{ch} , reducing the thickness of the underlying dielectric from 300 nm to 90 nm to 15 nm leads to mobility reduction till a minimal dependence on the W_{ch} is observed. However, mobility degradation is still observed for all devices with channel widths of a few hundred nanometers. Thus, a given device has two parameters influencing the observed mobility, depending on W_{ch}, namely edge scattering and enhanced charge distribution at the edges.

In summary, the mobility behavior in graphene devices was studied as a function of channel dimensions. The extracted mobility in back gated graphene was found to depend on channel dimension. For a given channel width, the length dependence is attributed to the device operating in both the quasi ballistic and the diffusive regimes, as previously reported. The width dependence however is partially attributed to edge scattering and partially to enhanced conductivity in the channel as a result of electrostatically induced charge accumulation along the edges. This previously overlooked dependence could be a major contributing factor to the scatter in mobility values that have been reported.

Acknowledgment:

This work was supported by the SRC NRI SWAN center, the Texas Emerging Technology fund and a TI Diversity Fellowship.

References

[1].K.I. Bolotin, et al., Solid State Commun. 146, 351(2008);[2].K.S. Novoselov, et al., Nature. 438,197(2005);[3].J.-H. Chen, et al., Nat Nano. 3, 206(2008); S. Adam, et al., Physica E. 40, 1022(2008); W. Zhu, et al., Phys. Rev. B. 80, 235402(2009);[4]. F.T. Vasko, et al., Appl. Phys. Lett. 97, 092115(2010);[5].M.C. Lemme, et al., IEEE Electron Dev. Lett. 28, 282(2007); D.B. Farmer, et al., Nano Lett. 9, 4474(2009);[6]. S. Kim, et al., Appl. Phys. Lett. 94, 062107(2009);[7]. Z.Chen, et al. in *IEEE Tech. Dig.* (2008).

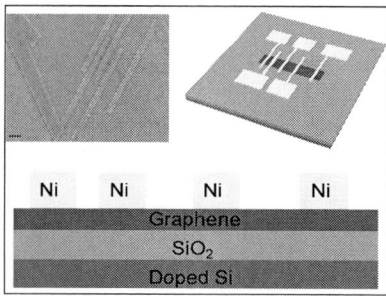

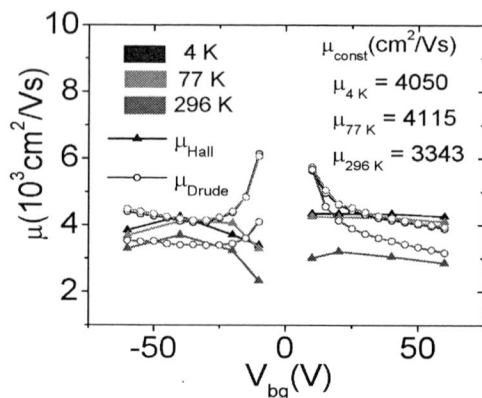

Fig. 1. SEM image, schematic of the top view and cross-section of a typical device structure used for the study (scale bar on SEM image is 2 μm) The contact lengths were fixed at 1 μm and the contact widths were defined by the width of the graphene flake.

Fig. 2. Plot comparing effective mobilities extracted using the constant mobility model and the Drude Model with the Hall mobility.

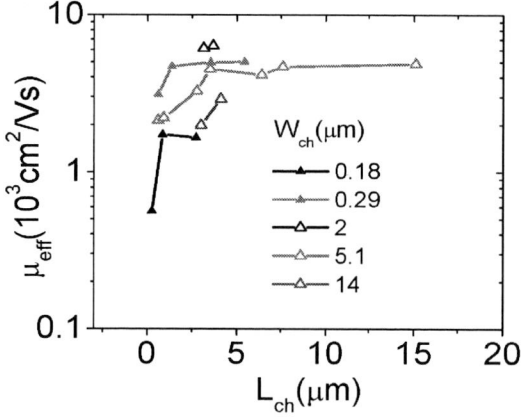

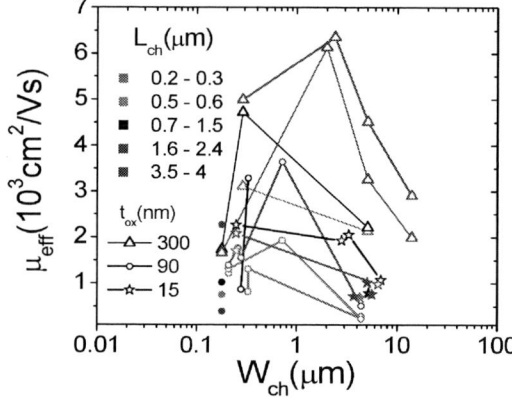

Fig. 3. Effective mobilities for devices on exfoliated graphene flakes, plotted as a function of channel length for different channel widths.

Fig. 5. Effective mobilities for exfoliated graphene devices as a function of L_{ch}, W_{ch} and dielectric thickness d.

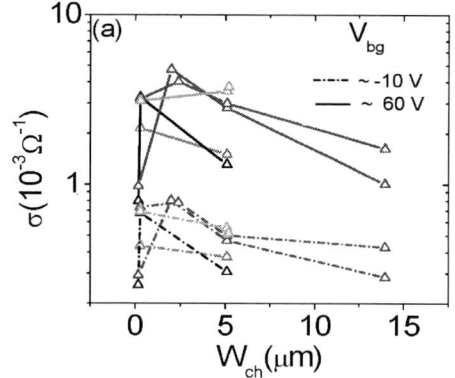

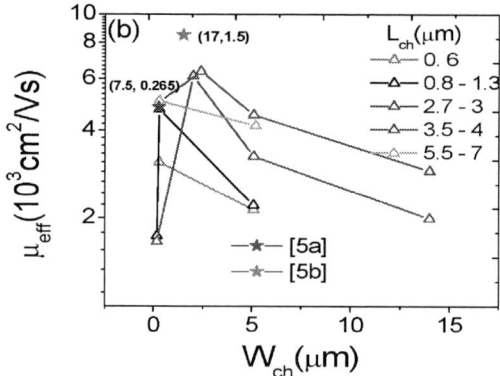

Fig. 4. Width dependence of (a) channel conductivity and (b) effective mobility as a function of different channel lengths. Coordinates indicate channel dimensions in the stated references.

978-1-61284-243-1/11 $26.00 © 2011 IEEE

Top-gated single-electron transistor in germanium nanowires

Sung-Kwon Shin, Shaoyun Huang, Naoki Fukata, Koji Ishibashi

Advanced Device Laboratory, RIKEN, Wako, Saitama 351-0198, Japan

Email: skshin@riken.jp, phone: +81-48-462-1111(ext.8434), fax: +81-48-462-4659

Germanium nanowires (GeNWs) of the group IV semiconductors could be one of the attractive candidates for electron-spin based quantum devices because of their long electron-spin coherence time. Besides, Ge has an advantage over Si in terms of the larger quantum effects due to the smaller effective mass. Single-electron transistors (SETs) are basic building blocks of such devices. To define the spin configuration in the dot, it is necessary to reach a few-electron regime or an even-odd regime where the single spin is realized for the odd number of electrons in the dot. So far, we have developed processes to fabricate SETs using n-type monocrystalline GeNWs with a back gate, and succeeded in observing the even-odd effect [1]. In this work, we have developed fabrication processes of the top-gate SETs to enhance the gating efficiency, and succeeded in reaching a few-electron regime.

The fabrication techniques of metal contacts to the GeNW are described in our previous report [1]. The source and drain electrodes were deposited with 80-nm thick nickel and were separated by 300 nm. Prior to the deposition of the top-gate electrode, 10-nm thick hafnium oxide (HfO_2) thin film was deposited on top of the GeNW by using atomic layer deposition (ALD) technique. Titanium / gold were deposited for the top-gate electrode. A schematic image of the top-gate device and a scanning electron microscopy image of the contacted device are shown in Fig. 1(a) and Fig. 1(b), respectively.

The electron transport characteristics were investigated in a dilution refrigerator with a base temperature of 25 mK. Source-drain current versus source-drain voltage (I_{ds}-V_{ds}) curves at -700 and -720 mV of V_{tg} are shown in Fig. 2, respectively. The coulomb blockade gap with suppressed I_{ds} could be modulated by tuning V_{tg}, demonstrating the SET operation. As shown in Fig. 3, pronounced Coulomb-oscillation peaks were observed in gate voltages larger than -850 mV. A charge-stability diagram in a V_{tg} region from -950 to -600 mV is shown in Fig. 4. V_{tg} positions of the Coulomb peaks occur at gate voltages where two adjacent diamonds meets. The first diamond did not close even below -1100 mV of V_{tg} and opened up to ±100 mV of V_{ds}. These facts mean the present quantum dot is completely depleted and the number of electrons can be reduced down to zero below V_{tg}=-850 mV. In our previous work, the number of electrons residing on the dot could not be reduced down to zero even at back-gate voltage of -10 V because the NWs could be highly degenerated and the gate-conversion factor (α) was much smaller [1]. The charging energy (E_c) and the average peak spacing (ΔV_g) are obtained to be 10 meV and 20 mV from the dimensions of the coulomb diamonds in Fig. 4, respectively. α is thus calculated to be 0.5. The α-value is two orders of magnitude than that of the back-gate device. The charging energy appears to be larger than the expected value when the whole NW between the source and drain is assumed to form a single dot. The reason is not clear, but it could be because the dot is formed with a single impurity or unexpected small dot formed by the potential fluctuations.

References

[1] S. Y. Huang et al., J. Appl. Phys. 109, 036101 (2011)

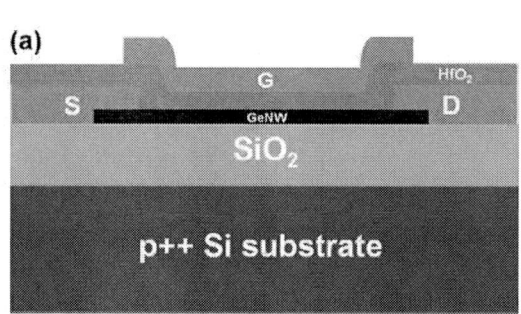 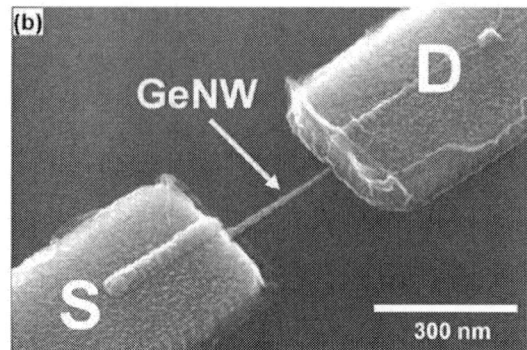

Fig. 1. (a) Schematic image of the top-gate device. (b) SEM image of the nickel-contacted device.

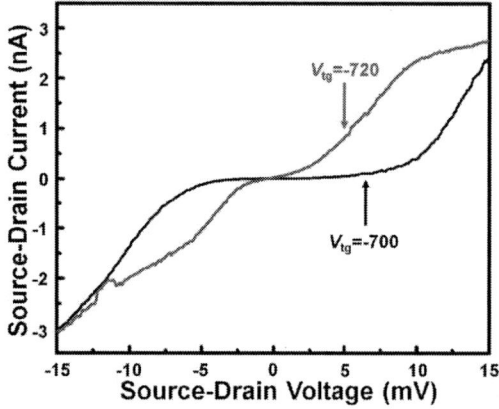 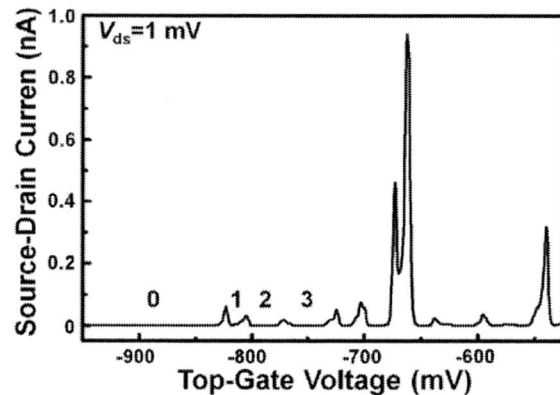

Fig. 2. Source-drain current versus source-drain voltage curves.

Fig. 3. Coulomb oscillation peaks at a fixed source-drain bias of 1 mV. The numbers indicate the number of electrons in the dot

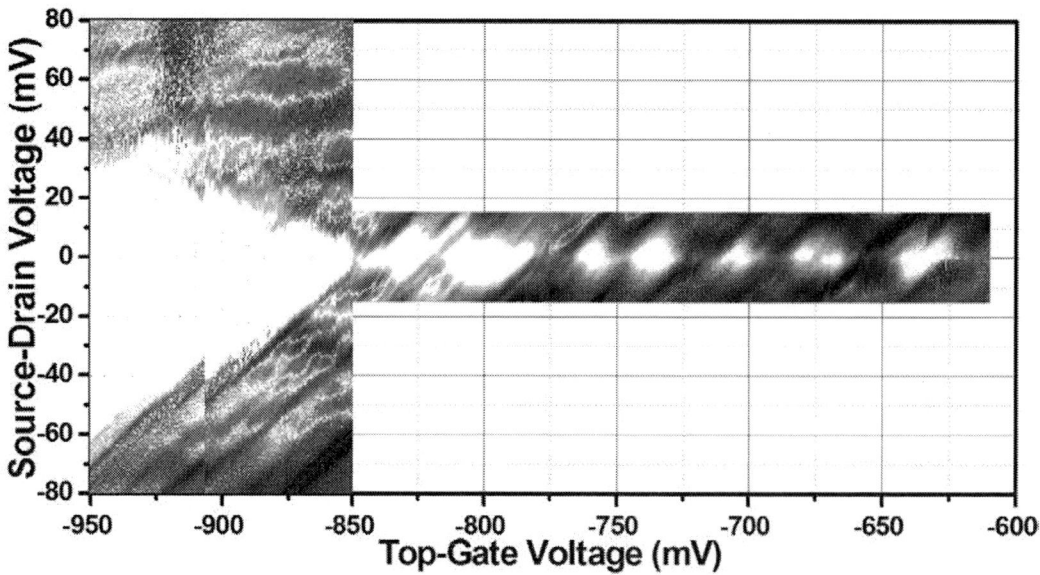

Fig. 4. Charge stability diagram of the device.

978-1-61284-243-1/11 $26.00 © 2011 IEEE

Reliability of Ambipolar Switching Poly-Si Diodes for Cross-Point Memory Applications

M. H. Lee[a,*], C.-Y. Kao[a], C.-L. Yang[a], Y.-S. Chen[b], H. Y. Lee[b], F. Chen[b], and M.-J. Tsai[b]

[a] Institute of Electro-Optical Science and Technology, National Taiwan Normal University, Taipei, Taiwan
[b] Electronics and Optoelectronics Research Laboratories, Industrial Technology Research Institute, Hsinchu, Taiwan
* phone: +886-2-77346747; fax: +886-2-86631954; e-mail: mhlee@ntnu.edu.tw

Cross-point memory framework provides high capacity, low power consumption, and low cost in nonvolatile-memory (NVM) technology [1,2]. Resistive cross-point memory structure is one of the potential candidates with scaling down beyond the flash memory [3]. In order to increase density for cross-point architecture, the vertical diode is integrated for the controller (Fig. 1) without planar MOSFET or BJT. The metal oxide diode has been reported on the switching devices with high leakage current [4]. The p/n diode has higher ON-current and uni-polar operation for PCM (Phase Change Memory) [5,6], which is compatible with IC process. The characteristic of bipolar programming in RRAM makes the requirement of bi-directional turn-ON behavior for the switching driving device [7]. In this work, the poly-Si n/p/n diode with ambipolar operation for RRAM applications and the stress reliability for programming will be demonstrated.

Standard 6-inch MOS-based line was employed in this study. Firstly, the n^+ Si substrate was prepared and 200nm-thick n^+ poly-Si was deposited by LPCVD. Then, the 100nm-thick undoped poly-Si was deposited and performed BF_2 implantation (30 keV, 5×10^{14} cm^{-2}). The annealing was carried out by RTA spike 800°C for p^+ activation. The sandwich structure was obtained by 200nm-thick n^+ poly-Si deposited on p^+ layer sequentially. After that, the diode area was defined by photolithography and etched mesa structure. Finally, the contact hole etching and metallization process were performed for the poly-Si n/p/n diode (Fig. 2 &3).

The current-voltage characteristics of the poly-Si n/p/n diode showed the bi-directional turn-ON behavior for bipolar memory applications (Fig. 4). The $J_{ON} \sim 0.1$ MA/cm^2 was obtained to provide memory programming for area = 2.25×10^{-8} cm^2. The currents were increased for small area cell due to less grain boundary. With noise margin consideration, maximum ratio between J_{ON} and J_{OFF} was ~ 7000 for 1/3 margin, and J_{OFF} was defined as 1/3 V_{ON} (Fig. 5). Note that 1/3 is for unselected lines left OPEN, while 1/2 is for unselected lines grounded with sneak path leakage effect [2]. The J_{ON}/J_{OFF} ratio is increasing with smaller cell size (Fig. 6). For a lower drive voltage (<1 V) would be able to increase the ratio, however, it is tradeoff for the lower J_{ON}. Both negative and positive biases made the positive shift of the J-V curve (Fig. 7). The reverse bias of pn junction was operated for the ambipolar poly-Si n/p/n diode (Fig.8). Both thermionic emission (TE) and trap-assisted tunneling (TAT) lead to increasing the number of deep (trap) states in the band gap, which is dangling bond formation at the grain boundary. The deep traps act as acceptor-like defect [8]. It results in positive shift of V_{ONSET}. The shift of V_{ONSET} increased with stress time, and the slope was almost constant with area (Fig. 9). The trap formation density with stress time which was proportional to ΔV_{ONSET} showed independently on area. J_{ON} showed increasing and decreasing for negative and positive stress, respectively, and the slopes were steeper for small area (Fig. 10). Note that the 10^{-7} sec for programming time and 10^9 for endurance were estimated, so that total stress time of 100 sec was for DC bias. The AC stress was also performed for comparison (Fig. 11). The increasing ΔV_{ONSET} was obtained with frequency increase for ±4V bias. Note that the ON time was half of the period so that AC stress 200 sec was the same as ON time with that of DC stress 100 sec. The breakdown voltage was estimated for 10 years (Fig. 12).

The ambipolar switching diodes with poly-Si n/p/n ware successfully demonstrated for cross-point memory applications, such as RRAM. The high $J_{ON} \sim 0.1$ MA/cm^2 was obtained to provide memory programming for area = 2.25×10^{-8} cm^2. Both negative and positive biases made the positive shift of the J-V curve with DC stress 100 sec and were contributed by acceptor-like defect formation. However, the reliability is sufficient for >10^6 cycles of 100 ns operation (~0.1 s total stress at 4V). V_{BD} of 2.25×10^{-8} cm^2 was estimated > 1V for 10 years. Finally, the authors are very grateful for the support and funding provided by the National Science Council (NSC 98-2221-E-003-020-MY3), for carrying out the process by National Nano Device Laboratories (NDL).

REFERENCES

[1] J. Liang, et al., in Interconnect Technology Conference (IITC), 2010, 6.3. [2] J. Liang, et al., IEEE Trans. Electron Devices., 2010, vol. 57, pp. 2531-2538.[3] S. Hanzawa, et al., in IEEE IEDM Tech. Dig., 2002, pp. 150-153.[4] M.-J. Lee, et al., in IEEE IEDM Tech. Dig., 2007, pp. 772-775.[5] J. H. Oh, et al., in IEEE IEDM Tech. Dig., 2006, pp. 49-52.[6] Y. Sasago, et al., in Symp. on VLSI Tech. Dig., 2009, pp. 24-25.[7] H. Y. Lee, et al., in IEEE IEDM Tech. Dig., 2008, pp. 297-300.[8] G. Fortunato, et al., Appl. Phys. Lett., 1986, vol. 49, pp. 1025-1027.

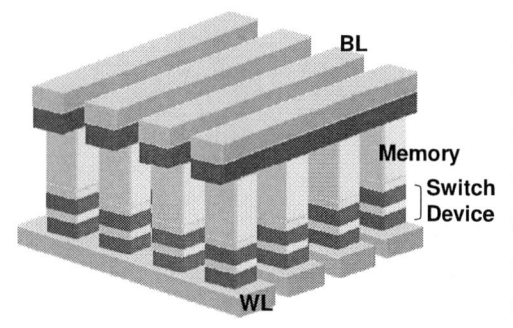

Fig. 1. The schematic diagram of cross-point memory. The vertical diode is integrated for the controller.

PROCESS FLOW

- n+ Si Substrate prepare
- n+ Poly-Si 200nm deposited
- Poly-Si 100nm deposited
- BF₂ I/I (30keV, 5×10¹⁴cm²) & Activation (800 °C spike)
- n+ Poly-Si 200nm deposited
- MESA etching
- Contact Hole & Metallization

Fig. 2. The process flow of the poly-Si n/p/n diode for cross-point memory applications.

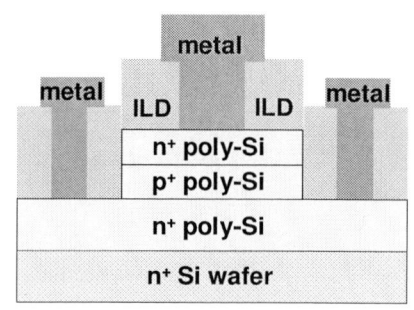

Fig. 3. The structure of poly-Si n/p/n didoe for cross-point memory applications.

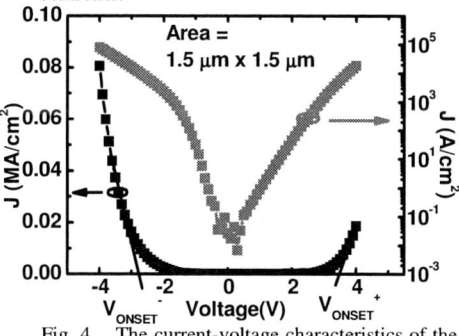

Fig. 4. The current-voltage characteristics of the poly-Si npn diode. The red and black dots are the logarithm and linear scale J vs. applied voltage, respectively. It shows the bi-directional turn-ON behavior for bipolar memory applications

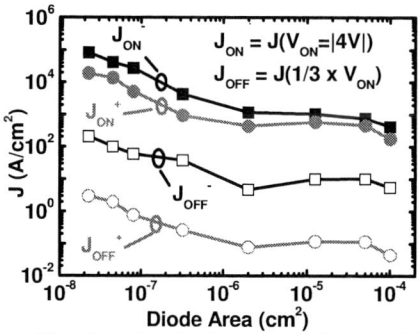

Fig. 5. J_{ON} & J_{OFF} vs. diode area. The currents are increased for small area cell due to less grain boundary.

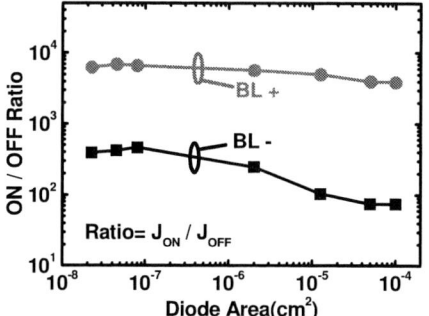

Fig. 6. The ratio of J_{ON} & J_{OFF} vs. diode area. The J_{ON}/J_{OFF} ratio is increasing with smaller cell size

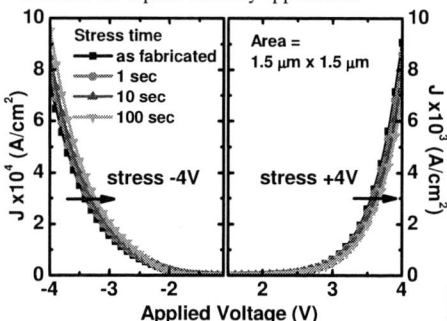

Fig. 7. The current-voltage characteristics of the poly-Si npn diode with DC stress ±4V. Both negative and positive biases make the positive shift of the J-V curve.

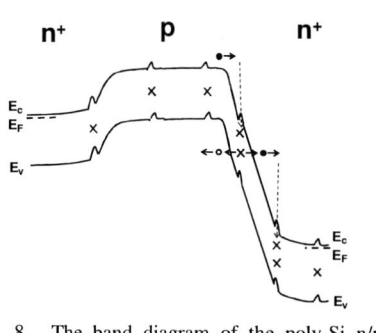

Fig. 8. The band diagram of the poly-Si n/p/n diode with operation. Both TE and TAT lead to increasing the number of deep (trap) states in the band gap, which is dangling bond formation at the grain boundary. The deep traps act as acceptor-like defect.

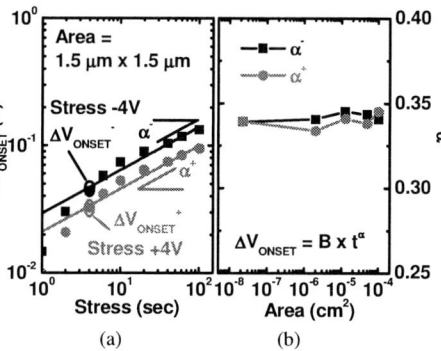

Fig. 9. (a) ΔV_{ONSET} vs. stress time. (b) The exponential of time vs. diodes area. The shift of V_{ONSET} increases with stress time, and the slope is almost constant with area.

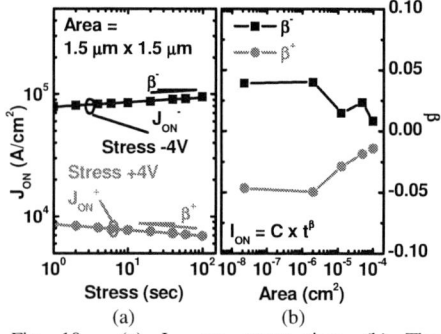

Fig. 10. (a) J_{ON} vs. stress time. (b) The exponential of time vs. diodes area. J_{ON} shows increasing and decreasing for negative and positive stress, respectively, and the slopes are steeper for small area.

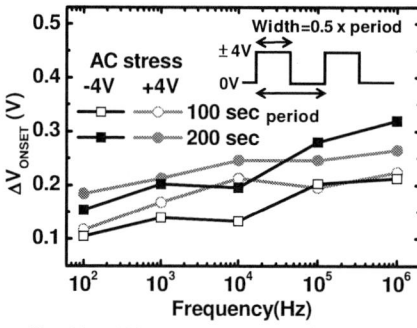

Fig. 11. ΔV_{ONSET} vs. operation frequency. The increasing ΔV_{ONSET} is obtained with frequency increase for ±4V bias. The inset shows the waveform.

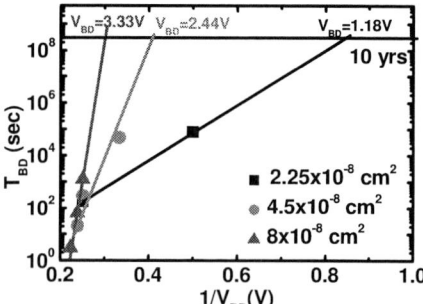

Fig. 12. The breakdown voltage is estimated for 10 year. V_{BD} of 2.25 x 10⁻⁸ cm² is estimated > 1V for 10 years.

978-1-61284-243-1/11 $26.00 © 2011 IEEE

90

Electrochemical supercapacitor based on flexible pillar graphene nanostructures

Jian Lin,[1] Jiebin Zhong,[1] Duoduo bao,[2] Jennifer Reiber-kyle,[3] Wei Wang,[4] Valentine Vullev,[2] Mihrimah Ozkan,[3] Cengiz S. Ozkan[1,4]

[1]Department of Mechanical Engineering,[2]Department of Bioengineering, [3]Department of Electrical Engineering, [4]Materials Science and Engineering Program, University of California, Riverside, CA92521 (USA), phone: 951-827-5016, email:cozkan@engr.ucr.edu

Here we report the fabrication of high conductive and large surface-area 3D pillar graphene nanostructures (PGN) films from assembly of vertically aligned CNT pillars on flexible copper foils and directly employed for the application in electrochemical double layer capacitance (EDLC) supercapacitor. The fabricated supercapacitor based on PGN films with excellent mechanical flexibility and electrical conductivity has high energy storage capability. The PGN films which were one-step synthesized on flexible copper foil (25 um) by CVD process exhibit high conductivity with sheet resistance as low as 1.6 ohm per square and high mechanical flexibility. The fabricated EDLC supercapacitor based on high surface-area PGN electrodes ($563m^2/g$) shows high performance with high specific capacitance of 330F/g and energy density as high as 45.8Wh/kg. All of these make this 3D graphene/CNTs hybrid carbon nanostructures highly attractive material for high performance supercapacitor and other energy storage material.

The surface morphology and PGN films is shown in figure 1. The transparent graphene layers and the dark black PGN film are clearly distinguishable from figure 1a and 1b. The diameters of CNT pillars are ranging from 5nm to 15nm, which indicates the formation of multi-wall carbon nanotubes (MWCNTs) and the height of CNT pillars is around 20 um (from figure 1c and 1d). The measured sheet resistance by Van der Pauw method reaches as low as 1.6 ohm/sqr from I-V curve shown in figure 1e. Plus PGN films show excellent mechanical flexibility when bent to different radium (see figure 1f).

The typical Raman characteristics shown in figure 2a collected from the graphene region shows that the graphene is few-layer graphene (FLG), which is similar to the reported CVD graphene. The Raman spectrum shown in figure 2b was collected from the dark area representing the CNT pillars. Unlike the splitting G+ and G- for the single wall carbon nanotubes (SWCNT) the predominant G peak located at 1582 cm^{-1} is usual the result of MWCNTs with large and various diameters.

The fabricated test cell comprising two copper foils with PGN films covered and separator is exhibited in figure 3a. Base on the cyclic voltammetry (CV) curves shown in figure 3b, the specific capacitance of PGN films after functionalization with HNO_3/IPA solution (67% HNO_3: IPA=1:4) was calculated to be 330F/g at scanning rate of 100mV/s using $C_s = \frac{\int i\,dV}{2 \times m \times \Delta V \times S}$.

This relatively higher specific capacitance than 2D graphene/CNTs film made from spincoating of CNTs on reduced graphene oxide or graphene is because of that the vertically uniformly-spacing aligned CNTs provides more effective surface area for electrolyte access into the pores. The energy density of PGN supercapacitor was calculated to be 45.8Kw/h using $E = \frac{1}{2}C_s(\Delta V)^2$.

In conclusion, a novel three dimensional architecture named pillar graphene nanostructures consisting of high dense and vertical aligned MWCNTs on large-area graphene layer were synthesized on flexible copper foil. The fabricated supercapacitor based on PGN electrodes shows high energy storage capability with high specific capacitance (330F/g) and high power density (45.8 Wh/g). This unique 3D architecture for supercapacitor electrodes would provide new pathway of achieving high energy storage technology.

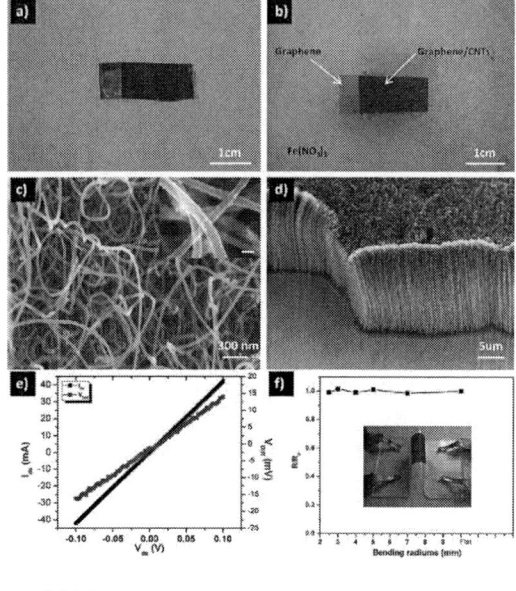

Figure 1. a) Optical micrograph of PGN grown over copper foil. b) Optical micrograph of PGN film floating in the aqueous $Fe(NO_3)_3$ solution. c) Top-view SEM image of PGN film. Inset is the top-view high magnification SEM of PGN (scale bar: 10nm). d) 45° side-view SEM image of PGN film. e) Two probe and four probe I-V characteristic of PGN film. Sheet resistance: $R_s = (\pi / \ln 2) \times (V_{diff} / I_{ds})$. f) Sheet resistance change of PGN film transferred on PET substrate under various bending radium. Inset shows the bending

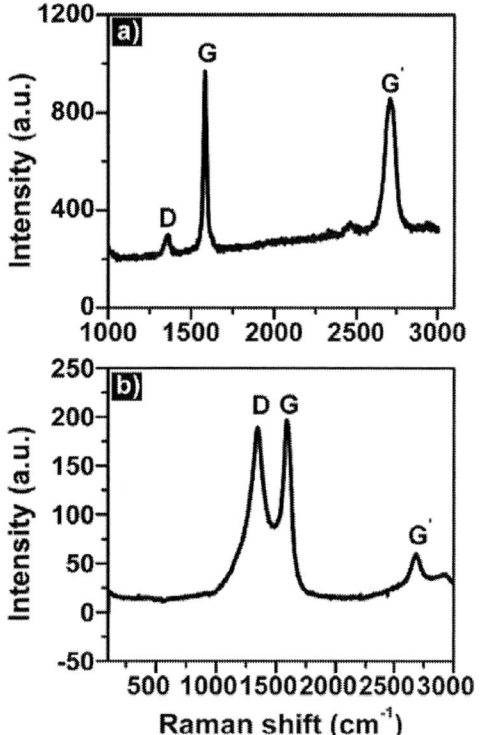

Figure 2. a) Raman spectra of few-layer graphene region. b) Raman spectra of PGN film from CNT pillars region. The excitation laser: 532 nm.

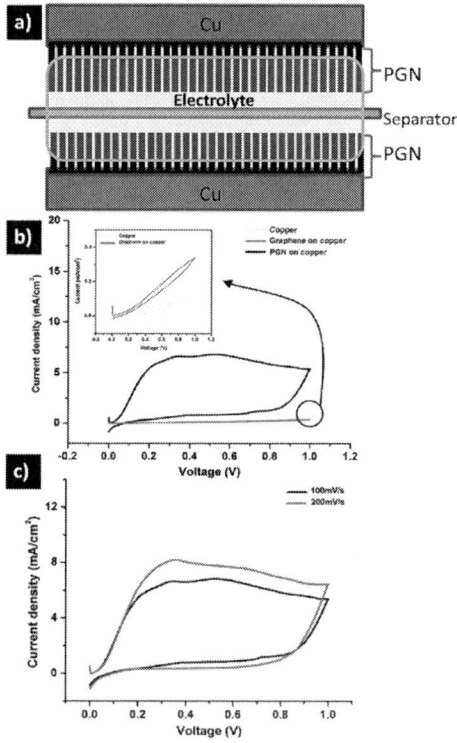

Figure 3. a) Schematic of test cell assembly. b) Comparison CV curves of capacitors based on copper foils, graphene layers on copper foils and PGN films on copper foils. Scanning rate: 100mV/s. inset is the zoom-in CV curves of capacitors based on copper foils and graphene layers on copper foils. c) CV curves of supercapacitor based on PGN films at scanning rate of 100mV/s and 200mV/s.

"Zero" Drain-Current Drift of Inversion-Mode NMOSFET on InP (111)A Surface

Chen Wang, Min Xu, Robert Colby, Eric A. Stach and Peide D. Ye

School of Electrical and Computer Engineering and Birck Nanotechnology Center,
Purdue University, West Lafayette, IN 47907, USA

InP is a commonly used compound semiconductor with wide applications in electronic, optoelectronic, and photonic devices. Compared to GaAs, InP is widely believed to be a more forgiving material with respect to Fermi level pinning and has a higher electron saturation velocity (2.5×10^7 cm/s) as well. It could be a viable channel material for high-speed logic applications if a high-quality, thermodynamically stable high-k dielectric could be found. [1] It is of great importance for the understanding of high-k/InP interfaces since InP is identified as a transition layer for ALD high-k/InGaAs quantum well transistor in state-of-the-art devices. [2] Motivated by previous work on surface orientation studies of GaAs [3] and InGaAs [4], we have systematically studied NMOSFETs, MOSCAPs, and interfacial chemistry on two different crystalline surfaces: InP (100) and (111)A (In-rich). With ALD Al_2O_3 in direct contact as gate dielectric, a record high drain current of 600 μA/μm is obtained for an InP inversion-mode MOSFET on the (111)A surface with a gate length of 1μm, which is a factor of 2.6 enhancement compared to the (100) surface at the same V_G-V_T condition. The smoother Al_2O_3/(111)A interface and a shift of the charge-neutrality-level (CNL) [5] on InP(111)A toward the conduction band edge is identified as the origin of this drain current enhancement in spite of the extracted interface trap density (D_{it}). [6] In this paper, we report on "zero" drain-current drift on InP (111)A MOSFETs which is a major issue to prevent commercializing InP MOSFET technology on (100) surface in 1980s. [7]

Figure 1 shows the schematic cross section of an InP inversion-mode MOSFET. ALD Al_2O_3 as gate dielectric was grown *directly* on semi-insulating InP substrates with the two different surface orientations (100) and (111)A after a 10 min. $(NH_4)_2S$ dip and DI water rinse. A 300°C TMA/H_2O chemistry was employed. A well-behaved I_{ds}-V_{ds} characteristic of a 1μm-gate-length inversion-mode InP NMOSFET on (111)A is demonstrated in Figure 2 with a maximum drain current of 600 μA/μm, which is *a factor of 3.5 larger* than that on (100), and demonstrates a record high drain current on inversion-mode InP NMOSFET. Sub-threshold characteristics of devices with L_g=2μm are shown in Figure 3. The subthreshold swing (*S.S.*) is ~150 mV/dec for (100) and ~230 mV/dec for (111)A. The threshold voltage (V_{th}) shifts from 0.5V for (100) to 0.03V for (111)A. I_{on}/I_{off} is higher for InP than that for InGaAs due to the wider bandgap. Combining with temperature dependent multi-frequency CV measurements performed on NMOSCAPs (not shown), it confirms that (111)A has a better interface near E_C but higher D_{it} in the midgap compared to (100). With CNL level shift toward conduction band edge on (111)A, more donor-like interface states contribute to the negative V_{th} shift.[8] The TEM images of Figure 4 show that the Al_2O_3/(111)A interface is smoother. This might explain the higher electron mobility and thus higher drain current on (111)A.

However, the drain-current degradation due to the interface and oxide bulk traps is a major issue for the practical application of InP MOSFETs. [7] To study the stability of on-currents of these InP MOSFETs on two different crystal orientations, a setup as illustrated in Figure 5(a) is used to measure drain-current drift versus time. For (111)A InP MOSFET, no obvious current drift is observed after 1500 s stress with a gate voltage of 3V and a drain voltage of 3V. In contrast, the drain current of InP (100) MOSFET goes down about 6.9% under the same bias conditions as shown in Figure 5(b). As I_{ds}-V_{gs} shown in Figure 6(a), the drain current degrades consistently about 6.9% after stress with a larger subthreshold swing, compared to the original drain current without stress. However, significant improvement of the drain current reliability is realized on (111)A surface as seen in Figure 6(b). After 1500 s stress with 3V at both gate and drain biases, I_{ds}-V_{gs} characteristic is kept almost as same as the virgin behavior for InP MOSFETs on (111)A surface.

Combined with CNL level shift, "zero" drain-current drift of (111)A surface can be interpreted through the band diagrams as illustrated in Figure 7. Supposing both (111)A and (100) surface are in strong inversion, more net negative charges from acceptor interface states on (100) surface make stronger electrical field on the oxide layer. Larger band bending leads to increase the possibility of tunneling from inversion to oxide bulk traps, which is often believed to be the main reason for the drain current degradation. InP MOSFETs on (111)A surface have not only better on-state performance but also "zero" drain-current drift and better reliability.

References:
[1] Y.Q. Wu et al., *Appl. Phys. Lett.* 91, 022108 (2007). [2] M. Radosavljevic et al., *IEDM Tech Dig.*, 319 (2009). [3] M. Xu et al., *IEDM Tech Dig.*, 865 (2009). [4] H. Ishii et al., *Appl. Physics. Express* 2, 121101 (2009). [5] P.D. Ye et al., *J. Vac. Sci. Technol. A* 26, 697 (2008). [6] M. Xu et al., in preparation. [7] D.L. Lile et al., *J. Appl. Phys.* 54, 260 (1983). [8] F Gozzo et al., *Solid State communications*.7, 81(1992).

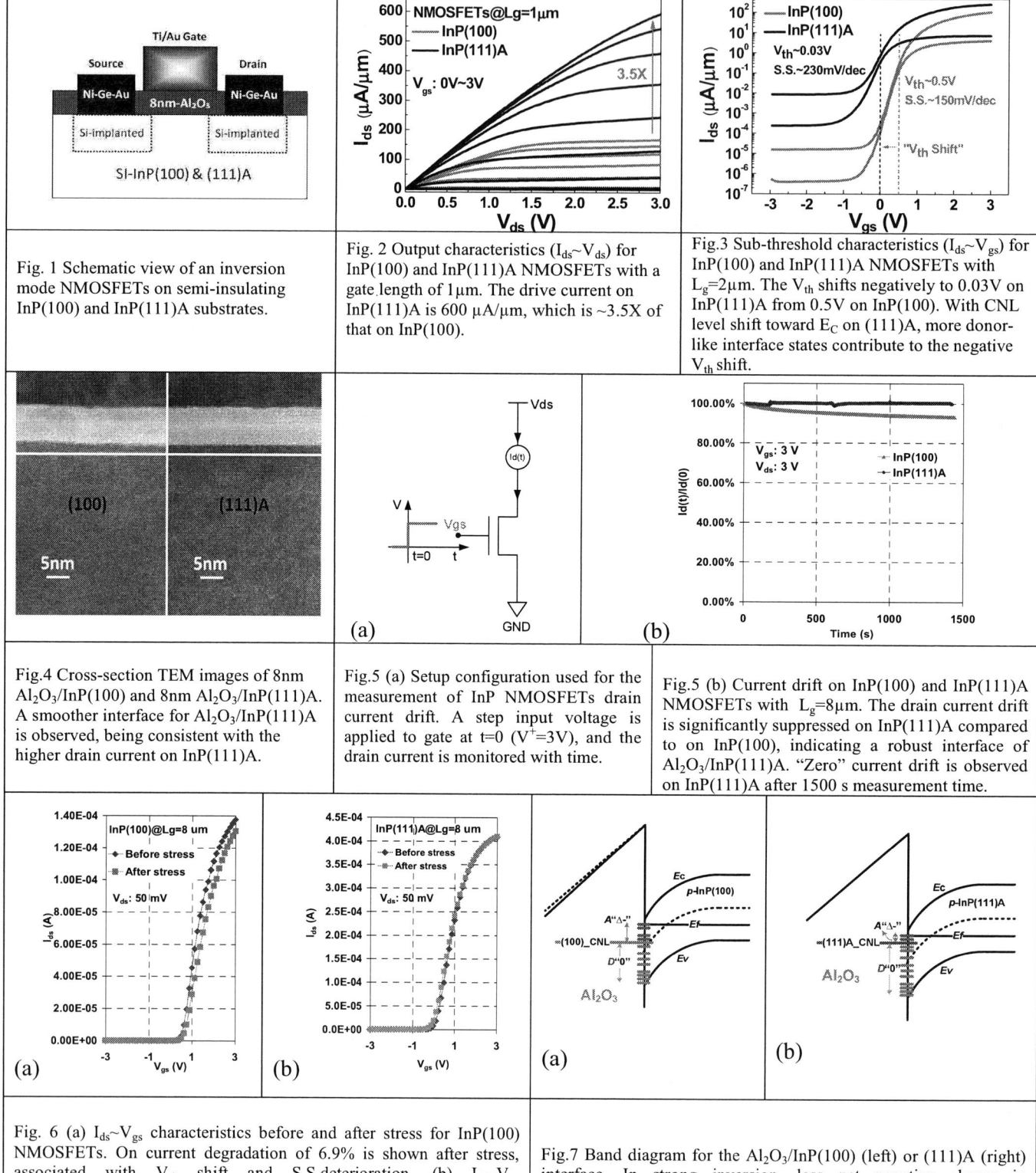

Fig. 1 Schematic view of an inversion mode NMOSFETs on semi-insulating InP(100) and InP(111)A substrates.

Fig. 2 Output characteristics ($I_{ds} \sim V_{ds}$) for InP(100) and InP(111)A NMOSFETs with a gate length of 1μm. The drive current on InP(111)A is 600 μA/μm, which is ~3.5X of that on InP(100).

Fig.3 Sub-threshold characteristics ($I_{ds} \sim V_{gs}$) for InP(100) and InP(111)A NMOSFETs with L_g=2μm. The V_{th} shifts negatively to 0.03V on InP(111)A from 0.5V on InP(100). With CNL level shift toward E_C on (111)A, more donor-like interface states contribute to the negative V_{th} shift.

Fig.4 Cross-section TEM images of 8nm Al_2O_3/InP(100) and 8nm Al_2O_3/InP(111)A. A smoother interface for Al_2O_3/InP(111)A is observed, being consistent with the higher drain current on InP(111)A.

Fig.5 (a) Setup configuration used for the measurement of InP NMOSFETs drain current drift. A step input voltage is applied to gate at t=0 (V^+=3V), and the drain current is monitored with time.

Fig.5 (b) Current drift on InP(100) and InP(111)A NMOSFETs with L_g=8μm. The drain current drift is significantly suppressed on InP(111)A compared to on InP(100), indicating a robust interface of Al_2O_3/InP(111)A. "Zero" current drift is observed on InP(111)A after 1500 s measurement time.

Fig. 6 (a) $I_{ds} \sim V_{gs}$ characteristics before and after stress for InP(100) NMOSFETs. On current degradation of 6.9% is shown after stress, associated with V_{th} shift and S.S.deterioration. (b) $I_{ds} \sim V_{gs}$ characteristics before and after stress for InP(111)A NMOSFETs. No obvious current degradation and V_{th} shift are found after the same stress.

Fig.7 Band diagram for the Al_2O_3/InP(100) (left) or (111)A (right) interface. In strong inversion, less net negative charge in Al_2O_3/(111)A interface results in less oxide band bending and less possibilities of tunneling from interface traps to border traps.

978-1-61284-243-1/11 $26.00 © 2011 IEEE   94

Improvement of efficiency in inverted bottom-emission white OLEDs by doping the hole transport layer

Hyunkoo Lee[1], Jeonghun Kwak[2], Jaehoon Lim[3], Kookheon Char[3], Seonghoon Lee[4] and Changhee Lee[1*]

[1]School of Electrical Engineering and Computer Science, Inter-University Semiconductor Research Center, Seoul National University, 599 Gwanak-ro, Gwanak-gu, Seoul 151-744, Korea
phone: +82-2-880-9093, fax: +82-2-877-6668 email: chlee7@snu.ac.kr
[2]Department of Electronics Engineering, Dong-A University, 840 Hadan2-dong, Saha-gu, Busan 604-714, Korea
[3]School of Chemical and Biological Engineering, Intelligent Hybrids Research Center, Seoul National University, 599 Gwanak-ro, Gwanak-gu, Seoul 151-744, Korea
[4]Department of Chemistry, Seoul National University, 599 Gwanak-ro, Gwanak-gu, Seoul 151-744, Korea

Recently, it is a critical issue to make larger display panels with low costs in organic light-emitting diodes (OLEDs). However, the uniformity and the cost of low temperature poly Si (LTPS) thin film transistors (TFTs) based backplanes for driving panels as well as the fine metal mask for pixel-patterning obstruct the realization of large-size OLED displays. To overcome these problems, using amorphous silicon (a-Si) TFTs which have high uniformity and cost-efficiency as the backplanes and white OLEDs (WOLEDs) which do not requiring any fine metal mask have been suggested previously. The inverted structure of OLEDs is much suitable rather than the conventional structure for a-Si TFTs because most a-Si TFTs have n-type channel.[1] Here, we demonstrate highly efficient inverted bottom-emission WOLEDs by controlling the balance of electrons and holes injected from electrodes. The maximum external quantum efficiency (E.Q.E.) of inverted WOLEDs was increased from 6.4% to 8.6% (about 34% improvement). To the best of our knowledge, this value is the highest E.Q.E. without other optical light extraction techniques in inverted WOLEDs reported to date.

Figure 1 shows the device structure of the inverted bottom-emission WOLEDs. We used three phosphorescent dopants to emit red, green, and blue, respectively. To control the electron-hole balance, we changed the hole transport layer (HTL); we doped 2,9-dimethyl-4,7-diphenyl-1,10-phenanthroline (BCP) to reduce hole current or molybdenum trioxide (MoO_3) to increase hole current into 1,1-bis[(di-4-tolyamino) phenyl]cyclohexane (TAPC).

Figure 2 and inset show the absorption spectra and current density-voltage (J-V) characteristics of different HTLs, respectively. The absorption spectrum of BCP doped TAPC film is similar with undoped TAPC film but the absorption spectrum of MoO_3 doped TAPC films is slightly different from undoped TAPC film. The absorption spectrum of MoO_3 doped TAPC film shows additional absorption peak at about 703 nm, which indicates the formation of charge transfer (CT) complexes. Therefore, the current increases drastically in the hole-only device (ITO/MoO_3 (5 nm)/HTL (100 nm)/MoO_3 (5 nm)/Al (100 nm)) with MoO_3 doped TAPC compared to the undoped device. On the other hand, the current decreases significantly in the device with BCP doped TAPC compared to the other devices. It indicates that MoO_3 acts as p-dopant in the TAPC, while BCP acts as traps for the hole transport in TAPC.

Figure 3 shows the luminance-voltage (L-V) and J-V characteristics for the inverted bottom-emission WOLEDs with different HTLs. The L-V and J-V characteristics are similar tendency with the J-V characteristics of hole-only devices.

Figure 4 and 5 show the E.Q.E-current density and the luminous current efficiency (L.C.E.)-current density characteristics for the devices with different HTLs, respectively. The efficiencies are increased for the device with the BCP doped HTL compared to the undoped device; the maximum E.Q.E. and L.C.E of the device with BCP doped HTL were 8.6% and 13.5 cd/A at 5.5 mA/cm^2, while those of undoped device were 6.4% and 10.5 cd/A at 4.2 mA/cm^2. On the other hand, the efficiencies are decreased for the device with MoO_3 doped HTL compared to undoped device. These results indicate that the electron-hole balance is a very critical factor for improving electroluminescence (EL) efficiency in OLEDs.

Figure 6 shows the normalized EL spectra of the devices with different HTLs at the same current density. The Commission Internationale de l'Éclairage (CIE) color coordinates of the devices with MoO_3 doped, undoped, and BCP doped HTLs are (0.34, 0.42), (0.36, 0.43) and (0.38, 0.42), respectively. BCP doped device has more redish spectrum, while MoO_3 doped device has more bluish spectrum compared to the device with undoped HTL. This is concerned with the recombination zone shift which is caused by different injected hole amounts.

This work was financially supported by the grant from the Industrial Source Technology Development Program (A1100-0901-1750) of the MKE of Korea and also supported partly by the MEST through the BK21 Program.

[1] H.-H. Hsieh, T.-T. Tsai, C.-Y. Chang, H.-H. Wang, J.-Y. Huang, S.-F. Hsu, Y.-C. Wu, T.-C. Tsai, C.-S. Chuang, L.-H. Chang and Y.-H. Lin, *SID Symposium Digest*, vol. 41, p.140 (2010).

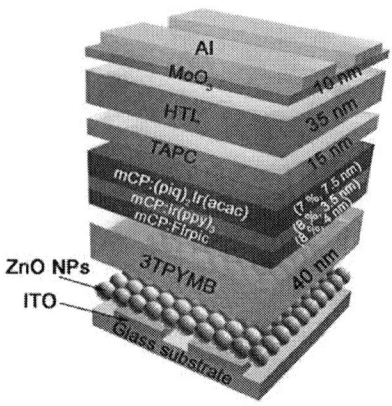

Fig. 1. Device structure of the inverted bottom-emission white OLEDs.

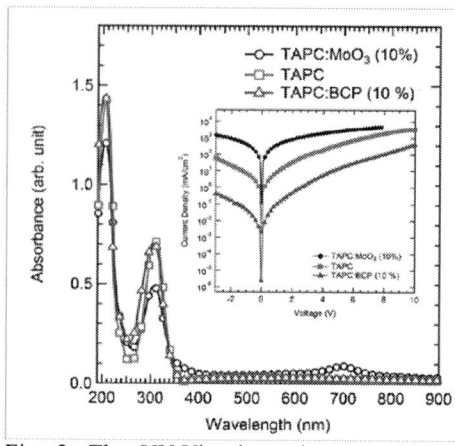

Fig. 2. The UV-Vis absorption spectra of MoO_3, undoped, and BCP doped TAPC films. Inset: Current density-voltage (J-V) characteristics of hole-only devices with different HTLs.

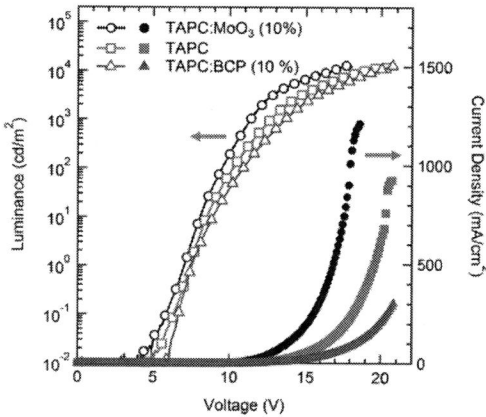

Fig. 3. Luminance-voltage (L-V), current density-voltage (J-V) characteristics for the devices with different HTLs.

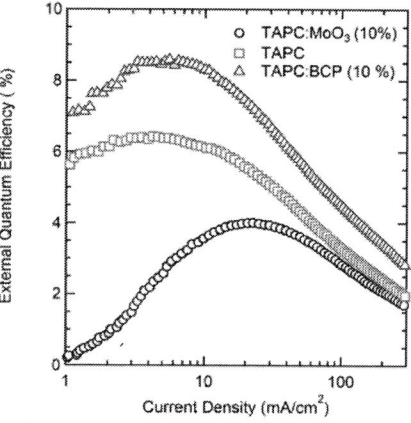

Fig. 4. External quantum efficiency (E.Q.E.)-current density characteristics for the devices with different HTLs.

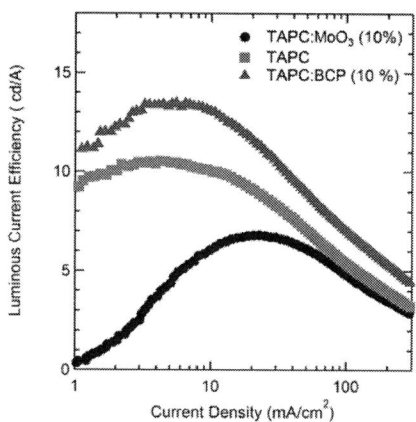

Fig. 5. Luminous current efficiency (L.C.E.)-current density characteristics for the devices with different HTLs.

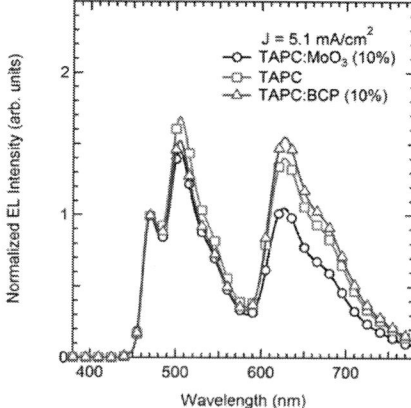

Fig. 6. Normalized EL spectra of the devices with different HTLs at the same current density (normalized at around 470 nm).

978-1-61284-243-1/11 $26.00 © 2011 IEEE

Vertical Organic Field-Effect Transistor Array Fabrication Based on Laser Holography Lithography Process

Donghyun Kim and Yongtaek Hong[*]

Department of Electrical Engineering and Computer Science, Seoul National University
599 Gwanak-ro, Gwanak-gu, Seoul, Korea
Phone: +82-2-880-9567, Fax: +82-2-871-5974, E-mail: yongtaek@snu.ac.kr

Due to the handiness in obtaining a short channel length, vertical organic field-effect transistors (VOFETs) have been pointed as an alternative form of conventional organic thin-film transistor (OTFT). With VOFET structure, it is relatively simple to obtain an short channel length and a large channel width-to-length ratio (W/L) value in a restricted device area, so a large current driving capability which can hardly be achieved with organic semiconductor, can be realized. Moreover, VOFETs can be utilized as a platform for many kinds of electrical applications associated with other various functional devices such as organic light-emitting diodes (OLEDs) and sensors.

We have recently established a laser-holography lithography method to apply in patterning higher-order nano-structures for VOFETs, to increase the controllability by altering gate voltages. It has been shown by other researchers that higher on/off ratio could be obtained with tinier gate electrode structures. [1] To realize more accurate gate electrode structures which have nanoscale channel openings, e-beam lithography or colloidal lithography techniques have been introduced. Compared with these methods, a laser-holography lithography technique applied in VOFET fabrications have many advantages because it is a simple and cost-competitive process and the exact lateral device structures can be estimated. In this work, we have made a VOFET array on a quartz substrate to verify feasibility of the laser-holography lithography for display applications.

A brief device schematic diagram of VOFETs is shown in Fig. 1. On a ITO sputtered quartz substrate, a vertical stack consisting of silicon nitride (SiN_x) insulating layer, aluminum (Al) gate electrode layer, another SiN_x layer, and chrome (Cr) hardmask layer, has been deposited through plasma enhanced chemical vapor deposition (PECVD) (SiN_x) and e-beam evaporation (Al, Cr) process. On this vertical stack, two-dimensional nanoscale patterns have been made through a laser-holography lithography process after spin-coating of a photoresist. With this nanoscale patterns, each layers have been etched away through reactive ion etching (RIE) process. To further enhance on/off ratio by depressing the gate leakage current, Al gate electrodes were oxidized with an oxygen plasma during etching process. [2] Defined channel regions were filled with pentacene and finalized with gold (Au) electrode through thermal evaporation. The finalized VOFET device fabricated on quartz substrate is shown in Fig. 2.

The transfer characteristics and output characteristics are shown in Fig. 3 and Fig. 4, respectively. In these results, a VOFET device fabricated on a ITO sputtered quartz substrate has showed the on/off ratio greater than 10^3 because of the improved gate voltage controllability originated from reduction in size of the lateral gate electrode patterns to the higher-order nanoscale patterns. More stable and higher drain current level was obtained even at low source-drain and gate-source voltages in comparison with our reference devices that were fabricated on heavily doped silicon substrate acting as a drain electrode. Sufficiently high current density levels which have a capability of driving OLED devices (~ 10 mA/cm^2) have been achieved. We think that this enhancement in current density level is originated from the reduction of process deviation sensitivity during RIE process and increased hole injection efficiency between pentacene and drain electrode (heavily-doped silicon and ITO), which can be affected by a different growth behavior of pentacene on each electrode during deposition.

In conclusion, we have made a VOFET array on a ITO sputtered quartz substrate using a laser holography lithography technique during lateral patterning of VOFET structures. Holographically generated higher-order nanoscale patterns were uniform in size, location and shape, hence more accurate device analysis would be possible through VOFETs patterned with a laser holography lithography process. We believe that this simple and cost-competitive patterning technique which enables uniform nanoscale patterning for a large area, would improve the process reliability of VOFETs, in addition to device performances of VOFETs.

This work was supported by the Industrial Strategic Technology Development Program (KI002104, Development of Fundamental Technologies for Flexible Combined-Function Organic Electronic Device) funded by the Ministry of Knowledge Economy (MKE, Korea).

[1] H. Yamauchi, Y. Watanabe, M. Iizuka, M. Nakamura and K..Kudo, IEICE Trans. Electron. E91-C, 1852 (2008)

[2] D. Kim, J. Jeong and Y. Hong, IEICE Technical Report, ED2010-59, SDM2010-60 (2010)

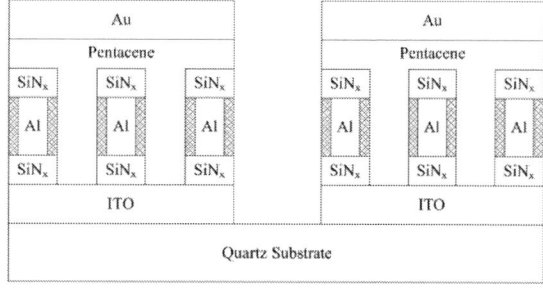

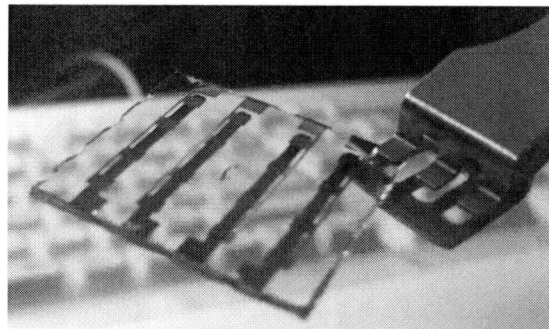

Fig. 1. A brief schematic diagram of a VOFET array. A vertical stack of SiN_x, Al, SiN_x, Cr layer were deposited through PECVD or e-beam evaporation. Cr hardmask was firstly patterned with photolithography and chemical wet-etching process, to have gate electrode and interconnection pattern of VOFET array. After the photoresist was spun-coated again, the sample was exposed to the holographic image that a laser holography setup generates. With a holographically generated nanoscale patterns, the vertical stack was etched sequentially. During RIE, Al layer was intentionally exposed to oxygen plasma to make oxidized gate electrode surfaces. Cr hardmask was removed through the same RIE process, after VOFET structure was formed. Finally, pentacene active layer and Au source electrode were deposited through a thermal evaporation.

Fig. 2. Finalized VOFET array fabricated on an ITO sputtered quartz substrate. Diffraction patterns can be observed in gate electrode patterns.

Fig. 3. Transfer characteristics of a VOFET fabricated on a quartz substrate. The gate voltage (V_G) was swept from 2 V to -2 V with -0.02 V steps. The drain voltage (V_D) was increased from 0 V to -1 V with -0.2 V steps. The measured pixel area of VOFET array was 1 mm × 1 mm. The on/off ratio was measured up to over 10^3 in low V_D region. (inset) A log-scale plot of transfer characteristics.

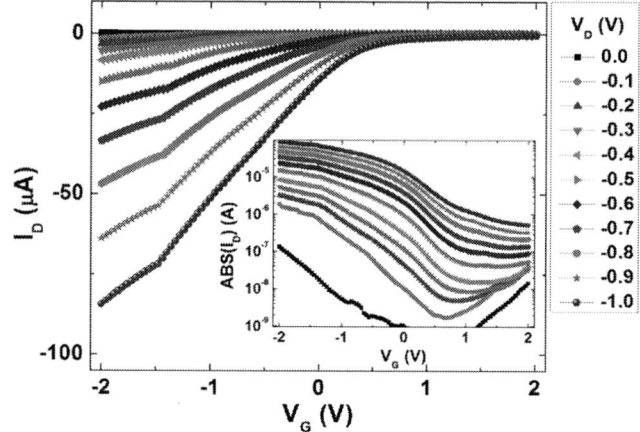

Fig. 4. Output characteristics of a VOFET fabricated on a quartz substrate. The same voltage ranges as in the transfer characteristics were applied. The drain voltage (V_D) was swept form 0 V to -1 V with -0.01 V steps and the gate voltage (V_G) was altered from 2 V to -2 V with -0.4 V steps. Distinct on/off behavior according to the applied gate voltage is shown here, too. However, the saturation behavior is not observed as many other types of vertical organic transistor devices have shown.

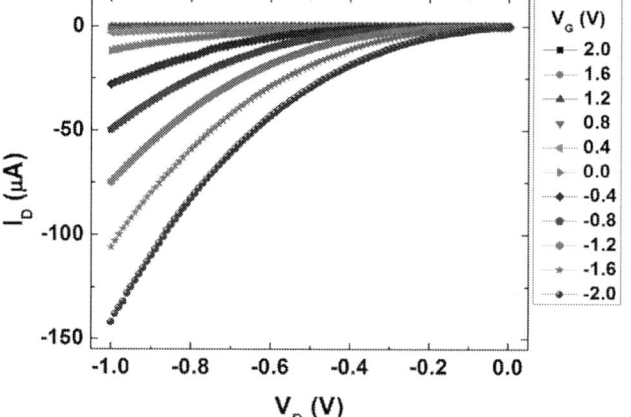

Ambipolar Nano-crystalline-silicon TFTs with Submicron Dimensions and Reduced Threshold Voltage Shift

Anand Subramaniam[1], Kurtis D. Cantley, Richard A. Chapman, Bhaswar Chakrabarti,
and Eric M. Vogel[2]

Department of Electrical Engineering, The University of Texas at Dallas, Richardson, TX 75080, USA
Phone: (480) 203 0304, Email: asubram3@utdallas.edu[1], eric.vogel@utdallas.edu[2]

Hydrogenated nano-crystalline-silicon (nc-Si) thin-film transistors (TFTs) are primary candidates for use in neuromorphic circuits and systems [1]. Such devices can be fabricated at low temperatures and over large areas, allowing cheap processing and three-dimensional integration with CMOS structures. The major drawbacks of nc-Si TFTs include low carrier mobility, threshold voltage (V_T) shift under bias stress and lack of p-channel operation due to unintentional n-type doping by oxygen impurity present in the nc-Si layer [2]. We have fabricated nc-Si TFTs that minimize all the above drawbacks, and are thus well suited for use in neuromorphic applications.

Top-gated TFTs with Cr source/drain and Al gate were fabricated with a staggered structure as shown in Fig. 1. The use of Cr as source/drain metal is to ensure low contact resistance [3]. All TFTs used 200 nm SiO_2 as the gate dielectric. Devices with sub-micron dimensions were fabricated by patterning the source/drain metal and nc-Si channel with electron-beam lithography. The nc-Si channel material (80 nm thick) was deposited using PECVD, at 13.56 MHz, 125 W RF power, 0.9 Torr chamber pressure, and 250° C deposition temperature in a 1:100 mixture of SiH_4 and H_2. A turbo pump was fitted to the PECVD to reduce the background vacuum pressure to 6×10^{-6} Torr. XRD measurements were performed and the average crystal size in the nc-Si layer was calculated using Scherrer's formula [4] to be approximately 7 nm (Fig. 2). For device reliability comparison, we also fabricated another set of samples with similar processing except that the turbo-pump was not used with the PECVD in this case. Such samples had a much higher oxygen impurity concentration in the nc-Si layer. SIMS results from both types of samples confirm the reduction in oxygen concentration when the turbo-pump was used (Fig. 3).

The electrical characteristics of the fabricated TFTs were measured and are shown in Fig. 4. The devices operate in both n- and p-channel regimes. In the n-channel regime, the devices have a threshold voltage of 3 V and subthreshold swing of 0.6 V/decade, whereas the corresponding values in the p-channel regime are −5 V and 0.65 V/decade. The field-effect mobility values in both regimes are plotted in Fig. 5. The maximum mobility for electrons is 20 cm^2/V·s and that for holes is 3 cm^2/V·s. The high hole mobility is ascribed to the low oxygen content in the nc-Si layer. Table 1 displays various device parameters at device lengths of 10 μm and 200 nm. The threshold voltage, subthreshold swing, and the peak electron mobility are not degraded at nanoscale dimensions [5]. Two of the fabricated ambipolar transistors were connected to form an inverter circuit. Fig. 6 shows the output of such an inverter at different operating voltages. The W/L ratio of the pull-up transistor is five times that of the pull-down device to compensate for the lower hole mobility of the TFTs. The voltage gain when operating at 6 V is 12, which is the highest value that has been shown using ambipolar nc-Si TFTs [6]. We attribute the high gain to the low subthreshold swing and low leakage current in the TFTs.

We also investigated how the on-current of the TFTs scale with the device dimensions. The results (Fig. 7) indicate that the on-current (measured at V_G=5 V, V_D=5 V) scales fairly linearly with device width for both n-channel and p-channel operation down to a channel width of 200 nm. The on-current also scales linearly with channel length down to 200 nm. Thus scaling the device dimensions is not detrimental to the device performance, allowing such devices to be used in architectures that require high density.

Finally, the shift in threshold voltage of the TFTs was measured after the application of different voltage stresses. The results (shown in Fig. 8) indicate that the degradation is significantly reduced for the samples with low oxygen concentration. Of the two accepted V_T degradation mechanisms in nc-Si TFTs [7], charge trapping in the gate dielectric is reversible. The observed V_T shift did not recover after a rest period of several hours or under the application of negative gate bias, suggesting that the shift is irreversible. Thus, under the applied bias conditions, defect creation in the channel layer is the principal V_T degradation mechanism. The results also suggest that the oxygen impurity present in the nc-Si layer plays an important role in the defect creation.

[1] K. D. Cantley et al., *IEEE Trans. on Nanotech.*, in press.
[2] T. Kamei et al., *J. Appl. Phys.*, vol. 96, pp. 2087, Aug. 2004.
[3] C-H. Lee et al., *J. Non-Crystalline Solids*, vol. 352, pp. 1732-1736, Apr. 2006.
[4] K. Bhattacharya et al., *Nanotechnol.*, vol. 18, pp. 415704, Sept. 2007.
[5] K. D. Cantley et al., *Appl. Phys. Lett.*, vol. 97, pp. 143509, Oct. 2010.
[6] K-Y. Chan et al., *Solid State Electronics*, vol .53, pp. 635-639, 2009.
[7] M. J. Powell et al., *Appl. Phys. Lett.*, vol. 52, pp. 1242, Aug. 1987.

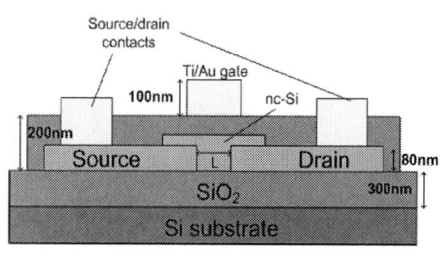

Fig. 1. Schematic cross-section of a staggered top-gated nc-Si TFT.

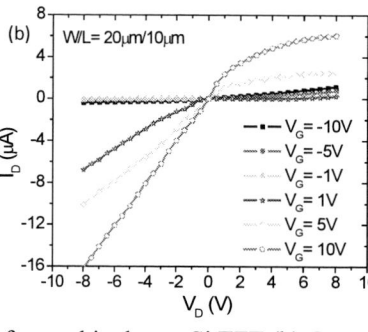

Fig. 2. XRD measurement on nc-Si film showing the crystalline peaks, used to calculate the crystallite size.

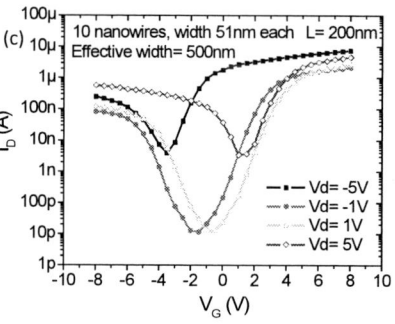

Fig. 3. SIMS data showing oxygen concentration in the nc-Si layer with and without the turbo-pump.

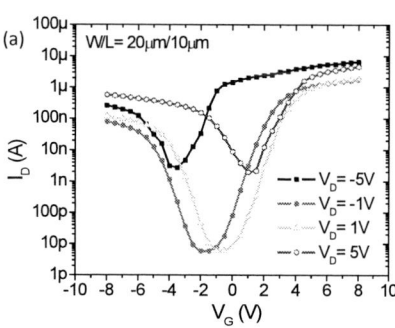

Fig. 4. (a) Transfer characteristics of an ambipolar nc-Si TFT (b) Output characteristics of the same device (c) Transfer characteristics of an nc-Si TFT with sub-micron dimensions.

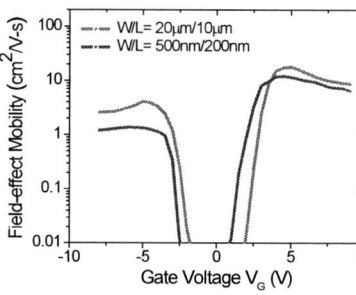

Fig. 5. Field-effect mobility of electrons and holes calculated from nc-Si TFTs.

W= 20μm	L= 10μm	L= 200nm
Threshold Voltage at V_D= 1V (V)	3.0	2.4
Subthreshold Swing (mV/dec.)	600	450
Peak Electron Mobility (cm²/V·s)	20	11

Table 1. Summary of device characteristics for regular and short-channel nc-Si TFTs.

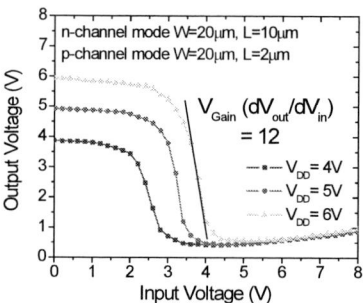

Fig. 6. Inverter characteristics when two ambipolar nc-Si TFTs are connected together.

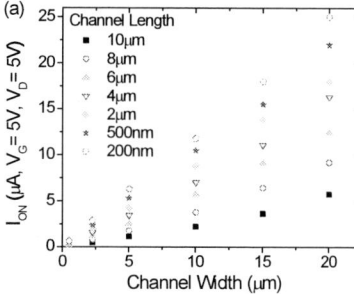

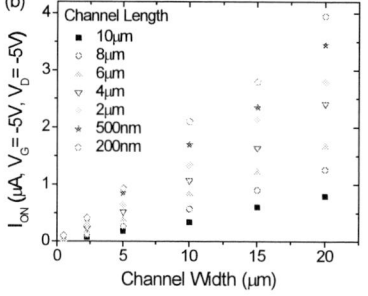

Fig. 7. The variation of on-current with device dimensions in (a) n-channel and (b) p-channel operation of nc-Si ambipolar TFTs.

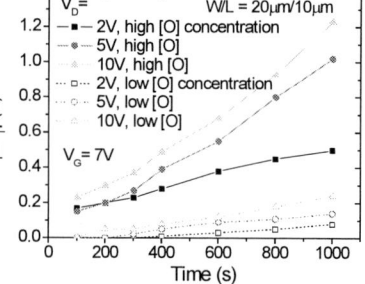

Fig. 8. Variation of the shift in threshold voltage with stress time for different applied bias values.

978-1-61284-243-1/11 $26.00 © 2011 IEEE

Monolithically Grown In$_x$Ga$_{1-x}$As Nanowire on Silicon Tandem Solar Cells with High Efficiency

Jae Cheol Shin[1], Kyou Hyun Kim[2], Hefei Hu[2], Ki Jun Yu[1], John A. Rogers[2,1], Jian-Min Zuo[2], and Xiuling Li[1,2*]

[*]xiuling@illinois.edu

[1]Department of Electrical and Computer Engineering, [2]Department of Materials Science and Engineering,
University of Illinois, Urbana, IL 61801

Heteroepitaxial integration of III-V and Si has been researched for many years since the Si is the prevalent platform and III-V can be used for light emitting source (i.e., direct bandgap) [1]. Although vertical InAs nanowires (NWs) growth on Si substrate (11.6% lattice mismatch) without catalysts and patterning has been demonstrated by several groups, [2, 3], direct heteroexpitaxial growth of ternary In$_x$Ga$_{1-x}$As nanowires hasn't been systematically studied yet, in spite of its important spectral coverage in the near infrared range. In this paper, we report the one-dimensional heteroepitaxial growth of dislocation free In$_x$Ga$_{1-x}$As nanowires on silicon (111) substrate in the entire composition range and demonstrate monolithically grown axial p-n junction tandem solar cells consisting of In$_x$Ga$_{1-x}$As NWs on Si with an efficiency that well exceeds the planar Si single junction solar cell fabricated using identical process.

Fig. 1 shows the SEM images of In$_x$Ga$_{1-x}$As NWs on Si (111) substrate for compositions as indicated. Most NWs are vertically grown on the Si (111) substrate; and the tip bending for some NWs (e.g. in Fig. 1c) occurs for ternary NWs only and is attributed to asymmetric strain distribution due to phase separation. The top and side surfaces are extremely flat with (111) hexagonal facet (Fig. 1e inset) on top and surrounded by six $\{1\bar{1}0\}$ facets. Shown in Fig. 2 are TEM images and electron diffraction patterns of InAs and In$_{0.30}$Ga$_{0.70}$As NWs. All images clearly show ordered crystalline planes which correspond to a mixture of zinc-blende (ZB) and wurtzite (WZ) structures often alternating every few monolayers. No misfit dislocation has been observed through the entire observed NW length and for all compositions. The arrangements of WZ and ZB layers can be determined directly from atomic resolution HAADF-STEM images where the atomic models are superimposed directly on the experimental images, as shown in Fig. 2d and e and 2i and j. Shown in Fig. 3 is the detailed growth and fabrication procedure of the tandem cell consisting of an axial p-n junction In$_{0.85}$Ga$_{0.15}$As NW on top of a planar Si p-n junction. Note that the composition of the InGaAs NW is labeled according to initial XRD peak position, and probably lower in In% after the annealing step. Fig. 4 shows the schematic and the I-V characteristics of the tandem photovoltaic device in dark and under air mass 1.5 (AM 1. 5, 100 mW/cm^2) with solar simulator at room temperature. V_{oc}, I_{sc}, and fill factor (FF) are 0.58 V, 22.1 mA/cm^2, and 0.66, respectively. Note that the area coverage of the InGaAs NWs is only 4 – 5%. The energy conversion efficiency is calculated to be 8.5%.

In summary, the direct 1D heteroepitaxy of ternary In$_x$Ga$_{1-x}$As NWs in a wide composition/bandgap range reported here enables monolithic tandem solar cells on Si substrate. Further improvement in energy conversion efficiency is expected due to the direct bandgap, enhanced light trapping, as well as bandgap and current matching by composition, doping, and height engineering.

[1] D. L. Huffaker and D. G. Deppe, Appl. Phys. Lett. 73 (4) 1998
[2] Wei, W.; Bao, X.; Soci, C.; Ding, Y.; Wang, Z.; Wang, D. Nano Lett., 9, 2926– 2934 2009
[3] K. Tomioka, J. Motohisa, S. Hara, and T. Fukui, Nano. Lett., 8, (10) 3475-3480, 2008

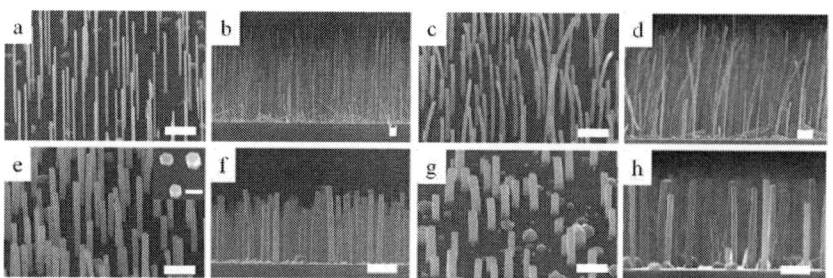

Figure 1. SEM images of $In_xGa_{1-x}As$ nanowires (NWs) on Si (111) substrates. (a) and (b) are InAs, (c) and (d) are $In_{0.85}Ga_{0.15}As$, (e) and (f) are $In_{0.3}Ga_{0.7}As$, (g) and (h) are $In_{0.2}Ga_{0.8}As$ NWs, respectively. (a), (c), (e), and (g) are the 45^0 tilted view images with same magnification (20 K) while (b), (d), (f), and (h) are the side view images with different magnifications as labeled. The scale bar represents 1 μm. Inset in (e) is the high magnification top view image and the scale bar represents 200 nm.

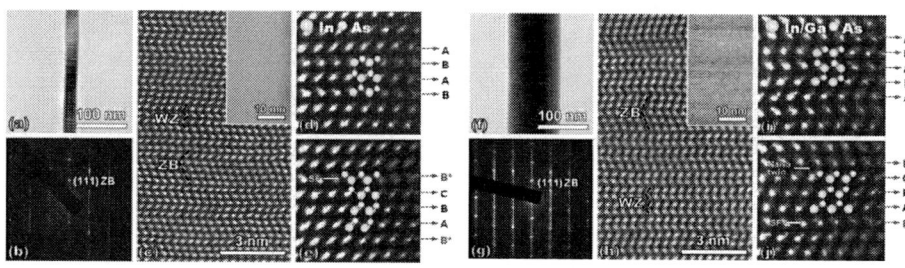

Figure 2. Electron microscopy characterization of $In_xGa_{1-x}As$ NWs. a – e are for InAs NW and f - j are for $In_{0.30}Ga_{0.70}As$. Both NW samples show intermixing of WZ (AB* stacking) and ZB (ABC stacking). Nano twins and stacking faults (SF) are indicated in the figures. The atomic structure models are superimposed on the atomic resolution Z-contrast STEM images in d, e, i and j. The stacking are labeled with * denotes layers with the In(Ga)-As bonds toward upper left directions. The mixture of WZ and ZB and stacking faults leads to streaky lines in the recorded diffraction patterns in b and g and contrast modulations in bright field TEM images in a and f.

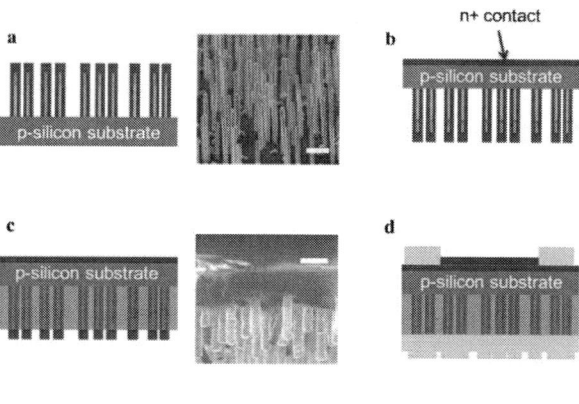

Figure 3. Growth and fabrication process of the tandem solar cell consisting of a p^+-p-n-p-n^+ InGaAs NW-Si structure.

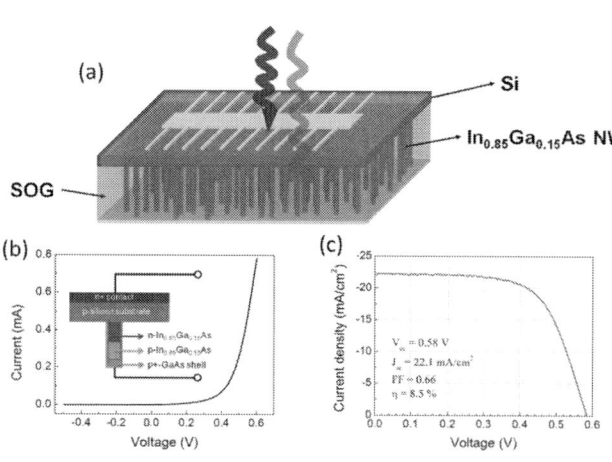

Figure 4. Tandem photovoltaic device with p-n $In_{0.85}Ga_{0.15}As$ NW array (sidewalls passivated with p+-GaAs) on p-n Si (111) substrate. Note the In% in the NW after fabrication is a, schematic illustration of a fully fabricated tandem photovoltaic device using p-n InGaAs NWs grown on p-n Si substrate. b, I-V characteristics of the p^+-p-n-p-n^+ structure in dark at room temperature. c, I-V characteristics of the photovoltaic device under AM 1.5 spectrum illumination.

978-1-61284-243-1/11 $26.00 © 2011 IEEE

3D Simulation of Electrical Characteristic Fluctuation Induced by Interface Traps at Si/high-κ Oxide Interface and Random Dopants in 16-nm-Gate CMOS Devices

Hui-Wen Cheng, Yung-Yueh Chiu and Yiming Li[*]

Department of Electrical Engineering, National Chiao Tung University, 1001 Ta-Hsueh Road, Hsinchu 300, Taiwan
[*]Corresponding author. Phone: +886-3-571-2121 ext. 52974; Fax: +886-3-5726639; E-mail address: ymli@faculty.nctu.edu.tw

The random dopant (RD)-induced threshold voltage fluctuation (σV_{th}) was explored recently [1-4]. RD fluctuation (RDF) has been one of challenges in nano-CMOS technologies; consequently, high-κ/metal gate (HKMG) approach is adopted to suppress intrinsic parameter fluctuation and leakage current for sub-45-nm generations. However, random interface traps (ITs) appearing at Si/high-κ oxide interface results in a new fluctuation source [2]. Effects of ITs and RDs on electrical characteristic fluctuation have been explored yet. In this work, we study influences of random ITs and RDs on 16-nm CMOS devices using an experimentally calibrated 3D device simulation [1-4]. Devices with totally random ITs, RDs, and "ITs+RDs" (i.e., 3D device simulation with considering random ITs and RDs simultaneously) are generated and simulated to assess the device variability.

The validated performance of studied HKMG devices, according to ITRS roadmap for low operating power, is experimentally quantified in our recent study [4]. The threshold voltages of 16-nm N-and P- MOSFETs are 250 and -250 mV, respectively. The devices we examined are the 16-nm N- and P-MOSFETs (width: 16 nm) with amorphous-based TiN/HfO_2 gate stacks, as shown in Fig. 1(a). For the simulation of IT fluctuation (ITF), we first generate 753 acceptor-like traps (pink dots) in a large 2D plane, as shown in Fig. 1(b), where the trap's concentration in the plane is around 1.5×10^{12} cm^{-2} based on experimental characterization, and the total number of traps follows the Poisson distribution. Then, the whole plane (with the statistically generated random ITs) is partitioned into many sub-planes, where the number of traps in each sub-plane (area: 16 x 16 nm^2) may vary from 0 to 8 and the average is 4. Density of ITs follows random ITs' energy [5-9], where the density of ITs varying from 5×10^{11} to 5×10^{12} cm^{-2}. The procedure above is repeated until all sub-regions are assigned at Si/HfO_2 oxide interface. Notably the approach enables us to examine ITs' influence simultaneously capturing the random traps' position and number effects over 2D interface. Second, the simulation of RDF follows the reported technique in [1-4], as shown in Fig. 1(c). For the device simulation with "ITs+RDs", we consider random ITs and RDs at the same time. Fig. 2(a) shows that the drain current (I_D) versus the gate voltage (V_G) for N- and P-MOSFETs, respectively, in which the solid line shows the nominal case with channel doping of 1.48×10^{18}-cm^{-3} and work function of 4.52 eV, the dashed lines are fluctuated cases, and the symbol line shows averaged result. The P-MOSFETs possess smaller σV_{th} than that of N-MOSFETs. "ITs+RDs"-induced current fluctuation diminishes as the screen effect occurs for both N- and P-MOSFETs. Selecting from one of simulated transistors (total 216 3D transistors are simulated), Fig. 2(b) shows the off-state (EOT = 0.8 nm, $V_G = 0$ V and $V_D = 0.8$ V) potential distributions and the on-state ($V_G = V_D = 0.8$ V) current densities for the device with "ITs+RDs". The surface potential results from two strongly interacted local spikes which are 0.11 and 0.12 eV induced by RD and IT, respectively. Cutting from the source (S) to drain (D), the 1D shapes of potential passing through the peaks of spike are also shown. Note the range of localized spikes in Fig. 2(b) is broadened owing to potential's short-range interaction resulting from the "ITs+RDs". Consequently, the whirlpool-like on-state current density spreads apart from S to D, as shown in Fig. 2(b). For two random variables $V_{th,ITs}$ and $V_{th,RDs}$, we have observed that $\sigma V_{th,ITs+RDs} < (\sigma^2 V_{th,ITs} + \sigma^2 V_{th,RDs})^{0.5}$ for N- and P-MOSFETs because $V_{th,ITs}$ and $V_{th,RDs}$ are not independent and identically distributed random variables. Thus, their statistical sum $(\sigma^2 V_{th,ITs} + \sigma^2 V_{th,RDs})^{0.5}$ overestimates the random "ITs+RDs"-induced V_{th} fluctuations. The black and grey bars are for N- and P-MOSFETs, respectively, for example, $\sigma V_{th,ITs+RDs} = 46.4$ mV and $(\sigma^2 V_{th,ITs} + \sigma^2 V_{th,RDs})^{0.5} = 50.4$ mV, and the correlation coefficients is equal to -0.171 for N-MOSFET, as shown in Fig. 2(c). Notably, not only RDs [1-4], but also the random ITs have random position and number effects on device characteristics. Fig. 3(a) shows I_{on}-I_{off} plot, Figs. 3(b)-3(e) disclose four different discrete "ITs+RDs" channels that have similar values of I_{on} or I_{off} but with various "ITs+RDs" distributions, where their corresponding cross-sectional on-state current density and off-state electrostatic potential at 1 nm below the gate oxide are presented, as shown in Figs. 3(b')-3(c') and 3(d')-3(e').

In summary, we have studied ITs and RDs of 16 nm N- and P-MOSFETs. Totally random devices with 2D ITs at Si/high-κ oxide interface and 3D RDs inside channel were examined. ITs and RDs result in localized spikes of potential barrier and induced characteristics are correlated to each other. The results of this study indicate that effect of random ITs and RDs on device variability should be counted simultaneously for HKMG CMOS devices.

This work was supported in part by National Science Council, Taiwan under Contract No. NSC-99-2221-E-009-175 and by TSMC, Hsinchu, Taiwan under a 2010-2011 grant.

[1] Y. Li, C.-H. Hwang, T.-C. Yeh, and M.-H. Han, "Simulation of Electrical Characteristic Fluctuation in 16-nm FinFETs and Circuits," in: *Proc. DRC*, 2009, pp. 139-140.

[2] H.-W. Cheng, F.-H. Li, M.-H. Han, C.-Y. Yiu, C.-H. Yu, K.-F. Lee, and Y. Li, "3D Device Simulation of Work-Function and Interface Trap Fluctuations on High-κ/Metal Gate Devices," in: *IEDM Tech. Dig.*, 2010, pp. 379-382.

[3] Y. Li, C.-H. Hwang, T.-Y. Li, and M.-H. Han, "Process-Variation effect, Metal-Gate Work-Function fluctuation, and random-dopant fluctuation in emerging CMOS technologies," *IEEE Trans. Electron Devices*, vol. 57, pp. 437-447, 2010.

[4] Y. Li, S.-M. Yu, J.-R. Hwang, and F.-L. Yang, "Discrete dopant fluctuations in 20-nm/15-nm-gate Planar CMOS," *IEEE Trans. Electron Devices*, vol. 55, pp. 1449-1455, 2008.

[5] G. Panagopoulos and K. Roy, "A physical 3-D analytical model for the threshold voltage considering RDF," in: *Proc. DRC*, 2009, pp. 81-82.

[6] P. Andricciola, H.P. Tuinhout, B. De Vries, N.A.H. Wils, A.J. Scholten, and D.B.M. Klaassen, "Impact of interface states on MOS transistor mismatch," in: *IEDM Tech. Dig.*, 2009, pp. 711-714.

[7] A. Appaswamy, P. Chakraborty, and J. Cressler, "Influence of Interface Traps on the Temperature Sensitivity of MOSFET Drain-Current Variations," *IEEE Electron Device Lett.*, vol. 31, pp. 387-389, 2010.

[8] T.G. Pribicko, J.P. Campbell, and P.M. Lenahan, "Interface defects in Si/HfO_2-based metal-oxide-semiconductor field-effect transistors," *Appl. Phys. Lett.*, vol. 86, p. 173511, 2005.

[9] M. F. Bukhori, S. Roy, and A. Asenov, "Simulation of Statistical Aspects of Charge Trapping and Related Degradation in Bulk MOSFETs in the Presence of Random Discrete Dopants," *IEEE Trans. Electron Devices*, vol. 57, pp. 795-803, 2010.

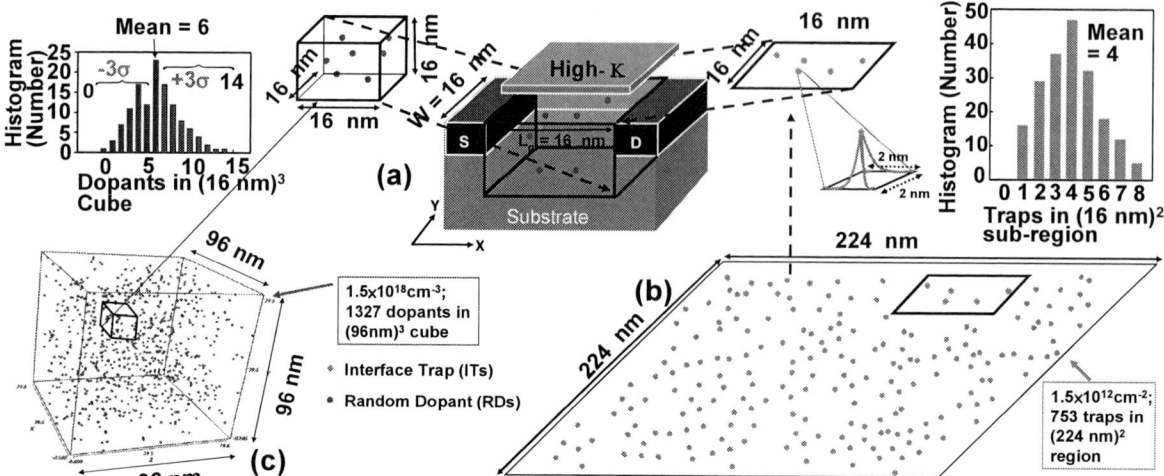

Fig. 1. (a) The source of randomness (pink dots are interface traps and blue dots are discrete dopants) and simulation settings for fluctuations of random ITs and RDs. (b) We first generate 753 acceptor-like traps in a large plane, where the trap's concentration in the plane is around 1.5×10^{12} cm^{-2} and the total number of generated traps follow the Poisson distribution. The energy of each trap on the plane is assigned according to distribution of trap's density. Then the entire plane is partitioned into sub-planes (size: 16 nm × 16 nm), where the number of traps in all sub-planes may vary from 1 to 8 and the average number is 4. (c) Discrete dopants randomly distributed in (96 nm)^3 cube with the average concentration of 1.5×10^{18} cm^{-3}. There will be 1327 dopants within the cube and dopants vary from 0 to 14 (the average number is 6) for all 216 sub-cubes. The size of each sub-cube is (16 nm)^3. The total sub-cubes and sub-planes are then mapped into device's 3D channel and 2D surface for RDs and ITs' position/number-sensitive simulation (a).

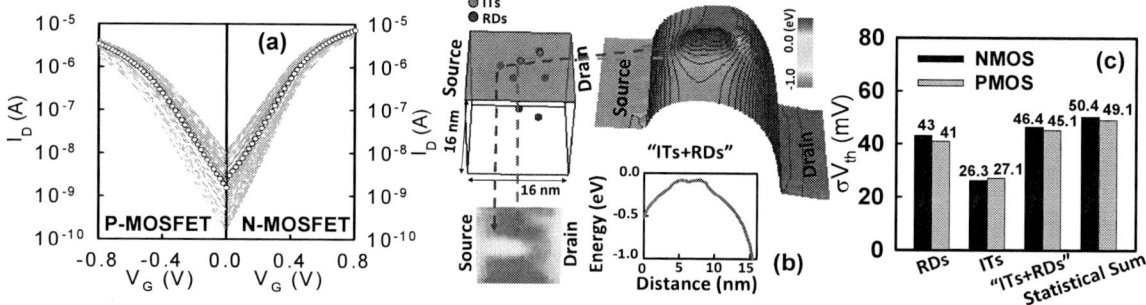

Fig. 2. (a) I-V characteristic for both N- and P-MOSFETs, respectively, in which the solid line shows the nominal case with channel doping of 1.48×10^{18} cm^{-3} and work function of 4.52 eV, the dashed lines are fluctuated cases, and the symbol line shows averaged result (b) The on-state ($V_G = V_D = 0.8$ V) current density and the off-state ($V_G = 0$ V and $V_D = 0.8$ V) potential distribution of the channel surface from one of simulated 16 nm transistors, where EOT = 0.8 nm. The device fluctuated by 8 random "ITs+RDs" simultaneously. The interactions of ITs and RDs on the band profile and current density are also shown, measuring from the source (S) to drain (D). (c) σV_{th} calculated from devices with RDs, ITs, "ITs+RDs" and statistical sum: $(\sigma^2 V_{th,ITs} + \sigma^2 V_{th,RDs})^{0.5}$.

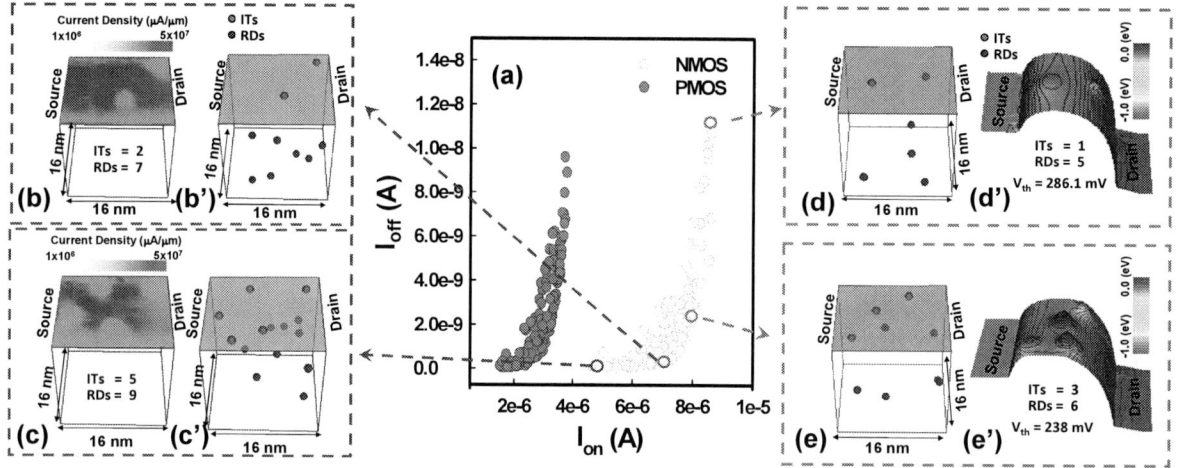

Fig. 3. (a) I_{on}-I_{off} plot of the total simulated 16-nm-gate transistors with "ITs+RDs" ($L_g = W = 16$ nm). (b) and (c) represent two cases of simulated devices with similar I_{on} but different I_{off}; (d) and (e) represent two cases of simulated devices with similar I_{off} but different I_{on}. The corresponding on-state current density and off-state potential contours of (b), (c), (d) and (e) are shown in (b'), (c'), (d') and (d'), respectively.

978-1-61284-243-1/11 $26.00 © 2011 IEEE

Creating dynamic nanowire devices using wrapped gates

Kristian Storm, Gustav Nylund, Magnus Borgström, Jesper Wallentin, Carina Fasth, Claes Thelander and
Lars Samuelson

Solid State Physics, the Nanometer Structure Consortium, Lund University, Sweden

Box 118, S-221 00, Sweden, Kristian.storm@ftf.lth.se

Semiconducting nanowires (NWs) constitute an interesting platform as building blocks for various types of devices as well as for studies of fundamental material transport properties in one dimension. Nanowires have many interesting properties, such as the ability to incorporate strongly lattice-mismatched material combinations along its axis due to radial relaxation of interface strain. Furthermore, its inherent cylindrical geometry makes it an ideal candidate for devices implementing gates wrapped around the nanowire channel; the optimal geometry for maximum gate to channel coupling.

Most conventional semiconductor devices are based upon doping for its operation. As device dimensions are decreased, the random position of a few incorporated impurity atoms may dominate device characteristics. Furthermore, the amount of surface in relation to the material volume increases, causing surface effects to be extremely important for nanoscopic devices. Since the Fermi level is often pinned at the material surface, this effect may also render doping less effective compared to bulk material.

Because of these problems of decreased doping effectiveness on the nanoscale, we are pursuing other means to control the potential landscape in nanoscopic systems than by doping alone. Nanowires provide an excellent system in which multiple sequential and fully wrapped gates along the nanowire axis may be used to induce various device behaviors. It may e.g. be possible to induce devices such as light-emitting diodes or Esaki diodes by gates alone, replacing doping in certain types of devices. Using gates to induce different device behaviors also enables dynamic control of material properties such as carrier concentration, as opposed to conventional devices in which the device properties are statically set by the doping levels.

In this work, we utilize InP nanowires to investigate how a wrapped gate may be used to tune the Fermi level across the bandgap of the semiconductor. We have previously demonstrated long-channel InP NW FETs in which it was possible to tune the Fermi level with efficiency near the thermal limit [1]. This was done using laterally placed, omega-gated p- and n-type NWs (Fig. 1), where the n-type devices exhibited a subthreshold slope below 70 mV/dec and the p-type devices showed ambipolar behavior.

Due to the promising results of the lateral InP NW transistors, dual-gated devices incorporating nominally intrinsic InP NWs were fabricated (Fig. 2), with each gate covering roughly half of the NW. By implementing two separate gates, sequential n- and p-type regions may be induced in a nominally intrinsic nanowire, and diode-like behavior is observed. It is clear that by varying the voltage on one of the gates, while keeping the other at a fixed potential, the device can be switched from ohmic to rectifying, with diode characteristics. The device exhibits an ideality factor of ~3 in its rectifying mode. This serves as the first step towards producing dynamic light emitting diodes, where the n- and p-type regions are induced and individually tuned by external gates. By incorporating a graded bandgap material between the two gates, it may also be possible to tune the wavelength of the emitted light by changing the voltage configuration of the gates.

Lateral single NW devices are interesting for proof of concept measurements, but in order to increase the current levels and investigate how the device properties scale with the number of NWs, the next step is to create vertical devices. Matrices of either 100 or 400 InP NWs were fabricated into vertical n-type field-effect transistors (Fig. 3) and their scaling properties investigated. The NWs were grown n-type with a short intrinsic segment defining the channel. Devices with four different channel lengths were fabricated (25nm, 50nm, 100nm and 200nm), and the nanowires in each device had diameters varying between 40 nm and 50 nm. By varying the NW diameter for each channel length, scaling properties could be mapped out and investigated. A comparison of the 50 nm channel with the 200 nm channel is shown in Fig. 3. It is clear that the subthreshold slope increases for decreasing channel length (see also inset of Fig. 3), and the channel control is lost for the 25 nm channel length, i.e. when the channel becomes small compared to the NW diameter, in accordance with MOSFET scaling theory.

[1] K. Storm, G. Nylund, M. Borgström, J. Wallentin, C. Fasth, C. Thelander and L. Samuelson,
Accepted to Nano Letters, DOI: dx.doi.org/10.1021/nl104032s

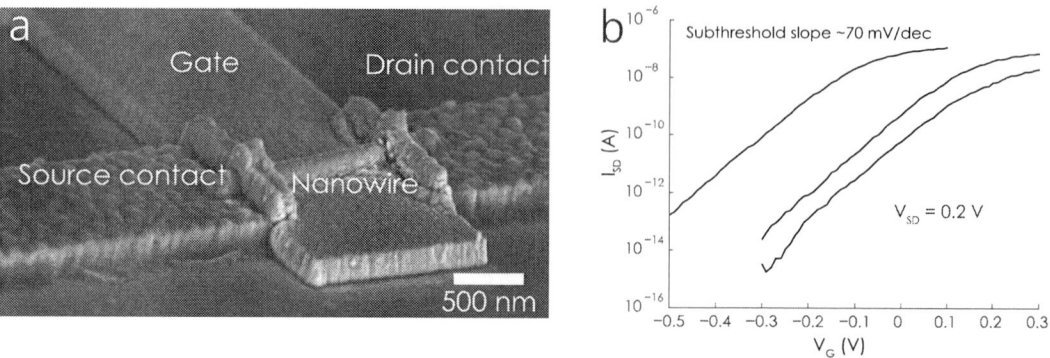

Fig 1 Lateral InP nanowire FETs. Scanning electron micrograph (a) of one device and transfer characteristics (b) of three typical devices.

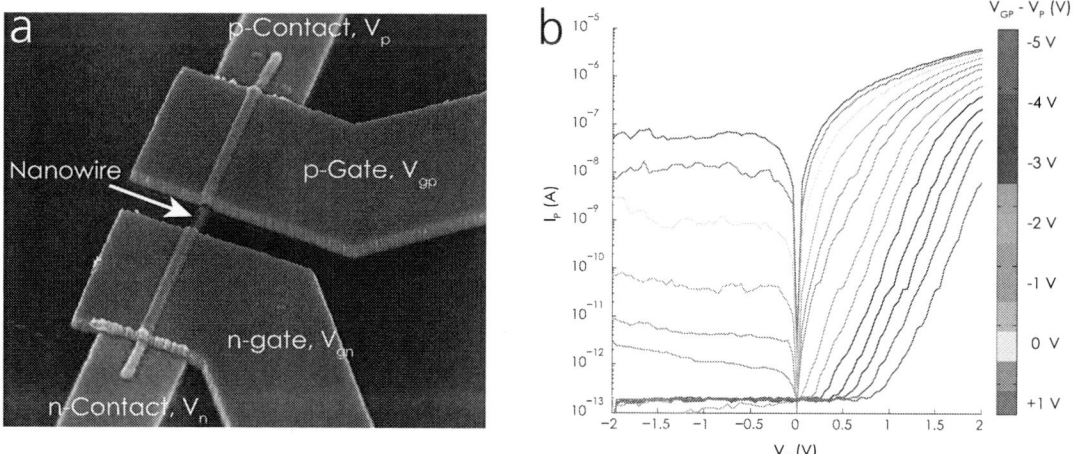

Fig 2 Gate-induced rectifying diode. A scanning electron micrograph of the dual-gated device is shown in (a) and its current-voltage characteristics in (b).

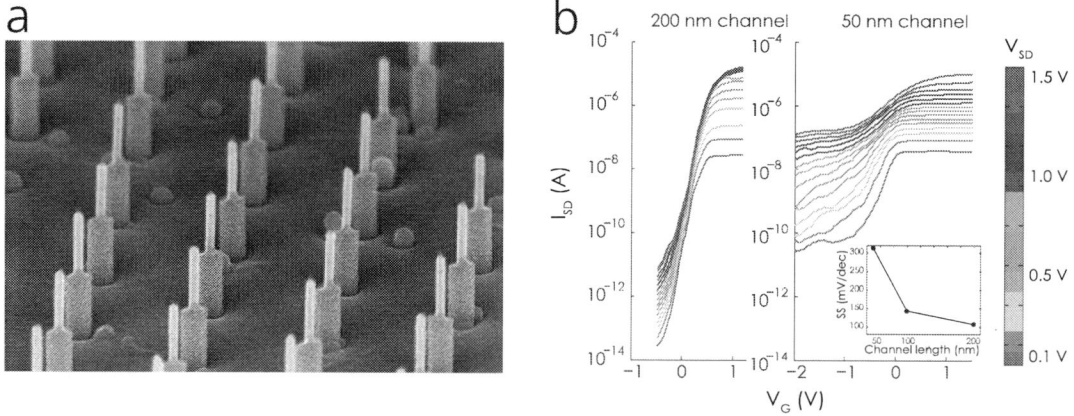

Fig 3 Vertical InP nanowire wrap-gate transistors. A scanning electron micrograph of the device (a) before the top contact is deposited, together with a comparison of the transfer characteristics of two different channel lengths (b). The inset of (b) shows the scaling behavior of the subthreshold slope as the channel length is varied.

978-1-61284-243-1/11 $26.00 © 2011 IEEE   106

InAlAs/InGaAs Metamorphic HEMT and MOS-HEMT with Regrown Source/Drain by MOCVD

Xiuju ZHOU, Qiang LI, Kei May LAU*

Department of Electronic and Computer Engineering,
Hong Kong University of Science and Technology, Clear Water Bay, Kowloon, Hong Kong.
Tel: (852) 2358-7049, Fax: (852) 2358-1485, Email: eekmlau@ust.hk

As scaling technologies are being stretched harder and harder in the roadmap of Si based CMOS, III-V compounds have become competitive alternative channel materials for the next generation high speed and low power transistors. Among various device structures, InGaAs HEMT has been intensively researched in the past few years because of its excellent carrier transport properties [1-3]. However, conventional HEMT structures requiring recessed gate technology may be difficult for digital VLSI applications due to their large footprint and higher parasitic capacitances [4]. Moreover, the gate recess process raises serious concerns in threshold voltage uniformity caused by variations in recess etching depth [5]. Selective Source/Drain (S/D) regrowth, which has been implemented in advanced Si pMOSFET, is an easier and scalable approach to facilitate ohmic contact in HEMT structures, with the benefits of eliminating reliability issues related to gate recess and parasitic reduction. In this paper, we describe the process and preliminary device results of metamorphic HEMTs (mHEMTs) and MOS-HEMTs on GaAs substrates with highly doped $In_{0.53}Ga_{0.47}As$ S/D by selective regrowth using MOCVD.

InAlAs/InGaAs metamorphic HEMT structures were grown on (100) oriented GaAs substrates in an Aixtron AIX-200/4 MOCVD system. Fig.1 shows the layered structure. From Hall measurements, an electron mobility of $7230 cm^2/V \cdot s$ with a sheet carrier density of $3.9 \times 10^{12}/cm^2$ at 300K was obtained. After the HEMT structure growth, a 1000 Å SiO_2 layer was used to pattern regions for S/D recesses etching down to the InGaAs channel layer. The sample was then loaded into the MOCVD system for $In_{0.53}Ga_{0.47}As$ regrowth in the etched regions at $670°C$. Good selectivity was achieved. The SiO_2 mask was removed by BOE subsequently. An AFM image in the regrowth region is given in Fig.2, showing a rms value of 1.0 nm over a scanned area of 3×3 μm^2.

Both metamorphic HEMTs and MOS-HEMTs featuring regrown S/D were fabricated. Fig.3 lists the major process flow. Firstly, mesa isolation was formed by wet etching down to the InAlAs buffer. A 12nm thick Al_2O_3 was deposited by ALD for the MOS-HEMT sample after immediate pre-treatment using $HCl:H_2O$ (1:10) for 3mins. Non-alloyed S/D ohmic contacts were formed using a six-layer metal scheme (Ni/Ge/Au/Ge/Ni/Au). Finally, gate electrodes were defined by electron beam evaporation of Ti/Pt/Au and lift-off. Fig.4 illustrates the cross-sectional schematic of the devices after processing. From TLM measurements, a low specific contact resistivity of 1×10^{-6} $\Omega \cdot cm^2$ was achieved for the non-alloyed S/D ohmic contacts. Fig.5 and Fig.6 show the output and transfer characteristics, respectively. 1-μm gate-length HEMT exhibits threshold voltage $V_T = -0.25$ V, maximum drain current $I_{dss} = 168$ mA/mm, and extrinsic peak transconductance $G_{max} = 302$ mS/mm, while the MOS-HEMT shows $V_T = -3.8V$, $I_{dss} = 186$ mA/mm, and $G_{max} = 76$ mS/mm. Fig.7 depicts gate leakage characteristics of both devices. The gate leakage for MOS-HEMT is five orders of magnitude lower compared with HEMT. Fig.8 illustrates the multi-frequency Capacitance-Voltage(C-V) response of MOS-HEMT. The sharp transition from accumulation to depletion region and the small frequency dispersion in the accumulation region indicate good Al_2O_3/InAlAs interface quality.

The DC performance of both HEMT and MOS-HEMT with regrown S/D is believed to be limited by the large Gate-to-Source separation L_{GS} (1.5μm) and Gate-to-Drain separation L_{GD} (1.5μm). In conventional HEMT structure, as shown in Fig.9, the access resistance in S/D-to-Gate region is dominated by the highly conductive n-InGaAs cap layer, which is small enough compared with the intrinsic channel resistance. However, for HEMT and MOS-HEMT featuring regrown S/D described in Fig.4, the current flows through 2DEG in S/D-to-Gate region, which results in a much larger access resistance. By minimizing L_{GS} and L_{GD}, and further thinning the InAlAs barrier and ALD-Al_2O_3, improved performance of the regrown devices is expected.

Reference:
[1] D. H. Kim and J. A. del Alamo, IEDM Tech. Dig., 2006, pp. 837-840.
[2] D.-H. Kim and J. del Alamo, IEEE Trans. Electron Devices, vol. 55, no. 10, pp. 2546–2553, Oct. 2008.
[3] T.-W. Kim, D.-H. Kim, and J. del Alamo, IEDM Tech. Dig., 2009, pp. 483–486.
[4] S. Oktyabrsky and P. D. Ye, Fundamentals of III-V Semiconductor MOSFETs. New York: Springer, 2010.
[5] Iain Thayne et al, Electrochem. Soc. Transaction, vol.25, no.7, pp.385-389. 2009.

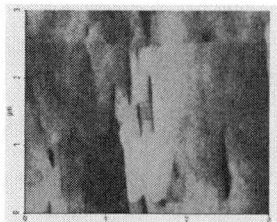

Undoped In$_{0.42}$Al$_{0.58}$As, 20nm.	Barrier
Si δ-doping, 4×10^{12}/cm^2	Delta doping
Undoped In$_{0.42}$Al$_{0.58}$As, 5nm.	Spacer
Undoped In$_{0.6}$Ga$_{0.4}$As, 30nm	Channel
Undoped HT- In$_{0.42}$Al$_{0.58}$As, 100nm	Buffer 5
Undoped LT- In$_{0.45}$Al$_{0.55}$As, 200nm	Buffer 4
Undoped HT- InP, 650nm	Buffer 3
Undoped LT- InP, 110nm	Buffer 2
Undoped GaAs, 100nm	Buffer 1
Semi-Insulating GaAs substrate	

Fig.1. mHEMT epitaxial layer structure

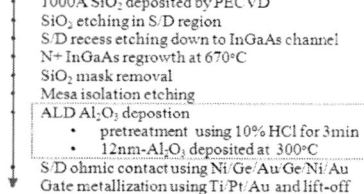

Fig.2. AFM image of the regrowth S/D

1000Å SiO$_2$ deposited by PECVD
SiO$_2$ etching in S/D region
S/D recess etching down to InGaAs channel
N+ InGaAs regrowth at 670°C
SiO$_2$ mask removal
Mesa isolation etching
ALD Al$_2$O$_3$ deposition
  • pretreatment using 10% HCl for 3min
  • 12nm-Al$_2$O$_3$ deposited at 300°C
S/D ohmic contact using Ni/Ge/Au/Ge/Ni/Au
Gate metallization using Ti/Pt/Au and lift-off

Fig.3. Process flow for mHEMT and MOS-HEMT. The only difference is that there is no ALD step in mHEMT

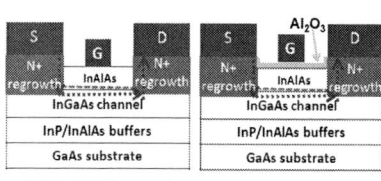

Fig.4. Schematic of mHEMT and MOS-HEMT. (blue dashed line: simplified drain current path)

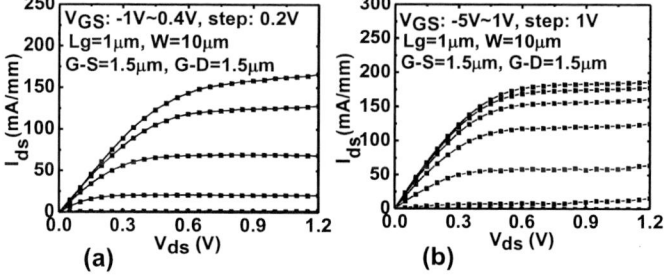

Fig.5. Output characteristics of mHEMT **(a)** and MOS-HEMT **(b)** with regrown S/D

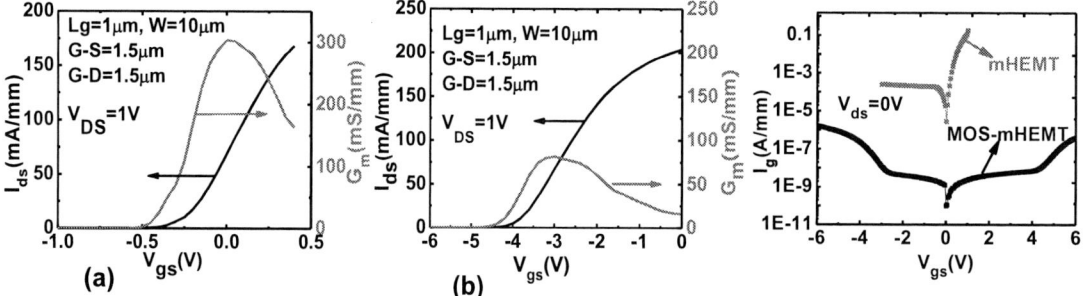

Fig.6. Transfer characteristics of mHEMT **(a)** and MOS-HEMT **(b)** with S/D regrowth

Fig.7. Gate leakage characteristics of mHEMT and MOS-HEMT

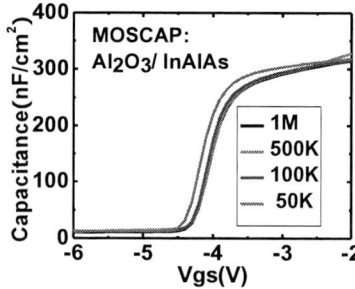

Fig.8. C-V curve of Al$_2$O$_3$/InAlAs interface for MOS-HEMT with regrown S/D

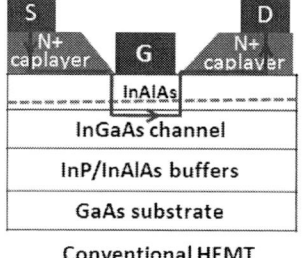

Conventional HEMT

Fig.9. Schematic of conventional HEMT (blue lines: simplified drain current path)

978-1-61284-243-1/11 $26.00 © 2011 IEEE

Numerical Study of Electronic Transport Through Bilayer Graphene Nanoribbons

K. M. Masum Habib[*] and Roger K. Lake

Department of Electrical Engineering, University of California, Riverside, CA 92521-0204
[*]Phone: (951) 827 4515, Email: khabib@ee.ucr.edu

Abstract – In graphene, a sheet of carbon atoms arranged in a honeycomb structure, charge carriers behave as massless Dirac fermions and move with extremely high speed leading to exotic electronic properties. However, lack of a band-gap reduces its utility for conventional electronic device applications. A tunable bandgap can be induced in bilayer graphene by application of a potential difference between the two layers.

Many proposed graphene based devices have multiple gates making them relatively complex device structures. In this abstract, we report a numerical study of electronic transport through the simplest possible geometry consisting of two graphene nanorobbons (GNRs). Each of our model geometry consists of two 14-C atomic layer wide armchair type GNRs (14-AGNRs) with one placed on top of the other. We consider two arrangements. In Scheme-I, GNRs are placed in a straight line alignment with an AB–stacked overlap, and in Scheme-II, GNRs are placed at right angles forming a crossbar. Each GNR is then independently contacted such that one is held at ground while the other has a bias applied to it.

The effects of bias on electronic structure of AB–stacked bilayer GNR was studied using semi-empirical π-bond tight binding model where the intra-layer coupling (-2.66 eV) and the inter-layer coupling (-0.266 eV) parameters were taken from Ref[1]. The applied bias was modeled by shifting the energy of the biased GNR by $U = -eV$ amount where, V is the applied bias voltage. It has been found that the bandgap initially increases with bias and then saturates. The maximum bandgap for 14-AGNR AB bilayer has been found to be ~ 0.25 eV. Since the bandgap increases with applied bias, we hypothesized that negative differential resistance (NDR) would occur in Scheme-I.

To test this hypothesis, we used a π-bond tight binding model coupled with the non-equilibrium Green's function (NEGF) formalism to study the electronic transport. The simulated current voltage $(I - V)$ characteristics indeed demonstrated NDR confirming our initial hypothesis. The peak and valley currents are $25\mu A$ and $1.38\mu A$ at 0.5 V and 1.1 V bias voltages respectively with a peak-to-valley currents ratio of ~ 18. To understand the $I - V$ characteristics the transmission coefficients are plotted as a function of electron energy. In agreement with and as discussed in [1,2], the transmission shows a Fabry-Perot resonant feature at low energy and both resonances and antiresonances at more excited energies.

The coherent current at any bias is proportional to the area under the transmission curve bounded by the chemical potential of the contacts. Initially, with the increasing bias, the transmission coefficient within the chemical potential energy window decreases which in turn results in NDR. As the bias of the bottom layer is increased to 1.1 V, the the transmission from hole states to electron states between 0 and -1.1 eV is strongly suppressed due to the large wavevector mismatch[2,3] of the states inside the top and bottom contacts leading to the valley current. Beyond 1.1 V bias, the transmission within the energy window begins to increase as the first excited subbands of the top and bottom GNR leads are pulled into the energy window, and the current begins to increase.

Encouraged from the results of the AB bilayer (Scheme-I), we considered the crossbar (Scheme-II). As the GNRs in a crossbar geometry can not be placed either in AA– or in AB– stacking sequences, the simple π-bond tight binding model is no longer applicable. Calculations, based on *ab-initio* density functional theory (DFT) and NEGF reveal that NDR also occurs in crossbar. In this case, however, two peak and two valley currents are found. The first peak and valley currents, 0.94 μA and 0.07 μA, occur at 0.2 V and 0.4 V respectively and the second peak and valley currents, 1.33 μA and 0.19 μA, occur at 0.6 V and 0.8 V respectively.

In conclusion, we have performed π-bond, *ab-initio* DFT, and NEGF based calculations to study the $I - V$ characteristics of two bilayer GNR structures where bias is applied between the GNRs by independently contacting each layer. It has been found that both the AB bilayer and the crossbar structures exhibit NDR and thus have potential for use in a crossbar architecture.

---

[1] J. W. González, H. Santos, M. Pacheco, L. Chico, and L. Brey, Phys. Rev. B **81**, 195406 (2010).
[2] J. W. González, H. Santos, E. Prada, L. Brey, and L. Chico, arXiv:1008.3255v1[cond-mat.mes-hall] (2010).
[3] V. N. Do and P. Dollfus, J. Appl. Phys. **107**, 063705 (2010).

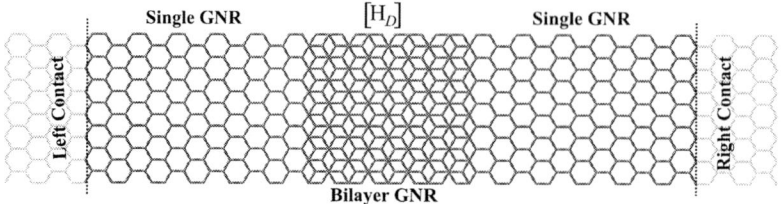

Figure 1: *Scheme-I* – Atomistic geometry of the bilayer graphene nanoribbon (GNR) model structure. Two overlying armchair type GNRs are placed in AB stacking sequence to form the bilayer region. The extended parts of the left and right GNRs are used as contacts. A bias is applied by independently contacting each GNR such that the left GNR is held at ground while the right has a potential applied to it. The electronic properties and transport are calculated using semi-empirical tight binding π-bond model and the non-equilibrium Green's function (NEGF) formalism respectively.

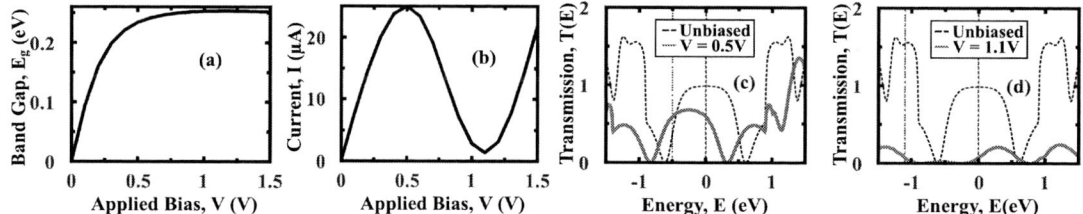

Figure 2: *Scheme-I results* – (a) Bandgap as a function of applied bias for infinite AB–stacked bilayer GNR. (b) Simulated current voltage ($I - V$) characteristics of the bilayer structure exhibiting negative differential resistance. The peak and valley currents occur at 0.5 V and 1.1 V respectively. Transmission as a function of energy for the bilayer structure: (c) at bias, $V = 0.5$ V and (d) at bias, $V = 1.1$V, superimposed on transmission at no bias. The vertical lines at the lower and upper energies represent the chemical potentials of right and left contacts, respectively. The chemical potential of the left contact is set at 0.

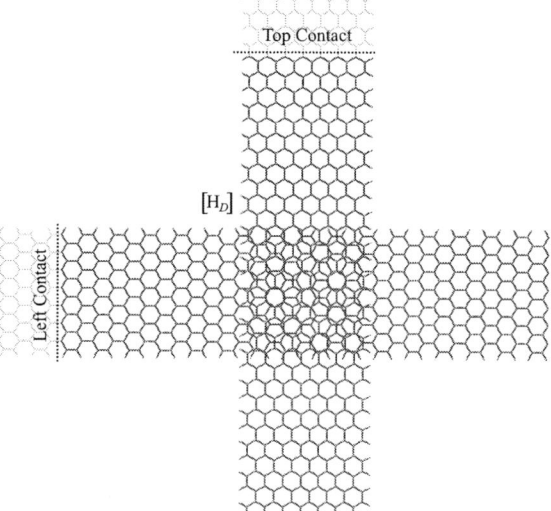

Figure 3: *Scheme-II* – Atomistic geometry of the GNR crossbar. Two hydrogen passivated relaxed armchair type GNRs are placed on top of each other at a right angle with a vertical separation of 3.35 Å. The relaxation was done using *ab-initio* density functional theory (DFT) code, FIREBALL. The extended parts of the GNRs are used as contacts. A bias is applied by independently contacting each GNR such that one is held at ground while the other has a potential applied to it. Calculations based on first principle DFT and the non-equilibrium Green's function (NEGF) formalism show that negative differential resistance occurs in this structure revealing potentials for use in a crossbar architecture.

978-1-61284-243-1/11 $26.00 © 2011 IEEE

Tunnel-FET Architecture with Improved Performance due to Enhanced Gate Modulation of the Tunneling Barrier

L. De Michielis[1], L. Lattanzio[1], P. Palestri[2], L. Selmi[2], A. M. Ionescu[1]

[1]: Nanoelectronic Devices Laboratory (Nanolab), Ecole Polytechnique Fédérale de Lausanne, EPFL, 1015, Switzerland
[2]: Department of Electrical, Managerial and Mechanical Engineering, University of Udine, 33100, Italy
Phone: +41 21 693 69 72; Fax: +41 21 693 36 40; e-mail: Luca.DeMichielis@epfl.ch

The Tunnel-FET (TFET) device is a gated reverse biased p-i-n junction whose working principle is based on the quantum mechanical Band-to-Band Tunneling (B2BT) mechanism [1]. The OFF-ON transition can be much more abrupt than for conventional MOSFETs, thus allowing a reduction of the supply voltage and power consumption in logic applications [2]. Several TFETs with *point* Subthreshold Swing (SS) lower than 60mV/dec have been experimentally demonstrated with different architectures as conventional single gate Silicon-on-Insulator (SOI), Double Gate (DG) and Gate-All-Around (GAA) [3,4]. Unfortunately in all cases a relatively large *average* SS and a poor on-current have been observed.

To trace back the cause of this behavior, we look at the potential distribution of a conventional n-TFET operating in the subthreshold regime (see Fig. 1-left), where the black lines are the equipotential surfaces at the tunneling region. Although the tunneling region (i.e. the source/channel junction) has been designed in order to be parallel to the transport direction, the potential coming from the gate is modifying the electrostatic configuration. For this reason, in a tunneling-path based picture of the B2BT mechanism, as the one implemented in TCAD tools, the electrons in the source region that are about to tunnel from the valence band to the conduction band are experiencing a force such that the resulting tunneling path is not parallel to the source-drain direction. In the specific case of Fig. 1, the direction of the tunneling path (assumed by the TCAD [6] to be the direction of the maximum valence band gradient) at the spatial coordinate corresponding to the maximum tunneling generation of holes is deflected by 40 degrees compared to the S/D direction. Then, as the gate voltage is increased and the device enters the on-state regime, the vertical component of the electric field increases and the direction becomes even more vertical.

In a well-designed TFET, the gate should effectively modulate the tunneling barrier width. It is clear that in order to fulfill this requirement the tunneling junction must be designed having the same orientation of the component of the electric field modulated by the gate: only in this case the optimal control of the carrier injection mechanism will assure the best modulation on the drain current levels.

Different TFET architectures based on this principle can be implemented and attempts in this direction with a tilted source junction have been demonstrated recently [5]. In this work we propose the new structure reported in Fig.1-right, (i.e. an embedded p+ source region covered with an epitaxial intrinsic channel layer) and compared its performance with the conventional structures. The simulations have been performed with the latest release of the commercial simulator Sentaurus TCAD 2010.12E [6]. The dynamic non-local path B2BT model has been employed together with physical models as doping dependent Shockley–Read–Hall recombination, doping and field dependent mobility (accounting also for the velocity saturation) and density gradient corrections (to account for quantization).

Figure 2 shows the transfer and the output characteristics of the new architecture: a steep average SS and high on-current levels can be observed. Fig. 3 shows the proposed process flow for the device. Thermal activation of dopants could result in different source junction abruptness: since TFETs are extremely sensitive to the smearing of the doping on the tunneling region, we have compared in Figs. 4, 5 the effect of different doping decays corresponding to typical flash annealing techniques, without finding relevant differences in the device behavior. Figure 6 compares a conventional SOI TFET with the new structure featuring the correct alignment between the tunneling junction and the transport direction. Although the conventional TFET features a sub-60mV/dec *point* SS, this is observed only over a very small range of V_{GS}: in fact its *average* SS evaluated from 20fA/μm to 1nA/μm is 89mV/dec compared to the 38mV/dec of *average* SS of the new architecture in the same current range. Also the higher on-current level and the smaller intrinsic delay of the proposed architecture with respect to the conventional structure can be observed: both are the result of the combined action of a proper alignment of the tunneling junction and a larger area where B2BT takes place.

In conclusion with this work we have shown that although commonly fabricated TFETs feature source/channel interfaces normal to the transport direction, in a well-designed TFET the tunneling junction should have the same orientation of the component of the electric field modulated by the gate: only in this case the gate can effectively modulate the tunneling barrier, resulting in a steeper *average* SS and higher I_{ON}. An example of TFET structure fulfilling this design rule has been proposed and simulated. It provides an improvement of the average SS (234%), of the I_{ON} (x18) and of the intrinsic speed (x13) compared to the reference SOI TFET.

[1] Y. Lu et al, DRC 2010, pp 17-18
[2] S. Datta et al, SNW 2010, pp1-2
[3] K. E. Moselund et al, DRC 2009, pp 23-24
[4] T. Krishnamohan et al, IEDM 2008, pp 1-3
[5] C. Hu et al, IEDM 2010, pp387-390
[6] Synopsys Sentaurus Device user's manual

The research leading to these results has received funding from the European Community's Seventh Framework Program (FP7/2007-2013) under grant agreement n° 257267.

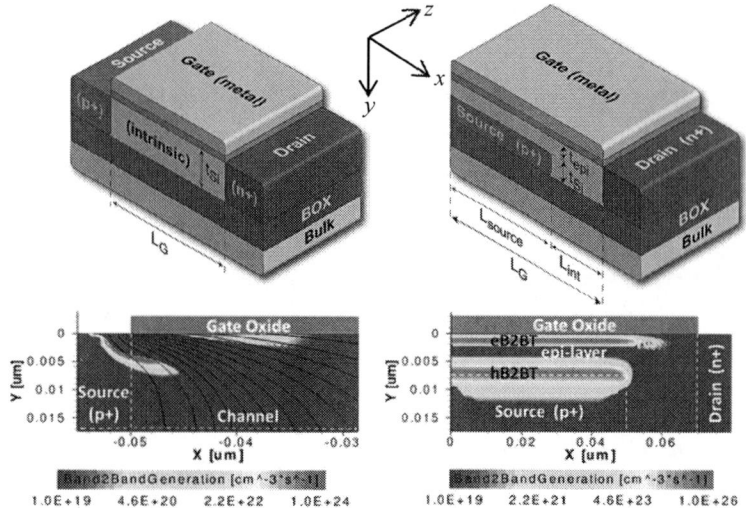

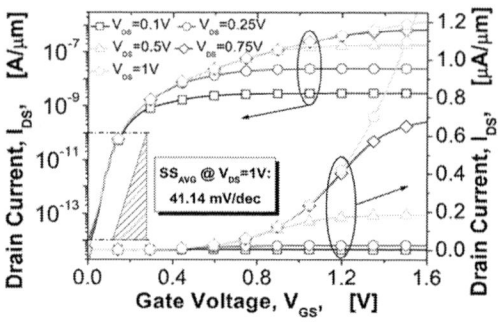

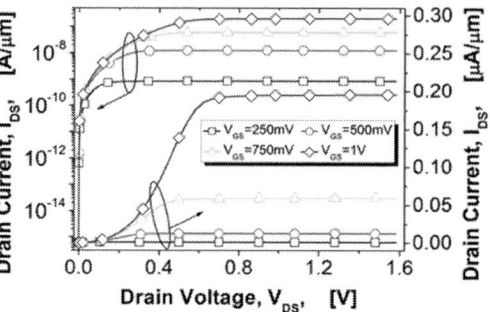

Figure 1. Top: sketch of a conventional SOI TFET with t_{Si}=20nm (left) and of the proposed structure with t_{Si}=10nm, t_{epi}=7.5nm (right). The gate workfunction is 3.9eV, the doping is $N_{A, source}$=10^{20}cm^{-3}, $N_{D, drain}$=3·10^{18}cm^{-3}, while N_D=10^{15}cm^{-3} has been used for the bulk and the intrinsic regions; all the junctions present a doping decay of 5nm/dec. 3nm of HfO$_2$ have been used as gate oxide (ε_r=22); t_{BOX}=145nm, t_{bulk}=1μm. Bottom: (left) simulation showing the e/h B2BT generation (upper/lower colored contours) and the equipotential surfaces (black lines) in a conventional SOI TFET in the subthreshold regime (V_{GS}=100mV, V_{DS}=1V). Right: B2BT generation in the proposed architecture (V_{GS}=1.6V, V_{DS}=100mV).

Figure 2. Transfer (top) and output (bottom) characteristics of the proposed TFET architecture. The gate length is L_G=100nm.

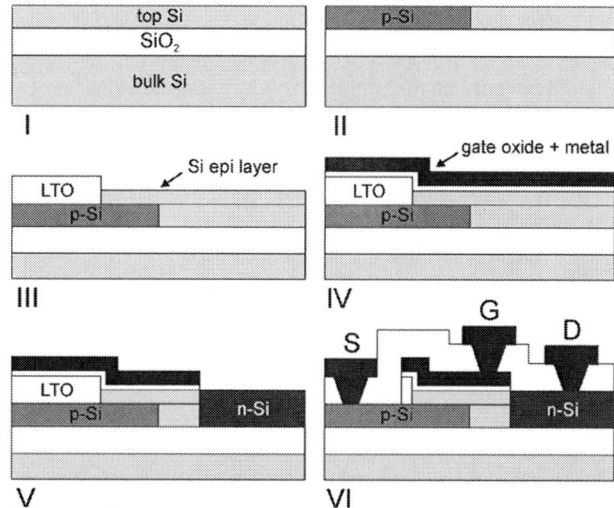

Figure 3. Proposed process flow. (I) Substrate: Silicon-on-insulator (SOI) wafer. (II) BF$_2$ ion implantation (source). (III) Low-thermal oxide (LTO) deposition + patterning and selective epitaxial growth (silicon vertical channel). (IV) Gate stack (oxide + metal) deposition. (V) Gate patterning + As ion implantation (drain) and activation (flash lamp annealing). (VI) Gate patterning + passivation and metal contacts formation (source/drain/gate).

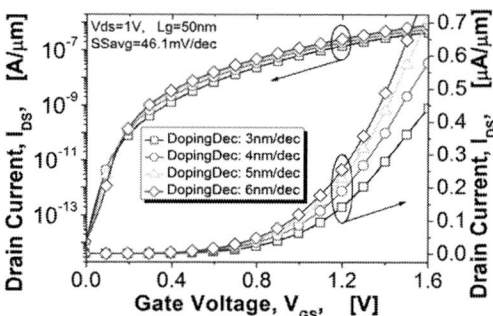

Figure 4. Transfer characteristic of the proposed TFET structure for different doping decays between the source region and the epitaxial layer.

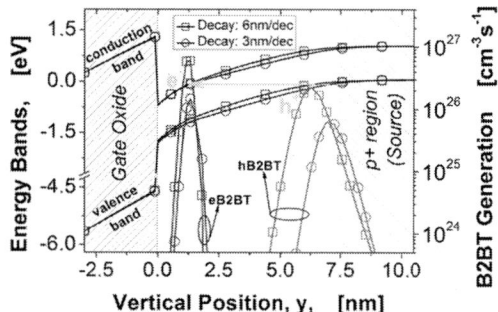

Figure 5. Band diagram and tunneling generation rates for two different doping decays at the source/epitaxial layer. V_{GS}=1.6V, V_{DS}=100mV, cut @ x=L$_{source}$/2=25nm.

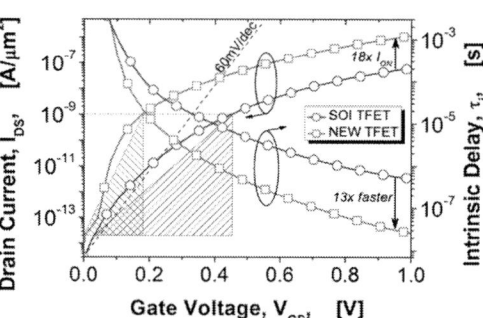

Figure 6. Comparison between a conventional SOI TFET and the proposed structure for same gate area. We observe a better average SS, higher Ion and faster device speed (smaller intrinsic delay τ_i=CV/I). V_{DS}=500mV.

978-1-61284-243-1/11 $26.00 © 2011 IEEE

Orientation dependent complex bandstructure of $Si_{1-x}Ge_x$ alloys

Arvind Ajoy[1], Kota V. R. M. Murali[2], S. Karmalkar[1], S.E. Laux[3]

[1] Indian Institute of Technology Madras, India,

[2] IBM SRDC, Bangalore, India, [3] IBM SRDC, T. J. Watson Center, USA

Phone: (91)44-22575415, Fax: (91)44-22574402, Email: arvindajoy@iitm.ac.in

Over the last decade, $Si_{1-x}Ge_x$ has increasingly been used as a channel material in MOSFETs. Though many studies have dealt with the real bandstructure of $Si_{1-x}Ge_x$, the effect of germanium mole fraction x on complex bandstructure has been unexplored. Complex bands fundamentally determine band to band tunneling (BTBT) current. For example, using the orientation dependent complex bandstructure of silicon [1], it has been shown [2] that the BTBT current in the [110] direction is an order of magnitude larger than that along the [100] direction. BTBT contributes significantly to off-current I_{off} in conventional MOSFETs, via the mechanism of gate induced drain leakage (GIDL). Additionally, BTBT determines the on current I_{on} in tunneling FETs, which have been suggested as next generation devices. Further, BTBT is more dominant in $Si_{1-x}Ge_x$ than silicon, owing to a narrower bandgap. In this work, we determine the orientation dependent complex bandstructure of $Si_{1-x}Ge_x$ along common crystallographic directions and predict trends in BTBT current.

$Si_{1-x}Ge_x$ is a substitutional alloy. The idea of bandstructure is hence only approximate [3]. Nevertheless, following [4], [5], we view relaxed $Si_{1-x}Ge_x$ as a weighted average of hydrostatically strained Si, Ge and implement a virtual crystal approximation (VCA) based $sp^3d^5s^*$ tight binding (TB) method, using parameters from [6], [7], [8]. To compute complex bands along a given transport direction n, we first determine primitive lattice vectors u_1, u_2, u_3 that are adapted to the plane perpendicular to n, i.e. $u_1 \cdot n \geq 0$ and u_2, u_3 are restricted to be within this plane [9]. Note that $k^\perp$ and $k^\parallel$ are respectively perpendicular and parallel to this plane (i.e. $k^\perp$ is along n). We then form layer Bloch sums $\xi_{\mu\varsigma m}(r; s, k^\parallel)$ [10], corresponding to orbital μ with spin ς on atom m in layer s, and write the wavefunction $\psi(r, k) = \sum_{\mu\varsigma m} \sum_s^{(M_1)} c_s^{\mu\varsigma m}(k^\perp) \xi_{\mu\varsigma m}(r; s, k^\parallel)$ over M_1 layers. Periodicity implies $c_{s+1}^{\mu\varsigma m}(k^\perp) = e^{ik^\perp \cdot u_1} c_s^{\mu\varsigma m}(k^\perp)$. Schrödinger's equation can now be written as a palindromic polynomial eigenvalue problem (1), which is solved using a companion linearization, to obtain $k^\perp(E)$.

$$\ldots [H_{s,s-1}][c_{s-1}] + ([H_{s,s}] - [1]E)[c_s] + [H_{s,s+1}][c_{s+1}] \ldots = 0 \qquad (1)$$

Fig. 1 compares the calculated energy gap E_g with experimental data [11]. Note [5] seems to have overlooked the temperature dependence of bandgap, and the effect of Harrison's strain parameters. We find our implementation to be in excellent agreement with experimental data, once a constant shift of 0.035 eV has been applied ([12] shows that this shift has been experimentally found to be independent of x). The conduction band minima transitions from being Si-like (i.e. along the $\langle 100 \rangle$) to being Ge-like (i.e. at L along $\langle 111 \rangle$), around $x = 0.85$. In order to calculate BTBT current, we need to determine complex bands $k^\perp(E)$ for k paths oriented along n, passing through various band extrema; these paths differ w.r.t $k^\parallel$. Fig. 2 shows the nature of $k^\perp(E)$ for the case of Si-like and Ge-like positions of the conduction minima. Paths having the same $k^\perp(E)$ are shown in the same colour. BTBT involves the transition of an electron from an imaginary valence band to a conduction band that is either imaginary or complex. The most probable tunneling event minimizes the area $\int_{E_g} Im(k^\perp)dE$, bounded by the imaginary parts of the valence and conduction bands involved in tunneling. This minimum area is denoted A_n (see inset of Fig. 3). The corresponding k paths through the conduction extrema (p_n in number) are shown in red in the Brillouin zone; the transverse momentum required to move from a black to a red k path is provided by a phonon. Assuming the probability of finding a phonon to be unity [2] and a uniform electric field F, a simple estimate for the transmission is $T_n \sim p_n exp(-2S)$, where the action $S = A_n/eF$. Thus, tunneling follows the path of least action. Finally, Fig. 3 shows that the trend in tunneling current changes from [111] > [110] > [100] to [110] > [111] > [100] as the conduction valleys change from being Si-like to Ge-like. Additionally, the dependence of BTBT current on orientation weakens with increasing Ge concentration. Our results should be useful for BTBT calculations in the design and optimization of $Si_{1-x}Ge_x$ devices.

The authors wish to thank Rajan Pandey, IBM SRDC, Bangalore for useful discussions.

[1] S. Laux, in *IEEE IWCE*, p. 1, 2009.
[2] R. Pandey et al., *IEEE TED*, 57(9), p. 2098, 2010.
[3] T. Boykin et al., *JPCM*, 19, p. 036203, 2007.
[4] S. J. Lee et al., *PRB*, 42(2), p. 1452, 1990.
[5] A. Paul et al., *IEEE EDL*, 31(4), p. 278, 2010.
[6] T. Boykin et al., *PRB*, 66(12), p. 125207, 2002.

[7] T. Boykin et al., *PRB*, 69(11), p. 115201, 2004.
[8] T. Boykin et al., *PRB*, 76(3), p. 35310, 2007.
[9] P. Aravind, *Am. J. Phys.*, 74, p. 794, 2006.
[10] T. Boykin, *PRB*, 54(11), p. 8107, 1996.
[11] J. Weber, *PRB*, 40(8), p. 5683, 1989.
[12] L. Yang et al., *Semicond. Sci. Technol.*, 19, p. 1174, 2004.

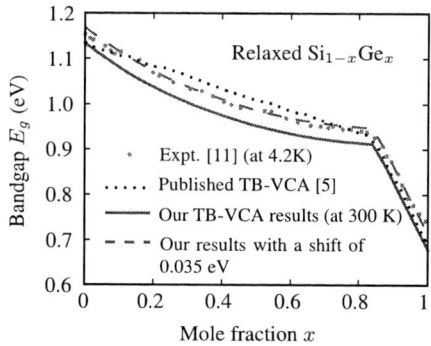

Fig. 1 Comparison of bandgaps calculated using TB-VCA with experimental data. The shift of 0.035 eV can be attributed to a temperature dependence of bandgap.

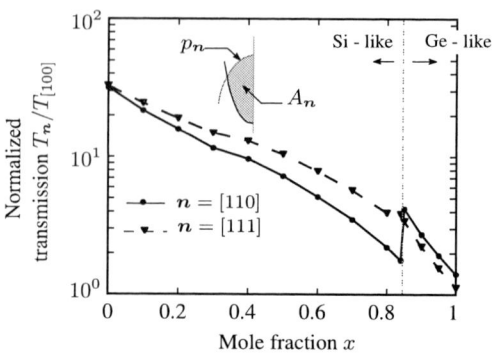

Fig. 3 Transmission probabilities for transport along [110] and [111], normalized with respect to [100] (electric field $F = 1$ MV/cm). Notice the change in trends when the bandgap becomes Ge-like, around $x = 0.85$. The abrupt change is possibly because of the approximations made in calculating T.

(a) $Si_{0.7}Ge_{0.3}$ (Si - like conduction valleys)　　　(b) $Si_{0.1}Ge_{0.9}$ (Ge - like conduction valleys)

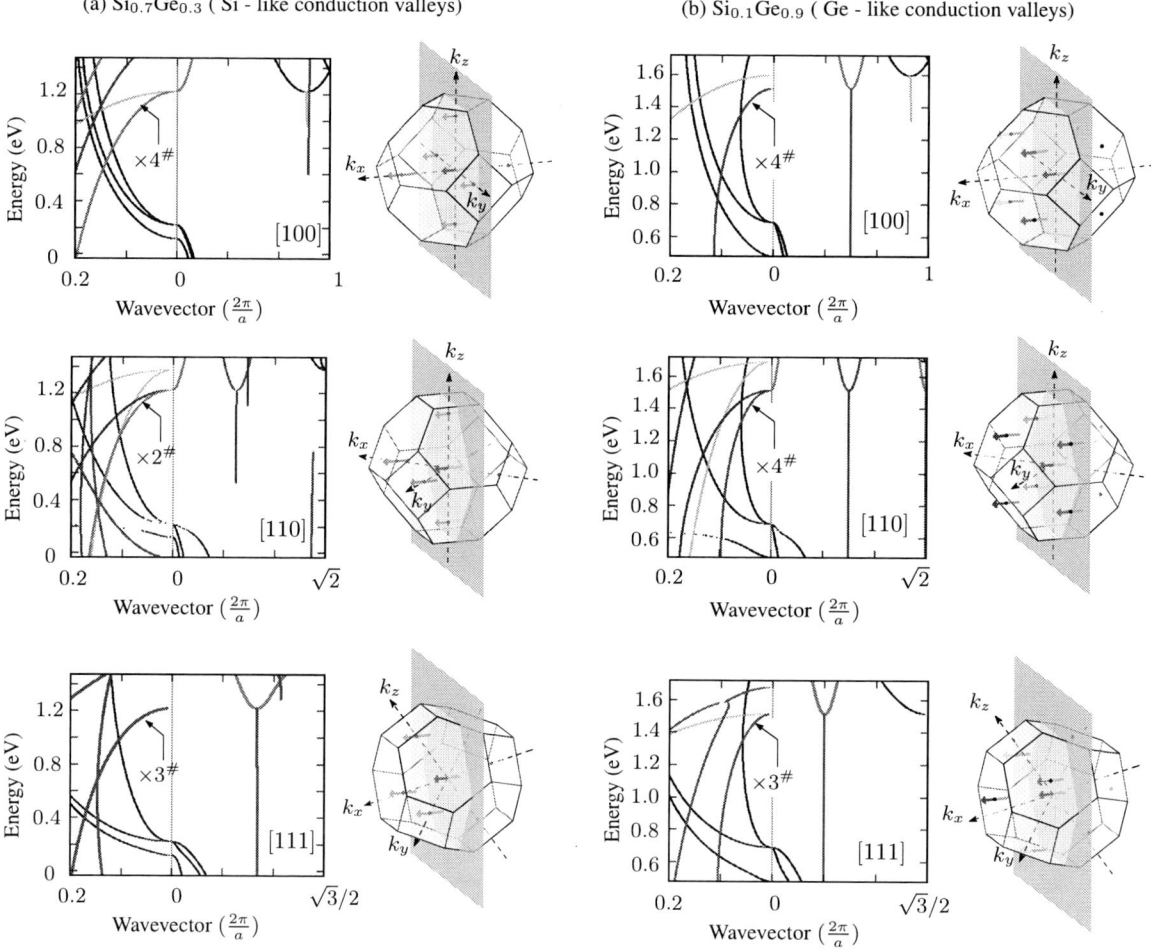

Fig. 2 Complex bandstructure $k^{\perp}(E)$ of $Si_{1-x}Ge_x$ with (a) Si-like ($x = 0.3$) and (b) Ge-like ($x = 0.9$) positions of conduction valleys, for transport along [100], [110] and [111]. In each case: (i) the Brillouin zone shows the plane with $k^{\perp} = 0$, and colour coded k paths passing through different band extrema; (ii) real and imaginary parts of the wavevector are shown on the right and left panels of the bandstructure plot respectively; (iii) pure real and pure imaginary bands are in black, red, blue, whereas complex bands in grey, dark red, dark blue respectively for the Black, Red and Blue paths; and (iv) the multiplicity p_n associated with paths is shown with a #.

978-1-61284-243-1/11 $26.00 © 2011 IEEE　　114

Towards Planar GaAs Nanowire Array High Electron Mobility Transistor

Xin Miao, and Xiuling Li*

Department of Electrical and Computer Engineering, Micro and Nanotechnology Laboratory
University of Illinois at Urbana-Champaign, 208 N. Wright St., Urbana, Illinois, 61801
phone: (217) 265-6354; *e-mail: xiuling@illinois.edu

We have demonstrated the first planar III-V nanowire (NW) based high electron mobility transistor (NW-HEMT) with self-assembled <110> planar GaAs NWs capped with Si-doped AlGaAs sheath as the channel. In contrast to vertical NW-FETs and post-growth aligned NW-FETs, self-assembled planar NW-FETs are well compatible with planar processing and can be deterministically positioned by patterning the Au catalysts [1][2]. The core-shell NW-HEMT heterostructure leaves the GaAs NWs undoped which avoids the issue of NW doping non-uniformity [3][4]. Therefore, device electrical uniformity can be achieved from uniform NWs because the epitaxial growth of AlGaAs thin film is well understood and controllable both in thickness and doping concentration.

The planar NW-HEMT structure was grown monolithically in an Aixtron 200 MOCVD reactor and fabricated using conventional optical lithography. Planar GaAs NWs were first grown at 460 °C using Au-catalyzed vapor-liquid-solid (VLS) mechanism on semi-insulating GaAs (100) substrate. They are self-aligned in the [0-11] or [01-1] directions and epitaxially pinned to the substrate [1]. Then by raising the temperature to 680 °C, we manage to switch from VLS NW growth mode to epitaxial thin film growth mode. 3nm intrinsic $Al_{0.35}Ga_{0.65}As$ was grown as spacer followed by 42nm Si-doped ($2 \cdot 10^{18}$ cm^{-3}) $Al_{0.35}Ga_{0.65}As$ as electron supply layer. At last, 5nm n^+ GaAs thin film was grown as source and drain ohm contact layer. Source and drain metal deposition was done right after growth using Ge/Au/Ni/Au stack followed by annealing at 400°C for 15s. Gate recess is done by selectively etching off the n^+ GaAs over $Al_{0.35}Ga_{0.65}As$. Ti/Au gate metal was evaporated thereafter without annealing.

The cross section of the planar GaAs nanowire is trapezoidal with {111} sidewall facets and (100) top and bottom facet. After growth, a sample is cleaved and dipped in the selective etchant Citric: H_2O_2 (4:1) to leave contrast between $Al_{0.35}Ga_{0.65}As$ thin film sheath and GaAs NW. As shown in Figure 1, a void created from the removed GaAs nanowire end can be clearly seen underneath the top $Al_{0.35}Ga_{0.65}As$ layer which conforms to the same crystal facet as the GaAs nanowire. Care must be taken to ensure the good crystal quality at the interface which determines the density of the two dimension electron gas (2DEG) for device operation.

Figure 2 shows SEM images of three fully processed AlGaAs-planar-GaAs-NW-HEMTs with single, double and triple NWs as conducting channels respectively. The separation between source and drain is 5 μm and gate length is ~1 μm. These nanowires are uniform in size so that uniform and scalable electrical property should be expected. Figure 3 shows the measured two-terminal source to drain I-V curves before gate deposition for single, double and triple NWs. Current level scales almost exactly with the number of NWs in the channel. After gate deposition, multiple-channel planar NW-HEMTs are probed. Figure 4 plots the transfer characteristics of all three devices, which shows peak Gm of 84-95 mS/mm at 2.5V drain bias. Output characteristics at 0V and 0.5V gate bias are plotted in Figure 5. The current scaling of the three-terminal devices is also evident. The device performances of multiple-channel planar NW-HEMT are summarized in Table I. Note that the scaled drive current and peak Gm are calculated by dividing GaAs nanowire's sidewall and top facets total length (450nm). The small difference of scaled drive current and peak Gm among three devices is due to processing related variations of gate length. Single-channel planar NW-HEMT has gate length of 940nm, while the other two have gate length of 1.1um. The excellent scaling properties demonstrated here will enable planar NW array based HEMTs with high output current.

An enhance-mode planar NW-HEMT with 3 μm long gate and 9 μm source to drain separation has also been fabricated using the same heterostructure mentioned above but with deeper gate recess. Figure 6 shows the device characteristics from which subthreshold slope of 79 mV/dec, scaled peak Gm of 50 mS/mm (at 0.1V drain bias) and I_{ON}/I_{OFF} ratio of 10^7 can be extracted. The dimension of NWs is determined by the size of Au catalyst. Therefore, we can scale both NW size and quantity in channel for planar NW-HEMTs. Benefit from small W/L ratio, enhanced gate controllability, high electron mobility and scalability of current levels, high speed-low power mixed signal circuits can be built out of this planar NW-FETs technology.

[1] Fortuna, S.A. et al., *Nano Lett.* **8**, 4421-4427(2008).
[2] Fortuna, S.A, and Li, X., *IEEE Elec. Dev. Lett.* **6**, 593-595 (2009).
[3] Li, Y. et al., *Nano Lett.* **6**, 1468-1473(2006).
[4] Perea, D. E. et al., *Nat Nano* **4**, 315-319, (2009).

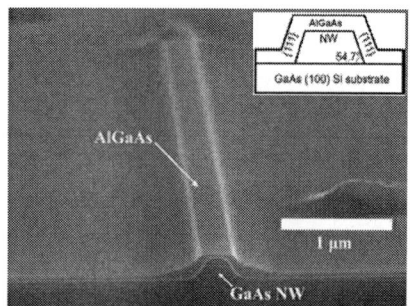

Fig. 1 Planar GaAs NW with AlGaAs coverage film where GaAs NW is selectively etched. Inset shows the schematic cross-section

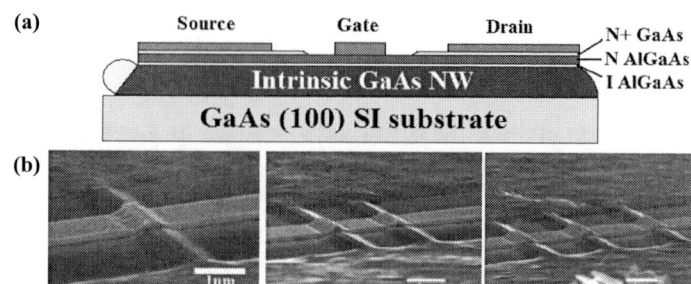

Fig. 2 (a) Schematic of planar NW-HEMT. (b) Images of single, double and triple channel planar NW-HEMTs; the scale bar is 1μm long

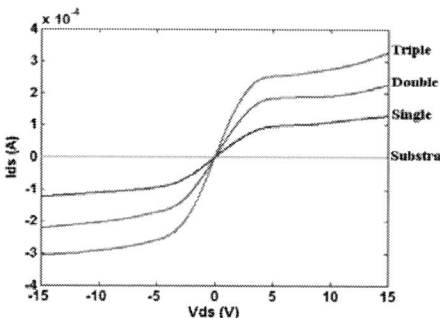

Fig. 3 Two-terminal source to drain I-V curves before gate deposition for single, double and triple NWs

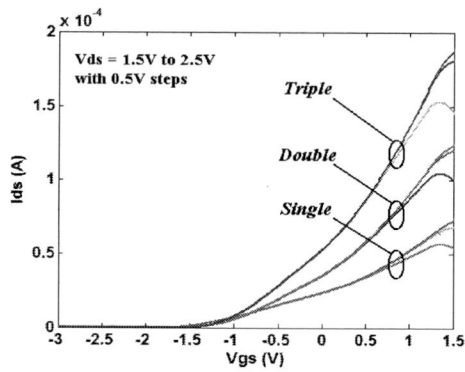

Fig. 4 Ids-Vgs transfer characteristics of single, double and triple channel planar NW-HEMTs for Vds = 1.5 to 2.5V with 0.5V steps

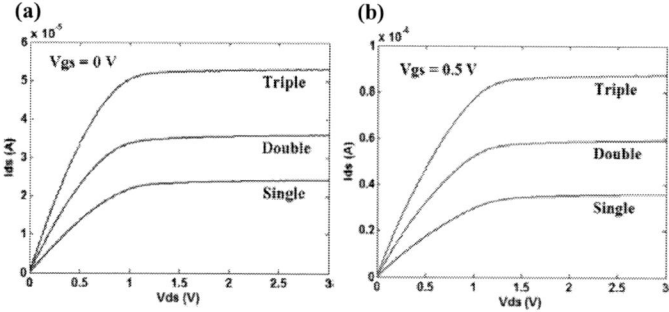

Fig. 5 Ids-Vds output characteristics of single, double and triple channel planar NW-HEMTs for (a) Vgs = 0V and (b) Vgs = 0.5V

Table I Multiple-Channel planar NW-HEMT summary

	single	double	triple
Drive current (μA) (at Vg=1V)	53.6	91.9	138.2
Scaled drive current (mA/mm)	119	102	102
Peak Gm (μs) (at Vd=2.5V)	42.7	75.5	116
Scaled peak Gm (ms/mm)	95	84	86
V_T (V)	-1.5	-1.5	-1.5
SS (mV/dec)	134	162	126
I_{ON}/I_{OFF}	10^4	10^4	10^4

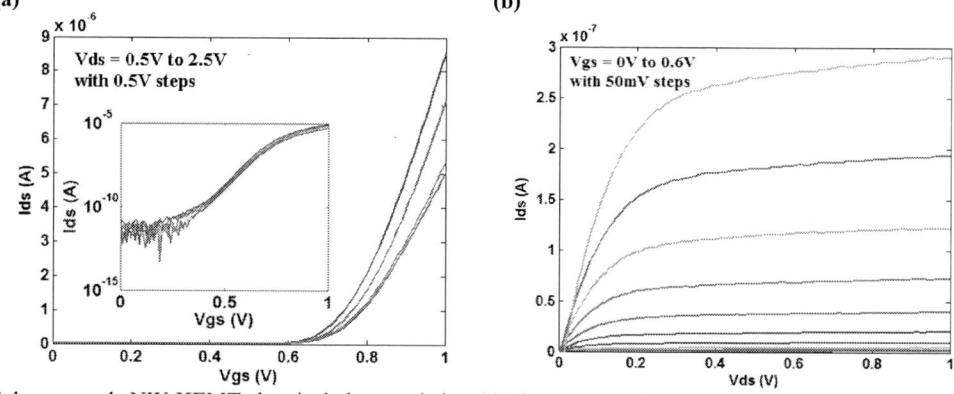

Fig. 6 Enhance-mode NW-HEMT electrical characteristics. (a) Ids-Vgs transfer characteristics for Vds = 0.5 to 2.5V with 0.5V steps; Inset shows Ids-Vgs plot on a semilog scale. Ids-Vds family of curves for Vgs = 0 to 0.6V with 50mV steps

Interface States at high- κ /InGaAs interface: H_2O vs. O_3 based ALD Dielectric

H. Madan[1,2], D. Veksler[1], Y.T. Chen[1,3], J. Huang[1], N. Goel[1], G. Bersuker[1] and S. Datta[2]

[1]SEMATECH, Albany, New York 12203, USA, [2]Penn State University, University Park, Pennsylvania 16802, USA

[3]University of Texas at Austin, Texas, USA. E-mail: himanshu@psu.edu

1. Introduction

Interface states at the high-κ/III-V interface are considered to be one of the major showstoppers for the implementation of III-V channel MOSFETs in VLSI technology. Due to the large interfacial state density, D_{it}, comparable with density of states of free carriers the standard, high-low freq., Terman, and conductance methods become unreliable. Recently, [1,2] proposed obtaining the parameters of the equivalent admittance circuit (including substrate capacitance, D_{it}, trap time constant, channel resistance and gate leakage) by fitting the experimental frequency dispersion of the capacitance and conductance curves in a self consistent manner.

In this study, we identify the D_{it} distribution and the traps characteristic time constant vs. the trap energy by *combining the above mentioned method [1,2] with the low-high frequency [3] and Terman techniques [4].* We apply the technique to study the defects in the water (H_2O) based ALD Al_2O_3/ $In_{0.53}Ga_{0.47}As$ and in the ozone (O_3) based ALD Al_2O_3/$In_{0.53}Ga_{0.47}As$ stacks. H_2O-based ALD allows reduction in the formation of the native oxide at the high-κ/IIIV interface, while O_3-based oxide is known to contain less OH groups within the high-κ resulting in less bulk trapping of carriers. Comparing the extracted trap capture cross-section dependences vs. temperature and trap energy, we conclude that: (i) water-based ALD allows reducing the number of electrically active traps (ii) the traps in the water-based ALD high-κ film respond (recharge) by more than an order of magnitude slower that those in O_3-based high-κ film.

2. Device Fabrication

N and P doped $In_{0.53}Ga_{0.47}As$ was epitaxially grown on a InP substrate. After the ammonia based surface clean the Al_2O_3 films were deposited on $In_{0.53}Ga_{0.47}As$ using either a H_2O-based or O_3-based atomic layer deposition (ALD) followed by post anneal. The TaN/TiN metal was deposited and patterned as top electrode. The AuGe alloy was deposited to form a backside ohmic contact.

3. Results

D_{it} extraction: The admittance characteristics of the fabricated capacitors were measured for 200K to 425K temperature range. Fig. 1 and 2 show the measured capacitance at RT for the H_2O and O_3-based ALD high-κ. Fig. 3 shows the calculated ideal C-V dependency for the 8nm Al_2O_3 high-κ/ $In_{0.53}Ga_{0.47}As$ stack [5] taking into account the conduction band non-parabolicity and carrier distribution in Γ, L and X valleys. We have used low temperature and high temperature C-V data sets in order to accurately evaluate the effects of interface and border traps. At 425 K the traps are fast enough ($2\pi f\ \tau >> 1$, see Fig. 5(b)) for the 1kHz C-V to be considered as a true low

frequency C-V. In this case the trap response is quasi-static and the trap capacitance, $C_{it}=qD_{it}$. The high temperature low frequency C-V and the "stretch-out" of the low temperature high frequency C-V characteristic (a "true" high frequency C-V) were theoretically reproduced. Fig. 4 shows the result of this iterative fitting exercise for both H_2O-based and O_3-based ALD high-κ. The C-V with non-parabolic (NP) correction including all valleys was used for this exercise. The simulated C-V with C_{it} represents the quasi-static case and C-V without C_{it} represents the high frequency case (includes the stretch-out in gate bias due to D_{it}). The evaluated D_{it} from p and n type samples are in good agreement (Fig. 5(a)). The trap density for the O_3-based ALD is ~1.5 times higher than that of H_2O-based ALD sample.

Interfacial layer analysis: An XPS study on these samples shows the presence of the As-O bonds in the O_3-based ALD sample. The use of O_3 as oxidant for the ALD growth of Al_2O_3 on $In_{0.53}Ga_{0.47}As$ has resulted in excessive interfacial oxidation consistent with [6]. Due to the presence of a native oxide at the oxide-substrate interface the measured oxide capacitance for the O_3-based ALD is ~ 10% lower than that of the H_2O-based sample.

Trapping kinetics: The characteristic traps capture times (for the mid gap traps) obtained from the conductance peaks (Fig. 5(b)) shows an order of magnitude faster response time for the O_3-based ALD high-κ. The extracted capture cross-section of the mid gap traps in both samples were found to be weakly depend on temperature (Fig. 5(b)) and for the O_3-based sample to be ~1 orders of magnitude larger than that for the H_2O-based sample. It is opposite to the expected trend if the electrons were to tunnel through the native oxide in the O_3-based sample before they can get trapped by the high-κ defects, leading to a decrease in the capture cross-section. This observation indicates that the traps may have a significantly different atomic structure, and chemical bonds. It is worth noting that the energy dependence of the capture time is rather weak in both samples (Fig. 6(a) and 6(b)). Fig. 7 shows the comparison of the mid-gap D_{it} compared with other works reported in literature.

4. Conclusion

By combining the capacitance and conductance analysis techniques, we obtained the D_{it} distribution throughout the band gap of $In_{0.53}Ga_{0.47}As$ capacitors with H_2O-based and O_3-based ALD oxides. The choice of appropriate temperature to obtain the quasi-static C-V and the DC voltage sweep rate is an essential for the correct extraction of D_{it}. Simultaneously we obtained the trap kinetics characteristics. We claim that: (i) the H_2O-based ALD deposition results in a fewer traps in the lower portion of $In_{0.53}Ga_{0.47}As$ band gap, (ii) is

978-1-61284-243-1/11 $26.00 © 2011 IEEE

related to the formation of the thicker native oxide in the O_3-based samples; (iii) the mid gap traps in the H_2O-based samples are significantly slower than those in the O_3-based samples, which indicate their different nature.

References

[1] A. Ali et al, IEEE TED 57 (2010), 742-748
[2] A. Ali et al, Appl. Phys. Lett. 97 (2010), 143502-143504
[3] www.ieeesisc.org/tutorials/2010_SISC_Tutorial.pdf
[4] L. Terman, Solid-State Electron. 5 (1962), 285-299
[5] V. Ariel-Altschul et al, IEEE TED 39 (1992), P1312
[6] B. Brennan et al, ECS let. 12 (2009), H205-H207
[7] G. Brammertz et al, Appl. Phys. Lett. 95, 202109 2009.
[8] H. C. Chiu et al, Appl. Phys. Lett. 93, 202903 2008.
[9] H. C. Lin et al, Microelectron. Eng. 86, 1554 2009.
[10] Y. Xuan et al, IEEE Electron Device Lett. 28, 935 2007.
[11] Y. Hwang et al, Appl. Phys. Lett. 96, 102910 2010.

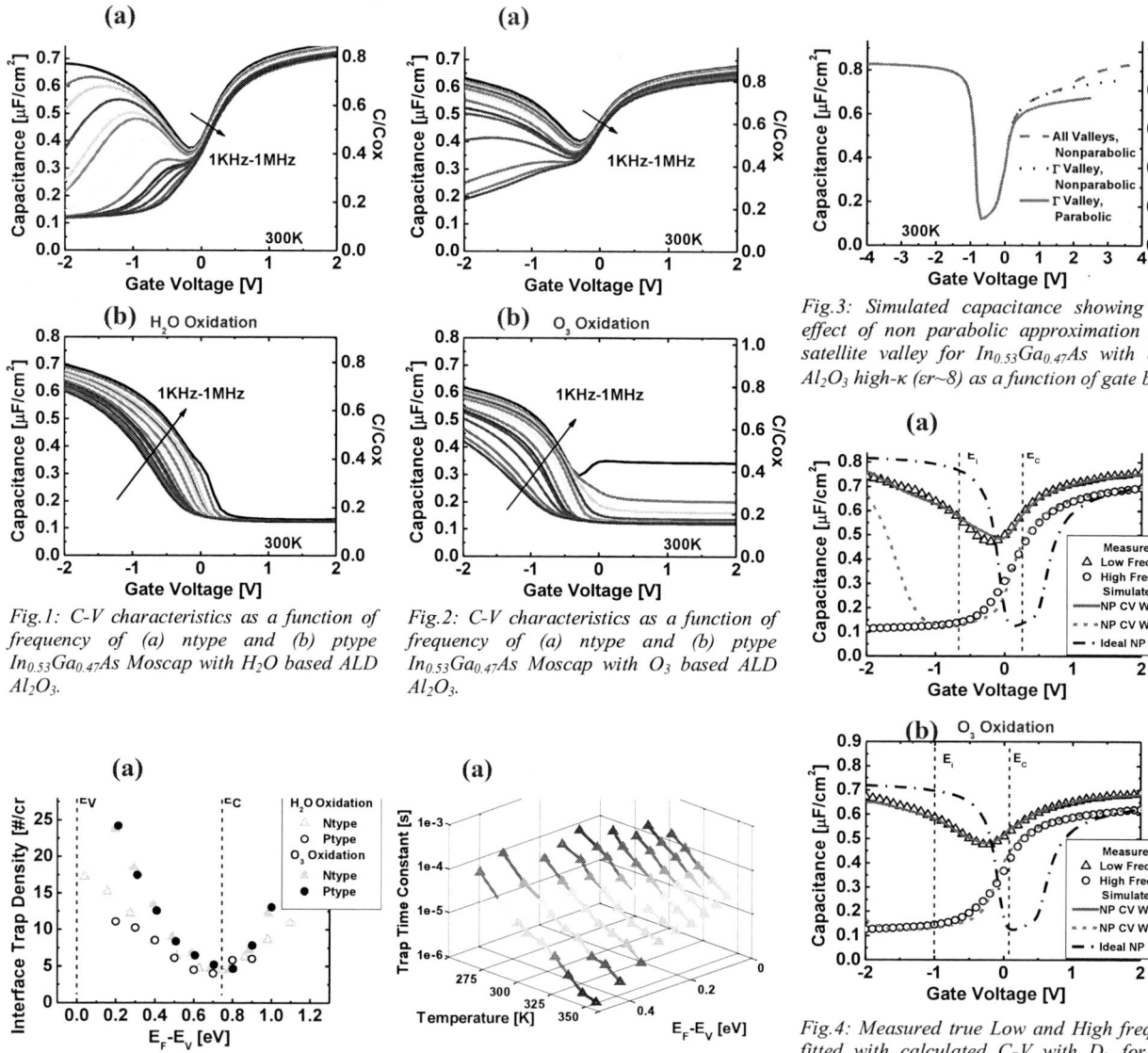

Fig.1: C-V characteristics as a function of frequency of (a) ntype and (b) ptype $In_{0.53}Ga_{0.47}As$ Moscap with H_2O based ALD Al_2O_3.

Fig.2: C-V characteristics as a function of frequency of (a) ntype and (b) ptype $In_{0.53}Ga_{0.47}As$ Moscap with O_3 based ALD Al_2O_3.

Fig.3: Simulated capacitance showing the effect of non parabolic approximation and satellite valley for $In_{0.53}Ga_{0.47}As$ with 8nm Al_2O_3 high-κ (εr~8) as a function of gate bias.

Fig.4: Measured true Low and High frequency fitted with calculated C-V with D_{it} for ntype $In_{0.53}Ga_{0.47}As$ Moscap with (a) H_2O and (b) O_3 based ALD Al_2O_3.

Fig.5: Extracted (a) interface trap density and (b) interface trap time constant and capture cross section at E_F~Eg/2 for the water and ozone based ALD Al_2O_3.

Fig.6: Trap time constant as a function of energy and temperature for (a) H_2O and (b) O_3 based ALD Al_2O_3.

Fig.7: Midgap D_{it} evaluated by conductance method at room temperature compared with other work reported in literature.

C-V Measurements of Single Vertical Nanowire Capacitors

P. Mensch, K. E. Moselund, S. Karg, E. Lörtscher, M. T. Björk, H. Schmid and H. Riel.

IBM Research – Zurich, Säumerstrasse 4, CH-8803 Rüschlikon, Switzerland.[1]

Email: kmo@zurich.ibm.com / Phone: +41 44 824 8111

The density of interface states, D_{it}, is important for the device performance in view of the fact that it limits the inverse subthreshold slope in both, MOSFETs and TFETs [1]. This poses particular challenges for nanowire (NW) devices, because the measured D_{it} is expected to increase due to the extensive processing and the various crystallographic orientations of the surface, which differ from the ideal (100) orientation. For a detailed investigation of the D_{it} of NWs it is best to analyze single NW MOS capacitors. However, the capacitance of a single NW MOS capacitor lies in the fF regime which is very challenging to measure. To date, very few capacitance measurements on single NWs have been reported, e.g., on lateral devices based on InAs [2], Ge [3], and Si [4]. D_{it} analysis of NWs has been demonstrated, however, based on capacitance measurements only of large arrays of InAs NWs [5]. In the present work, we report on the capacitance measurement and D_{it} analysis of *vertical* silicon MOS capacitors based on *single NWs*.

Silicon NWs with a diameter around 300 nm were fabricated by reactive ion etching from a hard mask; see Fig. 1 for a schematic of the process flow. In order to be able to isolate the response of the NW from that of the substrate a much higher doping of the substrate is required. In this case, the substrate contributes solely with a fixed parasitic capacitance, which does not depend on the bias level and, hence, can easily be corrected for. This doping differentiation was achieved by diffusion doping using a doped PECVD oxide selectively patterned to cover only the substrate and not the NWs, see [6] for details. Thus, the doping in the NWs is p-type with a concentration of $\sim 10^{16}$ cm^{-3}, whereas that in the substrate is at least 10^{19} cm^{-3}, also p-type. BenzoCyclobutene (BCB) was used as the isolating spacer layer. The gate stack consisted of 9.5 nm Al$_2$O$_3$ deposited by atomic layer deposition, and Ti/Pt was used as the gate metallization. Both individual NWs and arrays of 2x2, 5x5 and 10x10 were fabricated, see Fig. 2.

The dielectric response of single NWs was measured by a highly accurate capacitance bridge (Andeen Hagerling) at frequencies between 1 kHz and 19.5 kHz (Fig. 3). The measurements were done at room temperature with 1 aF resolution and a peak-to-peak noise level of less than ± 20 aF at an ac bias of 100 mV. In addition to careful calibration of the set-up, a combination of electro-static and electro-magnetic shielding of sample and probes against external noise sources was required to achieve a low noise level, see Fig. 4. Reference pads and leads which are not connected via NWs to the substrate were used to determine the parasitic capacitance.

C-V measurements of single NW capacitors as well as arrays of NWs are shown in Fig 5 (normalized to the number of NWs). We obtain reproducible values of gate capacitance of 1.8 fF for single NWs. For larger NW arrays (5x5 to 10x10), a capacitance of 3.4 fF/NW is determined. The difference is due to proximity effects in the etching process where isolated NWs become thinner than the average NW in arrays. The D_{it} of a single NW was extracted by the quasistatic method and found to be $(4\pm2)\times10^{12}$ cm^{-2} eV^{-1} in the lower half of the bandgap (Fig. 6). This is about one order of magnitude higher than the D_{it} of planar capacitors fabricated on Si (111) using the same gate stack. The increase in D_{it} is expected for NWs due to the effects of the etching process.

In summary, we have fabricated arrays of etched vertical silicon NW capacitors and measured the capacitance of arrays and single NWs with a noise level of ± 20 aF at room temperature. The single NWs have a total C_{ox} on the order of 1.8-3.4 fF at room temperature.

[1] W. Kao et al. Solid-State Electronics, **2010**, 54, 1665-1668
[2] A. Ford et al. Nano Letters, **2009**, 9, 360-365
[3] R. Tu et al. H. Nano Letters, **2007**, 7, 1561-1565
[4] H. Zhao et al. Electron Device Letters, **2009**, 30, 526-528
[5] S. Roddaro et al. Applied Physics Letters, **2008**, 92, 253509
[6] K. E. Moselund et al. Nanotechnology, **2010**, 21, 435202

"The research leading to these results has received funding from the European Union Seventh Framework Programme (FP7/2007- 2013) under grant agreement n° [257267]".

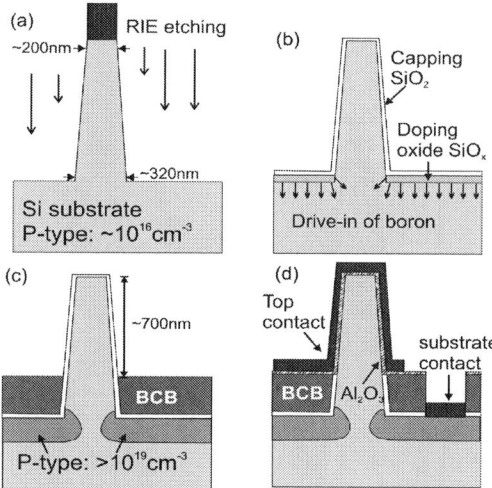

Figure 1. Process flow. (a) NWs etched by RIE. (b) Doping of substrate at base of NWs to ≥10¹⁹ cm⁻³ by diffusion doping. (c) BCB isolation layer is deposited.

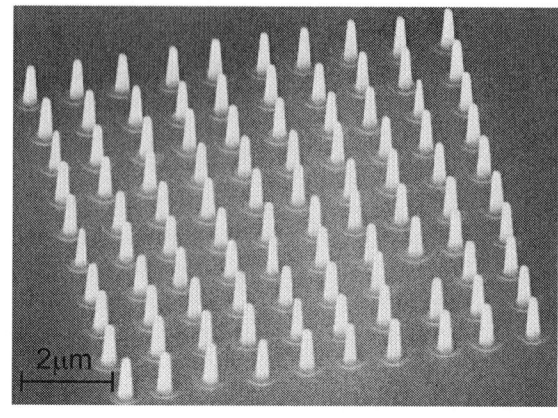

Figure 2. SEM image of 10x10 NW array including BCB layer. NW diameter ranges from ~320 nm at the base to ~200 nm at the top. The height is ~1.1 μm. Occasionally a NW is missing which is accounted for in the normalization.

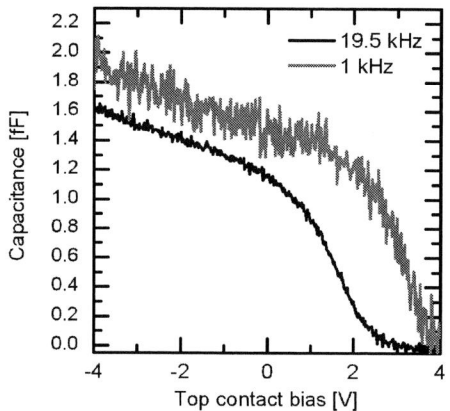

Figure 3. C-V curves of a single nanowire wrapped gate Si MOS capacitor at two different frequencies of 1 kHz and 19.5 kHz.

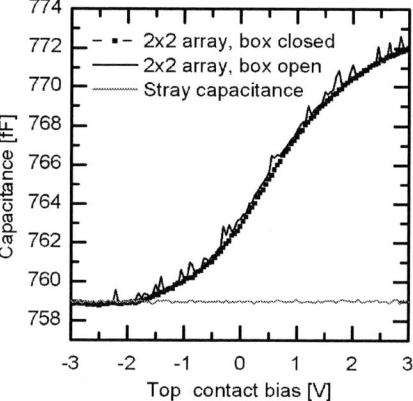

Figure 4. C-V characteristics of a 2x2 array of Si NW MOS capacitors measured with (solid) and without (dashed) noise shielding. The noise reduction due to shielding measures is one order of magnitude. Also shown is the parasitic capacitance level (light grey).

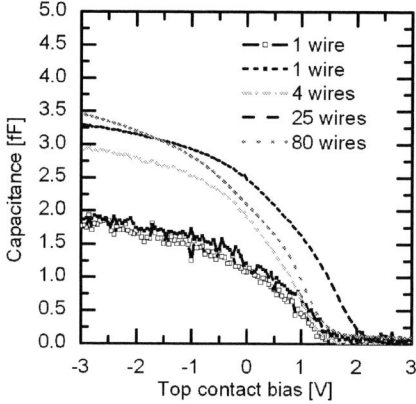

Figure 5. C-V curves of 2 different single NWs and of arrays of NWs divided by the number of nanowires measured at a frequency of 19.5 kHz.

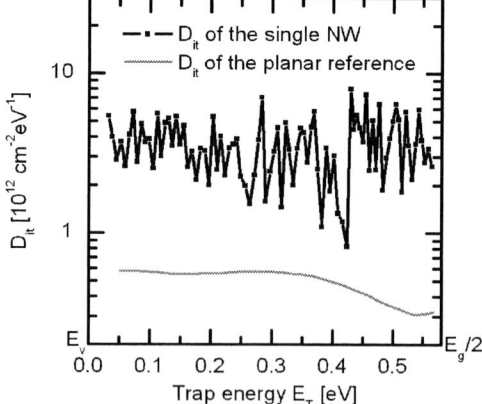

Figure 6. D_{it} extracted for a single p-type silicon NW, compared with the D_{it} of a planar Si (111) sample.

Barrier Height, Interface Charge & Tunneling Effective Mass in ALD Al$_2$O$_3$/AlN/GaN HEMTs

Satyaki Ganguly, Jai Verma, Guowang Li, Tom Zimmermann, Huili Xing, Debdeep Jena[*]

University of Notre Dame, Department of Electrical Engineering

275 Fitzpatrick, Notre Dame, IN 46556, [*]Email: djena@nd.edu, [*]Phone: 574-631-8835

Atomic layer deposited (ALD) high band gap (~6.5eV) [1], high k (~9.1) Al$_2$O$_3$ has emerged as an attractive candidate to support vertical scaling of AlN/GaN HEMTs [2] and its variants owing to its outstanding dielectric, thermal, and chemical properties. Integration of ALD oxides with GaN will enable lower gate leakage currents, high breakdown voltages, and surface passivation. In this work we present a comprehensive characterization of AlN/GaN MOS-HEMT gate stacks with ALD Al$_2$O$_3$ of various thicknesses. Through capacitance-voltage and Hall-effect measurements, we find the presence and propose an origin of benign donor-type interface charge (Q_{int}) at the AlN/Al$_2$O$_3$ junction, and relate its presence to the polarization charges in AlN. By studying tunneling transport in corresponding (Ni/Al$_2$O$_3$/Ni) M-I-M diodes, we extract the Ni/Al$_2$O$_3$ surface barrier height (Φ_B), the electron tunneling effective mass in Al$_2$O$_3$, and discuss the resulting HEMTs.

ALD/AlN Interface: AlN/GaN HEMT structures with ~4nm barrier were grown by MBE on semi insulating 0001 GaN templates (2μm) on sapphire. Mesa isolation was followed by Ti/Al/Ni/Au stack source/drain alloyed ohmic metallization. The same sample was then cleaved into four parts, and four different Al$_2$O$_3$ thicknesses (t_{ox}=2nm, 4nm, 6nm, 8nm) were deposited on the AlN surface by ALD with TMA and H$_2$O as the precursors. Finally, Ni/Au (50/150 nm) gate metal stacks were deposited simultaneously. The layer structure of the sample is shown in Fig. 1(a). Figure 1(b) & (c) shows the cross-section TEM image of gate stack of the (t_{ox}=4nm) sample, confirming the thicknesses, and shows thickness variations within an acceptable window. Fig. 2(a) shows the capacitance-voltage (C-V) measurement (@1MHz) data for the four samples (they show negligible hysteresis). The pinch-off voltage V_p increases with t_{ox} from -3.5 V for t_{ox}=2nm to -8.8V for t_{ox}=8nm. The carrier profile extracted [3] from the C-V measurement is shown in Fig. 2(b), indicating the varying depth of the 2DEG channel from the gate metal. Fig. 2(c) shows the decreasing gate leakage current with increasing t_{ox}, the property that will enable high breakdown voltages in AlN/GaN HEMTs. The increase in V_p with t_{ox} over the 4 samples can be *quantitatively* explained by the existence of a fixed *positive* sheet charge Q_{int}~6x10^{13} cm^{-2} at the (Al$_2$O$_3$/AlN) interface as shown in Fig. 3(a). Self-consistent 1-D Poisson-Schrödinger simulation [4] for Q_{int}=0 cannot explain the experimental data as shown in Fig. 3(a). The Ni/Al$_2$O$_3$ surface barrier height used is the measured (see later) value Φ_B=2.9eV. The interface density Q_{int}~6x10^{13} cm^{-2} is remarkably close to the surface polarization charge of a strained AlN layer. We argue that since the AlN surface is metal (Al)-face, in ALD oxygen atoms attach to Al and can be viewed as substituting the nitrogen site [5], thus acting as donor dopants by electron counting rules. The picture is essentially identical to modulation doping: the positive sheet charge at the Al$_2$O$_3$/AlN interface [inset, Fig. 3(a)], neutralizes negative polarization charges of the AlN surface, increasing the 2DEG density at the AlN/GaN heterojunction. Fig. 3(b) shows that the increase in the experimental 2DEG density n_s (from C-V) with t_{ox} can be explained if Q_{int} = 6x10^{13} cm^{-2} is assumed. This finding is verified with another AlN/GaN structure with a thinner AlN barrier. Fig. 3(c) shows the Hall-effect measured charge before and after ALD deposition (t_{ox}=5nm). Charge density loci predicted by Poisson-Schrödinger simulations for various Q_{int} (after ALD) and Φ_B are shown in Fig. 3(c), confirming that Q_{int} = 6x10^{13} cm^{-2} and Φ_B =2.9eV match the experimental data.

Ni/ALD Interface: Four M-I-M (Ni/ALD Al$_2$O$_3$/Ni) diode structures were fabricated with t_{ox}=8, 9, 10, 12nm for Fowler-Nordheim (FN) type tunneling studies. Fig. 4(a) shows the measured tunneling current density (J) vs. the oxide electric field (E_{ox}) for the four samples, and Fig. 4(b) shows the FN plots for the t_{ox}=10 and 12nm samples. From the slope of these plots Φ_B = 2.9eV and an electron tunneling effective mass of m_T^* ~ 0.16m_0 are extracted. This tunneling effective mass is similar to an earlier report [6].

The dc I-V characteristics of the depletion mode HEMT (t_{ox}=6nm) is shown in Fig. 4(c), and Fig. 4(d) shows the transfer characteristics with the gate capacitance. The subthreshold slope SS and V_p extracted from the transport measurement (transfer characteristics) is SS~285mV/decade and V_p=-7.6V are in close agreement with those extracted from the static (no transport) C-V plot (SS~320 mV/decade and V_p=-7.7V). This indicates that the SS is not a transport/short channel effect, but related to the gate stack itself.

The extraction of the polarization-related ALD/AlN interface charge, the Ni/ALD barrier height, and the tunneling effective mass of electrons through ALD Al$_2$O$_3$ into AlN/GaN heterojunctions reported here is expected to accelerate the choice of optimal gate stacks for nitride HEMTs. In addition, the role of ALD oxygen layers as possible modulation dopants can be cleverly exploited for novel purposes in nitride electronic devices.

[1]   N. V. Nguyen *et al*, *Appl. Phys. Lett.*, 93, 082105, (2008).
[2]   H. Xing *et al*, *ECS transactions*, *Vol.*11, (2007).
[3]   N. Onojima *et al*, *J. Appl. Phys.*, 101, 043703, (2007).
[4]   I. H. Tan *et al*, *J. Appl. Phys.*, 68, 4071, (1990).
[5]   T. Mattila *et al*, *Phys. Rev B*, 54, 23, (1996).
[6]   M. D. Groner *et al*, *Thin Solid Films*, 413, (2002).

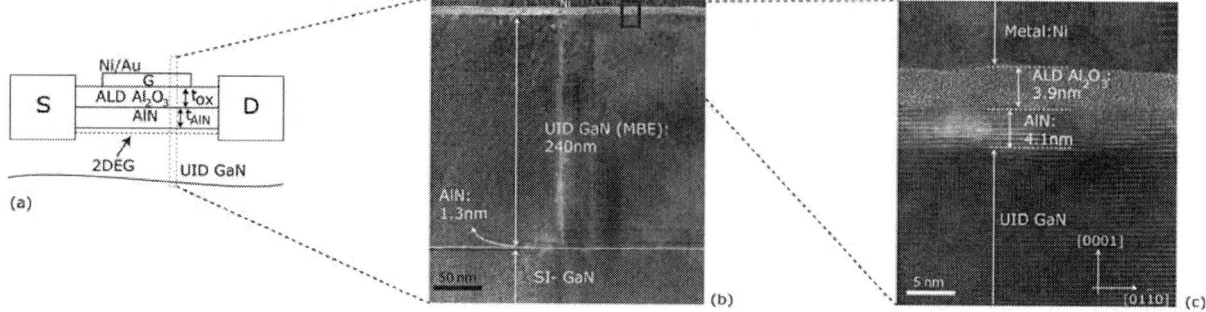

Fig.1 (a) Schematic layer structure of the sample; (b) High Resolution Transmission Electron Microscope image along zone axis [100] showing the layer structure of the sample; (c) High-resolution lattice image of the gate stack showing the crystalline AlN barrier layer, the amorphous ALD Al_2O_3 and Ni as gate metal.

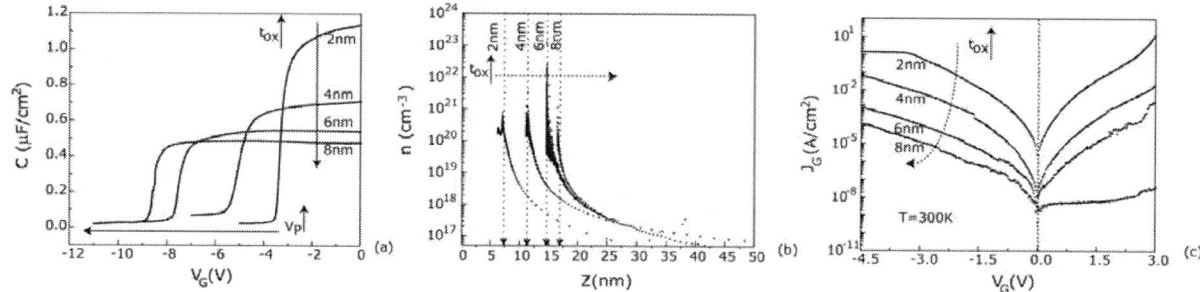

Fig. 2 (a) C-V plots (@1MHz) of the four samples with different t_{ox}, showing V_p increases with increasing t_{ox}; (b) Plots showing the charge profile and varying depth of 2DEG channel from the gate metal for samples with different t_{ox}; (c) Gate leakage current density vs. gate voltage plots indicating the reduction of J_G with increasing t_{ox}.

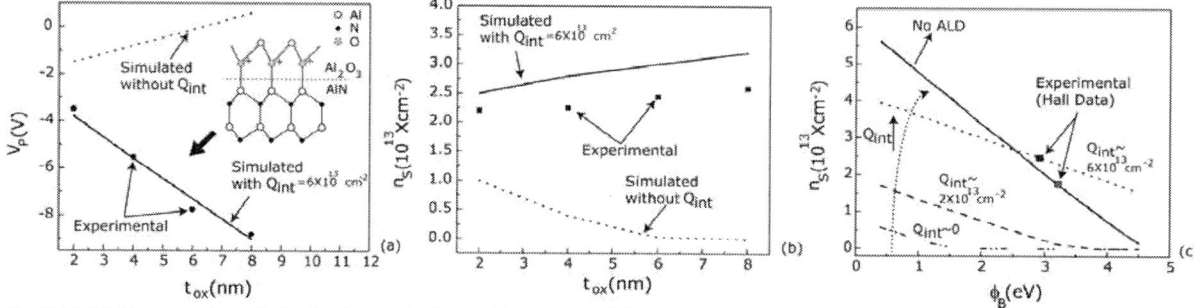

Fig. 3 (a)-(b) Experimental and simulated (w and w/o Q_{int}) V_p and n_s for different t_{ox}. The atomic arrangement at the Al_2O_3/AlN interface and the positive donor dopants giving rise to Q_{int}, are shown in the inset(a); (c) Simulation data plot showing the variation of n_s with Φ_B before and after 5nm ALD (for various Q_{int}). The Hall measured n_s before and after ALD (t_{ox}=5nm) are compared with the simulation result.

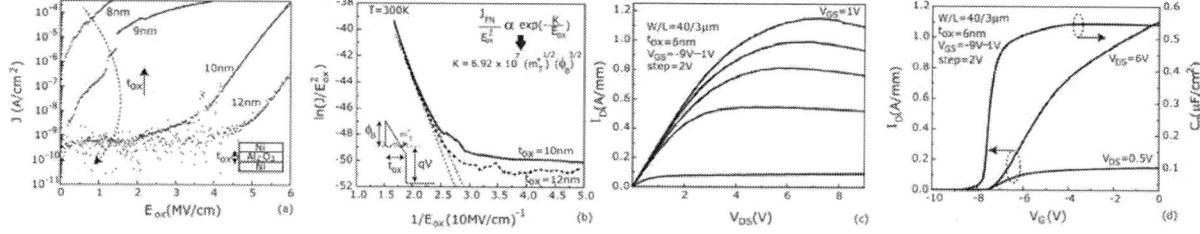

Fig. 4 (a) J versus E_{ox} for the M-I-M structures with different t_{ox}. Inset: Schematic layer structure of the M-I-M diode; (b) FN plots for the samples with t_{ox}=10nm, 12nm. The expression in the inset is used to extract Φ_B & m_T^* from the slope. Schematic band diagram for FN tunneling (@ $qV > \Phi_B$) is shown in the inset; (c) dc I-V characteristics of Al_2O_3/AlN/GaN (6/4/235nm) depletion mode HEMT; (d) Transfer characteristics (for V_{DS}=0.5, 6V) and gate capacitance (vs V_G) plot for the Al_2O_3/AlN/GaN (6/4/235nm) HEMT structure.

978-1-61284-243-1/11 $26.00 © 2011 IEEE

Graphene Field-Effect Transistors Using Large-Area Monolayer Graphene Grown by Chemical Vapor Deposition on Co Thin Films

M. E. Ramón[1], A. Gupta[1], C. Corbet[1], D. A. Ferrer[1], H.C.P. Movva[1], G. Carpenter[2], L. Colombo[3], G. Bourianoff[4], M. Doczy[4], D. Akinwande[1], E. Tutuc[1], and S.K. Banerjee[1]

[1]Microelectronics Research Center, The University of Texas at Austin, Austin, TX-78758, USA,
[2]IBM Research, 11501 Burnet Rd, Austin, TX 78758, United States, [3]Texas Instruments Incorporated, Dallas, TX 75243, United States, [4]Intel Corporation, Austin, TX 78746, United States
Phone: (512) 232-9159, Fax: (512) 471-5625, Email: mramon@mail.utexas.edu

There has been great interest in methods for the synthesis of high-quality, large-area graphene films, as required for practical applications in the electronics industry. In particular, recent developments in chemical vapor deposition (CVD) methods have shown a promising approach to grow large-area graphene on metal substrates by catalyzed CVD growth [1]. Reports of CVD growth on Cu and Ni are common [1-3]; however, there have been few efforts to grow graphene on Co [4], and attempts to grow graphene on $Co/SiO_2/Si$ resulted in very small domains of predominantly multilayer graphene that were not suitable for transistor fabrication. Unlike Ni, Co is attractive due to the low lattice mismatch ($< 2\%$) between graphene and the Co (0001) surface, and Co exhibits greater compatibility with Si than Cu, which is a deep trap in Si and a fast diffuser. Here we have demonstrated graphene field-effect transistors (GFETs) fabricated using large-area monolayer graphene grown by catalyzed CVD on Co films.

Graphene films were synthesized in a Low-Pressure Rapid Thermal Chemical Vapor Deposition (RTCVD) furnace using acetylene (C_2H_2) as the carbon source. Initially, Co films on SiO_2/Si substrates were annealed in H_2 at ~650°C to optimize the Co grain size (Fig. 1), followed by graphene growth at ~800°C and subsequent cooling to room temperature. After CVD growth, the Co was etched, and the graphene was transferred to another SiO_2/Si substrate. To obtain large-area monolayer graphene, the Co film thickness was also optimized (Fig. 2). The thickness, quality, and uniformity of the graphene films were evaluated by Raman spectroscopy (Fig. 3).

For GFET fabrication, CVD graphene films grown on Co were transferred to a low resistivity Si substrate, upon which a 285nm thick SiO_2 layer was grown by thermal oxidation. The low resistivity Si substrate allows for its use as a global back-gate. Monolayer graphene regions used for device fabrication were identified and selected by a combination of optical microscopy [5] and Raman spectroscopy [6]. An active region was defined by electron beam lithography (EBL) and oxygen plasma etching. A second EBL step was performed to define metal contacts for a four-point structure, followed by a 50 nm thick Ni evaporation and lift-off process. Fig. 4(b-inset) shows an example optical microscope image of a fabricated device. From the transfer characteristics (Fig. 4(a)), we observe an I_{ON}/I_{OFF} ratio of ~3 and a positive Dirac point ~15V, due to unintentional extrinsic doping caused by adsorbants that may have been introduced during device processing [7]. The hysteresis observed is indicative of trapped charges at the SiO_2/graphene interface [8], and the number of trapped charges per unit area N (i.e., charged impurities) was estimated to be ~1.2×10^{11} cm^{-2}, as determined from the shift in the Dirac point (~3V) and the capacitance (C_{ox}) of the SiO_2 bottom dielectric [9], where C_{ox} is ~12 nF/cm^2. Despite the possibility of unintentional doping during device fabrication, this value of trapped charges is indicative of a relatively clean sample [10].

The four-point resistance ($R_{4-Point}$) versus back-gate voltage (V_{BG}) (Fig. 4(b)) was fit to the model for GFET resistance described elsewhere [11]. In particular, we have compared electron and hole mobility as shown in Fig. 4(b). Specifically, the data was separately fit for holes, $V_{BG} - V_{DIRAC} < 0$, and for electrons, $V_{BG} - V_{DIRAC} > 0$. The resulting extracted mobility is ~1600 cm^2/V-s for holes and ~1000 cm^2/V-s for electrons, indicating preferential hole-conduction over electron-conduction, which may be due in part to the extrinsic doping [12]. As confirmation of the preferential hole conduction in our GFETs, Fig. 4(c) shows the output characteristics for holes and for electrons at various values of V_{BG}. As shown, the modulation of the drain current is much more significant when the graphene is electrostatically doped p-type ($V_{BG} - V_{DIRAC} < 0$), as compared to when the graphene is comparably electrostatically doped n-type ($V_{BG} - V_{DIRAC} > 0$). This is expected given the higher mobility for holes observed in our devices. Indeed, the I_{DS} at $V_{DS} = 100$mV for $V_{BG} - V_{DIRAC} = -15$V is ~1.8 times greater than the I_{DS} at $V_{BG} = V_{DIRAC}$, while the I_{DS} at $V_{DS} = 100$mV for $V_{BG} - V_{DIRAC} = +15$V is ~1.3 times higher than the I_{DS} at $V_{BG} = V_{DIRAC}$.

[1] Li, X. *et al.*, Science **324**, 1312, 2009; [2] Kim, K.S. *et al.*, Nature **457**, 706, 2009; [3] Reina, A. *et al.*, Nano Lett, **9**, 30, 2009; [4] Ago, H. *et al.*, ACS Nano **4**, 7407, 2010; [5] Blake, P. *et al.*, APL **91**, 063124, 2007; [6] Ferrari, A.C. *et al.*, PRL **97**, 187401, 2006; [7] Cao, H. *et al.*, APL **96**, 122106, 2010; [8] Lemme, M.C., Solid State Phenom., **156-158**, 499, 2010; [9] Wang, H. *et al.*, ACS Nano **4**, 7221, 2010; [10] Adam, S. *et al.*, PNAS **104**, 18392, 2007; [11] Kim, S. *et al.*, APL **94**, 062107, 2009; [12] Lemme, M.C. *et al.*, Solid State Electronics **52**, 514, 2008.

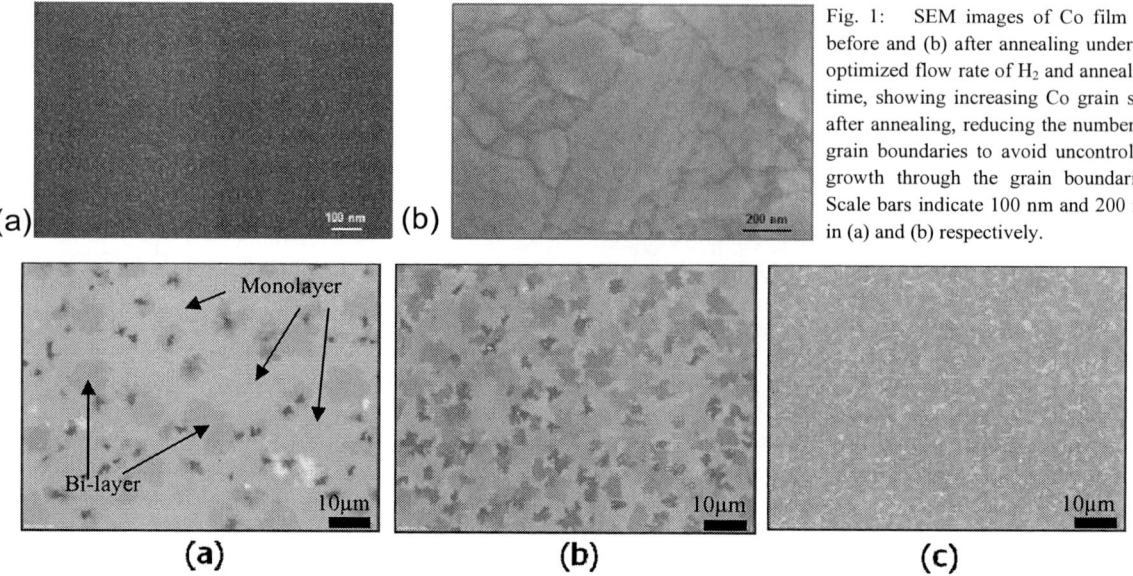

Fig. 1: SEM images of Co film (a) before and (b) after annealing under an optimized flow rate of H$_2$ and annealing time, showing increasing Co grain size after annealing, reducing the number of grain boundaries to avoid uncontrolled growth through the grain boundaries. Scale bars indicate 100 nm and 200 nm in (a) and (b) respectively.

Fig. 2: (a) Optical micrographs of graphene grown on (a) 100 nm Co film, showing large monolayer area along with small domains of bi- and multilayer area; (b) 200 nm and (c) 300 nm Co films. The domain size of monolayer graphene decreases with increasing Co film thickness.

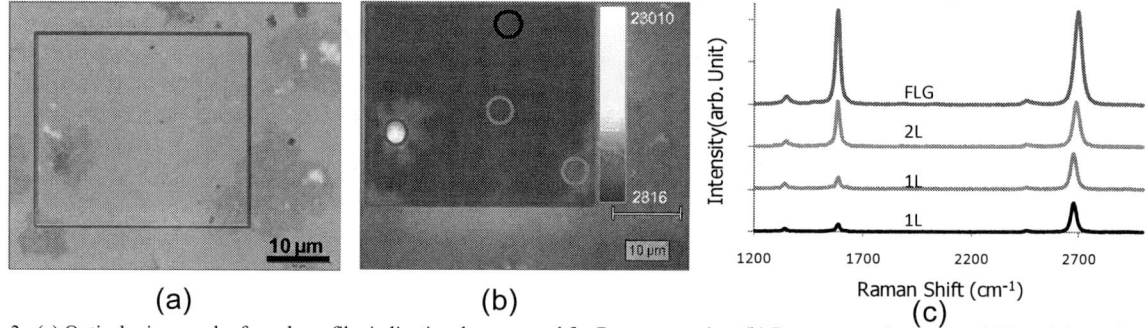

Fig. 3: (a) Optical micrograph of graphene film indicating the area used for Raman mapping. (b) Raman mapping image of 2D peak intensity. (c) The Raman spectra taken from the regions marked with corresponding colored circles as shown in (b). Raman spectrum corresponding to the black-circled (pink-circled) region shows a 2D peak position at 2675 cm^{-1} (2675 cm^{-1}) with a FWHM ~29 cm^{-1} (~30.5 cm^{-1}) and I$_{2D}$/I$_G$ ~3.89 (~3.1), indicating monolayer graphene [1]. In contrast, the Raman spectrum associated with the green-circled (blue-circled) region yields a 2D peak position at 2688 cm^{-1} (~2695 cm^{-1}), a FWHM ~35 cm^{-1} (~42 cm^{-1}) and I$_{2D}$/I$_G$ ~1 (~1), which reveals the formation of bilayer (multilayer) graphene [1]. These results support that most of the mapped area of the film (~95%) consists of monolayer graphene. All scale bars are 10 μm.

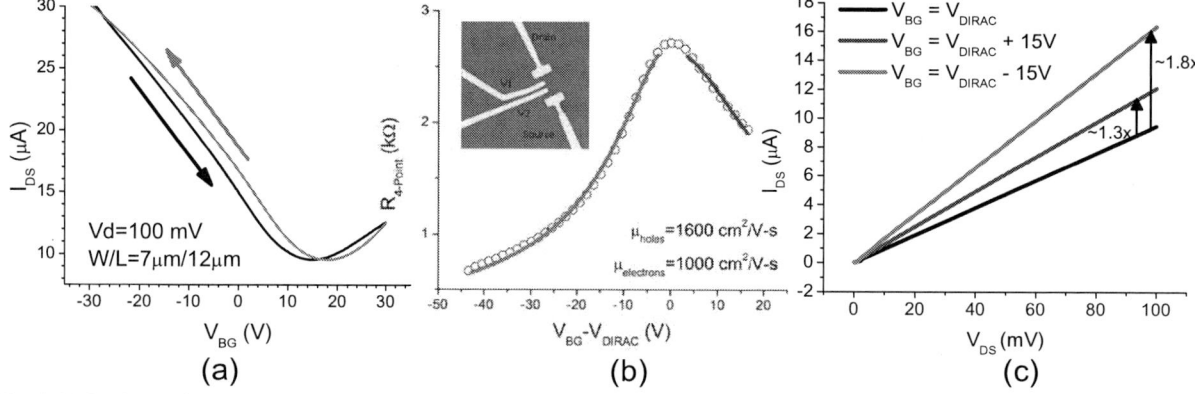

Fig. 4: (a) GFET transfer characteristics showing I$_{ON}$/I$_{OFF}$ ~ 3 and Dirac point ~15V, indicating unintentional doping during processing. From the hysteresis (~3V), the number of trapped charges per unit area was found to be ~1.2x10^{11} cm^{-2}. (b) R$_{4\text{-Point}}$ versus V$_{BG}$ - V$_{DIRAC}$ and extracted hole and electron mobility, indicating preferential hole-conduction over electron-conduction. (b-inset) Optical microscope image of a 4-point GFET. (c) Output characteristics for holes and electrons at V$_{BG}$ = V$_{DIRAC}$ (black), V$_{BG}$ = V$_{DIRAC}$ -15V (red), V$_{BG}$ = V$_{DIRAC}$ + 15V (blue). Drain current modulation is much greater when the graphene is electrostatically doped p-type (V$_{BG}$ - V$_{DIRAC}$ < 0), as compared to when the graphene is comparably electrostatically doped n-type (V$_{BG}$ - V$_{DIRAC}$ > 0), which is expected given the higher mobility for holes as compared to electrons (b).

978-1-61284-243-1/11 $26.00 © 2011 IEEE

Modeling of Dielectric Breakdown-Induced Time-Dependent STT-MRAM Performance Degradation

Georgios Panagopoulos, Charles Augustine and Kaushik Roy
School of ECE, Purdue University, West Lafayette, IN 47907, Email: gpanagop@purdue.edu

In recent years, spin-transfer torque magnetoresistive random access memory (STT-MRAM) has gained a lot of interest as a promising memory candidate for future embedded applications. STT-MRAM possesses desirable memory attributes such as excellent readability, writability, stability, non-volatility, and unlimited endurance. Moreover, ITRS reports that STT-MRAM can endure 10^{15} cycle operations before breakdown [1] thus meeting 10 yrs life-time. As shown in Fig. 1, STT-MRAM bitcell consists of one access transistor and one magnetic tunnel junction (MTJ) (1T-1R). One of the primary reliability concerns in STT-MRAM is the dielectric breakdown of the tunnel junction MgO in the MTJ known as time-dependent dielectric breakdown (TDDB). The thickness of MgO is on the order of 1nm and the voltage across the MTJ during write operation is approximately 0.7V resulting in electric field of ~10MV/cm across it which can induce TDDB [2-3]. Thus, such high stress conditions can lead to lower breakdown time (T_{BD}) which can go even lower with further MgO thickness scaling. In addition to the hard breakdown (HBD) in MTJ which results in very low MTJ impedance and inability to function as memory, experimental results show that soft breakdowns (SBD) also exists [7,8]. SBDs cause minor degradation in the MTJ resistance and they have shorter average time to appear compared to HBDs. In this paper, we explore in detail the physical mechanism behind both HBD and SBD, and using percolation model we estimate the time dependent degradation in the MTJ performance parameters such as tunneling magneto-resistance (TMR), write current (J_C), write-time (T_{WR}) and lifetime (T_{LIFE}).

In order to model the effect of TDDB in MTJs, we follow the same approach as gate dielectric breakdown modeling in MOSFETs [4]. Specifically, during SBD, a short-circuit path across the MgO is formed as shown in Fig. 1. Under continuous application of voltage across MTJ (V_{MTJ}), more paths are formed resulting in successive increase in tunneling current as presented in Fig. 2. It is obvious that the increase in current due to the percolation path is not Ohmic but follows a power law: $I_{SBD}=G_0 V^\gamma$ (Simmons model [2-4]) and this enables the prediction of MTJ current levels at various voltages. G_0 and γ are fitting parameters and they are determined by experimental data as shown in Fig. 2. Note that this current is added with the original MTJ tunneling current (Δ_1 coherent band tunneling for parallel MTJ (P) and direct tunneling current for anti-parallel case (AP)). Since the formation of a percolation path is a probabilistic phenomenon, the time-to-breakdown is a random variable which follows the Weibull distribution. Moreover, the multiple percolation paths are uncorrelated and independent of the pre-existing paths (both spatially and temporally), which further simplifies the data analysis and prediction of T_{BD}. Due to such independent nature, with data for one breakdown (say the 1st breakdown) we can predict the distributions for all the higher order soft breakdowns (i.e. 2nd, 3rd, 4th etc) using the fitted Weibull parameters α and β (see Table 1). Fig. 3 shows the benchmarked data from [2] for hard breakdown (4th SBD in our case) and produced distributions for the breakdowns 1st to 3rd. As described earlier, the current through the MTJ increases in successive breakdowns resulting in MTJ performance degradation such as TMR reduction, and increase in J_C and T_{WR}.

At the circuit level, we can describe the effect of SBD on STT-MRAM as a voltage dependent current source in parallel to the MTJ (see Fig. (1)). The percolation path current does not depend on the state of the MTJ (P or AP) and hence the difference between P and AP currents decreases due to leakage through that path. We have modeled the net resistance of the MTJ after successive SBDs and for different applied voltages as is presented in Fig. 5. It can be noted that resistance successively decreases at different time points and the difference in resistance between P and AP states also decreases resulting in lower TMR (Fig. 6). The impact of SBD on write current (J_{WR}) is also indicated in Fig. 7 and it can be observed that J_{WR} increases as the time progresses for iso-write time (T_{WR}) requirement. Higher J_{WR} is required to compensate for the current lost through the R_{PERC} in parallel shown in Fig. 7. Fig. 8 shows the plot of bitcell TMR under continued application of nominal voltage of 0.5V (determined) after extrapolation using Eqs in Table 1) for a period of 10 yrs. TMR continuously degrade over time and the critical TMR for correct STT-MRAM read operation (to distinguish between P and AP states) is also indicated in the figure. When the bitcell TMR gets lower than the critical TMR, the STT-MRAM bitcell fails to function, thus shortening the memory life-time ($T_{LIFE,RD}$). In addition to the critical TMR constraint, bit-cell also requires write current larger than critical current (J_C) to switch the MTJ from P to AP and vice-versa. Fig. 9 shows that J_C increase for 10 year period and the bitcell current for the write-operation is indicated in the figure. As the J_C requirement increases, the cell will fail to switch within the given write time (T_{WR}) resulting again in cell failure and finite lifetime ($T_{LIFE,WR}$). The net result of both read and write failures is the reduction of memory lifetime which is the minimum of $T_{LIFE,RD}$ and $T_{LIFE,WR}$ (Fig. 10). Note that, in addition to SBD, process variations in the MTJ, especially in the MgO thickness can also degrade the lifetime and are predicted by the percolation model as shown in Fig. 11. Hence, providing enough design margins for both read (providing higher TMR than required) and write in STT-MRAM is necessary to increase the lifetime of the memory bitcell and array. Note that, dynamically adjusting the write voltage can also be a solution to increase the STT-MRAM lifetime.

In conclusion, we have studied the performance degradation of a STT-MRAM cell and the predicted lifetime due to successive soft breakdowns. At the device level, the degradation is modeled using percolation theory for the formation of conducting paths across MgO which leads to TMR degradation at the circuit level (parameters are presented in Table 1). The device model is coupled with self-consistent LLG-NEGF simulation model [6] to predict the STT-MRAM performance over the memory life-time and can be used as a guide by memory designers for realizing high-performance, low-power, scalable and reliable STT-MRAMs.

References:
[1] http://www.itrs.net
[2] C. Yoshida et. al., *IEEE IRPS, 2009*, pp. 139-142.
[3] K. Hosotani et. al., *IEEE, IRPS*, 2007, pp. 650-651.
[4] M. A. Alam et. al., *IEDM*, 2002, pp. 151-154.
[5] Q. Chen et. al., *JAP, 2009*, **105**, 07C931
[6] C. Augustine et. al, IEDM, 2010, 22.7.1-22.7.4
[7] D. V. Dimitrov et. al., *APL*, 2009, **94**, 123110.
[8] D. Kim et. al., *JJAP, 2003*, 42, pp. 1242-1245.
[9] A. A. Khan et. al, *JAP*, 2008, **103**, 123705

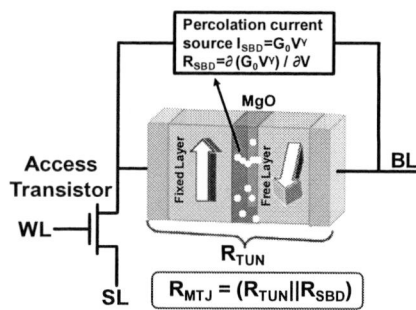

Fig. 1: 1T-1R STT-MRAM bitcell with TDDB induced percolation current source. After a SBD a resistance R_{SBD} is added in parallel to MTJ to emulate the effect of TDDB on MgO.

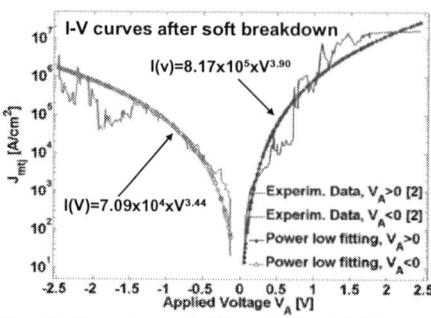

Fig. 2: Experimental [1] and fitted MTJ I-V curves after breakdown. The data follow the power law $(=G_0V^\gamma)$ known as Simmons's model [2-4].

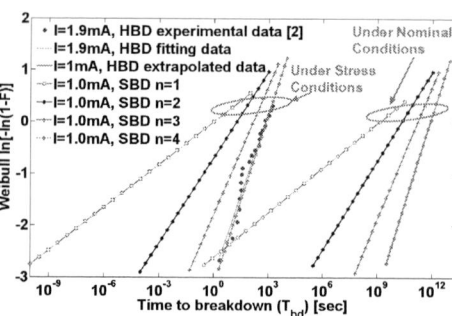

Fig. 3: HBD Weibull distributions for different MTJ currents (1mA and 1.9mA) and extrapolated Weibull distributions for SBDs using the equations presented in Table 1.

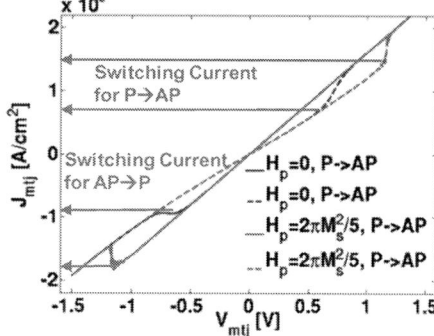

Fig. 4: MTJ I-V characteristics for PMA (perpendicular magnetic anisotropy Hp=0) and IMA (in-plane magnetic anisotropy, $Hp=2\pi Ms^2$).

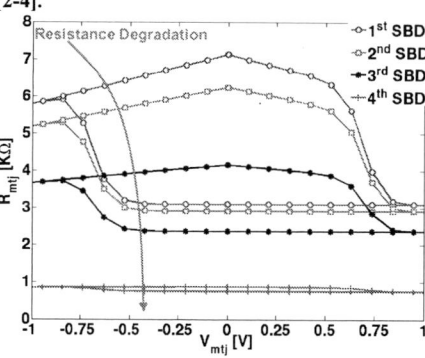

Fig. 5: R_A and R_P temporal degradation after successive SBDs due to TDDB.

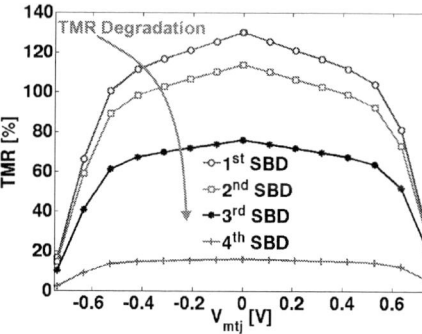

Fig. 6: TMR vs. V_{MTJ} curve after 0, 1st, 2nd, 3rd and 4th dielectric soft breakdowns.

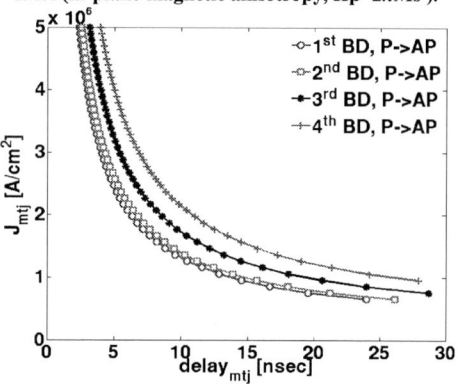

Fig. 7: MTJ current density (JC) vs. switching delay curves after 0, 1st, 2nd, 3rd and 4th dielectric breakdowns.

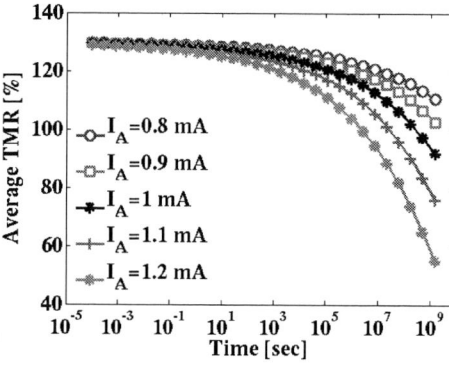

Fig. 8: Average MTJ TMR degradation as a function of time for different applied currents. For large currents ~1.2mA the TMR can be reduced to 77% after ~12 days.

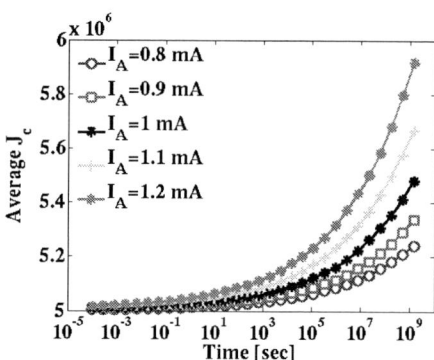

Fig. 9: Average MTJ critical write current (JC) as a function of time (@iso switching delay) after successive SDBs.

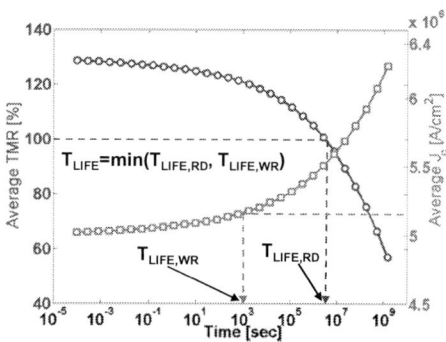

Fig. 10: Bitcell TMR and write current degradation after successive breakdowns at constant stress of 1mA. Critical TMR (100%) and maximum available write current available in the bitcell (J_{MAX}) is also shown in the figure, which determines the T_{LIFERD} and T_{LIFEWR} respectively.

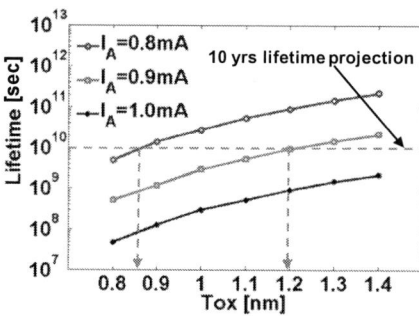

Fig. 11: Increasing the MgO thickness increases the MTJ lifetime drastically. This result is generated by changing the parameter Tox in the percolation model.

Parameter	Value
Magnetization Ms	800 emu/cm³
K_{u2}	4×10^4 erg/cm³
Dumping coefficient α	0.01
Free Layer Volume V [WxLxT]	200nm x 100nm x 2.1nm
MgO thickness	1.0nm
NMOS size [WxL]	1000nm x 110nm
Read/Write voltage	0.3V/1.0V

$$W_{HBD} = \beta_{HBD}\ln t - \beta_{HBD}\ln\alpha_{HBD}$$

$$W_{SBD}(n) = \frac{\beta_{HBD}}{N}\ln\left(\frac{t}{\alpha_{SBD}}\right) - \ln(n!)$$

$$T_{LIFE} = \min\left(T_{LIFE,RD}, T_{LIFE,WR}\right)$$

Table 1. STT-MRAM and percolation model parameters used in our analysis.

Introduction of ALD Beryllium Oxide Gate Dielectric for III-V MOS Devices

T. Akyol[1], J. H. Yum[1,5], D. A. Ferrer[1], M. Lei[2], M. Downer[2], C. W. Bielawski[3], T. W. Hudnall[4],
G. Bersuker[5], J.C. Lee[1], S. K. Banerjee[1]

[1]Microelectronics Research Center – Dept. of Electrical and Computer Eng., [2]Dept. of Physics, [3]Dept. of Chemistry, University of Texas at Austin, TX 78758, [4]Texas State University at San Marcos, [5]Sematech Austin, TX 78741 USA
Phone: (512)736-6859 / E-mail: tarikakyol@mail.utexas.edu / Fax: (512)471-8420

Using atomic layer deposited (ALD) Beryllium oxide (BeO) as a gate dielectric for the first time, we present improved surface channel MOSFETs on III-V substrates. We used a self-aligned gate-last process to fabricate MOSFETs on semi-insulating InP substrates with TaN gate electrode. The electrical characteristics of n-MOSFETs and MOS-Capacitors and physical characteristics of the BeO high-κ dielectric film were investigated and are summarized in this paper. BeO gate dielectric n-MOSFETs show excellent surface channel dc output characteristics, supporting high possibility of utilizing it in III-V CMOS technology.

The most challenging issue in III-V MOS devices is the lack of high quality interface like that between SiO_2 and Si substrate. Introduction of atomic layer deposition and high-κ gate dielectrics such as Al_2O_3, HfO_2 and ZrO_2 mitigated this barrier and made the III-V channels a component of the MOSFET menu [1-2]. Recently, by employing ALD BeO as a gate dielectric, excellent results such as smaller capacitance-voltage (C-V) hysteresis, lower frequency dispersion, interface defect density (D_{it}) and gate leakage current have been achieved compared to Al_2O_3 with same effective oxide thickness (EOT) [3]. In addition, BeO has high thermal stability on Si and III-V substrates, has large energy band-gap (10.6eV) and it is a strong diffusion barrier [4-5]. In this work, we report results of InP MOSFETs with ALD BeO as a high-k gate dielectric and its comparison with Al_2O_3.

Table 1 illustrates the self-aligned gate-last MOSFET fabrication procedure that we followed. The schematic vertical device structure of TaN/BeO/InP MOSFETs and the ring-type device layout used are shown in Figure 1. Semi-insulating (100) InP wafers were used for this process. We prepared the ALD BeO precursor (Dimethyl beryllium) for the first time by synthesizing $Be(CH_3)_2$ from $BeCl_2$ via Grignard metathesis [6]. Atomic force microscopy (AFM) results after ALD BeO deposition shows a very low RMS surface roughness of 0.19nm in 3 x 3 um scan area. In comparison, the surface roughness of ALD Al_2O_3 sample with same physical thickness was measured as 0.23nm (Figure 2).

C-V and I-V characteristics of MOS Capacitors with ALD BeO dielectric are superior to its Al_2O_3 counterpart in terms of lower frequency dispersion, hysteresis and leakage current (Figure 3). X-ray photoelectron spectroscopy (XPS) results in Figure 4 demonstrate the interface quality comparison between Al_2O_3 and BeO samples with and without post-deposition annealing (PDA). We see that BeO is indeed successfully deposited by ALD using our Dimethyl beryllium precursor. It has an efficient self-cleaning effect by removing the InP native oxide and has better thermal stability compared to ALD Al_2O_3.

Channel carrier mobility is generally regarded as one main figure of merit benchmarking the MOSFET performance. Calculated effective mobility of InP n-MOSFETs with BeO dielectric (using split C-V method) exhibit maximum electron mobility 762 cm^2/V.s, which is the highest reported value for a surface channel InP n-MOSFET with high-k dielectrics. Figure 5 compares dc output characteristics of n-channel MOSFETs (W/L = 600/10 um) fabricated with ALD Al_2O_3 and BeO gate dielectrics. BeO MOSFETs show better performance with lower threshold voltage, higher maximum transconductance, drive current and current on/off ratio. Even though our ALD BeO quality and fabrication process were not fully optimized, it can be concluded from these MOSFET characteristics results that ALD BeO has great potential for application in III-V technology.

In summary, for the first time, we have fabricated and characterized surface channel n-MOSFETs on SI-InP substrate with ALD BeO dielectric. These MOSFETs show greatly improved device characteristics compared to that for Al_2O_3. Introducing ALD BeO as a gate dielectric or interface passivation layer (IPL) may be the key solution to overcome the performance limit of current III-V MOS devices with high-κ dielectrics.

This work was supported in part by DARPA, Micron Foundation, Robert Welch Foundation grant F-1038 and NSF grant DMR-0706227.

[1] J. R. Kwo et al., *ECS Transactions*, v19, 2 (2009).

[2] R. J. W. Hill et al., *Solid State Tech.*, v53, 3 (2010).

[3] H. Yum et al., *J. App. Phys.*, v109, 4 (2011).

[4] V. A. Sashin et al., J. Phys.: Condens. Matter., v15, p. 3567 (2003).

[5] K. J. Hubbard and D. J. Schlom, J. Mater. Res. v11, p.2757 (1996).

[6] H. Gilman and F. Schulze, J. Chem. Soc., v49, p.2663 (1927).

- Surface cleaning in 1% HF
- ALD Al$_2$O$_3$ (100A°) capping layer deposition
- Align-mark formation (photo-litho. & etching)
- Photo-litho. and ion implantation for S/D (n-channel) Si 35 KeV, 3x10^{14} cm^{2}
- Removal of photoresist and S/D activation with RTA (at 750 °C, 15sec)
- Removal of Al$_2$O$_3$ capping layer (in BOE)
- Surface clean & passivation (1% HF and (NH$_4$)$_2$S)
- Deposition of ALD BeO/ Al$_2$O$_3$ and PVD TaN
- Gate patterning (photo-litho. & RIE)
- S/D contact formation (photo-litho. & BOE etch)

Table 1: Gate-last MOSFET fabrication process flow

Figure 2: AFM results after ALD BeO / Al$_2$O$_3$ deposition with surface roughness (Rq) values

Figure 4: XPS Analysis of ALD Al2O3 and BeO samples with and without post-deposition annealing

Figure 1: Schematic of device cross-section structure and ring-type layout

Figure 3: C-V and I-V characteristics of n-InP MOS-CAPs

Figure 5: DC output characteristics of n-channel InP MOSFETs (W/L = 600/10 μm) with BeO and Al$_2$O$_3$ dielectrics

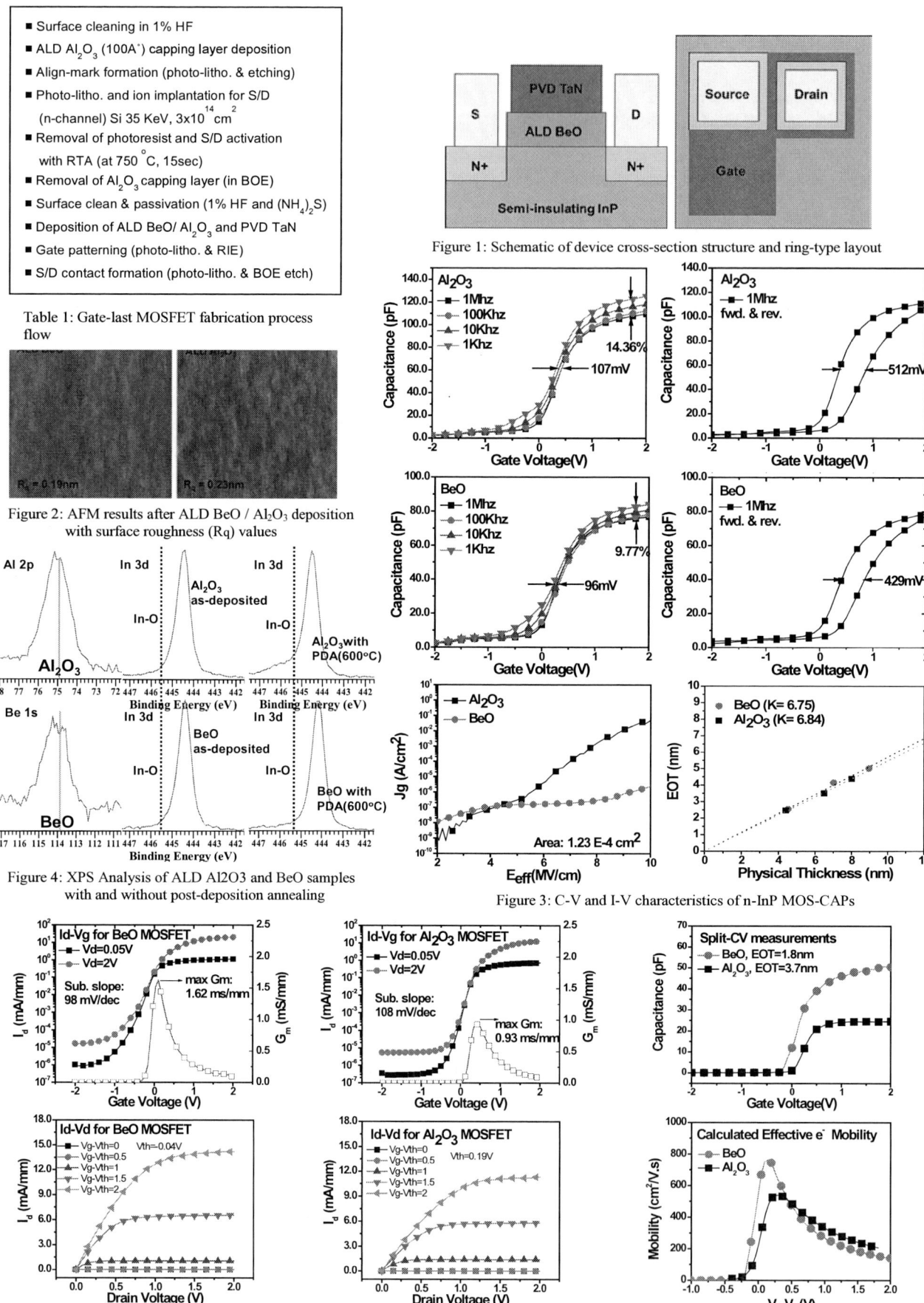

978-1-61284-243-1/11 $26.00 © 2011 IEEE

Transport Properties of CVD-Grown Graphene Nanoribbon Field-Effect Transistors

Austin S. Lyons, Ashkan Behnam, Edmond K. Chow, and Eric Pop

Dept. of Electrical and Computer Engineering, University of Illinois, Urbana-Champaign
Micro and Nanotechnology Lab, 208 N Wright St, Urbana IL 61801. E-mail: epop@illinois.edu

Graphene nanoribbons (GNRs) are promising candidates for nanoelectronics as interconnects or field-effect transistors (FETs) [1,2]. Previous GNR studies used chemically derived [1] or mechanically exfoliated [2] graphene, which are not practical for large scale fabrication. In this work we present a comprehensive analysis of GNR FETs obtained by chemical vapor deposition (CVD) [3], which is promising for creating wafer-scale circuits. We demonstrate low-bias, high-bias, and temperature-dependent measurements. We find that CVD GNRs have properties comparable to the best state-of-the-art GNRs obtained by other methods, suggesting that grain boundaries play a negligible role in sub-100 nm devices. This approach also serves to identify future challenges and represents a first step towards large-scale integration.

GNR Fabrication: The CVD graphene and GNR device process are illustrated in Figs. 1-2. CVD of methane on Cu foil results primarily in monolayer graphene growth (Fig. 1b) on both sides of the Cu foil. One graphene side is protected with PMMA, while the other is removed with an oxygen plasma etch. The Cu foil is removed using $FeCl_3$, and the PMMA/graphene film is transferred to SiO_2 (90 nm)/Si substrates. Fig. 2 shows GNRs defined by e-beam lithography in the channel region, but the CVD graphene extends under the Ti/Au (0.5/40 nm) contacts, which aids in lowering the contact resistance.

Low-Field Characterization: We present complete data sets of 22 GNRs with widths W = 15−50 nm and lengths L = 100−700 nm. Low-bias measurements in vacuum (~10^{-5} torr) at room temperature are fitted against a transport model [4], revealing mobility $\mu \approx$ 300-500 cm^{2}V^{-1}s^{-1} (at 5×10^{12} cm^{-2} carrier density) and contact resistance $R_C W \geq 500$ $\Omega \cdot \mu$m, as shown in Figs. 3 and 4. These properties are comparable to the best state-of-the-art chemically derived or exfoliated GNRs [1,2], but have not been demonstrated on CVD GNRs until now. This finding suggests that polycrystalline grain boundaries may play a weak (if any) role in such small CVD-grown devices. Fig. 4(b) shows the mobility is proportional to temperature for T $\geq$ 125 K, suggesting that low-field transport is limited by charged impurity scattering [4,5].

High-Field Characterization: We also performed high-field measurements on the same GNR FETs at 300 K to investigate maximum current density (I_{MAX}), up to GNR breakdown (Figs. 5-8). Some devices were capped by an AlO_x layer (Fig. 2) and showed average $I_{MAX} \sim$ 0.7 mA/μm, with the highest $I_{MAX} \sim$ 1.2 mA/μm. For other devices we removed the protective AlO_x layer and found *higher* peak current density $I_{MAX} \sim$ 1.3 mA/μm, with the highest value of ~3.4 mA/μm in air ambient for the widest device measured (W ~ 50 nm). Two control GNRs measured in vacuum both exhibited $I_{MAX} \sim$ 2.8 mA/μm, and all data are summarized in Figs. 6-7. Unlike in carbon nanotube FETs [6], current saturation was not observed in these GNR FETs (Fig. 6), although current densities are all relatively high. As shown in Fig. 6, no clear scaling between I_{MAX} and device dimensions (L/W ratio) is found. This suggests that sample-to-sample variability from impurities, edges, and contacts remains a concern and must be further addressed.

GNR Breakdown: We now turn to the behavior of such GNRs up to breakdown (BD), where no other data sets exist today. The GNR breakdown power P_{BD} scales approximately proportionally with GNR area (Fig. 7), indicative of the role of heat dissipation from the GNR to the SiO_2 under such high bias conditions [7]. At high fields our narrower GNRs broke cleanly and irreversibly, similar to carbon nanotubes [6], as shown in Fig. 5. However, our widest GNRs with W ~ 50 nm broke down in several stages, as shown in Fig. 8. This appears to suggest that breakdown may occur along grain boundaries which are more likely to occur in wider GNRs, although further study is needed to confirm this.

In summary, this work represents the first demonstration of GNR FETs using CVD-grown graphene, a key step towards large-scale integration. CVD GNRs are comparable to the best state-of-the-art GNRs obtained by other methods, and can support current densities up to ~3 mA/μm. Presenting data from 22 samples, this work also serves to identify key future challenges, such as device variability.

[1] X. Wang et al, PRL 100, 206803 (2008). [2] Y. Yang et al, EDL 31, 237 (2010). [3] X. Li et al, Science 324, 1312 (2009). [4] V. Dorgan et al, APL 97, 082112 (2011). [5] T. Fang et al, PRB 78, 205403 (2008). [6] A. Liao et al, PRB 82, 205406 (2010). [7] E. Pop, Nano Res. 3, 147 (2010).

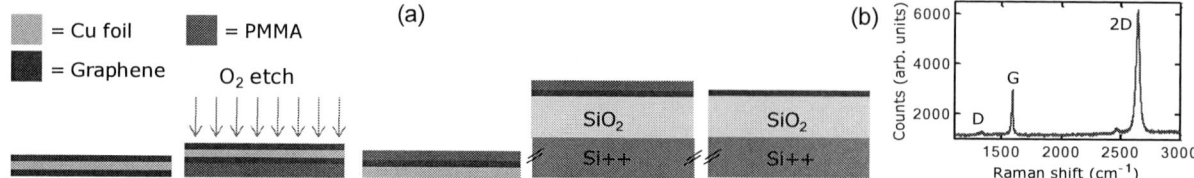

Fig. 1: (a) CVD growth on Cu and transfer process to SiO_2(90 nm)/Si substrates. The substrate is used as the back gate in electrical measurements. PMMA liftoff completes the CVD growth and transfer process. **(b)** Raman spectrum of CVD graphene, indicating good quality (small D peak) and predominantly monolayer coverage (2D:G > 2).

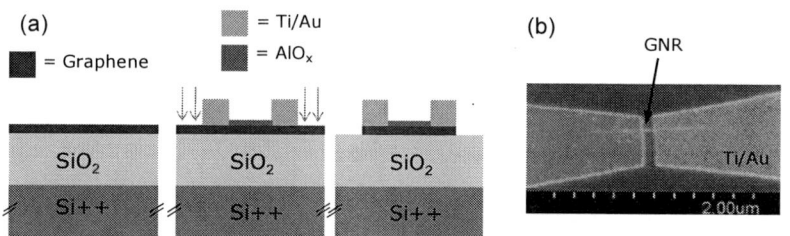

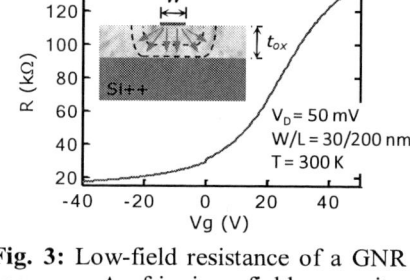

Fig. 2: (a) Ti/Au (0.5/40 nm) source-drain electrodes are patterned in PMMA. A second step patterns a thin (~2 nm) Al nanoribbon which then oxidizes in ambient. After liftoff, the AlO_x is used as an etch mask to form GNRs. **(b)** SEM image of finished device (AlO_x ribbon can be seen over the contacts). For some devices the AlO_x was etched away (batch b1 below).

Fig. 3: Low-field resistance of a GNR in vacuum. A fringing field capacitance model (inset) was used in the mobility calculations below. I_{ON}/I_{OFF} ~ 10 is comparable to other GNRs with W > 15 nm [1].

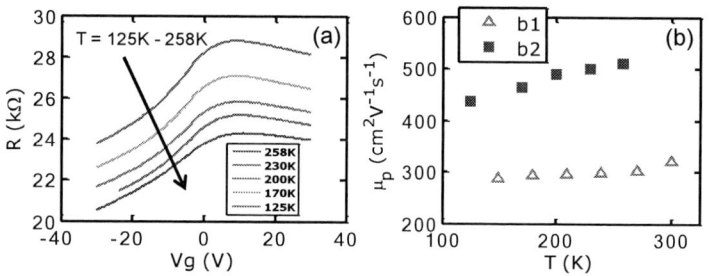

Fig. 4: (a) R-Vg for a GNR capped with AlO_x (W/L = 40/400 nm). **(b)** Extracted hole mobility increases with T for both an uncovered GNR (batch b1) and an AlO_x capped GNR (batch b2).

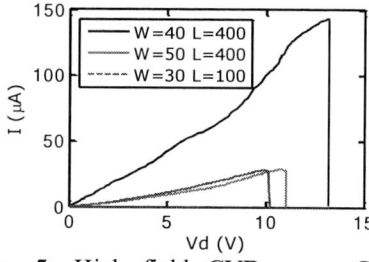

Fig. 5: High field CVD-grown GNR measurements. Unlike for CNTs [6], no current saturation is observed.

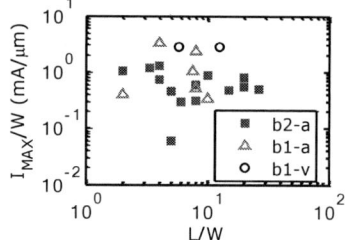

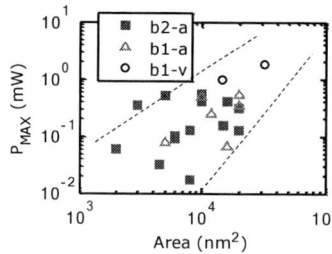

Fig. 6: I_{MAX} typically reaches >1 mA/μm but does not appear to scale with L/W. Most devices from batch b1 were broken in air (△) while two were broken in vacuum (o).

Fig. 7: Breakdown power tends to scale with device area, indicating the role of heat dissipation through the SiO_2 [7].

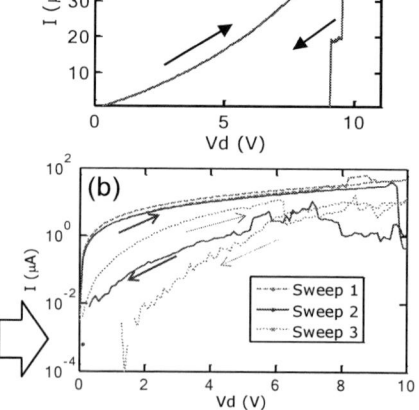

Fig. 8: Two wider (W = 50 nm) GNRs undergoing successive partial breakdowns. **(a)** W/L = 50/100 nm showing partial breakdown followed by complete breakdown. **(b)** W/L = 50/200 nm with successive partial breakdowns.

978-1-61284-243-1/11 $26.00 © 2011 IEEE

Protein Nanopore-gated Bio-transistor for Membrane Ionic Current Recording

Tae-Sun Lim, Dheeraj Jain, Peter J. Burke

Integrated Nanosystems Research Facility, Department of Electrical Engineering and Computer Science,
University of California Irvine, Irvine, CA, 92697 USA
E-mail: pburke@uci.edu, Phone: (949) 824-9326, Fax: (949) 824-3732

Although naturally occurring biological nanopores have shortcomings such as a relatively weak structural durability and a limited life-time, they are still intriguing candidates for nanobiosensing applications due to their sensitivity and specificity to analytes as well as various choices of ion channels depending on functionalities. In order to overcome limitations of biological nanopores, man-made solid-state nanopores have been explored. The fabricated solid-state nanopore is structurally durable and suitable for nanofabrication process yet it is still challenging to construct and a low throughput process, and lacks the chemical specificity of natural ion channels[1]. Can bionanotechnology be applied to improve this situation? Recent work has shown that nanomaterials (nanotubes, nanowires) can be gated by electrolyte, and even coated with lipid bilayers allowing charges of either the bilayer themselves[2]. These reports focus on time average changes in source/drain current due to gating by charges near the nanowire/nanotube. Thus, to date, no nanowire/nanotube device has been able to measure the time-dependent single ion channel recording.

Here we show direct interrogation of the dynamical opening and closing of ion channel pores with an integrated nano-bio system in a controlled microfluidic environment. Randomly oriented semiconducting CNT network transistors are covered by biomembranes (lipid bilayers) incorporated with transmembrane proteins (ion channels). This forms a robust and direct probe of ion channel currents with millisecond temporal resolution and sub-pico amp current resolution. The effect of both biomembranes and ionic current through transmembrane proteins on the electrical characteristic of transistors has been evaluated. We also demonstrate for the first time electrophysiological recording of single ion channel events from individual ion pores using our CNT transistors. The device developed here could offer a nano-electrophysiology system to study interactions between biomolecules and bionanoelectronics and bioelectrical activities of the cellular membrane.

Fig. 1(a) shows schematics of experimental set up of CNT device showing p-type I-V curves depending on source-drain bias. The continuous and defect-free supported lipid bilayers were formed on top of fabricated CNT FET as shown in Fig.1(b,c). Electrical characteristic modulations of CNT FETs were clearly observed by introduction of biomembranes and reconstitution of ion channel into biomembranes (Fig. 1(d)). This result indicates that CNT FETs can be gated by tiny ionic current flowing into local environment of CNTs in which the type of ions can be selected and controlled by the ion channel used. For ion channel recording with CNT electrophysiological system, time-dependent ionic current flow of specific ions was recorded (Fig. 1(e)). I-V responses of devices were measured to obtain the conductivity of devices and the ability of the biomembranes to isolate CNT device from the medium solution Fig 1(f). For the first time, single ion channel recordings using a CNT device were successfully recorded with high signal to noise ratio. Each current spike with the range of 0.5 -10 pA indicates ionic current flow when an individual ion channel is open.

This approach, significantly different than any prior approach to the study of ion channel trans-membrane currents, will find broad applications in electrophysiology, nanopore based sequencing, and membrane protein studies.

This work was supported by NIH National Cancer Institute Grant (CA143351-01).

[1] D. Branton, et al., "The potential and challenges of nanopore sequencing," Nature Biotechnology, vol. 26, pp. 1146-1153, 2008.

[2] X. Zhou, et al., "Supported lipid bilayer/carbon nanotube hybrids," Nat Nanotechnology, vol. 2, pp. 185-90, Mar 2007.

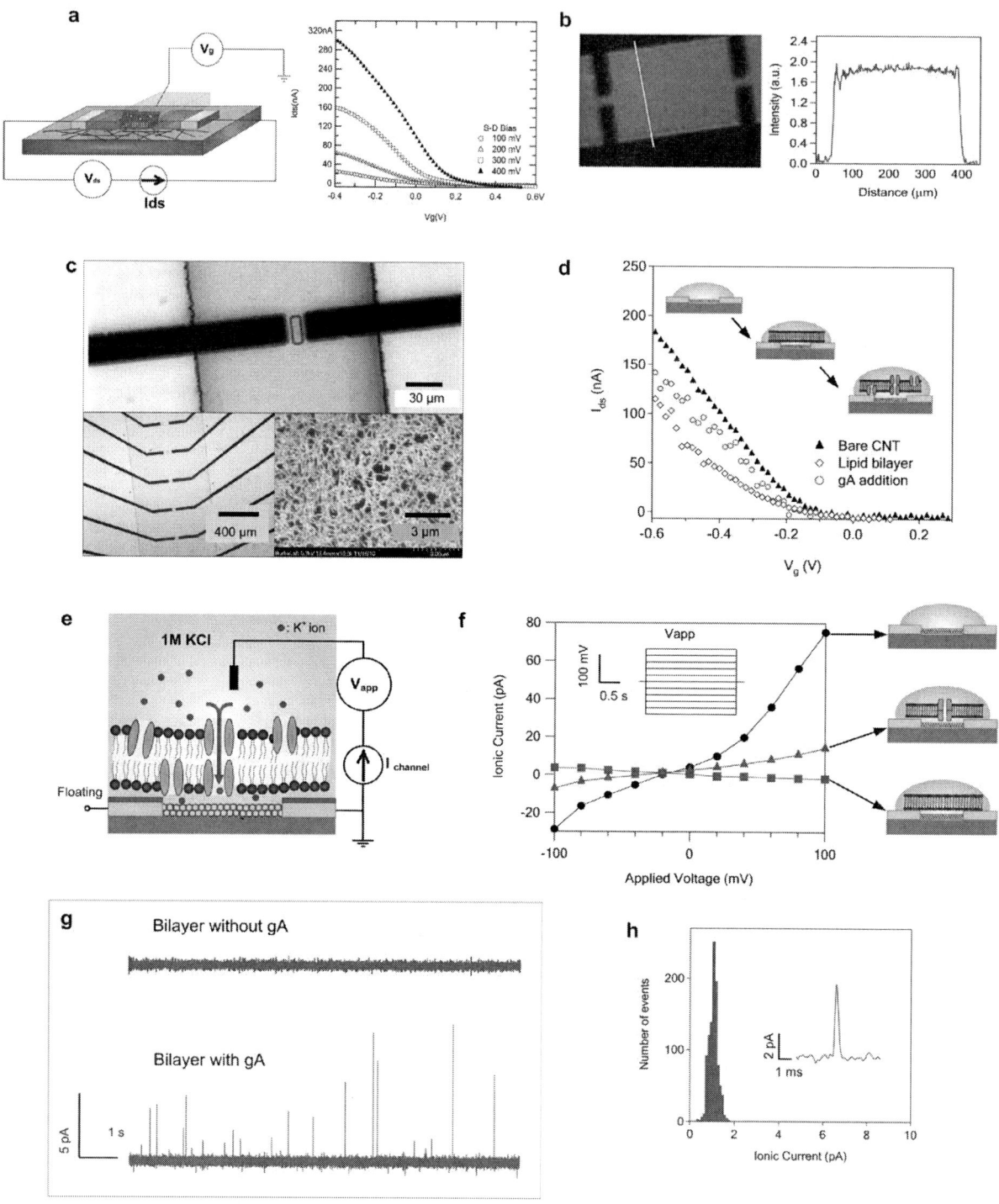

Figure 1 (a) Experimental set-up of CNT FET and typical p-type depletion curves (b) Fluorescence image of lipid bilayer and intensity profile (c) Micrograph of fabricated device and a SEM of CNT networks (d) Depletion curves after introduction of biomembranes (blue) and ion channels(red) (e) Cartoon of ion channel recording showing ionic current flow (f) I-V responses of CNT device of bare(black), biomembrane(blue), ion channel(red) (g) Individual ion channel current trace (h) Histogram of single ion channel events and typical event dwell-time.

Tunnel FET-Based Pass-Transistor Logic for Ultra-Low-Power Applications

Sung Hwan Kim, Zachery A. Jacobson, Pratik Patel, Chenming Hu, and Tsu-Jae King Liu

EECS Department, University of California, Berkeley, CA 94720-1770 USA

Phone: +1-510-529-1315, Fax: +1-510-642-2739, E-mail: shpkim@eecs.berkeley.edu

Abstract — **Germanium-source tunnel-FET-based pass-transistor logic gates are proposed and benchmarked against conventional CMOS logic gates via mixed-mode simulations, for 15 nm L_G. For low throughput applications (>100 ps gate delay), TPTL is advantageous for reductions in dynamic energy and leakage power.**

I. INTRODUCTION

The germanium-source silicon n-channel tunnel field effect transistor (TFET) has been demonstrated to achieve the highest on/off current ratio (I_{ON}/I_{OFF}) to date, for a supply voltage (V_{DD}) of 0.5 V [1]. Due to the valence-band offset between Ge and Si, it is difficult to leverage the lower energy bandgap of Ge to achieve high I_{ON} for a Si p-channel TFET. The lack of a complementary TFET design with commensurate performance motivates the study of nTFET-based pass-transistor logic (PTL) circuit topologies. In this work, two-dimensional device mixed-mode simulations are used to assess the performance of nTFET PTL gates and to compare their power/energy requirements against those of conventional CMOS logic gates under various delay constraints.

II. DEVICE DESIGN AND SIMULATION

Fig. 1(a) shows the optimized Ge-source nTFET design reported in [2]. A dynamic nonlocal band-to-band tunneling (BTBT) model calibrated to experimental data in [1] is used to perform dc, ac, and transient simulations [3]. Due to its steep switching behavior (Fig. 1(b)), the TFET can have lower threshold voltage (V_{TH}) and achieve lower on-state resistance than a MOSFET for a given I_{OFF} specification, at low supply voltage (V_{DD}). This provides for faster pass-transistor operation and improved pull-up (PU) behavior of the nTFET as compared against a fully-depleted (FD) nMOSFET [4] designed to meet ITRS design specifications at the same gate length [5] (Fig. 2).

III. TFET-BASED CIRCUIT DESIGN CONSTRAINTS

Because of its asymmetric structure, the nTFET is designed to conduct on-state current in only one direction (electrons flowing from the source to the drain). In addition to this constraint for TFET-based circuit design, care must be taken to consider the possibility of significantly forward-biasing the source-drain p-i-n diode since it results in a large parasitic diode current (I_S or I_{DIODE}) that flows from the source to the drain independently of the gate bias (Fig. 3(a)). This diode current is enhanced by the use of a small bandgap material (*i.e.* Ge) in the source region to maximize TFET I_{ON}, because it reduces the built-in potential [6]: In the Ge-source nTFET, electron injection into the p-type Ge source comprises the dominant component of forward-bias diode current (Fig. 3(b)). I_{DIODE} can be reduced by increasing the gate work function, with the tradeoff of increased V_{TH} and degraded TFET I_{ON}/I_{OFF} (Fig. 4).

IV. NTFET-BASED PASS TRANSISTOR LOGIC

2-input AND/NAND and OR/NOR gates implemented with nTFETs are shown in Fig. 5(a) and 5(b), respectively. As compared with nMOSFET implementations [7], the nTFET-based designs require one additional transistor for each output. The aforementioned design constraints are met by reducing the number of input signals to the source/drain terminals, by tying PU TFETs to V_{DD} (M2 and M3 in Fig. 5(a)) and tying PD TFETs to ground (M2 and M3 in Fig. 5(b)) to prevent I_{DIODE} from opposing the intended PD or PU operation, and by leveraging I_{DIODE} to assist the intended PU or PD operation (M1 in Fig. 5(a) and Fig. 5(b)). For example, for the NAND gate with A=0 and B=1, M3 pulls up the output node via BTBT current and is assisted by M1 which supplies current to the output node via I_{DIODE}. For the NOR gate with inputs A=1 and B=0, M3 discharges the output node via BTBT current and is assisted by M1 via I_{DIODE}. Level restoration to V_{DD} is needed in-between stages, and can be easily achieved with weak pMOSFET pull-up devices as shown in Fig. 6(a). The corresponding transient response of the nTFET NAND gate is shown in Fig. 6(b).

The leakage power and switching energy requirements for nTFET PTL *vs.* CMOS NAND gates are compared for various delay constraints in Fig. 7. For delays down to 100 ps, the TPTL implementation is more energy efficient.

V. CONCLUSION

Pass-transistor logic gate designs for Ge-source Si n-channel TFET technology are proposed and benchmarked against conventional CMOS technology implementations. For delays down to 100 ps, TPTL-based design is projected to have better energy efficiency and hence it is promising for ultra-low-power logic applications.

978-1-61284-243-1/11 $26.00 © 2011 IEEE

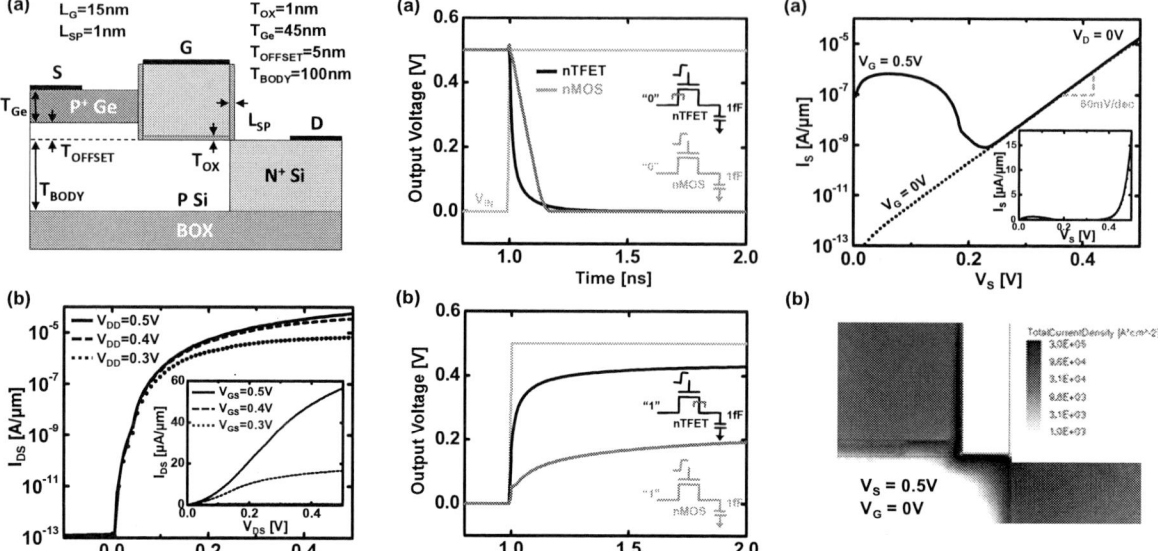

Fig. 1. (a) Schematic cross-sectional view of the raised Ge-source nTFET and device parameters. (b) Transfer characteristics (inset: output characteristics) for various V_{DD}.

Fig. 2. Comparison of the (a) discharging and (b) charging characteristics of the nTFET *vs.* nMOSFET. Inset: symbols and the input signals for the two devices.

Fig. 3. (a) Forward-bias source-drain current of a TFET for V_G=0 and 0.5V. (b) Current density contour plot showing electron injection from the N^+ drain to be the predominant source of diode current (surface element).

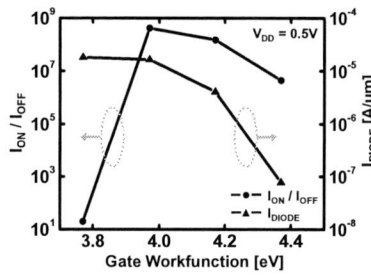

Fig. 4. I_{ON}/I_{OFF} and I_{DIODE} vs. gate work function.

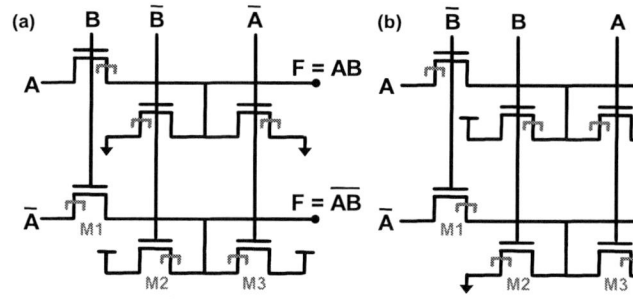

Fig. 5. 2-input nTFET PTL schematics for (a) AND/NAND and (b) OR/NOR.

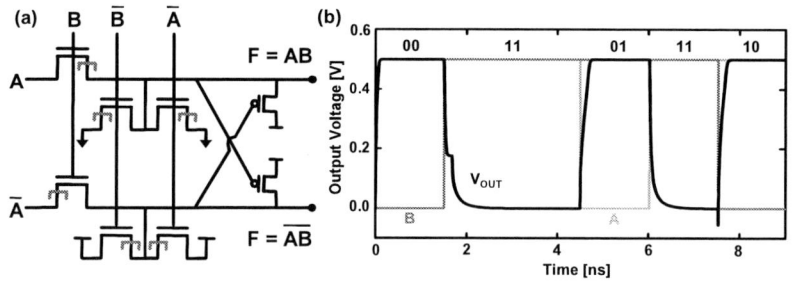

Fig. 6. (a) Minimum-sized weak pMOSFET pull-up devices can be used to restore high outputs to V_{DD}. (b) Transient response of the nTFET NAND for various combinations of the input signals A and B. Gate work function = 4eV, $I_{OFF, nTFET}$ = 0.1pA/μm, and $I_{OFF, PMOS}$ = 10pA/μm.

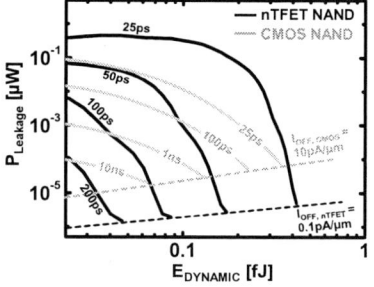

Fig. 7. Average $P_{Leakage}$ *vs.* $E_{DYNAMIC}$ for a nTFET NAND gate (with minimum-sized PMOS level restorers as in Fig. 6(a)), compared against a conventional CMOS NAND gate (W_{PMOS}/W_{NMOS} = 1.2), for various delay constraints.

REFERENCES [1] S. H. Kim *et al.*, *Symp. VLSI Tech.* Digest, 178, 2009. [2] S. H. Kim *et al.*, *IEEE Elec. Dev. Lett.* **31**, 1107, 2010. [3] *Sentaurus User's Manual*, Synopsys, v.2010.03. [4] C. Shin *et al.*, *Int'l SOI Conf.*, 2009. [5] *ITRS*, 2010. http://public.itrs.net. [6] K. Bhuwalka *et al.*, *Jpn. J. Appl. Phys.* **43**, 4073, 2004. [7] J. Rabaey, *Digital Integrated Circuits*, Prentice Hall, 2003.

978-1-61284-243-1/11 $26.00 © 2011 IEEE

Metal/III-V Effective Barrier Height Tuning using ALD High-κ Dipoles

Jenny Hu, Krishna Saraswat, H.-S. Philip Wong

Department of Electrical Engineering, Stanford University, Stanford, CA 94305, USA, Email : jennyhu@stanford.edu

INTRODUCTION

As we continue to scale silicon CMOS to its limit, there is an increased interest in III-V compounds as a high mobility alternative material [1]. To use III-V materials for CMOS, it is critical to develop a method to controllably tune the effective metal/III-V barrier height ($\Phi_{B,eff}$) for FET structures that use Schottky barrier and scalable non-alloyed ohmic contacts. In our previous work [2-3], we demonstrated the ability to shift the metal/n-GaAs Fermi level towards the conduction band by inserting a thin Si_xN_y or Al_2O_3, and reduced $\Phi_{B,eff}$ from 0.75 to 0.17eV. **In this work, we expand our investigated dielectrics to include ALD HfO_2, TiO_2, and ZrO_2. A similar reduction in $\Phi_{B,eff}$ is observed in all cases. We also discover that the presence of high-κ/high-κ dipoles in HfO_2/TiO_2 will further reduce the $\Phi_{B,eff}$ and contact resistance (R_C) beyond that of a single dielectric, despite an overall thicker dielectric.**

SCHOTTKY BARRIER MODULATION

Fermi level pinning in metal/semiconductor junctions causes Schottky barrier heights to be roughly independent of the metal (Fig.1A). By inserting an ultrathin insulator, we create a metal insulator semiconductor (MIS) contact, whereby the creation of a positive dielectric dipole in accordance with the bond polarization theory (Fig.1B) can shift the Fermi level. MIS contacts can be viewed as two resistances in series: a tunneling resistance through the dielectric (R_T) and a resistance associated with the barrier (R_{SB}), where there exists an optimal insulator thickness (t_{INS}) to minimize R_C as illustrated in Fig. 2.

FABRICATION

To emphasize the thermionic barrier effect, we used lightly doped MBE grown n-GaAs ($2x10^{16}cm^{-3}$) and p-GaAs ($3.5x10^{16}cm^{-3}$). Samples underwent organics degrease, HCl native oxide removal, and $(NH_4)_2S$ sulfur passivation immediately prior to ALD deposition (Fig.3). Diodes were then defined by metal evaporation through shadow masks.

HIGH-κ MIS RESULTS

Back-to-back diode current (Fig.4A) illustrates an initial increase in the reverse current with the insertion of HfO_2, indicating a reduction in $\Phi_{B,eff}$. Upon further increase in HfO_2 thickness, current begins to be tunneling limited. The optimal thickness is roughly 2nm, with a reverse current 2 orders of magnitude higher than the Schottky case. Fig.4B shows the effect of the ALD deposition temperature on $\Phi_{B,eff}$. HfO_2 deposited at 250°C results in a lower $\Phi_{B,eff}$ than the 150°C case, highlighting the MIS sensitivity to material properties.

Fig. 5 illustrates the R_c vs. t_{INS} tradeoff for HfO_2, with the Schottky case as a reference. Metals with band edge work functions, Al (Φ_M = 4.1 eV) for n-GaAs and Pt (Φ_M = 5.65 eV) for p-GaAs, were chosen for an ideally near zero Schottky barrier height. The existence of a minimum R_C for t_{INS} > 0 on n-GaAs confirms there is a reduction in $\Phi_{B,eff}$. However, in the p-GaAs case, there is a monotonic increase in R_c, indicating there is no reduction in $\Phi_{B,eff}$ and R_{SB}. This suggests the $\Phi_{B,eff}$ tunning in III-V MIS contacts is due to a shift in the Fermi level through electronic dipole formation [4], rather than the Fermi level depinning cited for Si and Ge MIS contacts [5-6]. If Fermi level depinning through the reduction of MIGS were the reason

behind the observed n-GaAs $\Phi_{B,eff}$ reduction with Al, then there should also be a similar reduction in the p-GaAs $\Phi_{B,eff}$ with Pt.

MIS contacts using ALD TiO_2 and ZrO_2 also indicate the creation of a dipole, with their R_C vs. t_{INS} summarized in Fig.6. Of the investigated dielectrics, TiO_2 achieves the lowest R_C, with over a magnitude reduction in R_C from our previously reported SiN contact [2]. The ideal dielectric for MIS contacts would be one that forms a large dipole and has $\Delta E_C \leq 0$ to reduce the tunneling penalty.

HIGH-κ/HIGH-κ MIS RESULTS

In the continued search for a near Ohmic MIS contact, we investigated the use of two-layer high-κ dielectrics, based on reports of SiO_2/high-κ dipoles created by an equalization of the areal oxygen density (σ) at the interface [7]. High-κ oxide materials have different oxygen densities. As a result, electronic dipoles should also be formed at high-κ/high-κ interfaces. Al/8.5Å HfO_2/13Å TiO_2/n-GaAs diodes were fabricated, and by inserting 8.5Å HfO_2, the reverse current was increased 20 times over that of Al/13Å TiO_2/n-GaAs (Fig.7). This indicates an additional $\Phi_{B,eff}$ reduction beyond that introduced at the TiO_2/n-GaAs interface, because otherwise additional HfO_2 would only reduce the current. This result is counter-intuitive, in that adding a larger bandgap insulator for an overall thicker dielectric, can actually increase current and reduce $\Phi_{B,eff}$. However, this can be explained by the presence of dipoles (Fig.1C).

A comparison of R_C (Fig.8) verifies that the HfO_2/TiO_2 contact achieves a lower R_C than both the TiO_2 and HfO_2 contacts. For a fixed 13Å TiO_2, we discover that varying the HfO_2 thickness results in a R_C vs. t_{INS} tradeoff very similar to that of a single dielectric MIS, with 8.5Å as the optimal HfO_2 thickness. This implies a similar underlying mechanism, with an additional contributing dipole at the HfO_2/TiO_2 interface.

The exact reasoning for the observed reduction in $\Phi_{B,eff}$ in the high-κ/high-κ MIS is still being studied. However, in looking at the calculated areal oxygen density of various dielectrics (Fig.9), $\sigma_{HfO2} > \sigma_{TiO2}$ indicates under the oxygen density reasoning, a dipole pointing towards HfO_2 would be formed [7], which agrees with our observations.

Fig.10 summarizes the minimum achieved R_C for all investigated dielectrics. Equipment limitation prevented the measurement of $\Phi_{B,eff}$, but from previous measurements of $\Phi_{B,eff}$ = 0.17eV for Al_2O_3 MIS contacts, it can be deduced from the lower R_C of HfO_2/TiO_2 MIS contacts that the $\Phi_{B,eff}$ would be much lower than 0.17eV.

CONCLUSION

In summary, we successfully demonstrate R_C and $\Phi_{B,eff}$ tuning of Al/n-GaAs junctions by a MIS diode structure using ALD HfO_2, TiO_2, and ZrO_2 dielectrics. We also introduce for the first time the use of HfO_2/TiO_2, two high-κ dielectrics in combination to further shift the Fermi level and reduce $\Phi_{B,eff}$. The underlying mechanism is believed to be the formation of a high-κ/high-κ dipole, which opens doors to the exploration of a multitude of other high-κ/high-κ dielectrics to ultimately achieve $\Phi_{B,eff} \leq 0$. This MIS structure provides much flexibility in the design of ideal source/drain contacts for III-V MOSFETs and Schottky Barrier FETs, where in real applications highly doped substrates would significantly reduce R_C and $\Phi_{B,eff}$. Further study of the dipole interaction and effective work function will lead to a better understanding of the physics behind metal/III-V contacts.

978-1-61284-243-1/11 $26.00 © 2011 IEEE

Acknowledgements: This work is supported in part by the Focus Center Research Program (MSD), Intel Corporation, and member companies of the Initiative for Nanoscale Materials and Processes (INMP) at Stanford. J. Hu is additionally supported by the Intel PhD Fellowship, Stanford Graduate Fellowship, and the National Defense Science and Engineering Graduate (NDSEG) Fellowship.

REFERENCES:
[1] ITRS 2009, http://www.itrs.net/reports.html.
[2] J. Hu et al., *JAP*, pp. 063712, 2010.
[3] J. Hu et al., *IEEE VLSI - TSA*, pp.123, 2009.
[4] B. E. Coss et al., *IEEE Symp. VLSI Tech.*, pp.104, 2009.
[5] D. Connelly et al, *IEEE Trans. on Nanotech.*, pp.98-104, 2004.
[6] M. Kobayashi et al., *JAP*, pp. 023702, 2009.
[7] K. Kita et al., *IEDM Tech. Dig.*, pp.29, 2008.

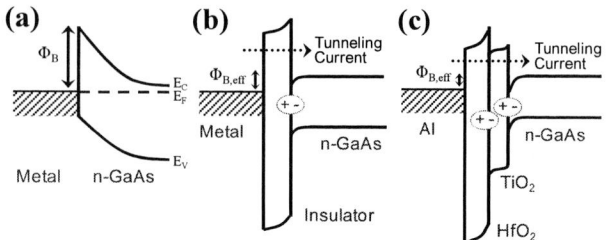

Fig. 1. (a) Schematic band diagram of a pinned Fermi level. (b) Band diagram showing a shift in the Fermi level through the formation of an electronic dielectric dipole. (c) Adding another high-κ dielectric further reduces $\Phi_{B,eff}$ by introducing an additional shift in the Fermi level caused by a dipole at the HfO₂/TiO₂ interface.

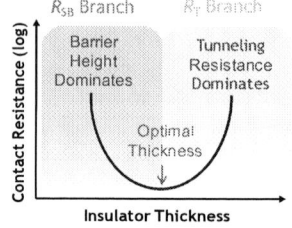

Fig. 2. Schematic of R_C vs. t_{INS}. There exists an optimal thickness for minimal R_C, arising from the tradeoff between a reduced barrier (R_{SB}) and an increased tunneling resistance (R_T).

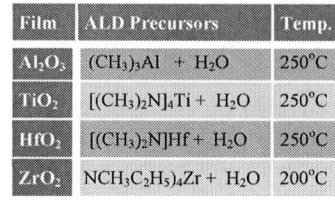

Fig. 3. ALD precursor and deposition temperatures.

Film	ALD Precursors	Temp.
Al₂O₃	(CH₃)₃Al + H₂O	250°C
TiO₂	[(CH₃)₂N]₄Ti + H₂O	250°C
HfO₂	[(CH₃)₂N]Hf + H₂O	250°C
ZrO₂	NCH₃C₂H₅)₄Zr + H₂O	200°C

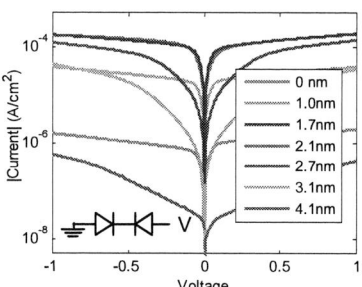

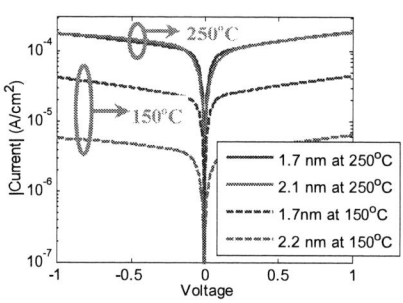

Fig. 4. Diodes were measured back-to-back to eliminate the back contact resistance contributions and to emphasize the reverse current, the dominant indicator of changes in $\Phi_{B,eff}$. (a) Al/HfO₂/n-GaAs illustrates a change from rectifying to increased conduction to tunneling limited current with thicker HfO₂. (b) Impact of the ALD deposition temperature on $\Phi_{B,eff}$ is shown, where changes in the film density and stoichiometry could explain the differences between the 250°C and 150°C films.

Fig. 5. Rc vs. t_{INS} tradeoff for n-GaAs and p-GaAs. Al (Φ_M = 4.1 eV) and Pt (Φ_M = 5.65 eV) were chosen for the n-type and p-type samples for their band edge metal workfunctions. The reduction of R_C in n-GaAs indicates the presence of an electronic dipole. The steady increase in R_C in p-GaAs indicates the mechanism is not a depinning, but rather a shift in the Fermi level.

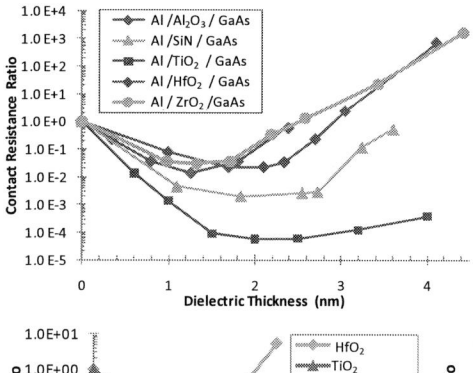

Fig. 6. Contact resistance vs. dielectric thickness (R_C vs. t_{INS}) tradeoff comparison for different dielectrics. R_C ratios are taken relative to the Schottky case. The tradeoff for each dielectric depends on the dipole magnitude and the ΔE_C.

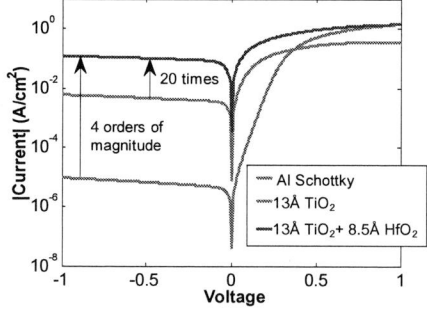

Fig. 7. MIS diode current of Al/TiO₂/n-GaAs and Al/HfO₂/TiO₂/n-GaAs. The increase in both the forward and reverse currents with the addition of HfO₂, a larger bandgap material, is counter intuitive and indicates the presence of a electronic dielectric dipole that reduces $\Phi_{B,eff}$.

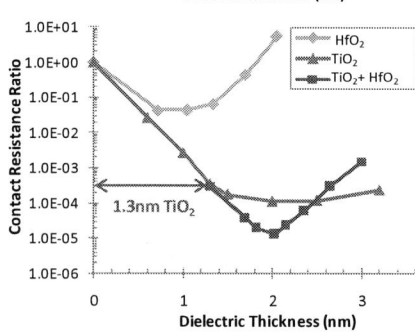

Fig. 8. R_C vs. t_{INS} for MIS contacts using TiO₂, HfO₂, and TiO₂+HfO₂. In Al/HfO₂/TiO₂/n-GaAs, R_C is reduced beyond that of just TiO₂, which indicates the presence of a high-κ/high-κ dipole.

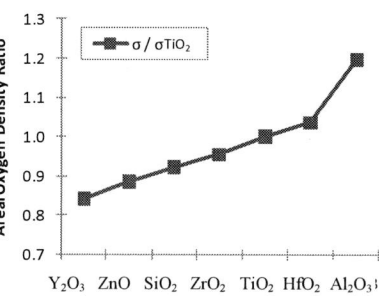

Fig. 9. Calculated areal oxygen densities (σ) relative to that of TiO₂ for a variety of high-κ dielectrics are plotted.

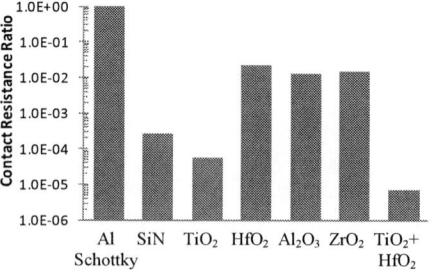

Fig. 10. The minimum achieved contact resistances for each material is shown.

978-1-61284-243-1/11 $26.00 © 2011 IEEE

Intrinsic DC Operation and Performance Potential of 50nm Gate Length Hydrogen-terminated Diamond Field Effect Transistors

David A. J. Moran[1], Oliver J. L. Fox[2], Helen McLelland[1], Stephen Russell[1], Paul W. May[2]

[1]The School of Engineering, The University of Glasgow, Glasgow, G12 8LT, U.K

[2] The School of Chemistry, The University of Bristol, Bristol, BS8 1TS, UK

The hydrogen-terminated diamond surface has demonstrated unique potential in the development of high power and high frequency field effect transistors (FETs) [1]. Further exploration into the intrinsic performance limitations and device operation as gate length is reduced however is essential in unveiling the potential of this exotic material system as a viable and competitive high power and high frequency device technology.

Progress and challenges involved with the scaling of hydrogen-terminated diamond FETs to sub-100nm gate dimensions have recently been reported [2,3]. In this work we describe the intrinsic operation and DC performance of devices at the 50nm node; the shortest gate length yet realised for an operational diamond FET.

50nm gate length devices were fabricated as illustrated in Fig. 1 using Element Six provided CVD grown (001) orientated single-crystal diamond. Devices were isolated electrically by selective removal of the hydrogen-termination with oxygen plasma. Sub-100nm gate length device yield and performance was improved substantially through fine optimization of the KI wet etch of the Au encapsulation layer which forms the ohmic contacts directly. The inherent ohmic contact edge roughness and large gate-ohmic separation associated with this process are therefore reduced. A Scanning Electron Microscope image of the resultant device structure is presented in Fig. 2.

Typical 50nm device output characteristics are presented in Fig. 3. Performance figures include a maximum drain current $I_{dmax} \sim 300$ mA/mm and a peak extrinsic transconductance $g_m \sim 100$ mS/mm. An increase in off-state output conductance is observed for larger magnitude V_{ds} as further verified upon inspection of device logarithmic transfer characteristics (Fig. 4). This short-channel type response results in an increase in sub-threshold swing (SS) and a shift in threshold voltage (V_{th}) from positive to negative as greater bias is applied between source and drain (Fig.5). Although electrostatic control of the drain current by the gate is reduced at more negative V_{ds}, a minimum I_{on}/I_{off} of $\sim 1.5 \times 10^4$ is maintained across the inspected bias range indicating the ability to pinch off the drain current at a gate dimension 50nm. Fig. 6 demonstrates the degradation of I_{on}/I_{off} for increased magnitude source-drain voltage and fixed gate bias close to device pinch off.

Device intrinsic transconductance was determined by extraction of parasitic access resistances from low V_{ds} device on-resistance (R_{on}) which in turn was extracted directly from the linear response of device output characteristics (as shown in Fig. 3). Given the symmetric positioning of the gate contact between source and drain and accounting for the channel resistance beneath the gate, the total source resistance (R_s) was determined from this figure and the intrinsic transconductance then extracted by a similar process to that described in [3] to be ~ 660 mS/mm. A value of 8.8 Ω.mm was determined for R_s, of which 4.0 Ω.mm is attributed to ohmic contact resistance (R_c) as measured by TLM test structures and the remaining 4.8 Ω.mm to the lateral resistance through the diamond between source and gate (R_{s-g}). Utilising the sheet resistance figure of 10 kΩ/sq for the exposed diamond surface as measured by TLM, a source-gate separation of $L_{s-g} \sim 480$nm was determined for these devices.

Comparison between the intrinsic and extrinsic transconductance values of 660 mS/mm and 100 mS/mm respectively emphasizes the need to greatly reduce the total access resistance within these devices to better enable the potential performance of this technology. Fig. 7 demonstrates the impact of R_s upon extrinsic transconductance g_m at this gate dimension as it is increased to its intrinsic value at $R_s = 0$ Ω.mm. The individual components of R_s i.e. R_c and R_{s-g} are highlighted in addition to the dependency upon the source-gate contact separation L_{s-g}. It is observed that as L_{s-g} approaches zero, an increase in extrinsic transconductance up to a value approaching 200 mS/mm for a similar value of $R_c = 4.0$ Ω.mm should be attained.

In summary, although short channel effects have been observed in the operation of 50nm hydrogen-terminated diamond FETs, a minimum I_{on}/I_{off} figure of $\sim 1.5 \times 10^4$ indicates pinch-off of the drain current is achievable at this gate dimension and across the inspected bias range. This is the first time the detailed intrinsic operation of diamond FETs with gate length below 100nm has been reported. Furthermore, deduction of the large intrinsic transconductance figure of 660mS/mm indicates substantial improvement in DC performance should result from further reduction to device access resistance and lead to substantial improvement in device high frequency performance for hydrogen-terminated diamond FETs with sub-100nm gate length.

[1] K. Ueda *et al, IEEE Electron Device Letters*, vol. 27, no. 7, pp. 570 – 572, July. 2006.
[2] D. A. J. Moran *et al*, doi:10.1016/j.mee.2010.11.029, *Microelectronic Engineering* (in press)
[3] David. A. J. Moran *et al, IEEE Electron Device Letters,* (accepted 9th Feb 2011 - in press)

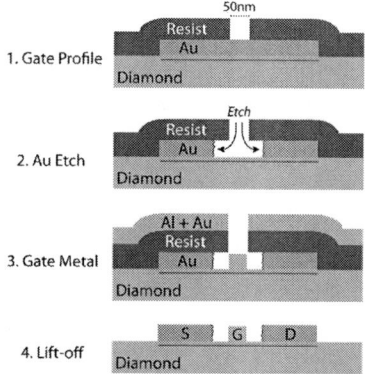

Fig. 1. Hydrogenated diamond FET fabrication process

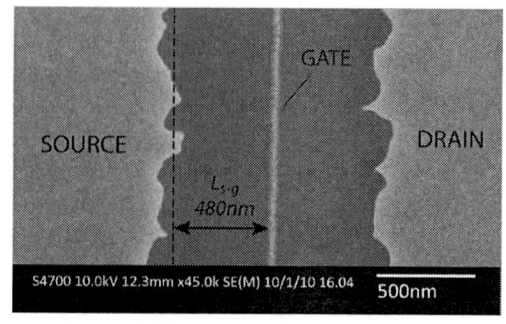

Fig. 2. SEM image of 50nm FET structure

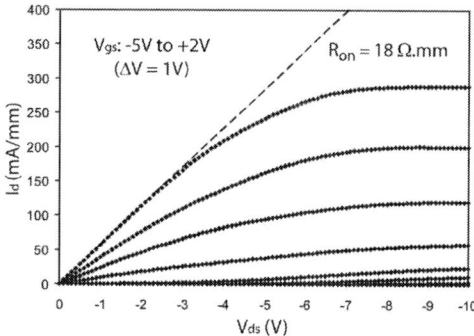

Fig. 3. 50nm device Output characteristics

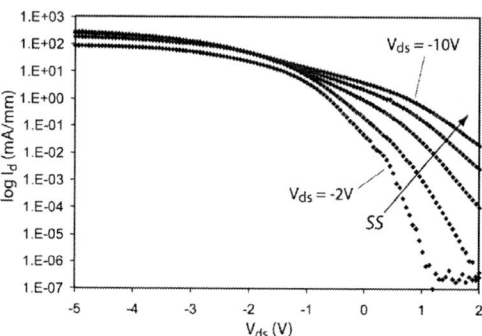

Fig. 4. 50nm device log Transfer characteristics

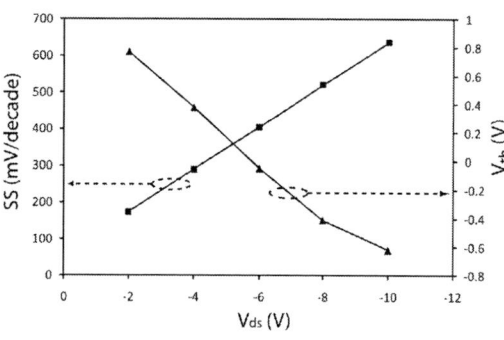

Fig. 5. Sub-threshold swing (SS) & threshold voltage (V_{th}) vs. V_{ds}

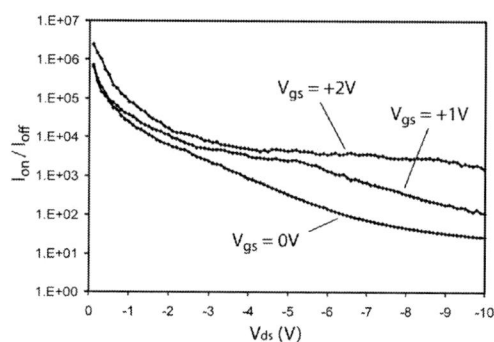

Fig. 6. I_{on}/I_{off} vs. V_{ds} for $V_{gs}=0V$, +1V & +2V

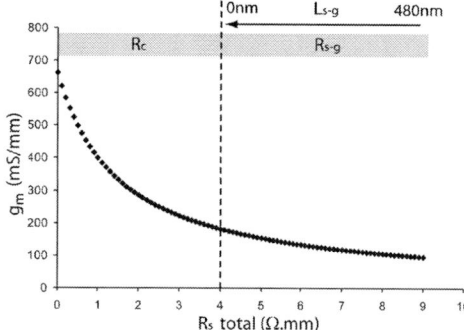

Fig. 7. Transconductance vs total source resistance. Contribution of contact resistance (R_c) and lateral access resistance (R_{s-g}) illustrated.

978-1-61284-243-1/11 $26.00 © 2011 IEEE

Improvement of f_T in InAl(Ga)N barrier HEMTs by Plasma Treatments

Ronghua Wang[1], Guowang Li[1], Tian Fang[1], Oleg Laboutin[2], Yu Cao[2], Wayne Johnson[2],
Gregory Snider[1], Patrick Fay[1], Debdeep Jena[1], Huili (Grace) Xing[1,*]

[1]Department of Electrical Engineering, University of Notre Dame, Notre Dame, IN 46556
[2]Kopin Corporation, Tauton, MA 02780, USA * Email: hxing@nd.edu

GaN-based high electron mobility transistors (HEMTs) have been developed for high-temperature, high-frequency and high-power applications. To improve the transistor speed, various techniques have been explored in addition to scaling down the gate length and top barrier thickness: ultrathin SiN passivation to reduce access resistance and parasitic capacitances [1]; re-grown ohmic contacts and self-alignment to minimize access resistances [2, 3]; O_2 plasma treatment in the gate region prior to the metal deposition to suppress rf transconductance collapse [4]; and dielectric-free passivation (DFP) by a O_2-containing plasma treatment in the access region to shorten the gate extension in InAlN HEMTs [5]. Here we report a comparative study on the impact of various plasma treatments in the access region (DFP) as well as under the gate for InAl(Ga)N barrier HEMTs, and propose a model for the observed f_T improvement.

The HEMT heterostructures (Fig. 1) consist of either a lattice-matched ternary $In_{0.17}Al_{0.83}N$ (10.3 nm) or a tensile-stained quaternary $In_{0.13}Al_{0.83}Ga_{0.04}N$ (10.8 nm) barrier, a 0.5 nm AlN spacer and a C-doped semi-insulating GaN buffer on SiC substrate grown by metal organic chemical vapor deposition at Kopin Corporation. A Si/Ti/Al/Ni/Au ohmic stack annealed at 860 °C in N_2 resulted in a contact resistance R_c of 0.38 for InAlN barrier HEMTs (Group A), 0.36 & 0.50 Ω-mm for InAlGaN barrier (Group B & C) HEMTs, respectively. Rectangular Ni/Au gates were defined by electron-beam lithography and lift-off. As-fabricated devices in Group A and B were treated with an O_2-containing plasma in the access region only for DFP, while devices in Group C were treated with O_2, CF_4, or BCl_3 plasma in the gate region prior to the metal deposition. 2D electron gas (2DEG) transport properties monitored on wafer are shown in Table 1. All the devices have a nominal source-drain separation of 1.5 µm and a gate width of 2 x 50 µm.

As an example of the effectiveness of DFP on f_T improvement, small signal rf performance of an InAlN barrier HEMT with L_g = 60 nm is shown in Fig. 2 (a) before and after DFP. The increase in f_T from 125 to 210 GHz after the DFP treatment is attributed to the passivation of surface states in the access region and thus a shortened gate extension. Since a virtual gate forms due to surface states near the gate trapping electrons under large drain-gate electric field, high output resistance was observed in the I-Vs characteristics before DFP; after DFP of the surface states, strong short channel effects were seen [5]. The f_T dependence on gate length L_g for the InAlN and InAlGaN HEMTs (Group A and B) processed in parallel is shown in Fig. 2 (b), which suggests it is promising to achieve f_T > 250 GHz with L_g < 50 nm by employing the DFP technique. Fig. 2 (c) shows the trend of f_T as a function of V_{ds} at a gate bias near peak g_m before and after DFP for an InAlGaN barrier HEMT in Group B. The f_T of post-DFP devices increases quickly with V_{ds} transiting from the linear region to the saturation region, reaches a maximum value of 220 GHz at V_{ds} = 4.8 V, and then slightly drops at higher V_{ds}, ascribed to the widened gate depletion region toward the drain. In contrast, the pre-DFP f_T gradually increases with increasing V_{ds}, which suggesting that for large V_{ds} the transit time increase across the lengthened gate extension region is overcompensated by the transit time reduction under the shortened effective gate with increasing V_{ds}.

Four different plasma treatments under the gate only were applied to the devices in Group C, including O_2, CF_4, BCl_3, and BCl_3 followed by O_2. It is found that the impact of the gate-only plasma treatment on the device performance, except for the BCl_3-only treatment, is similar to that of DFP – an access-region-only plasma treatment [5], and surprisingly the device rf performance stayed the same when subsequent DFP treatments were applied. An example of CF_4 plasma treatment on HEMTs with gate lengths of ~100 nm is shown in Fig. 3. After the plasma treatment, short channel effects became more severe; threshold voltage shifted by -0.7 V, peak $g_{m,ext}$ increased from 380 to 420 mS/mm; the reverse biased gate leakage dropped by one order of magnitude, and drain induced barrier lowing (DIBL) increased from 50 to 170 mV/V; f_T also increased from 90 to 130 GHz. We thus postulate that the 2DEG density increase after plasma treatment (Table 1) is not the major contributor for the observed electrical gate length reduction. Instead, a conduction path on or near the surface of the HEMT barrier connects the metal gate and the surface states in the virtual gate region; when this conduction path is cut off by the plasma treatment, the electrical gate length is shortened thus the transistor speed improved. The conduction path could be related to N-vacancies that could then be filled by F or O during the plasma treatment [4]. This work has been supported by DARPA-NEXT (John Albrecht, HR0011-10-C-0015), AFOSR (Kitt Reinhardt), AFRL/MDA (John Blevins) and AFOSR-YIP (Kitt Reinhardt).

[1] M. Higashiwaki, et al., *Appl. Phys. Express*, vol. 1, no. 2, p. 021103, 2008.
[2] Nidhi, et al., *IEDM Tech. Dig.*, p. 20.5, 2009.
[3] K. Shinohara, et al., *IEDM Tech. Dig.*, p. 30.1, 2010.
[4] J. W. Chung, et al., *IEDM Tech. Dig.*, p. 30.2, 2010.
[5] R. Wang, et al., submitted to *IEEE Electron Device Lett.*, 2011.

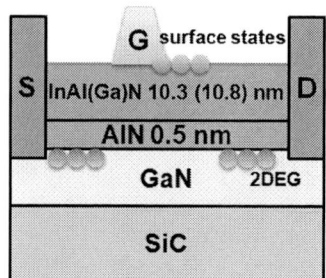

Table 1 Hall transport properties of HEMT structures before and after DFP.

	Before DFP			After DFP		
	R_{sh} (Ω/sq)	n_s (x 10^{13} cm^{-2})	μ (cm^2/ V.s)	R_{sh} (Ω/sq)	n_s (x 10^{13} cm^{-2})	μ (cm^2/ V.s)
InAlN (A)	290	1.62	1330	257	1.86	1300
InAlGaN (B)	227	1.45	1900	190	1.83	1790

Fig. 1 Schematic cross section of D-mode InAl(Ga)N/AlN/GaN HEMTs, showing surface states near the gate.

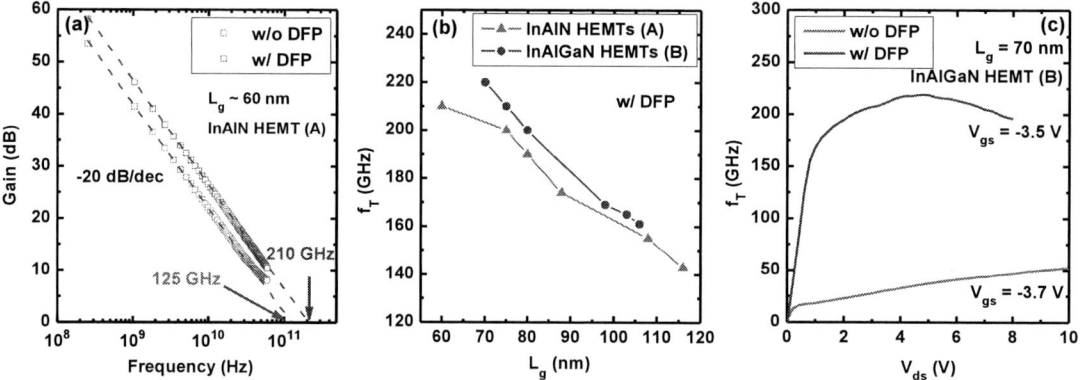

Fig. 2 (a) Current gain as a function of frequency for an InAlN barrier HEMT with L_g = 60 nm and L_{sg} = 400 nm, showing f_T increased from 125 to 210 GHz after DFP.   (b) Dependence of f_T on L_g, indicating the scalability of InAlN barrier HEMTs (Group A) and InAlGaN barrier HEMTs (Group B) after DFP.   (c) f_T as a function of V_{ds} near peak g_m gate bias of an InAlGaN barrier HEMT with L_g = 70 nm, and L_{sg} = 300 nm, before and after DFP, respectively.

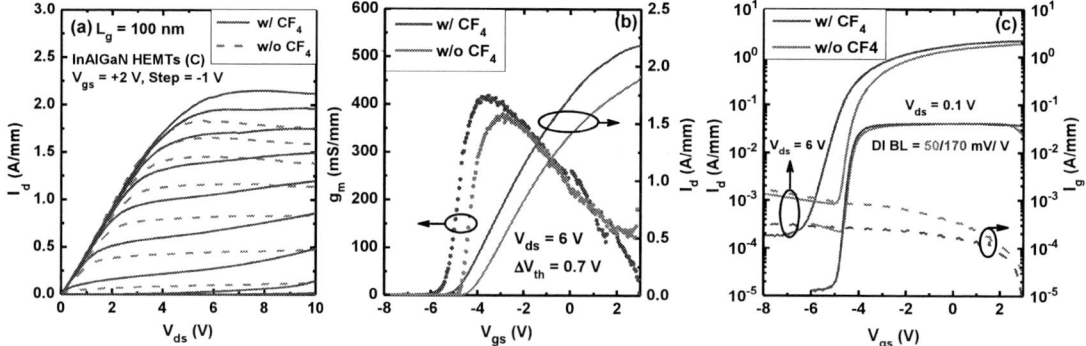

Fig. 3 InAlGaN barrier HEMTs (L_g = 100 nm) with and without CF$_4$ plasma treatment in the gate region only: (a) common source family of I-Vs; (b) linear-scale transfer characteristics at V_{ds} = 6 V, showing a negative V_{th} shift of 0.7 V and an extrinsic peak g_m increase from 380 to 420 mS/mm; (c) semi-log-scale transfer characteristics at V_{ds} = 6 and 0.1 V, showing CF$_4$ treatment under the gate reduces the gate leakage while DIBL increased from 50 to 170 mV/V.

978-1-61284-243-1/11 $26.00 © 2011 IEEE

Scaling behavior and velocity enhancement in Self-aligned N-polar GaN/AlGaN HEMTs with maximum f_T of 163 GHz

Nidhi[*], S. Dasgupta, D. F. Brown[1], J. S. Speck and U. K. Mishra

ECE Department, University of California Santa Barbara, Santa Barbara, CA 93106, USA
[1]Hughes Research Laboratories, Malibu, California

N-polar GaN/AlGaN HEMTs have recently been demonstrated as a potential technology for high frequency applications [1]. Excellent $f_T.L_G$ product of 16.8 GHz-μm has been achieved for $L_G = 130$ nm. As the devices are being scaled to further improve the performance, vertical scaling and hence, aspect ratio of the device becomes an important aspect of the device design. In this paper, we analyze self-aligned devices with varying gate-lengths fabricated on a 10 nm GaN channel with identical access regions to understand the effect of aspect ratio on the scaling of DC and RF device performance with gate-length scaling.

The device layer structure (Fig. 1a) consists of a graded AlGaN barrier with an AlN interlayer to reduce alloy scattering. It is capped with 10 nm GaN channel resulting in sheet resistance of ~450 Ω/□. The growth details are given in [2] and band diagram is shown in Fig. 1b. MOCVD SiN_x (5 nm) is used as the gate dielectric. The device fabrication consisted of blanket-deposition of gate-stack (W/Cr/SiO$_2$/Cr), e-beam lithography for gate definition and subsequent selective-etching of the gate-stack. PECVD SiN_x-spacers (50 nm) are formed around the gate before MBE-regrowth of highly-doped graded InGaN/InN to get very low ohmic contact resistance of 25 Ω-μm (Fig. 2). More details on the gate-first self-aligned device fabrication process are given in [3, 4].

The DC measurements resulted in saturated drain current density ($I_{D,sat}$) of 1.89 A/mm and 1.56 A/mm at $V_G = 2$ V for $L_G = 80$ nm and 150 nm respectively and peak extrinsic transconductance (g_m) of 337 mS/mm and 521 mS/mm at $V_{DS} = 2.5$ V for $L_G = 80$ nm and 150 nm respectively (Fig. 3). The small-signal measurements resulted in maximum f_T and f_{MAX} of 163 GHz and 20 GHz respectively for $L_G = 100$ nm (Fig. 4).

To understand scaling behavior of these devices, we need to study the variation of DC parameters and small-signal performance with the gate-length. Saturation current density ($I_{D,sat}$) is plotted a function of L_G in Fig. 5. It can be observed that the current remains relatively constant for $L_G > 150$ nm denoting the current limited by saturation velocity in the channel. But, as the gate-length is decreased further, the current tends to increase indicating enhancement in velocity above the saturation velocity in the channel. This kind of enhancement in velocity has not been well understood for GaN HEMTs and needs to be further investigated with analysis of small signal data and fabrication of transistors with shorter gate-lengths. Fig. 6 shows the variation of peak transconductance (g_m) as a function of gate-length. The transconductance increases slightly as L_G decreases from 250 nm to 150 nm due to slight enhancement in velocity, however as the gate-length is decreased further, the gate loses control of the channel and the transconductance drops. To further investigate the hypothesis that the gate-control is being compromised as the gate-length is reduced, we study the variation of threshold voltage as function of gate-length. As shown in Fig.7, threshold voltage remains relatively flat for larger aspect ratios and drops severely as aspect ratio falls below 5.5. This provides a design guideline that for good gate-control, aspect ratio has to be designed to be greater than 5.5.

We also studied the effect of aspect ratio of the small signal performance. As shown in Fig. 8, excellent $f_T.L_G$ product of 18 GHz.μm has been achieved for $L_G = 150$ nm. Also, due to loss of gate modulation and short channel effects, the $f_T.L_G$ product drops severely below an aspect ratio of 6. To understand how average channel velocity varies with gate-length, velocity was calculated from the slope of $(1/2\pi f_T)$ delay versus L_G plot with and without the 80 nm point (Fig. 9). The extracted average channel velocity was found to be 1.28×10^7 cm/s for $L_G \geq 100$ nm and 1.51×10^7 cm/s for $L_G \geq 80$ nm, supporting the DC data indicating enhancement in channel velocity as L_G decreases beyond 100 nm. A rough estimate of intrinsic velocity was also calculated using the method described by Jessen et. al. in [5] by using source resistance, $R_S = R_C + R_{sheet,InGaN} * 0.5$μm $+ R_{sheet,2DEG} * 50$ nm $= 0.18$ Ω-mm (Fig. 10). The intrinsic velocity is calculated to be 1.49×10^7 cm/s for $L_G \geq 100$ nm and 1.8×10^7 cm/s up to $L_G \geq 80$ nm.

To conclude, due to self-aligned nature of these devices, enhancement in channel velocity was observed for L_G as large as 80 nm. Critical aspect ratio of 5.5-6 is determined to be necessary to reduce short channel effects.

[1] Nidhi et. al., IEDM 2009, [2] S. Keller et. al., JAP, Mar 2011, [3] S. Dasgupta et. al., APL, 96, 143504 (2010), [4] Nidhi et. al., IEEE EDL 32, 33 (2011). [5] Jessen et. al., IEEE TED 54, 2589 (2007).

*Corresp. author: nidhi@ece.ucsb.edu, Phone: +1-805-893-3812-ext. 202

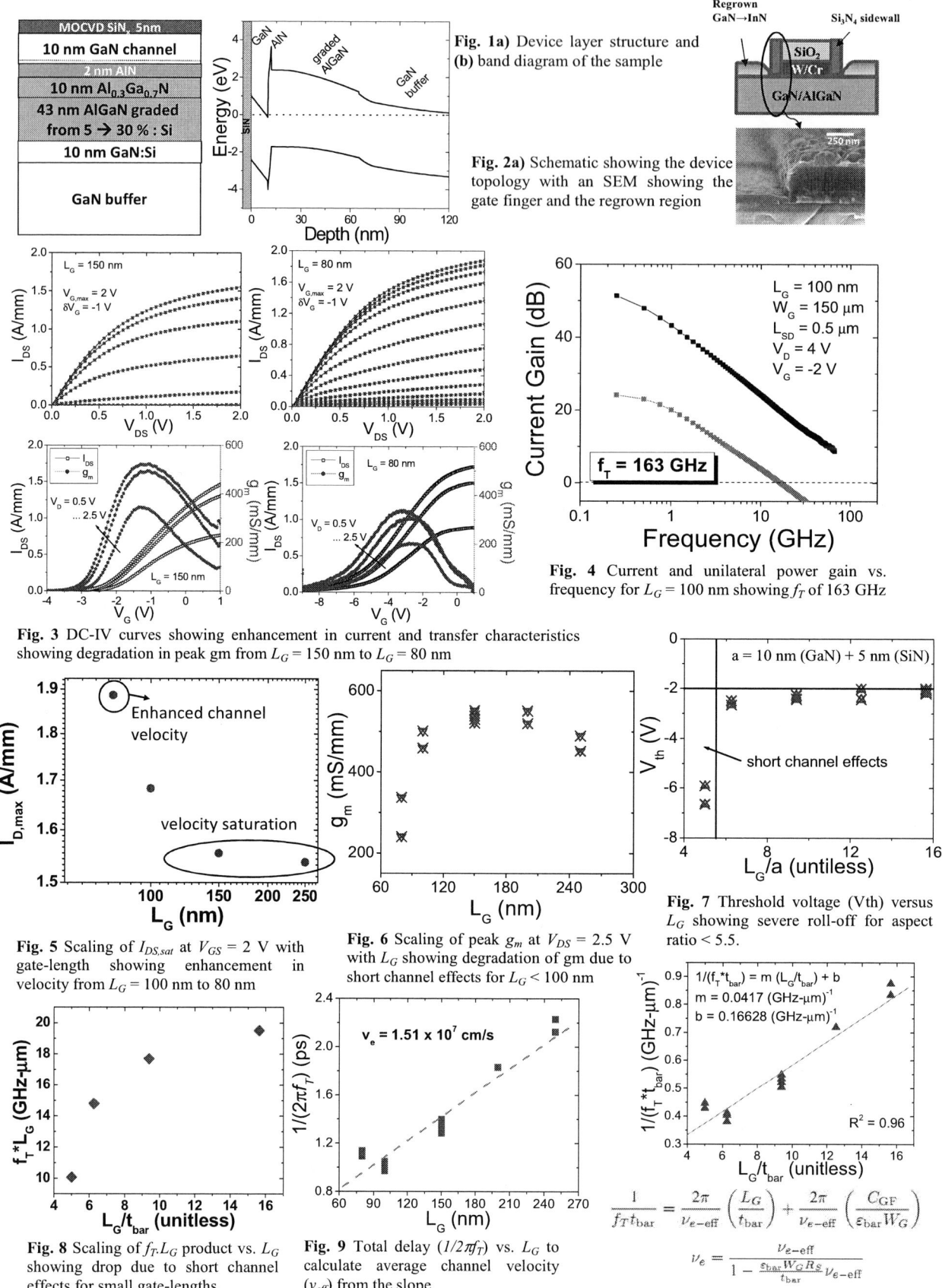

Fig. 1a) Device layer structure and **(b)** band diagram of the sample

Fig. 2a) Schematic showing the device topology with an SEM showing the gate finger and the regrown region

Fig. 4 Current and unilateral power gain vs. frequency for $L_G = 100$ nm showing f_T of 163 GHz

Fig. 3 DC-IV curves showing enhancement in current and transfer characteristics showing degradation in peak gm from $L_G = 150$ nm to $L_G = 80$ nm

Fig. 5 Scaling of $I_{DS,sat}$ at $V_{GS} = 2$ V with gate-length showing enhancement in velocity from $L_G = 100$ nm to 80 nm

Fig. 6 Scaling of peak g_m at $V_{DS} = 2.5$ V with L_G showing degradation of gm due to short channel effects for $L_G < 100$ nm

Fig. 7 Threshold voltage (Vth) versus L_G showing severe roll-off for aspect ratio < 5.5.

Fig. 8 Scaling of $f_T.L_G$ product vs. L_G showing drop due to short channel effects for small gate-lengths

Fig. 9 Total delay ($1/2\pi f_T$) vs. L_G to calculate average channel velocity (v_{eff}) from the slope

$$\frac{1}{f_T t_{bar}} = \frac{2\pi}{v_{e\text{-eff}}} \left(\frac{L_G}{t_{bar}} \right) + \frac{2\pi}{v_{e\text{-eff}}} \left(\frac{C_{GF}}{\varepsilon_{bar} W_G} \right)$$

$$v_e = \frac{v_{e\text{-eff}}}{1 - \frac{\varepsilon_{bar} W_G R_S}{t_{bar}} v_{e\text{-eff}}}$$

Fig. 10 Extraction of velocity (v_{eff}) as described in [5] from $1/(f_T t_{bar})$ vs. (L_G/t_{bar}) gives same v_{eff} as Fig. 9. Intrinsic velocity (v_e) is thus calculated from the above equation.

Fermi-level Pinning at Metal/Antimonides Interface and Demonstration of Antimonides-based Metal S/D Schottky pMOSFETs

Z. Yuan[1], A. Nainani[1], J.-Y. Lin[1], B. R. Bennett[2], J. B. Boos[2], M. G. Ancona[2] and K. C. Saraswat[1]

[1]Dept. Of Elec. Eng. Stanford University, CA 94305 USA [2]Naval Research Laboratory, Washington, DC 20375 USA

Phone: 650-995-3871, fax: 650-723-4659, email: zeyuan@stanford.edu

Introduction: III-V semiconductors are considered as promising candidates to replace silicon as the channel material in future technology nodes for transistors [1]. III-V n-channel MOSFETs have been extensively studied [2-4], showing high electron mobility. However, one of the most critical challenges in realizing high performance III-V MOSFETs is the difficulties in source/drain (S/D) design including parasitic resistance due to low solubility and poor activation of dopant and the "source starvation" effect due to low density of states [5-6]. Annealing of implant damage after S/D ion-implantation is also more problematic in III-V's due to the presence of 2 or more atomic species vs. group IV semiconductors (Fig.1). Use of Schottky-barrier (SB) metal S/D is a promising strategy to overcome these limitations [7]. Meanwhile, for III-V based CMOS logic, achieving a high mobility pMOSFET in a III-V channel remains a challenge. Antimony (Sb) based compound semiconductors have the highest electron and hole mobilities amongst all III-V materials. Recently, high performance strained channel InGaSb pMOSFETs [8] have been demonstrated. In this paper, we study the metal contact to antimonides compound. Good metal contact formed on p-type material and current suppression on n-type samples is attributed to the Fermi-level pinning at metal/antimonide interface and charge-neutral level being near the valence band edge. Schottky-barrier S/D p-MOSFETs is proposed and experimentally demonstrated which combines an $In_xGa_{1-x}Sb$ channel for good hole transport with metal S/D for low access resistance.

Metal contact to GaSb: To investigate metal contact on antimonides, we built Schottky diodes on moderately doped n- and p-type GaSb substrates of carrier concentration $\sim 10^{17} cm^{-3}$. Native oxides were removed by HCl clean. Metals were deposited by e-beam evaporation and patterned to make top electrode. Backside contact was formed using blanket evaporation of metal to measure diode IV. Ideally, by schottky-mott relation, low workfunction metals such as Al and Ti would give higher Schottky barrier for holes than for electrons. Strong Fermi-level pinning is observed at metal/GaSb interface. Fig. 2 (a), (b) shows the J-V characteristics of Al/n-, p-GaSb, and Ti/n-, p-GaSb respectively. Both Al and Ti show rectifying behaviors on n-GaSb and ohmic contact on p-GaSb. Lower (higher) Schottky barrier height for metal/p-(n-) GaSb is confirmed with temperature dependence measurement of Al/GaSb Schottky diode as shown in Fig. 3. Furthermore, contact resistivity for Ti on lightly doped p- and n-GaSb are measured to be $7-9 \times 10^{-5} cm^2$ and $3-6 \times 10^{-3} cm^2$ respectively using TLM test structures. Due to the intrinsic gap state distribution, the charge-neutral level (E_{CNL}) is found to be near valence band edge [9] as shown in Fig. 4. J-V characteristics, temperature dependence measurement and contact resistivity all confirm that antimonide (Sb binary and Sb rich ternary) compound exhibits Fermi-level pinning towards valence band.

Schottky-Barrier S/D pMOSFETs: The band line-up at metal/GaSb interface is favorable for SB pMOSFETs as it has potential to provide high on-current and low off-current simultaneously. For low bandgap material with low Schottky barrier, such as p-GaSb, on-current limitation is mitigated compared with high bandgap material e.g. Si [10]. The current level under forward/reverse bias for Ti contact on p-type material is sufficient for hole injection. The on/off ratio can be roughly estimated by the difference in current levels under reverse bias for contacts on n- and p-type materials. To further cut off the leakage, heterostructure design with Al containing compound as the barrier layer is adopted. The schematic of the fabricated device structure is shown in Fig. 5. Heterostructure stack is grown by MBE with top surface terminated by ultra-thin GaSb. Details on the growth conditions can be found in [11]. Source/Drain metals Ti/Pt was deposited by evaporation. ALD Al_2O_3 was then deposited at 300°C for use as the gate dielectric. Gate metal electrode was formed with Au lift-off to give an overlap between S/D and gate contacts, which allows the operation of SB p-MOSFETs [12] and avoids the use of ion implantation. Source/Drain contacts were opened by etching of Al_2O_3. Any thermal process after gate oxide deposition is avoided, which makes it a gate-late process. Less extrinsic resistance enables better intrinsic device performance evaluation. Temperature during the entire process flow never exceeds 300°C. I_{ds}-V_g and I_d-V_d curves for device of L_g=25um, T_{Al2O3}=14nm are plotted in Fig. 6 and Fig. 7 respectively. A room temperature I_{on}/I_{off} over 500X is achieved. Leakage in off-state comes from the leakage through the barrier layer, since undoped antimonides are of p-type due to intrinsic defect states. Ambipolar behavior, which is common for SB MOSFETs [13], has not been observed, meaning the conduction is highly favorable to hole as compared to electron at the contact. Fig. 8 shows the reduction in leakage current by the introduction of barrier layer. The temperature-dependence measurement is shown in Fig. 9. Significant reduction in leakage at low temperature is observed due to the freeze out of p-type defect states in barrier layer. Fig. 10 shows the extracted mobility versus gate voltage without any correction, which gives a peak mobility of $350cm^2/Vs$.

[1] S. Takagi et al, IEEE Trans. on Elect. Dev., v. 55, pp.21, 2008. [2] J. Huang et al, IEDM Tech. Dig., p. 335, 2009. [3] H. Zhao et al., Appl. Phys. Lett. 94, 193502, 2009. [4] U. Singisetti et al., DRC, p. 253, 2009. [5] M. V. Fischetti et al., IEDM Tech. Dig., p. 109, 2007. [6] H. Tsuchiya et al., IEEE Elect. Dev. Lett., vol. 31, no. 4, p. 365, 2010. [7] S. H. Kim et al, IEDM 2010, 26.6.1. [8] A. Nainani et al., IEDM 2010, 6.4.1. [9] J. Robertson et al., J. Appl. Phys. 100, 014111, 2006. [10] S. Zhu et al., IEEE Elect. Dev. Lett., vol. 26, no. 2, p. 81, 2005. [11] B. R. Bennett, et al., Appl. Phys. Lett., 91, 042104, 2007. [12] T. Takahashi et al., IEDM Tech. Dig., p. 697, 2007. [13] J. M. Larson et al., IEEE Trans. on Elect. Dev., vol. 53, no. 5, p 1048, 2006.

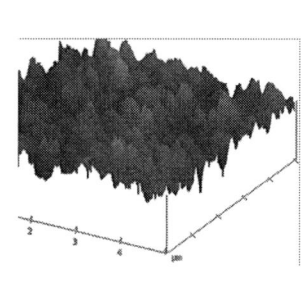

Fig. 1 AFM measured root mean square (rms) roughness of above 4.4nm on GaSb after ion implantation, 600°C RTA.

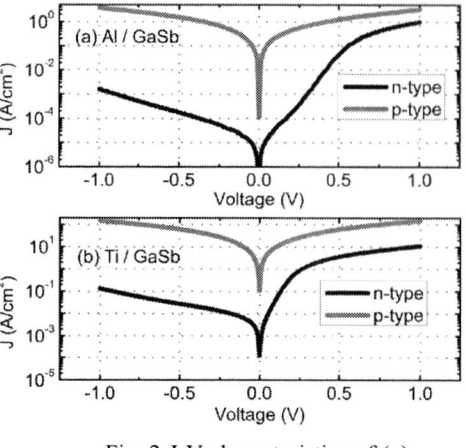

Fig. 2 J-V characteristics of (a) Al/GaSb (b) Ti//GaSb contacts

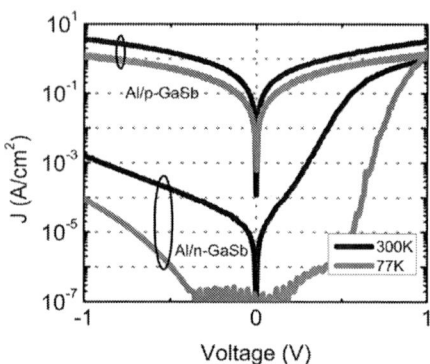

Fig. 3 Temperature dependence measurement of Al/GaSb Schottky diodes

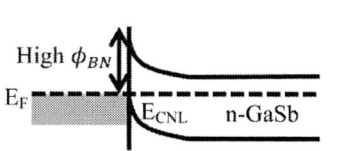

Fig. 4 Band diagram of metal contact to n-type and p-type GaSb

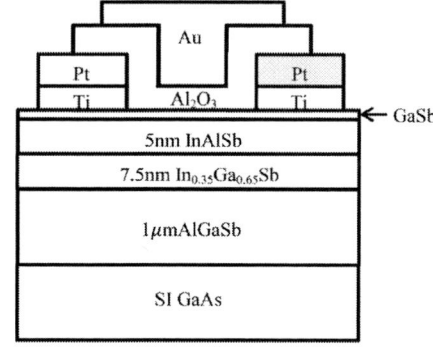

Fig. 5 Device structure of Schottky Barrier pMOSFET with barrier layer to cut off leakage between S/D through body.

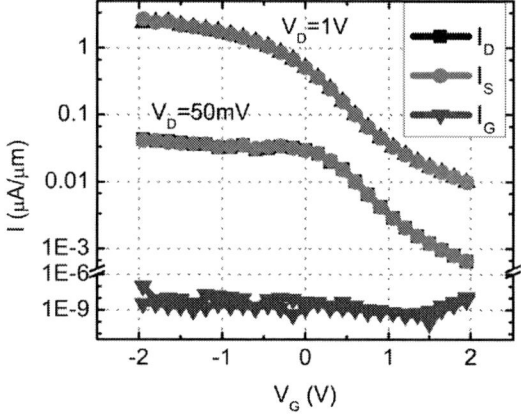

Fig. 6 Transfer characteristics of SB pMOSFETs, L_G=25μm, T_{Al2O3}=14nm

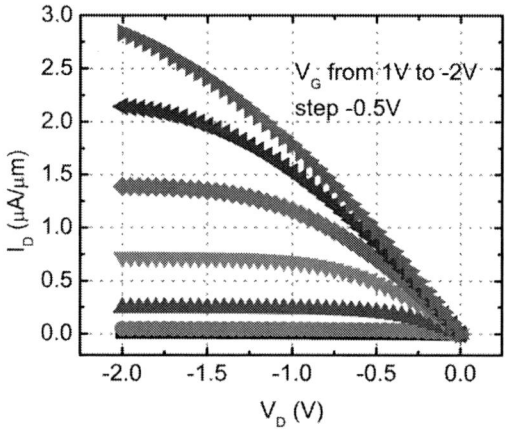

Fig. 7 Output characteristics of SB pMOSFETs, L_G=25μm, T_{Al2O3}=14nm

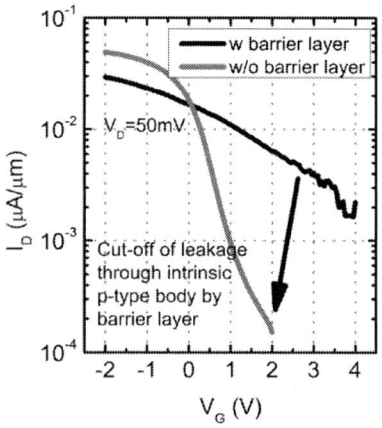

Fig. 8 Comparison of SB pMOSFETs, L_G=100μm, with and without barrier layer.

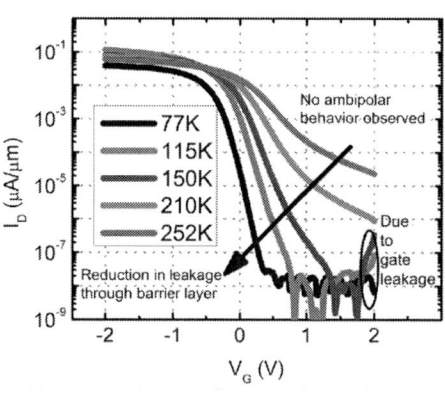

Fig. 9 Temperature dependence measurement of SB pMOSFETs, L_G=100μm

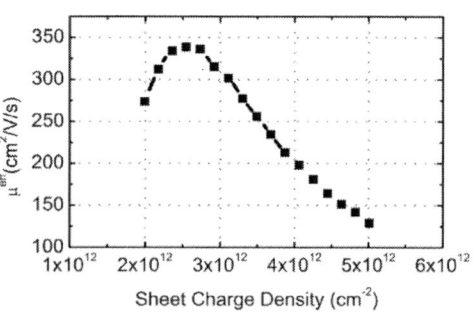

Fig. 10 Hole mobility versus sheet charge density extracted from SB pMOSFET.

Uniaxially Tensile Strained Accumulation-Mode Gate-All-Around Si Nanowire nMOSFETs

Mohammad Najmzadeh, Didier Bouvet, Wladek Grabinski, Adrian M. Ionescu

Nanoelectronic devices lab., Swiss Federal Institute of Technology (EPFL), CH-1015 Lausanne, Switzerland

phone: +41 21 693 5633, email: mohammad.najmzadeh@epfl.ch

In this work we report an experimental study on accumulation-mode (AM) gate-all-around (GAA) nMOSFETs based on silicon nanowires with uniaxial tensile strain. Their electrical characteristics are studied from room temperature up to ~400 K and carrier mobility, flat-band and threshold voltages are extracted and investigated.

Recently, highly single-type doped devices such as accumulation-mode (AM) and junctionless (JL) FETs have been proposed as straightforward switch architectures able to eliminate some of the limitations of nano-scale FETs such as the ultra-abrupt junctions and related processing challenges, allowing to fabricate even shorter channel devices [1-3]. Achieving a high I_{on} in a heavily doped channel is an engineering challenge since the carrier mobility in heavily doped devices is limited by ionized impurity scattering. In this paper we evaluate the impact of a local uniaxial tensile strain engineering method by local bending of nanowires to improve the mobility in highly doped AMOSFETs, without affecting their I_{on}/I_{off}.

Figure 1 represents the process flow to make dense array (~8 NW/µm) of deeply scaled sub-50 nm wide/thick Si nanowires on a Silicon-On-Insulator substrate. A sacrificial oxidation was formed in the presence of silicon nitride hard mask to shrink further the width of the Si nanowires, round the possible sharp corners as well as accumulate uniaxial tensile stress in the channel after stripping the hard mask and suspending the channel [4]. The gate stack includes 5 nm HfO$_2$ ALD deposition, RTA and afterward, 20 nm TiN deposition by sputtering to accumulate further uniaxial tensile stress in the suspended nanowires [5], causes local buckling of the suspended nanowires and induces clearly >1.5 GPa uniaxial tensile stress in the channel, measured from the arc length of the buckled NWs.

The electrical characterization was carried out at different temperatures using a Cascade prober and a HP 4155B Semiconductor Parameter Analyzer. Figure 2 depicts the I_D-V_G and I_D-V_D characteristics of a 2 µm long channel AMOSFET, including an array of 10 parallel Si nanowires, at 298 K. According to [6] and as described in Figure 3, the flat-band condition in an AMOSFET is pretty close to the threshold voltage and the accumulation layer will be created after reaching the flat-band condition while a heavily doped JL MOSFET, due to having the main current at the middle of the channel above threshold voltage, is mainly operating below the flat-band voltage and the flat-band condition will be reached while reaching the density of carriers on the surface to the doping concentration. Note that in the subthreshold region ($V_G < V_{TH}$) the conduction in both AMOSFET and JLFET is in the bulk of the NW.

The threshold voltage, V_{TH}, of the AMOSFET was extracted using the transconductance change (TC) method [7] which is quasi-independent of device series resistance and does not require the use of any accurate analytical model. Interestingly, as shown in Figure 4, the TC peak is appearing almost at the expected (theoretical) threshold voltage from the linear region of the I_D-V_G curve. By considering the current in the accumulation channel as the main dominant current in the GAA nanowire and using an earlier developed analytical model in [8], the flat-band voltage was extracted using the $I_D/g_m^{0.5}$ method [9] independent of series resistances and mobility attenuation factor and the results were plotted in Figure 4, providing systematically higher value than V_{TH} [6].

The transfer characteristics of the AMOSFET at 298-398 K are reported in Figure 5. The low field electron mobility at different temperatures was extracted using the slope of the $I_D/g_m^{0.5}$ vs. V_{GS} in the linear region (V_{DS}=100 mV), minimizing the influence of series resistances and mobility attenuation factor and plotted in Figure 6. As Figure 6 shows, the electron mobility in highly doped AMOSFET is reducing as $T^{-0.97}$ which is lower than for intrinsic or low n-doped Si and in agreement with prior reports [10], being explained by the dominant role of ionized impurity scattering in highly doped Si MOSFETs. Similar lower dependence on temperature is expected in JL FETs.

We report a threshold voltage reduction coefficient with temperature of -0.92 mV/K, a flat-band voltage coefficient with temperature of -1.73 mV/K and, finally, a subthreshold slope degradation of -0.43 %/K or -0.456 mV/dec.K (with a 106 mV/dec. subthreshold slope at 298 K), as reported in Figure 7.

As a conclusion, we have demonstrated the first AM gate-all-around nMOSFETs on SOI substrates with electron mobility enhanced by process-induced uniaxial tensile strain and reported their performance from 298 K to 398 K.

This work was supported by Swiss National Foundation (SNF). Thanks to CIME – EPFL for SEM observation.

[1] J.-P. Colinge et al., *Nature Nano.* (2010). [2] J.-P. Colinge et al., *Jpn. J. Appl. Phys.*, vol. 48, (2009). [3] J. Wu et al., *ECS Transactions*, vol. 33 (2010). [4] M. Najmzadeh et al., *Microelec. Eng.* (2009). [5] N. Singh et al., *IEEE EDL* (2007). [6] A. Nazarov et al., *Springer,* 1st ed. (2011). [7] H.-S. Wong et al., *SSE* (1987). [8] J.-P. Colinge et al., *IEEE TED*, vol. 37, (1990). [9] G. Ghibaudo, *Electronics Letters* (1988). [10] S. Reggiani et al, *IEEE TED*, Vol. 49, 2002.

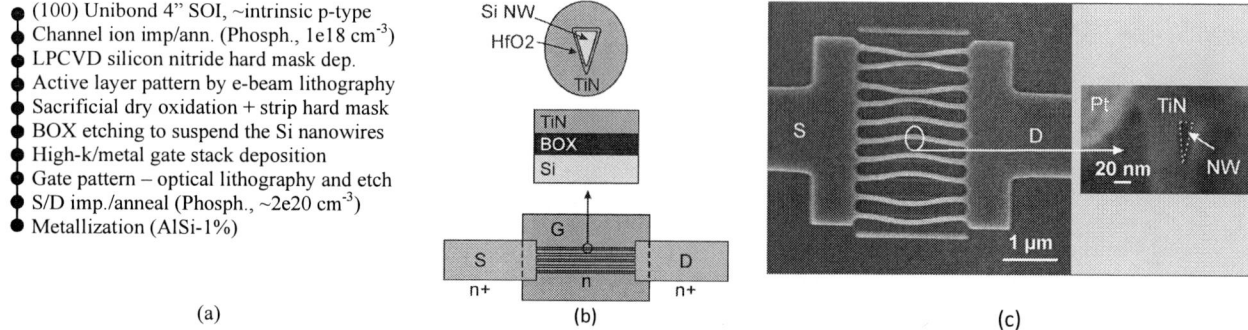

- (100) Unibond 4" SOI, ~intrinsic p-type
- Channel ion imp/ann. (Phosph., 1e18 cm^{-3})
- LPCVD silicon nitride hard mask dep.
- Active layer pattern by e-beam lithography
- Sacrificial dry oxidation + strip hard mask
- BOX etching to suspend the Si nanowires
- High-k/metal gate stack deposition
- Gate pattern – optical lithography and etch
- S/D imp./anneal (Phosph., ~2e20 cm^{-3})
- Metallization (AlSi-1%)

(a)　　　　　　　　(b)　　　　　　　　(c)

Fig. 1. Process flow to make GAA Si nanowire AMOSFET using a top-down Si NW platform (a), the schematic of the GAA Si NW AMOSFET (b), the SEM picture of an array of strained Si NWs after the gate stack step and the cross-section of a suspended GAA NW AMOSFET (right).

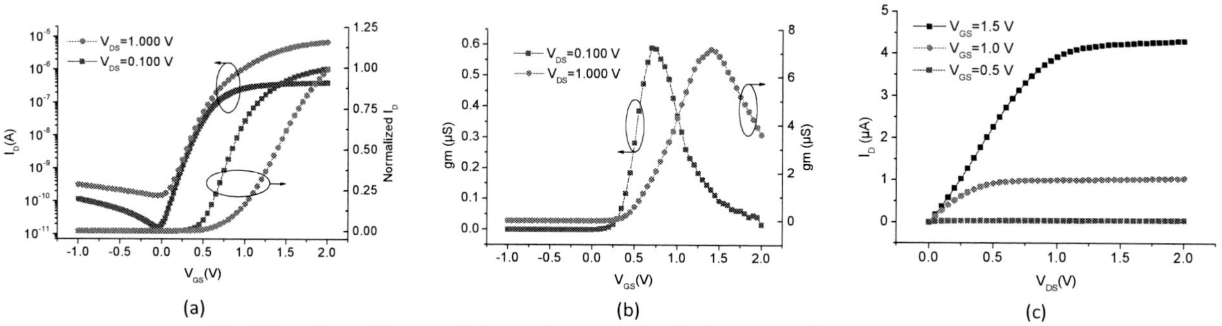

(a)　　　　　　　　(b)　　　　　　　　(c)

Fig. 2. Transfer and transconductance characteristics of a GAA *strained* AMOSFET at 298 K.

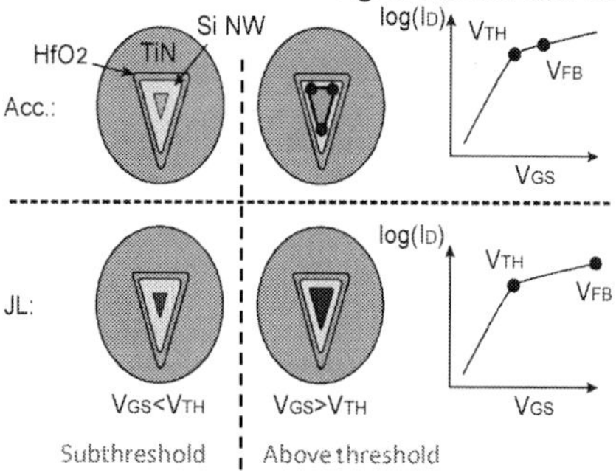

Fig. 3. Depiction of conduction path in GAA accumulation and junctionless MOSFETs vs. V_{GS}.

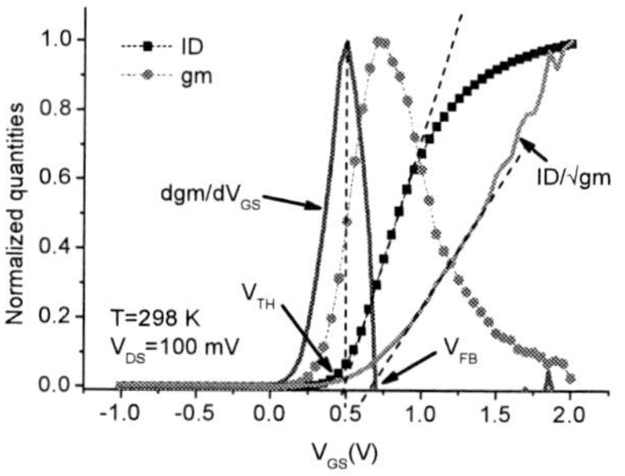

Fig. 4. Extraction of V_{TH} and V_{FB} of a GAA AMOSFET using TC and ID/gm$^{0.5}$ methods.

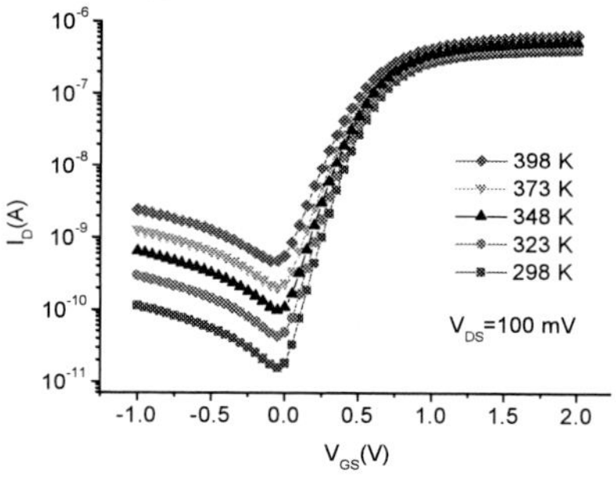

Fig. 5. Transfer charac. of a GAA *strained* AMOSFET at different temperatures.

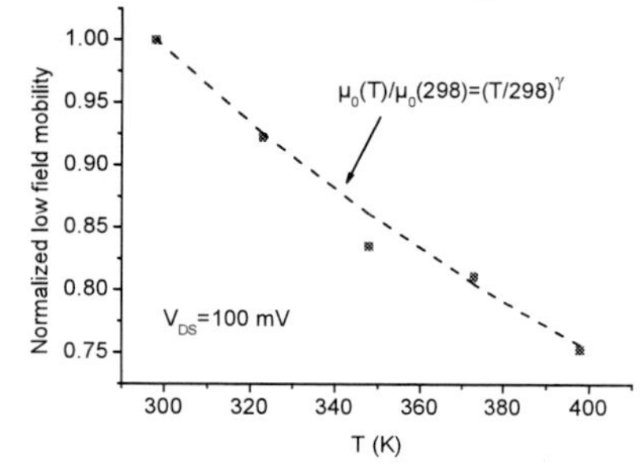

Fig. 6. Normalized low field electron mobility dependence on temperature for a GAA *strained* AMOSFET. The extracted γ is -0.966.

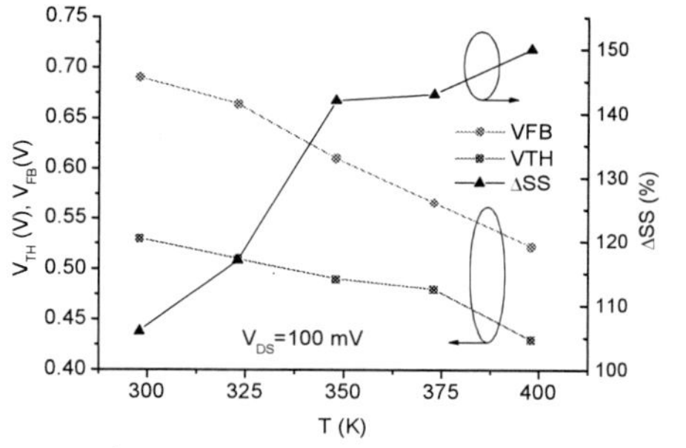

Fig. 7. Variation of extracted threshold voltage, flat-band voltage and subthreshold slope of a GAA *strained* AMOSFET with temperature.

978-1-61284-243-1/11 $26.00 © 2011 IEEE

Spintronics Search Engines

Hanan Dery, Berkehan Ciftcioglu, Yang Song, Hui Wu, Michael Huang, Roland Kawakami, Jing Shi, Ilya Krivorotov, Igor Zutic, and Lu J. Sham

We present a novel design concept for spintronic nanoelectronics that emphasizes a seamless integration of spin-based memory and logic circuits. The building blocks are magneto-logic gates [1,2] based on a hybrid graphene/ferromagnet material system. We use network search engines as a technology demonstration vehicle and present a spin-based circuit design with smaller area and lower energy consumption than the state-of-the-art CMOS counterparts. This design can also be applied in applications such as data compression, coding and image recognition. In the proposed scheme, over 100 spin-based logic operations are carried out before any need for a spin-charge conversion. Consequently, supporting CMOS electronics requires little power consumption. The spintronic-CMOS integrated system can be implemented on a single 3-D chip. These nonvolatile logic circuits hold potential for a paradigm shift in computing applications.

Figure 1 shows a universal and reconfigurable magneto-logic gate (MLG). Five magnetic terminals are deposited on top of a single layer graphene sheet. The spin accumulation profile in the sheet determines the logic result and it is governed by the magnetic directions of the biased sections (A-B and C-D). Using spin-transfer torque, the logic operands (magnetization directions of A-D) are encoded via the individual writing currents, $I_w(t)$, across the low resistive and all-metallic path (CoFe/Cu/Py/Cu/CoFe). The readout is triggered by the reading current signal, $I_r(t)$ that perturbs the magnetization of the middle contact. The binary logic output is the resulting on/off transient current, $I_M(t)$. Figure 2 shows the transient current across the middle contact in response to a 1 ns in-plane rotation of the magnetization direction of M. This modeled electrical behavior corresponds to matching operation between the stored (B & C) and search (A & D) bits. The bias is V_{dd}=1 V and the external capacitor is C_e=1 fF. The five contacts are 50 nm wide and 100 nm deep in the z direction. The spacing between contacts is 30 nm. The resistance and intrinsic capacitance of each contact, are respectively, 200 kΩ and 0.4 fF (areal conductance of 10^5 $\Omega^{-1}cm^{-2}$ and areal capacitance of 0.08 F/m^2). Figure 3 shows an optical microscope image of a lateral graphene sheet topped with ultrathin Ti-seeded MgO tunnel barrier and Co contacts [2]. The graphene flake is bounded by the dash line. This structure has been used to extract spin dependent graphene parameters.

To demonstrate the potential of MLG-based circuits, we use MLGs as building blocks to construct a spintronics search engine. The associative search of MLGs enables a highly scalable architecture with low power consumption. Figure 4(a) shows a circuitry for one m-bit word in a MLG-based search engine based on STT-MRAM technology. In each MLG, the current direction in one bit-line (vertical red wire) encodes a bit of the input search word at the outer contacts. The inner contacts (not the middle contact) hold the bit of the stored word (written by the horizontal cyan wires). If the stored and search magnetization directions are similar/dissimilar, the spin accumulation in the graphene channel is low/high, indicating a match or mismatch. For don't care bits the right and left search bit are opposite. The CMOS sensing circuit compares the match-line voltage output, Vt, with a reference level and generates a digital output voltage, V_k, indicating a match or mismatch in line k. Figure 4(b) shows the overall search engine architecture (Nxm bits). In the design example, N=25000, m=128.

[1] H. Dery et al., Nature 447, 573 (2007).
[2] H. Dery et al., ArXiv 1101.1497 (2011).
[3] W. Han et al., Phys. Rev. Lett. 105, 167202 (2010).

978-1-61284-243-1/11 $26.00 © 2011 IEEE

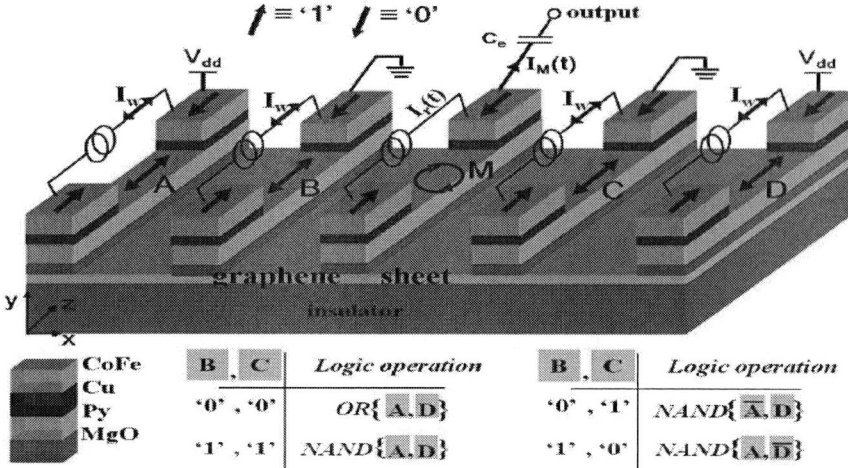

Fig. 1: Five-terminal Magneto-Logic Gate. In search application, the information bit is stored in B and C contacts and the search bit is stores in A & D. In case of don't care A and D are encoded with opposite bit values.

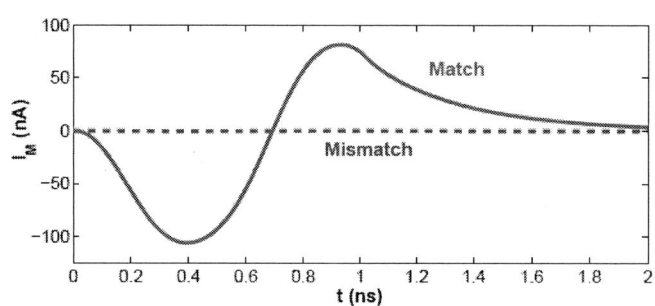

Fig. 2: Modeled electrical behavior of a magneto-logic gate set for matching between the stored (B & C) and search (A & D) bits.

Fig. 3: Optical image of graphene based lateral non local spin valve.

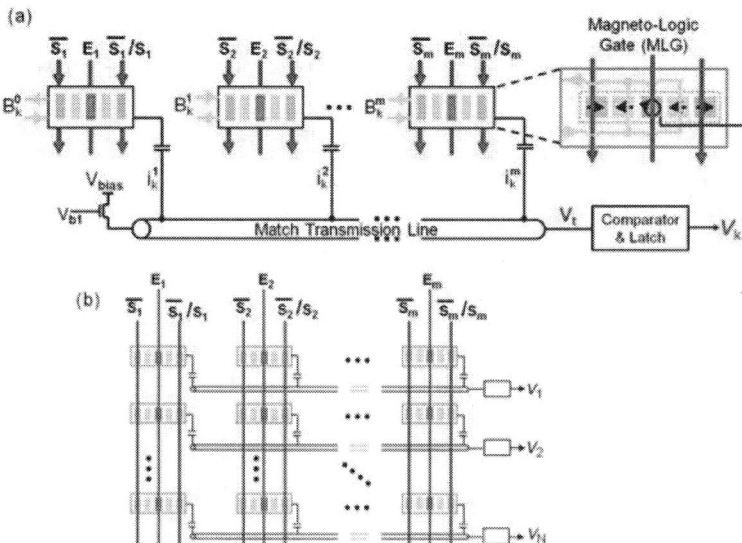

Fig. 4: (a) Spin-based search engine circuitry for one m-bit word using MLG cells and an AND sensing architecture. Each MLG consists of 5-FM terminals with an inherent nonvolatile XOR logic. (b) Overall search engine architecture (N_m bits). In the design example, N=25000, m=128. (See Figures 5 and 6 in Ref. [2] for details of the CMOS interface circuits.

978-1-61284-243-1/11 $26.00 © 2011 IEEE

Low Frequency Transconductance and Output Resistance Dispersion of Epitaxial Graphene Nanoribbon-based Field Effect Transistors

G. Aroshvili[1], N. Meng[2], D. Vignaud[2], D. Pavlidis[1] and H. Happy[2]

[1]Department of High Frequency Electronics, Darmstadt University of Technology, Merckstr.25, D-64283, Darmstadt, Germany
[2]Institute of Electronics, Microelectronics and Nanotechnology, CNRS and Univ.p Lille 1, Ave. Poincaré, F-59652 Villeneuve d'Ascq, France

Email: aroshvili@hfe.tu-darmstadt.de, pavlidis@hfe.tu-darmstadt.de Tel: +49 (6151)16-4162

Graphene-based devices have recently attracted strong attention due to very promising features such as two-dimensional material properties and high carrier mobility [1]. Significant effort has been placed on studies of high frequency characteristics of graphene transistors [2, 3] and first low-frequency noise studies have been reported [4]. However the low-frequency transconductance and output resistance dispersion of graphene FETs are less understood. These play a major role in determining the device performance and are the subject of the studies reported in this paper. The channel or the ungated region of the device is usually responsible for such effects. The Graphene Nano Ribbon Field Effect Transistors (GNRFETs) reported here have been fabricated using an array of parallel graphene nano ribbons, described in [2]. The devices were dual gate FETs fabricated with coplanar access structure for RF characterization. Ni/Au (50/300 nm) was used for source and drain contacts and the GNR array was defined by e-beam lithography. To achieve accurate ribbon width control, hydrogen silsesquioxane (HSQ) was used as mask material. The excess of graphene surface was then etched by O_2 RIE. After removing HSQ, the Al_2O_3 gate oxide was obtained by oxidation of a thin aluminium layer (about 2nm) in two steps leading to a final thickness of ~ 5 nm. Finally, the top gate (L_g=150 nm) was realized using Ni/Au 50/300nm. The photograph of the final device is shown in Fig. 1.

Fig. 2 presents typical DC I_{DS} (V_{DS}, V_{GS}) characteristics of dual gate GNRFET with gate length of 300 nm. The inset on Fig. 2 depicts the corresponding transfer characteristics. The presence of traps in a device is considered to be a major source of dispersion effects [5], which are usually more pronounced at low frequencies. The measurement of transconductance (g_m) frequency dispersion (10^2-10^5Hz) was made using a similar technique to the one described in [6]. As depicted in Fig. 3, at low drain bias (V_{DS} = 0.1V) a dramatic positive transconductance dispersion occurs during the first 10 kHz leading to a ~ 250% g_m-increase compared with DC values. The g_m decays then down to 60% dispersion at -0.5V gate voltage and is stable over the rest of the measurement frequency range. At high drain biases (V_{DS} = 0.8V) the increased g_m trend is still present in the beginning of the tested frequency range but manifests considerably lower dispersion (less than 40%). This is later on followed by a g_m decrease over the entire frequency range reaching 10% negative dispersion at 100 kHz. Fig. 4 shows the gate bias dependence of the transconductance dispersion under constant drain bias of 0.8V. Figs. 5 and 6 show the gate voltage dependent output resistance dispersion under small (0.1V) and large (0.8V) drain voltages. The output resistance was found to increase up to 30% under small drain bias, while the increase was limited to 15% in case of large bias.

Under small V_{DS} bias, the graphene transistors show minimum dispersion when V_{GS} is closer to pinch-off (Fig. 5). Conditions of this type are associated with current modulation and thus stronger frequency dispersion from the channel. This is not, however the case here and therefore the access region appears to play a major role in dispersion. The opposite trends are observed under large V_{DS} (Fig. 4), where the dispersion becomes stronger when moving to pinch-off, indicating therefore that the channel may indeed play a more important role for dispersion. The dispersion appears also to be lower under saturation rather than linear operation. The results indicate that both the access and channel regions are responsible for dispersion. Moreover the access region appears to contain a larger number of traps than the channel. The appearance of a positive g_m change followed by a negative one, suggests that the channel and access region traps have different time constants. R_{DS} was found to always increase with frequency. Dispersion was also found to be lower under higher V_{DS} bias. This is opposite to what is normally seen in most devices indicating that charge redistribution under the channel does not play a significant role. Figure 7 depicts the polar plot of S_{21}, which when extrapolated beyond the lowest frequency measured (40 MHz) allows one to estimate the transconductance at low frequency based on high-frequency and thus trap-independent data. A value of 0.27 mS was estimated at V_{GS} = 0.5 V and V_{DS} = 0.8 V. Since the DC value corresponding to the same bias was 0.35 mS, the high frequency g_m exhibits a ~ 23% dispersion compared with the DC value. This confirms the negative transconductance dispersion observed by low-frequency measurements and suggests that the dispersion continues beyond 100 KHz. Various fabrication steps may be responsible for the observed effects. These include (i) plasma etching leading in nanoribbons with rough edges; (ii) Al_2O_3 gate oxide directly deposited on graphene inducing degradation of electronic and structural properties of the material. Roughness of nanoribbons may affect channel conductance and access resistances, while the dielectric may limit channel mobility through phonon scattering, and structural defects. This agrees with reports on the effect of dielectric overlayers on graphene properties [7, 8]. Low temperature characterization made on the same devices [9] shows also the strong dependence of temperature on device performance and dispersion. Overall, we report here for the first time the frequency dependence of graphene device characteristics and discuss possible trap related reasons responsible for it.

[1] Yu-Ming Lin et al., *Nano Letters,* Vol.9, No.1, pp 422-426, 2009; [2] M. Meng et al., *European Microwave Integrated Circuits Conference (EuMIC),* pp 294-297, 2010; [3] Yu-Ming Lin et al., *Science,* Vol.327, No.5966, p 662, 2010; [4] Imam, S.A. et al., *Micro & Nano letters, IET,* pp.: 37-41, 2010   [5] S.H.Hsu et al., *Proceedings of International Semiconductor Device Research Symposium,* pp 315-317, 1999; [6] G. I. Ng et al., *IEEE Electron Device Letters,* Vol. 38, No. 4, pp 862 - 870, 1991; [7] S. M. Song and B. J. Cho, Nanotechnology 21 (2010) 335706; [8] J. A. Robinson et al., *ACS Nano,* Vol. 4, No. 5, 2667–2672, 2010; [9] N. Meng et al., *68th Device Research Conference,* Notre Dame, 2010.

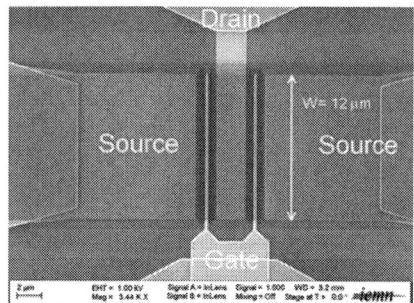

Figure 1. Graphene nanoribon FET structure. L_g=150nm, Two arrays of 120nm ribbons with width, spacing = 50nm. Source-to-Drain spacing of 1.6um

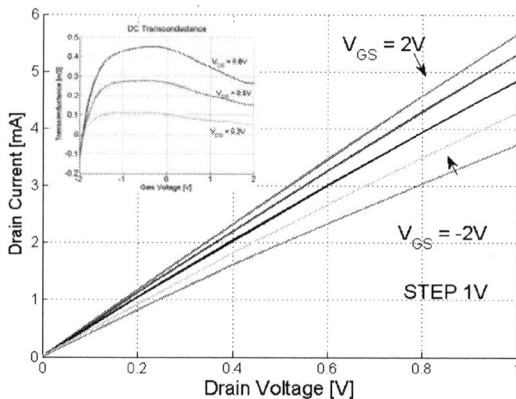

Figure 2. Typical DC output characteristics of two finger 300nm gate length GNRFET with gate voltage swept from -2 to 2 volts with 1V step. Inset: Typical DC transfer characteristics of the same device at various drain bias voltages.

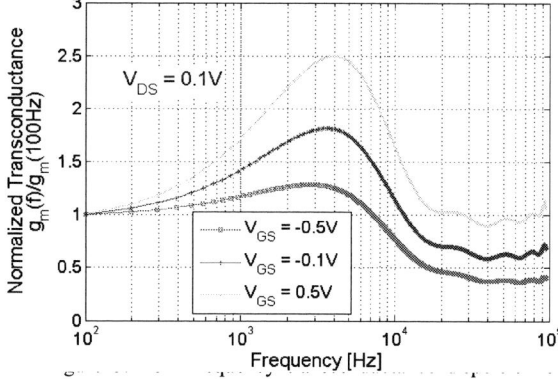

GNRFET at fixed drain bias of 0.1V for different gate bias

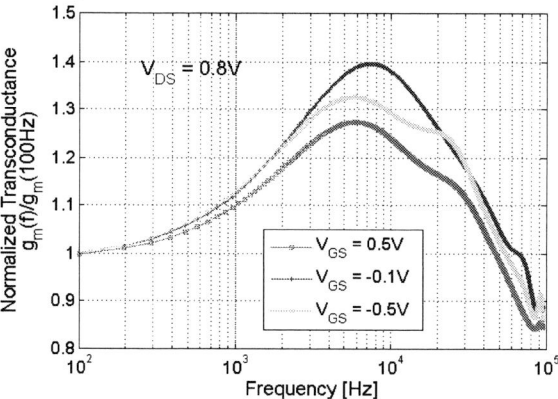

Figure 4. Low frequency transconductance dispersion of GNRFET at fixed drain bias of 0.8V for different gate bias voltages.

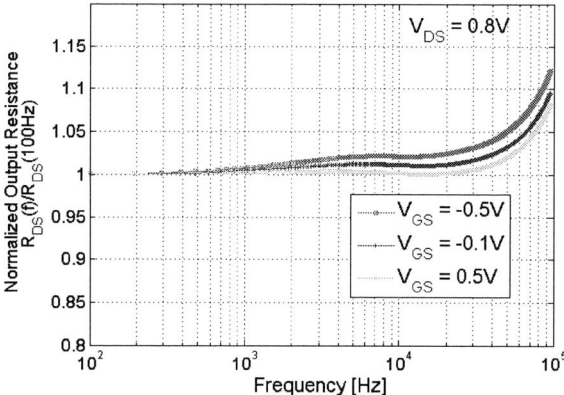

Figure 5. Low frequency output resistance dispersion of GNRFET at fixed drain bias of a. 0.8V for different gate bias voltage

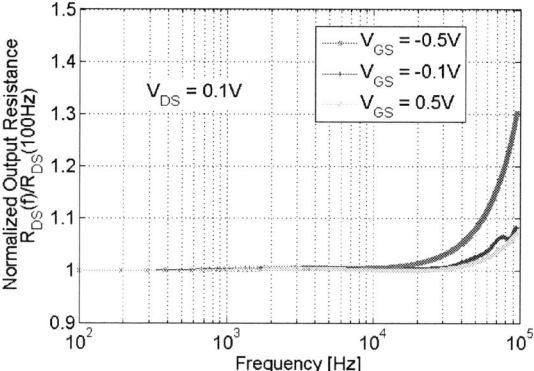

Figure 6. Low frequency output resistance dispersion of GNRFET at fixed drain bias of 0.1V for different gate bias voltages.

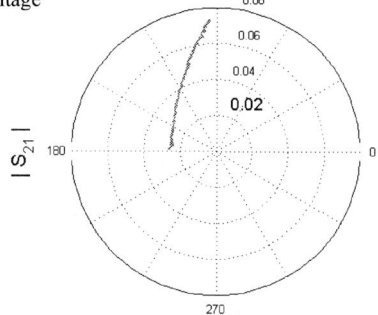

Figure 6. Polar plot of forward transmission S_{21}, when interpolated towards low frequencies is used to estimate DC transconductance of 0.27mS.

978-1-61284-243-1/11 $26.00 © 2011 IEEE

InAs/SiGe on Si Nanowire Tunneling Field Effect Transistors

C. Kshirsagar and S. J. Koester*

University of Minnesota-Twin Cities, 200 Union St. SE, Minneapolis, MN 55455
*Ph: (612) 625-1316, FAX: (612) 625-4583, Email: skoester@umn.edu

Introduction: Tunneling field-effect transistors (TFETs) are of tremendous interest for advanced logic applications due to their potential for sub-60-mV/dec subthreshold slope which could enable supply voltage scaling beyond what is practical for conventional MOSFETs. However, TFETs based upon tunneling in Si suffer from low on current, I_{ON}, and fail to provide steep slope at high current levels. III-V TFETs are more promising due to their potential for high drive current, but the poor gate oxide quality remains a significant challenge. Recently, a hybrid III-V-on-Si approach [1] has been proposed as a potential solution to this problem, whereby the small effective band gap, E_{geff}, of the InAs/Si heterojunction could increase I_{ON}, while preserving the high-quality Si/dielectric interface in the channel. Experimental demonstrations of nanostructured InAs-on-Si Esaki diodes and TFETs suggest this approach is feasible [1],[2]. However, InAs-on-Si heterostructures still exhibit relatively large E_{geff} (~ 0.4 eV in unconfined geometries) and quantum effects increase E_{geff} substantially in confined geometries. In this paper, we provide a simulation analysis of a new device structure, the InAs/SiGe/Si TFET that could overcome this problem by utilizing a compressively-strained SiGe layer to further decrease E_{geff}. We show that I_{ON} in these devices increases by 5x (at constant I_{off}) and further explore the various trade-offs and performance-limiting factors in these devices.

Device Description and Simulation Assumptions: Comprehensive 3D simulations were carried out using Synopsys Sentaurus Device. A nanowire p-TFET geometry with an InAs source, $Si_{1-x}Ge_x$ channel and Si drain was considered (**Fig.1**). The nominal gate length, L_g, and x-value were 20 nm and 50%, respectively, while the remainder of the structural parameters are shown in the **Fig.1** and **Fig. 2**. A non-local tunneling model was utilized. The InAs was assumed to be relaxed, while the SiGe was assumed to be under biaxial compressive strain. The relaxed state of InAs is reasonable considering the results in [1],[5] that show that InAs grown on Si produces an array of misfit dislocations at fixed lattice spacing. Nonparabolicity and quantum confinement effects were computed analytically and the band parameters adjusted accordingly for a given nanowire diameter. The SiGe was considered to be fully biaxially strained. Interface traps were not considered in this study, as improved understanding of the physical nature of the SiGe/InAs interface traps is necessary in order to properly model their effect on carrier transport.

Simulation Results and Discussion: The calculated values of the InAs/SiGe effective band gap as a function of diameter and composition are shown in **Fig. 3**. The I_d vs. V_{gs} curves for various source dopings (**Fig. 4**) indicate two limiting constraints for these devices: low doping reduces the I_{on} of the device and high doping degrades the subthreshold slope. **Fig. 5** shows the subthreshold slope vs. drain current at various source doping levels. It can be seen that, as the degeneracy is increased with doping, the subthreshold slope degrades from 18 mV/dec to the thermal limit of 60 mV/dec, at which point the Boltzmann "tail" of the Fermi-Dirac distribution dominates the tunneling current. **Fig. 6** and **Fig. 7** show the I_d-V_{gs} and subthreshold slope behavior for pure silicon and at various Ge compositions. It can be seen that increased Ge concentration is beneficial both in terms of better subthreshold slope and higher on currents. **Fig. 8** shows the on current at various I_{off} values at $V_{dd} = 0.5$ V. It can be seen that the drive current improves by 5x for $I_{off} = 1$ nA/μm as the Ge composition is increased from 0 to 50 %. **Fig. 9** shows the I_{on}-I_{off} characteristic for various doping levels in the source. Higher I_{on} is achievable with increased doping, though at the expense of poorer subthreshold slope. As V_{dd} is reduced, the trade-off between source degeneracy and depletion leads to an optimum source doping level, as shown in **Fig. 10**. A strong dependence of I_{on} on the diameter is also observed (**Fig. 11**), and the utilization of SiGe in the channel allows the reasonable drive currents to be maintained with increased confinement. Good short-channel behavior for the nominal devices (10 nm diameter) is observed, while the relatively-large band gap of SiGe suppresses direct source-to-drain tunneling for $L_g > 10$ nm (**Fig. 12**).

In conclusion, our results show that hybrid InAs/SiGe p-TFETs may lead to improved drive currents and steeper subthreshold slope compared to InAs/Si devices. On currents as high as 150 μA/μm at 1 nA/μm and $V_{ds} = 0.5$ V are possible using 50% SiGe. However, these devices have a fundamental trade-off between the source doping and the subthreshold slope which limits the maximum I_{on} and I_{on}/I_{off} ratios that can be obtained. The effect of defects in both InAs/Si and InAs/SiGe TFETs remain outstanding issues requiring further study.

---

[1] M.T. Bjork, et al., *Appl. Phys. Lett.* **97**, 163501 (2010); [2] K. Tomioka, et al, *Appl. Phys. Lett.* **98**, 083114 (2011); [3] M. V. Fischetti and S. E. Laux, *J. Appl. Phys.* **80**, 2234 (1996); [4] D. V. Lang, et al, *Appl. Phys. Lett.* **47**, 1333 (1985); [5] E. Ertekin, et al, *J. Appl. Phys.* **97**, 114325 (2005).

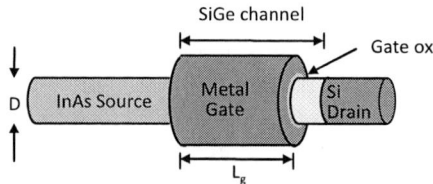

Parameter	Value
Source Doping (n)	5×10^{18} cm^{-3}
Channel Doping (p)	1×10^{15} cm^{-3}
Drain Doping (p)	1×10^{21} cm^{-3}
Gate dielectric thickness (t_{ox})	1 nm
Gate Length (L_g)	20 nm
Diameter (D)	10 nm
Composition (X)	0 – 50 %

Fig. 1. Schematic diagram of nominal device structure and structural parameters.

Material	Parameter	Value
InAs	$m_c(\Gamma) / m_0$	0.023
InAs	E_g (eV)	0.354
SiGe	m_v (perp.) / m_0	0.28 – 0.23
SiGe	m_v (ll) / m_0	0.21 - 0.1
SiGe	E_g (eV)	0.828 -1.108

Fig. 2. Calculated and extracted material parameters for non-local tunneling [3,4].

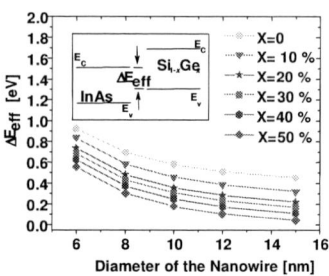

Fig. 3. Plot of InAs/Si$_{1-x}$Ge$_x$ effective band gap vs. diameter for different x values.

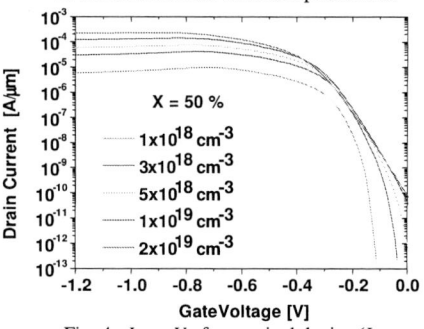

Fig. 4. I_d vs. V_{gs} for nominal device (L_g = 20 nm and x = 50%) for different source doping levels.

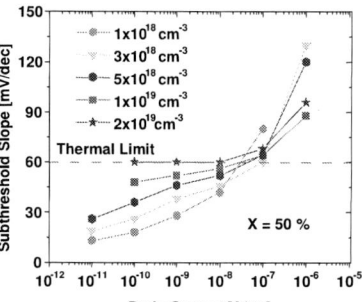

Fig. 5. SS vs.log(I_d) for different source doping levels. Strong degenerate doping limits subthreshold slope to ~ 60 mV/dec.

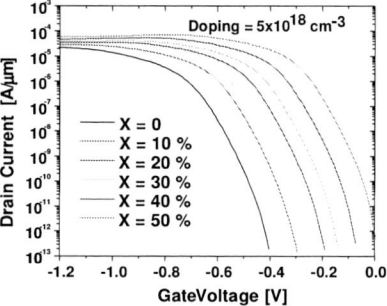

Fig. 6. I_d vs. V_{gs} for at different channel Ge concentrations. N_d(source) = 5 x 10^{18} cm^{-3}.

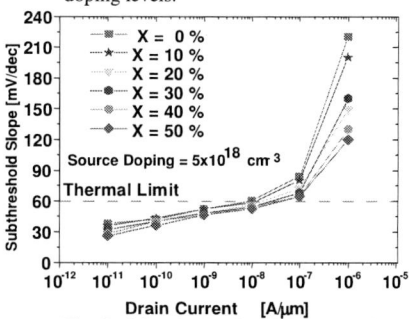

Fig. 7. SS vs.log(I_d) for different x-values. Increasing Ge composition improves SS.

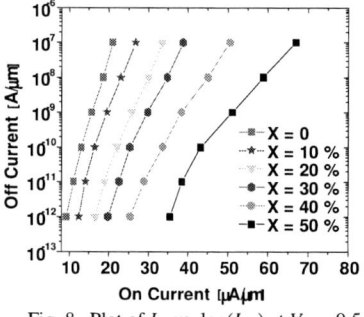

Fig. 8. Plot of I_{on} vs. log(I_{off}) at V_{dd} = 0.5 V for different x-values.

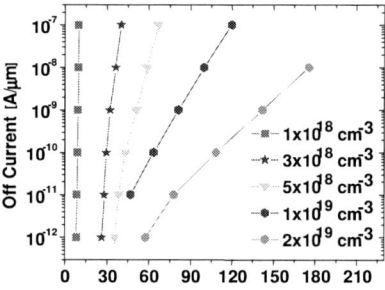

Fig. 9. Plot of I_{on} vs. log(I_{off}) at V_{dd} = 0.5 V for different Source doping

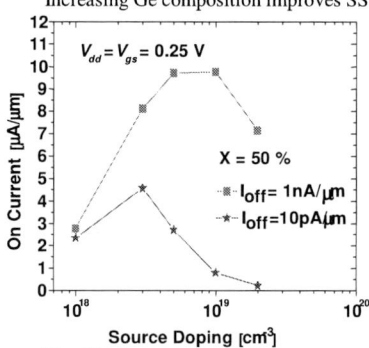

Fig. 10. On current at various source doping level when V_{dd} is scaled down to 0.25 V

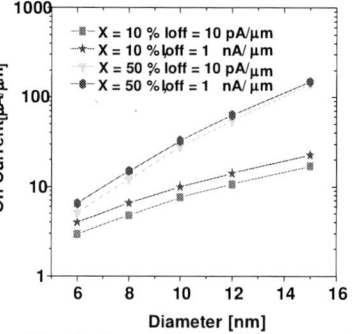

Fig. 11. On current for different nanowire diameter at Source doping = 3x 10^{18} cm^{-3} V_{dd} = 0.5 V.

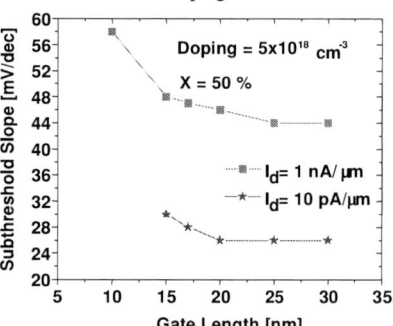

Fig. 12. Effect of gate length scaling on sub threshold slope at V_{dd} = 0.5 V.

Session IV.A (Corwin Pavilion East)

Spin/Memory

Tuesday AM, June 21st, 2011

Session Chair: Abram Falk, University of California, Santa Barbara and Brian Doyle, Intel Corporation

8:20 AM IV.A-1 Invited Paper
Electrical measurement of the spin Hall effects in Fe/In$_x$Ga$_{1-x}$As heterostructures
E. S. Garlid[1], Q. O. Hu[2], C. Geppert[1], M. K. Chan[1], C. J. Palmstrøm[2,3], and P. A. Crowell[1], [1]School of Physics and Astronomy, University of Minnesota, Minneapolis, Minnesota, USA, [2]Dept. of Electrical and Computer Engineering, University of California, Santa Barbara, California, USA, and [3]Dept. of Materials, University of California, Santa Barbara, California, USA

9:00 AM IV.A-2 Student Paper
Simultaneous Spin and Charge Transport in Gated Si Devices
J. Li and I. Appelbaum, Center for Nanophysics and Advanced Materials and Department of Physics, University of Maryland, College Park, Maryland, USA

9:20 AM IV.A-3 Student Paper
Unidirectional information transfer with cascaded All Spin Logic devices: A Ring Oscillator
S. Srinivasan[1,2], A. Sarkar[1,2], B. Behin-Aien[1,2], and S. Datta[1,2], [1]School of Electrical and Computer Engineering, Purdue University, W. Lafayette, Indiana, USA and [2]NSF Network for Computational Nanotechnology (NCN), W. Lafayette, Indiana, USA

9:40 AM IV.A-4
Proposal for piezoelectric-ferromagnet bilayer based microwave Oscillators without any external magnetic field or spin transfer torque
D. Bhowmik and S. Salahuddin, Department of Electrical Engineering and Computer Sciences, University of California, Berkeley California, USA

10:00 AM Break

10:20 AM IV.A-5
Orthogonal Spin Transfer MRAM
D. Bedau[1], D. Backes[1], H. Liu[1], J. Langer[2], P. Manandhar[3] and A. D. Kent[1], [1]Department of Physics, New York University, New York, New York, USA, [2]Singulus Technologies AG, Kahl am Main, GERMANY, and [3]Spin-Transfer Technologies, Quincy, Massachusetts, USA

10:40 AM IV.A-6
Thermal Effects and Instability in Unipolar Resistive Switching Devices
A. Chen and M.-R. Lin, Strategic Technology Group, GLOBAL FOUNDRIES, Sunnyvale, California, USA

11:00 AM IV.A-7 Student Paper
A Hybrid Ferroelectric and Charge Nonvolatile Memory
S. R. Rajwade, K. Auluck, J. Shaw, K. Lyon and E. C. Kan, School of Electrical and Computer Engineering, Cornell University, Ithaca, New York, USA

11:20 AM IV.A-8 Invited Paper
Spin-torque switchable perpendicular magnetic junctions for solid-state memory
J. Z. Sun[1,2], R. P. Robertazzi[1], J. J. Nowak[1], P. L. Trouilloud[1], G. Hu[1], M. C. Gaidis[1], S. L. Brown[1], D. W. Abraham[1], E. J. O'Sullivan[1], W. J. Gallagher[1], D. C. Worledge[1], and A. D. Kent[2], [1]IBM-MagIC MRAM Development Alliance, IBM T. J. Watson Research Center, Yorktown Heights, New York, USA and [2]Dept of Physics, New York University, New York, USA

978-1-61284-243-1/11 $26.00 © 2011 IEEE

978-1-61284-243-1/11 $26.00 © 2011 IEEE

Electrical measurement of the spin Hall effects in

Fe/In$_x$Ga$_{1-x}$As heterostructures

E. S. Garlid,[1] Q. O. Hu,[2] C. Geppert,[1] M. K. Chan,[1] C. J. Palmstrøm,[2,3] and P. A. Crowell[1]

[1]*School of Physics and Astronomy, University of Minnesota, Minneapolis, MN 55455*
[2]*Dept. of Electrical and Computer Engineering,*
University of California, Santa Barbara, CA 93106
[3]*Dept. of Materials, University of California, Santa Barbara, CA 93106*

There has been extensive theoretical discussion of the spin Hall effect (SHE) and the various ways that it could be exploited to generate or manipulate spin currents. However, only a handful of recent experiments have investigated this effect, and in semiconductor materials they have relied on optical techniques to either detect or generate spins [1,2].

We report on an all-electrical measurement of the spin Hall effect in epitaxial Fe/In$_x$Ga$_{1-x}$As heterostructures with n-type channel doping (Si) (n $\sim$ 5 $\times$ 10^{16} cm^{-3}) and highly doped Schottky tunnel barriers (n$^+$ $\sim$ 5 $\times$ 10^{18} cm^{-3}).[3] A transverse spin current generated by an ordinary charge current flowing in the In$_x$Ga$_{1-x}$As is detected by measuring the spin accumulation at the edges of the channel. The spin accumulation is identified through the observation of a Hanle effect in the Hall voltage measured by pairs of ferromagnetic (FM) contacts [see Fig. 1(a)]. We investigate the bias and temperature dependence of the resulting Hanle signal and we determine that both skew and side-jump scattering contribute to the total spin Hall conductivity.

Typical spin valve and Hanle effect curves (see Ref. [4] for a discussion) on a lateral spin valve device are shown in Fig. 1(b). These data establish that the FM contacts are sensitive to the spin polarization in the channel as well as its dephasing by precession in an applied magnetic field. We also establish that the Fe contacts, which have an easy axis along [110], show sharp and reproducible switching behavior as well as nearly perfect remanence. To measure the SHE, we apply a field B$_y$ to precess the spins into the FM easy axis direction. The resulting Hanle curve can be fit using a diffusion model similar to that used in Ref. [4]. For each bias current, a single set of parameters is used to fit the data sets obtained at different distances from the edge of the channel. The only free parameters in the fit are the magnitude of the spin polarization at the sample edge, P_0, and the spin lifetime, τ_s. The spin current j_s can be determined from the parameters extracted from the fit using the solution to the diffusion equation in zero magnetic field.

We find that the magnitude and sign of the spin Hall conductivity $\sigma_{SH} = j_s/E$ in GaAs is in agreement with models of the extrinsic SHE due to ionized impurity scattering [5, 6]. An analysis of the dependence of the SH signal on channel conductivity [Fig. 2(a)] shows both skew scattering and side jump components, and their ratio in the four samples is roughly constant, consistent with theoretical predictions, although the magnitude is too large by a factor of three [5, 6]. The temperature dependence of the spin Hall conductivity (Fig. 2(b)) is weak over the range of our experiment (T < 150 K), although our sensitivity at the highest temperatures is limited by the short spin lifetime in the channel.

The current status of measurements of the *inverse* spin Hall effect, in which a spin-polarized current is converted to a charge current, is more complicated. I will show that our devices, when operated in a regime of extremely large spin accumulation, show a spin-dependent Hall effect due to a combination of spin-orbit and hyperfine interactions. Although the microscopic mechanism is not clear, the size of the inverse spin Hall effect can be much larger than the direct effect. This serves as an example of some of the unique physics that can occur when a large non-equilibrium spin population exists in the semiconductor. This also has ramifications for the physics of multi-terminal devices in the presence of more than one source of spin current.

This work has been supported by NSF under DMR 08-040244, the NSF MRSEC and NNIN programs, and the Office of Naval Research.

[1] Y. K. Kato, R. C. Myers, A. C. Gossard, and D. D. Awschalom, Science **306**, 1910 (2004).

[2] J. Wunderlich, B. Kaestner, J. Sinova, and T. Jungwirth, Phys. Rev. Lett. **94**, 047204 (2005).

[3] E. S. Garlid, Q. O. Hu, M. K. Chan, C. J. Palmstrøm, and P. A. Crowell, Phys. Rev. Lett. **105**, 156602 (2010).

[4] X. Lou, C. Adelmann, S. A. Crooker, E. S. Garlid, J. Zhang, K. S. M. Reddy, S. D. Flexner, C. J. Palmstrøm, and P. A. Crowell, Nature Physics **3**, 197 (2007).

[5] H.-A. Engel, B. I. Halperin, and E. I. Rashba, Phys. Rev. Lett. **95**, 166605 (2005).

[6] W. K. Tse and S. D. Sarma, Phys. Rev. Lett. **96**, 056601 (2006).

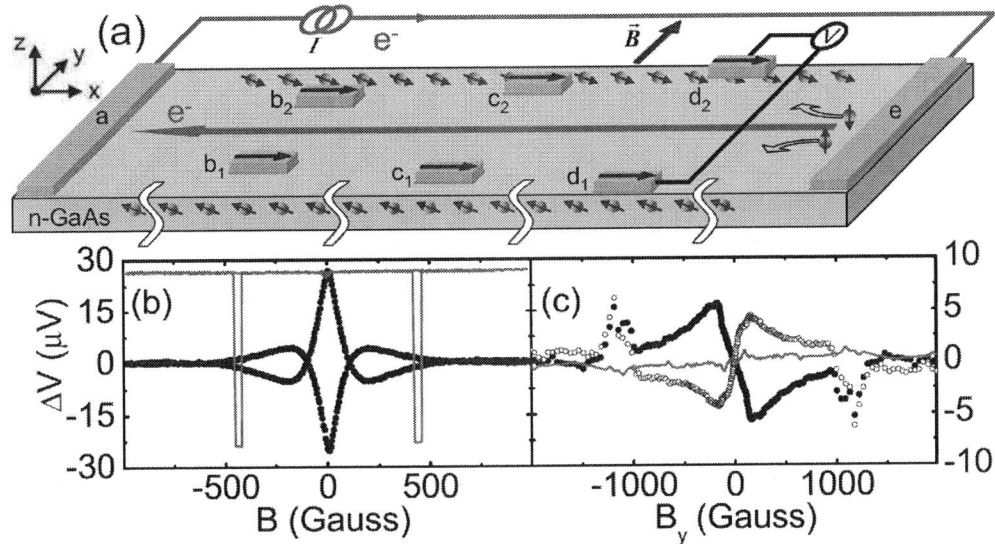

Fig. 1. (a) Schematic diagram (not to scale) of the device layout and SHE measurement. (b) Non-local spin valve (−) and Hanle (•) data obtained on a Fe/GaAs heterostructure at T = 60 K for $j_{Inj} = 8.2 \times 10^2$ A/cm^2. (c) SHE data obtained on a GaAs device at T = 30 K with the two Hall contacts in the parallel state for $j_x = \pm 5.7 \times 10^3$ A/cm^2 (• and ○) and antiparallel state (−).

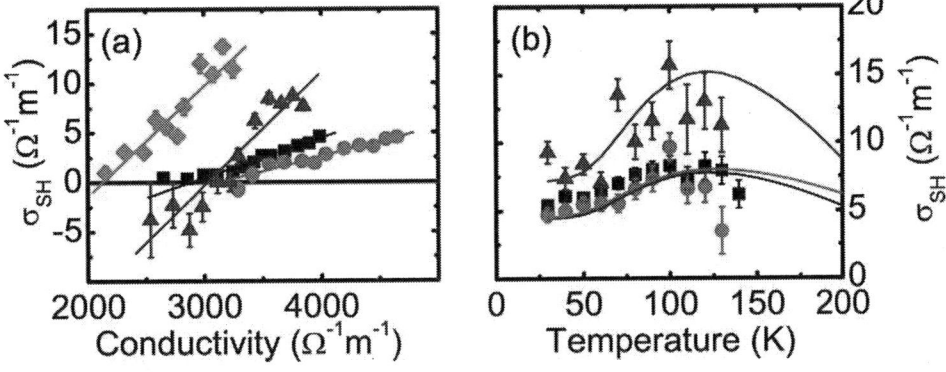

Fig 2. (a) Dependence of σ_{SH} on σ_{xx} at T = 30 K. A linear fit is used to extract γ and σ_{SJ}. (b) Temperature dependence of σ_{SH} from T = 30 to 130 K (solid points) and predicted temperature dependence using $\sigma_{xx}(T)$ and n(T) from sum of skew and side-jump contributions. The In concentrations x are 0.00 (black squares), 0.03 (red circles), 0.05 (blue triangles), 0.06 (magenta diamonds).

978-1-61284-243-1/11 $26.00 © 2011 IEEE

978-1-61284-243-1/11 $26.00 © 2011 IEEE   158

Simultaneous Spin and Charge Transport in Gated Si Devices

Jing Li and Ian Appelbaum

Center for Nanophysics and Advanced Materials and Department of Physics, University of Maryland, College Park, Maryland 20742, USA

Phone: (301)405-0890, fax: (301)405-3779, email: appelbaum@physics.umd.edu

Recent advances in the development of techniques for electrical injection and detection of spin-polarized electrons in silicon have aroused intensive research on exploiting devices and circuits that utilize the spin degree of freedom [1-3] as well as electron charge in this dominant material of the semiconductor integrated circuits industry.

In this work, lateral spin-transport devices employing ballistic hot-electron injection and detection methods are used to study temperature-dependent spin and charge transport controlled by an electrostatic back gate using native oxide (SiO_2) insulator. Spin transport is studied by the "Larmor clock" technique [4], where the ns-scale spin transit-time distribution is recovered from quasi-static precession measurements in a perpendicular magnetic field with potentially sub-ns resolution [see Fig. 1]. The empirical spin transit-time distribution can be directly derived via Fourier transform and later used to calculate spin mobility through $\mu_s = L^2/(V_{AC}\bar{t})$, where $\bar{t}$ is the mean of the spin transit-time distribution, L is the length of lateral transport channel, and V_{AC} is the longitudinal accelerating voltage across the channel from injector to detector. Simultaneous charge transport characteristics are investigated by Hall effect, where ballistic injection of electrons into an otherwise undoped silicon transport channel allows direct measurement of charge mobility (in obvious contrast to ohmic injection, which directly measures only carrier density). From the transverse voltage across the channel, charge mobility is derived through $\mu_c = V_H L/(WB\,V_{AC})$, where W is the width of transport channel, B is the perpendicular magnetic field, and V_H is the transverse voltage. The inset to Fig. 2 shows the linear relationship between Hall voltage and applied field, and the main panel of Fig. 2 shows that the Hall mobility is indeed independent of the current as expected for ballistically-injected currents greater than approximately 2 A (again, in direct contrast to ohmic injection, where the Hall voltage is directly proportional to current).

As in Fig. 3, it is found that both charge and spin mobility decrease as temperature rises, consistent with the role of phonon scattering in both processes. Spin mobility is lower than charge mobility under the same gate biases, but charge mobility decreases more rapidly than spin mobility when electrons are attracted to the Si/SiO_2 interface as shown in Fig. 4. The results indicate that enhanced momentum scattering near the Si/SiO_2 interface increases spin relaxation rate, consistent with the Elliott-Yafet spin relaxation mechanism.

[1] I. Appelbaum et al., *Nature*, vol. 447, p. 295 (2007)

[2] B. Huang et al., *Phys. Rev. Lett.*, vol. 99, p. 177209 (2007)

[2] H.-J. Jang et al., *Phys. Rev. Lett.*, vol. 103, p. 117202 (2009)

[4] B. Huang et al., *Phys. Rev. B*, vol. 82, p. 241202(R) (2010)

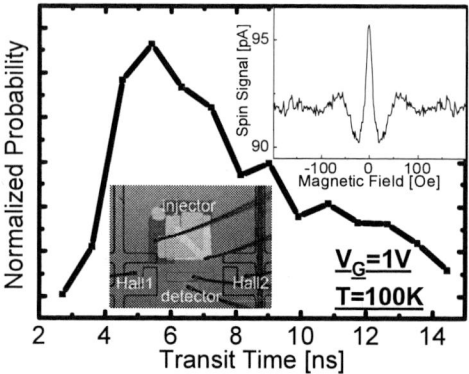

Fig. 1: Transit-time distribution for spin current under accelerating bias 16V and gate bias 1V at 100K. Inset 1 (upper right): symmetrized spin precession data, from which the distribution was calculated using Fourier transform. Inset 2 (bottom left): plan view of a wire-bonded silicon lateral spin-transport device.

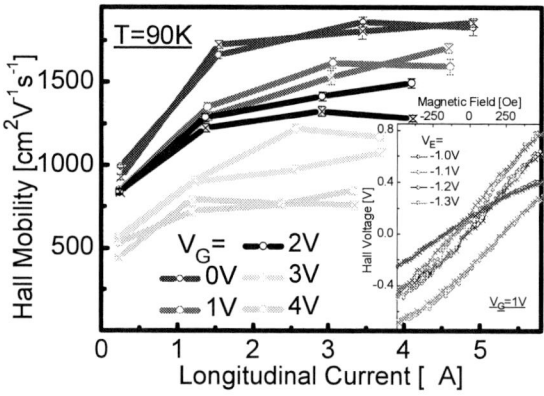

Fig. 2: Longitudinal current dependence of Hall mobility under different gate bias (0-4V) at 90K. Inset: Magnetic field dependence of Hall voltage, from which Hall mobility was calculated, under 1V gate bias at different injector bias (1.0-1.3V).

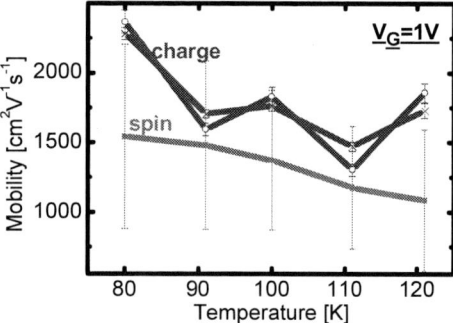

Fig. 3: Comparison between spin and charge mobility temperature dependence under gate bias 1V.

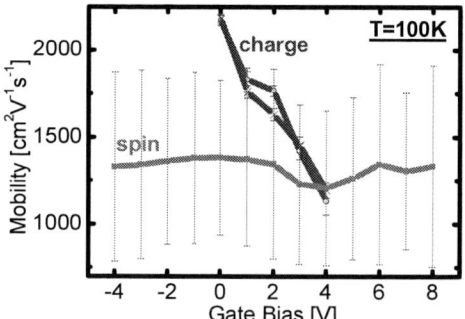

Fig. 4: Comparison between spin and charge mobility gate dependence at 100K.

978-1-61284-243-1/11 $26.00 © 2011 IEEE   160

Unidirectional information transfer with cascaded All Spin Logic devices: A Ring Oscillator

Srikant Srinivasan*[1,2], Angik Sarkar*[1,2], Behtash Behin-Aien*[1,2], and Supriyo Datta[1,2], *Fellow, IEEE*

[1]School of Electrical and Computer Engineering, Purdue University, W. Lafayette, IN 47907 USA.

[2]NSF Network for Computational Nanotechnology (NCN), W. Lafayette, IN 47907 USA

E-mail: srikant@purdue.edu; ph: (765) 496-1095. *These authors contributed equally to this work.

Magnet based logic devices have received much attention as potential alternatives [1] to charge based electronics in order to answer the ever growing concern [2] about the limits of CMOS scaling, especially since it has been shown that the energy required to turn a magnet could be as low as a few atto-joules [3]. The recently proposed All Spin Logic (ASL) device [4] is one such scheme whereby information is stored in the state of magnets and is communicated between magnets purely through spin currents, thus operating entirely within a new paradigm: using spin as a state variable.

One fundamental aspect for logic application that features naturally in CMOS – unlike interacting magnets – is that information transfer is unidirectional and that there is sufficient isolation to ensure that the input state is not affected by the output state. Nevertheless, we have shown [6] that in an ASL device the physics of spin torque can be utilized to engineer similar 'intrinsic' unidirectionality among equally sized communicating magnets. The inevitable question that must be addressed in this context is what happens when such devices are cascaded for logic operation. Can these devices really drive subsequent stages? Can there be sneak paths which degrade input-output isolation? Will the magnets switch in a deterministic fashion and will the outputs be stable? Indeed, in answering these questions our contribution in this paper is twofold:

- We present the first simulator for multi-magnet networks interacting via spin currents (Fig. 2) that can describe their simultaneous dynamics as well as their interaction through spin transport extending our earlier work in [5].
- We demonstrate through this simulator, the feasibility of unidirectional information transfer in cascaded ASL devices using the example of a Ring Oscillator (Figs. 1(b, c)) based on realistic device parameters (Table 1).

The behavior of the ASL ring oscillator is easily understood by first examining the simplest case in Fig. 1(a). Every magnet has a physical barrier beneath it which allows it to interact separately with the preceding and succeeding stages through a non-magnetic channel. When supply voltage (Vss) is applied on any magnet in the chain, a charge current flowing through the magnet into the ground terminal gets polarized as it traverses the magnet. When Vss<0 the magnet injects spins and when Vss>0 the magnet extracts spins. The resulting spin current then propagates to the other magnets through the channel and exerts a torque on them. When Vss is greater than a certain threshold value, the resulting spin current can actually flip the other magnets to a parallel (Vss<0) or anti-parallel (Vss>) state with respect to the injector. Ideally, for unidirectional information flow, we want one side of each magnet to behave as the input, from where it can receive information, and the other side to behave as the output from where it can pass it on. Among the several schemes [6] to enforce such directionality, here we use different injection polarizations for the two sides in Fig. 1(a) which ensures that the right side (output) of the magnet injects (solid arrow) more spin current than it can receive (dashed arrow) and vice-versa for the left side (input).

Fig 1(b) shows a schematic of a 3-inverter oscillator and Fig. 1(c) shows the equivalent ASL layout that we simulate. When a positive Vss is applied on all the magnets in this structure, they each extract spins and try to invert the next one in the chain (Fig. 3 region I) and this sets up a chain of periodic oscillations reminiscent of applying a constant 'Vdd' to a chain of CMOS inverters. Interestingly, when the Vss is made negative (Fig. 3 region III), they each inject spins and try to copy their state onto the next one thus settling to a constant stable value much like a buffer. It should be noted, however, that magnetic logic is inherently non-volatile. When the supply voltage is removed all the magnets simply remain in the state they were last in (Fig. 3 region II).

To summarize, we have presented the first simulator that simultaneously describes magnetization dynamics as well as spin transport in multi-magnet ASL networks and used it to demonstrate the possibility of large scale functional spin logic blocks through the example of an All Spin ring oscillator.

[1] S. A. Wolf et. al., Science 294, 5546 (2001)

[2] T.N.Theis *et.al.*,Science 327,1600 (2010).

[3] B. Behin-Aein et. al., IEEE Trans. Nanotech. 8, (2009).

[4] B.Behin-Aein *et.al., Nat. Nano.* **5**, 266(2010)

[5] B.Behin-Aein *et.al., App. Phys. Lett,* 98 (2011)

[6] S. Srinivasan *et.al*, submitted *IEEE. Trans. Mag.*

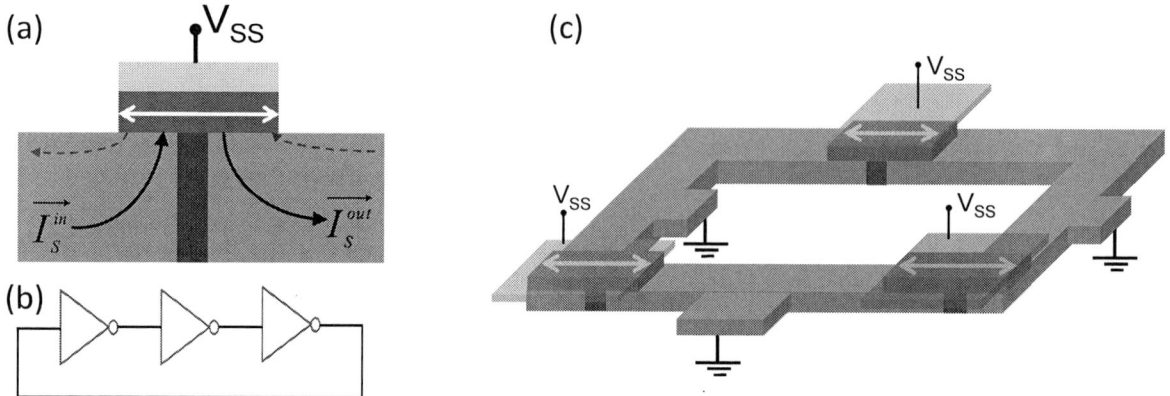

Fig. 1: (a) Side view of any one individual magnet showing that for ASL operation, the information content lies in the state of the magnets ($m_z=\pm1$) and each magnet has a "listening" (input) side physically separated from a "talking" (output) side. The propagation of information is purely through spin currents while Vss powers the circuit. **(b)** Schematic representation for a proposed ring oscillator. **(c)** Equivalent ASL oscillator layout with same Vss applied to all the magnets.

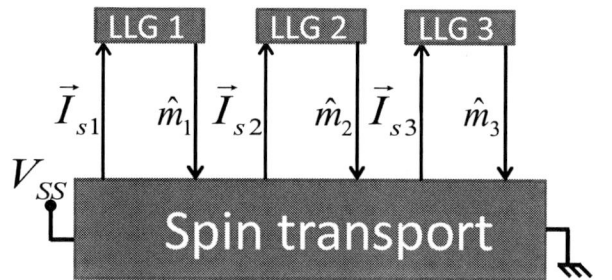

Fig. 2: Coupled Spin-transport/ Magnet-dynamics model used to simulate the ASL ring oscillator. The dynamics of each magnet is described by Landau-Lishitz-Gilbert(LLG) equation while the "Spin transport" section calculates the instantaneous vector spin currents ($\vec{I}_s$) incident on the magnet as a function of their magnetization ($\hat{m}$) and applied voltage Vss.

Simulation Parameter	Value
Magnet dimensions	$100 \times 40 \times 2$ nm^3
Saturation Magnetization (M_s)	780 emu/cc (Permalloy)
Uniaxial Anistropy Energy barrier ($K_{u_2}V$)	30 K_BT (T=300K)
Magnet/Interface polarization	Output side = 0.5 Input side = 0.2
Channel(Cu) dimensions	$100 \times 40 \times 100$ nm^3
Spin Diffusion Lengths	500 nm (Cu) 5 nm (Py)
Resistivity	6.9 ohm-nm (Cu) 171 ohm-nm (Py)
Supply Voltage (V_{SS})	20 mV

Table 1: Simulation Parameters used to obtain Fig. 3. The materials/values listed have been taken from actual experiments on lateral spin devices. They do not reflect the optimized parameter set for best performance but rather demonstrate feasibility of operation with realistic numbers.

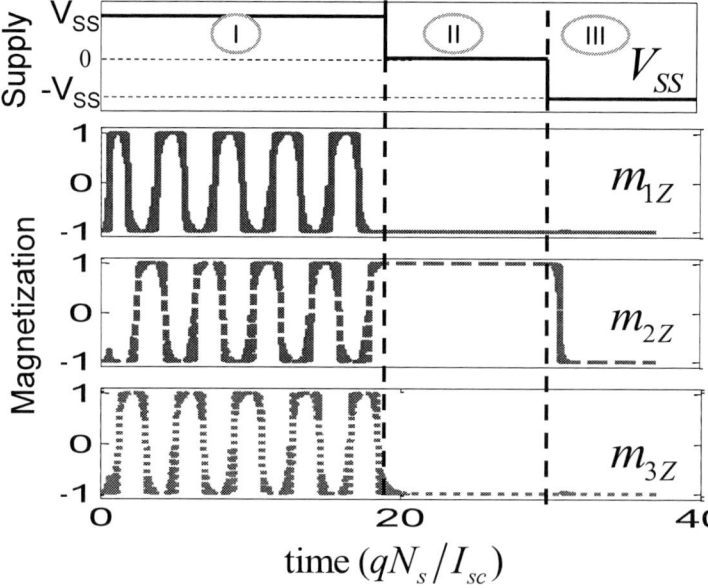

Fig. 3: When $V_{SS}>0$ is applied (region I), each magnet performs an 'invert' operation on the next one which eventually feeds back through the chain leading to oscillator characteristics. When Vss=0 (region II), the system is turned off and the magnets simply remain in whatever state they were in. On the other hand when $V_{SS}<0$ (region III), each magnet performs a 'copy' operation on the next one and they all settle into the same state where they remain stable. The horizontal axis is in time in the units of the ratio of number of Bohr magnetons (N_s) in the magnet to the critical spin current (I_{sc}) required to turn the magnet.

Proposal for piezoelectric-ferromagnet bilayer based microwave oscillators without any external magnetic field or spin transfer torque

Debanjan Bhowmik, Sayeef Salahuddin

Department of Electrical Engineering and Computer Sciences, University of California, Berkeley CA - 94720, USA.
Phone: +1 510 5014808, Email: debanjan@eecs.berkeley.edu

Introduction: We propose and simulate a magnetic device which, to the best of our knowledge, is the first of its kind that creates magnetic oscillations at rf/microwave frequency without an external high frequency magnetic field (used in conventional FMR and MRI) or spin transfer torque (used in STNO-s)[1,2]. In this scheme, the dc magnetic field around which magnetization precesses and r.f. magnetic field which causes the excitation in a conventional ferromagnetic resonance is replaced by the 'effective magnetic field'-s contributed by the shape and magnetoelastic anisotropy of the ferromagnet. Since the frequency of precession is directly dependent on the magnitude of the effective dc magnetic field, a frequency tunable oscillator is obtained by tuning the amplitude of dc voltage on the piezoelectric.

Device Operation: The key physics hinges on the fact that a strain caused in a piezoelectric (PZ) can be transferred to the ferromagnet (FM) above in a bilayer structure as a change in its energy anisotropy. Fig 1a shows the schematic of a device. Fig. 1 b shows the procedural flow of how the strain gets transferred to FM. Essentially, the change in energy anisotropy in the magnet looks like an effective magnetic field that can then be modulated by voltage applied on piezoelectric. This is shown in the hysteresis diagrams of Fig. 2b, which we obtained through magnetic energy calculations following the procedure highlighted in Fig 1b[3,4].

However, having a voltage controlled effective dc magnetic field is not enough for creating FMR since damping would cause the magnetization to finally align with the field. This damping torque is counteracted in our device by providing an additional high frequency strain to the magnet, as opposed to r.f. magnetic field in conventional FMR and MRI which needs r.f. antennas and is hence expensive, or spin transfer torque in STNO-s that needs relatively high input current(in mA)[1,2].Fig 3a shows the proposed device structure to achieve this purpose. Applying voltage of r.f. frequency on Inter Digital Transducers(IDT-s) creates surface acoustic waves on a piezoelectric substrate, which propagate along y direction causing strain of that frequency along the transverse dimension(z).Ellipsoidal nanomagnets are placed on the piezoelectric such that their major axes make 45^0 with both y and z axes. Because of shape anisotropy, the effective dc magnetic field acts along the major axis[4]. High frequency strain along z axis causes an effective r.f. magnetic field perpendicular to the effective dc magnetic field as magneto-elastic anisotropy term is directly proportional to strain (relation in Fig 1b). This makes the magnetization precess around the direction of effective dc magnetic field(major axis of the ellipsoid in this case)(Fig 3b)[4]. Fig 3c shows the frequency spectrum of the absorbed power obtained through Landau-Lifshitz-Gilbert (LLG) equation for a Ni nanomagnet of dimensions-105nm x 70 nm x 3 nm. The resonance frequency is found out to be 2GHz in the absence of any dc electric field on the piezoelectric below. As expected application of dc electric field on the piezoelectric changes the effective dc magnetic field of the magnet thereby changing the resonance frequency.

Finally, a feedback loop has to be provided to convert this ferromagnetic resonator to a self sustaining oscillator. To do this, we note that the precession of magnetization about the major axis would cause a sinusoidal temporal variation of magnetization component along the minor axis. Therefore, a spin-valve formed on the Ni nanomagnet such that pinned layer is along the minor axis creates a sinusoidally varying voltage through GMR effect that can be fed back to IDT through an amplifier, thereby completing the feedback loop (Fig. 4a, 4b). The frequency dependent amplitude and phase of the ferromagnetic resonator is shown in Fig 4c,4d where the gain and phase of the amplifier used in the feedback path can be chosen such that the Barkhausen criterion of net gain in the loop to be 1, necessary to have self sustained oscillations, is satisfied only for the resonant frequency of the ferromagnetic system[5]. We simulate this system through LLG. Using an amplifier that provides overall gain of -25dB and 255^0 phase lag in the feedback loop we have been able to create magnetic oscillations at the frequency of about 2GHz(Fig 4e) which corresponds to the resonant frequency of the system. By varying dc voltage on the piezoelectric, this frequency can be modulated as shown in Fig 3c.

Conclusion: To summarize, we proposed and simulated a nanomagnetic microwave oscillator that does not require any DC or r.f. magnetic field (unlike in conventional FMR), nor any spin torque current (unlike in STNO's). The frequency of oscillation can be effectively tuned by DC voltage in the microwave domain. The fact that magnetic fields and currents are not needed can make it a much cheaper, denser and less power hungry alternative for future electromagnetic oscillator applications.

References:[1]D.Gignous,M.Schlenker , Magn. Mat. and App. Springer-Chap,23,(2005)[2]S.E.Russek et al, Handbook of Nanophysics,Chap 38,CRC(1999)[3]Hu et. al. Phys . Rev. B. 80(2009) [4] R.Handley ,Mod. Mag. Mat, Wiley(1999) [5] Sedra Smith, Mic. Circ, Chap 12(1997).

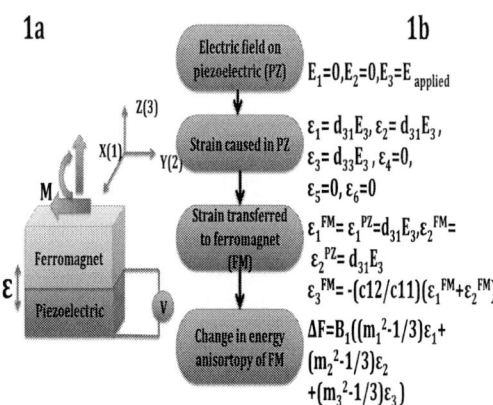

1a

1b

Electric field on piezoelectric (PZ)
$E_1=0, E_2=0, E_3=E_{applied}$

Strain caused in PZ
$\varepsilon_1=d_{31}E_3, \varepsilon_2=d_{31}E_3, \varepsilon_3=d_{33}E_3, \varepsilon_4=0, \varepsilon_5=0, \varepsilon_6=0$

Strain transferred to ferromagnet (FM)
$\varepsilon_1^{FM}=\varepsilon_1^{PZ}=d_{31}E_3, \varepsilon_2^{FM}= \varepsilon_2^{PZ}=d_{31}E_3$
$\varepsilon_3^{FM}=-(c12/c11)(\varepsilon_1^{FM}+\varepsilon_2^{FM})$

Change in energy anisortopy of FM
$\Delta F=B_1((m_1^2-1/3)\varepsilon_1+ (m_2^2-1/3)\varepsilon_2 +(m_3^2-1/3)\varepsilon_3)$

Fig1-Model-a. Piezoelectric-ferromagnet bilayer where voltage applied on piezoelectric causes strain(ε) in it which on being transferred to ferromagnet changes its magnetic anisotropy due to magnetoelastic coupling. **b.** Flowchart showing the governing equations for the model. The relationship between strain and applied electric field on piezoelectric is obtained for a Class 4mm crystal (like PZT). Relation between strain in piezoelectric and magnet is obtained considering ideal bonding at interface and zero perpendicular stress.

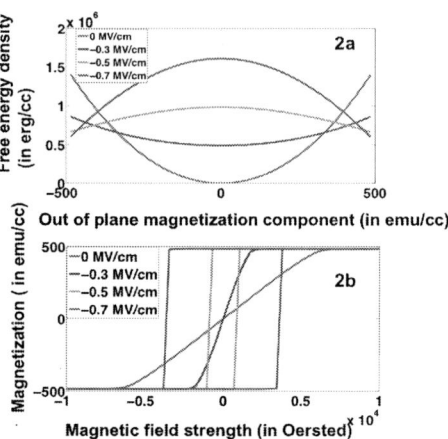

Fig2- Rotation of magnetization out of plane by electric field-a. For a Ni-PZT system , in absence of electric field on PZT(i.e. strain in Ni) ,magnetization(M) is in plane due to strong demagnetizing field but application of sufficient electric field on PZT (d33=410 pm/V,d31=-170pm/V – values at Phase Boundary)changes the easy axis to out of plane direction thereby rotating M out of plane. We show through magnetic energy calculations that a field of -0.5MV/cm (0.5V on 10nm PZT sample) should switch M out of plane. **b.** Out of plane hysteresis curves showing the same (obtained through LLG simulations)

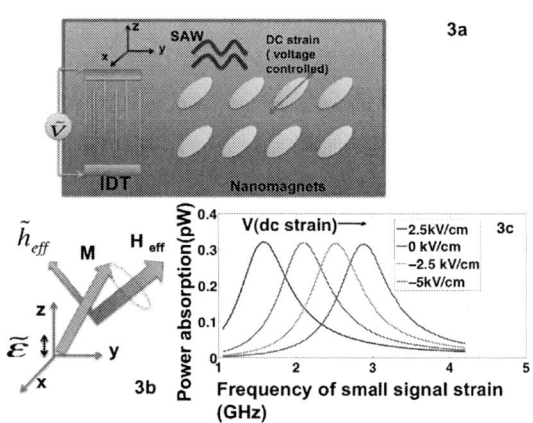

Fig3-Creating ferromagnetic resonance without any external magnetic field-a. Surface acoustic waves are created on the piezoelectric substrate along y dir. by applying voltage signal on Interdigital Transducer(IDT). They create high frequency strain along the transverse dir.(z) which gets transferred to the elliptical Ni nanomagnets. **b.** Effective field due to shape anisotropy(H_{eff}) replaces the dc magnetic field used for standard FMR. High frequency strain along z axis causes an effective r.f. magnetic field (h_{eff}) perpendicular to H_{eff} causing the magnetization to precess around H_{eff} at that frequency. **c.** FMR spectrum where peak in power absorption corresponds to the resonant frequency, which can be tuned by applying voltage on the piezoelectric below that causes d.c. strain along the major and minor axes of the nanomagnet thereby changing H_{eff} keeping h_{eff} unaffected. Thus frequency of our resonator is voltage controlled.(Elliptical Ni nanomagnets with major axis length 105nm,minor axis length 70nm and thickness 3nm on PZT substrate are used for LLG simulations) .

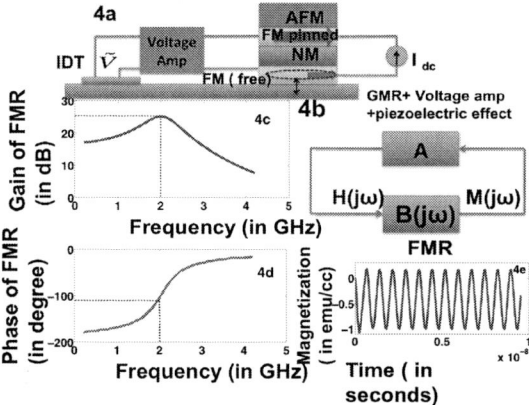

Fig4- Voltage controlled FMR based oscillator- a. The ferromagnetic resonator shown in Fig3 is connected to a feedback loop. A spin valve structure built on the nanomagnet of Fig3a picks up the precession of magnetization due to FMR in it by GMR. The GMR (voltage) signal is amplified and applied on the IDT to create strain on the nanomagnet. **b.** Block diagram representing the system **c,d.** The frequency response of the ferromagnetic resonator B(jω) . To create self sustained oscillations at the Larmor frequency of the ferromagnet(2GHz) in this case a voltage amplifier should be designed such that AB=1 at that freq. (Barkhausen criterion).From Fig 4c and 4d we find that an amplifier with creating overall gain of -25dB and phase lag of 255^0(independent of input frequency) would create such oscillation.**e.** LLG simulations with given FMR response and such feedback amplifier show self sustained oscillations.

978-1-61284-243-1/11 $26.00 © 2011 IEEE

Orthogonal Spin Transfer MRAM

D. Bedau[1*], D. Backes[1], H. Liu[1], J. Langer[2], P. Manandhar[3] and A. D. Kent[1]

[1]Department of Physics, New York University, New York, NY 10003, USA
[2]Singulus Technologies AG, 63796 Kahl am Main, Germany
[3]Spin-Transfer Technologies, One Adams Place, 859 Willard Street, Quincy MA 02169 USA
*phone: 212 998 7662, email: db137@nyu.edu

Spin-Transfer Magnetic Random Access Memory (ST-MRAM) devices hold great promise as a universal memory [Katine 2008]. ST-MRAM is non-volatile, has a small cell size, high endurance and may match the speed of SRAM. A disadvantage of the common collinearly magnetized ST-MRAM is their non-deterministic switching process, which leads to long switching times and broad switching time distributions [Devolder 2008, Koch 2004]. This delay is due to the fact that the torque is zero if the layers are either parallel or antiparallel [Slonczewski 1996] and hence switching cannot be initiated by the torque alone. Typically the process is started by an initial misalignment of the free layer stemming from thermal excitations. Relying on thermal initiation leads to incoherent reversal with an unpredictable incubation delay in the ns range [Devolder 2008] and broad switching time distributions [Koch 2004]

In contrast, orthogonal ST-MRAM (OST-MRAM) uses a polarizing layer that is magnetized perpendicular to the plane (Figure 1), thus providing instant torque and deterministic switching, with the potential for switching times below 50 ps [Kent 2004, Kent 2007]: OST-MRAM has a near maximal spin torque in both initial and final states. Spin transfer leads to a torque rotating the free layer magnetization out of the plane. The z component of the magnetization causes a large demagnetizing field $B_{demag} = -\mu_0 \, M_{eff} \, m_z$ around which the free layer magnetization precesses with the frequency $\omega = \gamma \, B_{demag}$. As the current pulse is switched off, the magnetization returns into the plane and the switching process is complete. By using the large demagnetizing field the switching process can be made very fast while keeping the power consumption low. To read out the state of the memory cell, a third magnetic layer with a pinned in-plane magnetization is placed on top of the free layer, separated by a tunnel barrier. This reference layer causes an additional spin torque that favors one free layer state for each current polarity, further stabilizing the two states and reducing the timing requirements of the switching pulse [Beaujour2009].

For the polarizer exchange coupled Co/Ni and Co/Pd multilayers that have both a high spin polarization and a high perpendicular anisotropy have been used. An ultrathin MgO tunnel barrier is deposited on top of the free layer, forming the readout together with the reference layer. The reference layer is exchange coupled to an antiferromagnet to increase the anisotropy. The full layer stack was deposited using a Singulus TIMARIS PVD module. The materials have been characterized by VSM, FMR and CIPT. Figure 2 shows the magnetization for in-plane and perpendicular fields: the free layer is very soft, and the exchange bias from the AFM is about 100 mT. The perpendicular polarizer has a coercive field $\mu_0 H$ of 26 mT.

OST-MRAM devices with sizes from 40 nm x 80 nm up to 80 nm x 240 nm were made by e-beam and optical lithography. The devices had more than 100% MR and resistance area products RA of about 2-10 $\Omega\mu m^2$. Figure 3 shows the two states of the free layer for a 60 nm x 180 nm hexagon with an enegy barrier of 40 kT at the measurement field of 10 mT. The dipole field has been mostly compensated by the synthetic antiferromagnet.

Reliable switching is observed at room temperature with 0.7 V amplitude for pulses of 500 ps duration (Fig. 4). Less than 450 fJ of energy is used to switch the device and no incubation delay was found [Liu 2010]. The switching is bipolar (Fig. 5), i.e. switching between AP and P states is possible with either pulse polarity. The same holds for switching between P and AP states. Without the reference layer we would expect a totally symmetric distribution of the switching probabilities, in contrast to the measured switching probabiltiy (Fig. 5). This asymmetry is caused by the additional torque from the reference layer which breaks this symmetry and favors one state. Device characteristics depend on the composition of the layer stack; devices with a modified stack not only switch at zero effective field, but also show unipolar switching, as shown in Fig. 6 for a 50 nm x 100 nm rectangular device. These OST-MRAM device characteristics will be presented and compared to those obtained with conventional ST-MRAM bit cells.

[Liu 2010] H. Liu et al., Appl. Phys. Lett. **97**, 242510 (2010)
[Beaujour 2009] J.-M Beaujour et al., SPIE, **7398**, 73908D (2009).
[Devolder 2008] T. Devolder et al., Phys. Rev. Lett **100**, 057206 (2008).
[Katine 2008] J. A. Katine et al., J. Magn. Magn. Mater. **320**, 1217 (2008).
[Kent 2004] A. D. Kent et al., Appl. Phys. Lett. **84**, 3897 (2004).
[Kent 2007] A. D. Kent, Nature Materials **6**, 399 (2007).
[Koch 2004] R. H. Koch et al., Phys. Rev. Lett. **92**, 088302 (2004).
[Slonczewski 1996] J. C. Slonczewski, J. Magn. Magn. Mater. 159, L1 (1996).

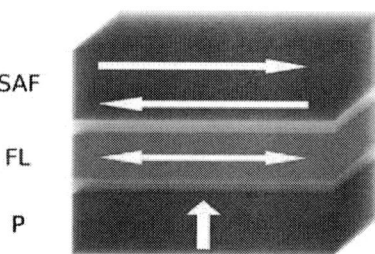

Fig. 1: Schematic depiction of an OST-MRAM stack. A perpendicular polarizer creates a torque on an in-plane free layer. The state of the free layer is read out by an in-plane reference layer pinned by a synthetic antiferromagnet on top.

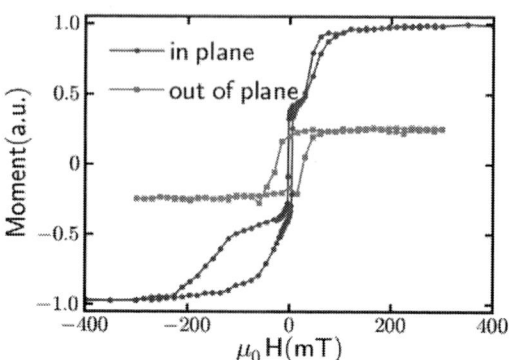

Fig. 2: VSM measurements of the magnetization of the layer stack demonstrating a soft free layer, about 100 mT exchange bias from the SAF and 26 mT coercive field for the perpendicular polarizer.

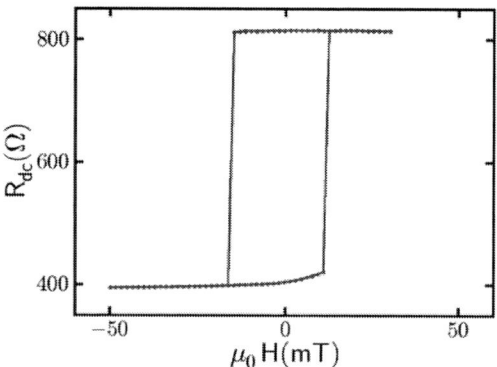

Fig. 3: Free layer hysteresis for a 60 nm x 180 nm hexagon-shaped device. The device has 107% MR, a coercive field of 14 mT, and -2 mT remaining dipole field.

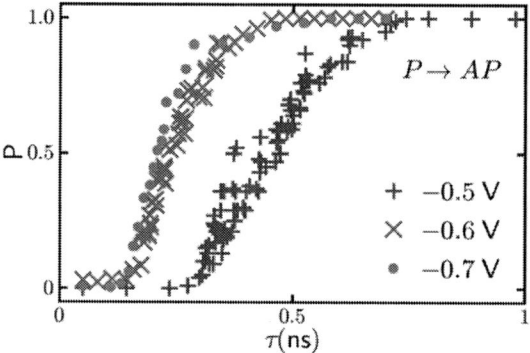

Fig. 4: P to AP switching for the device shown in Fig. 3. 100% switching probability is achieved for pulses of less than 500 ps duration. A field of 10 mT was applied; the measurement was carried out at room temperature.

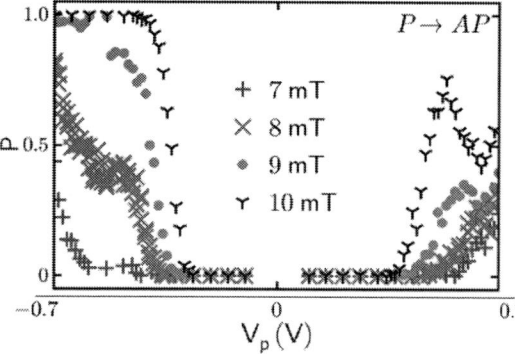

Fig. 5: Switching probability as a function of the pulse amplitude for P to AP switching, same device as in Fig. 3 and 4. The bipolar switching and the oscillation of P are an indication of the precessional nature of the switching process. P is not symmetric with respect to the polarity due to the influence of the reference layer.

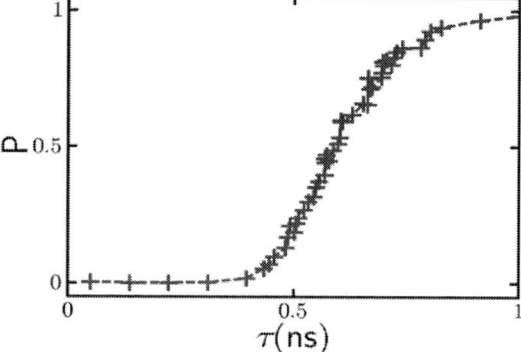

Fig. 6: Switching probability as a function of the pulse duration for a 50 nm x 100 nm rectangular device for a pulse amplitude of 0.75 V. The opposite polarity does not change the device state. The device had a coercive field of 8 mT. The measurements were done at zero effective field.

978-1-61284-243-1/11 $26.00 © 2011 IEEE

Thermal Effects and Instability in Unipolar Resistive Switching Devices

An Chen and Ming-Ren Lin

Strategic Technology Group, GLOBALFOUNDRIES, Sunnyvale, CA 94085; an.chen@globalfoundries.com; (408) 749-6352

Metal oxide based resistive switching devices have demonstrated promising characteristics for next-generation nonvolatile memory applications [1-4]. These devices can be electrically switched between a high-resistance state (HRS, or OFF-state) and a low-resistance state (LRS, or ON-state). The switching from HRS to LRS is called SET and that from LRS to HRS is RESET. Resistive switching may be bipolar (i.e., SET and RESET in opposite bias directions) or unipolar (i.e., SET and RESET in the same bias direction). Unipolar resistive switching devices are more compatible with two-terminal selection devices (e.g., diodes) for 3D stackable crossbar arrays. Although the resistive switching mechanisms are not yet clearly understood, it is generally believed that the switching is caused by mixed ionic/electronic effects involving some defects (e.g., mobile ions, charge traps, oxygen vacancies, *etc*). The RESET in bipolar resistive switching can be explained by the reversal of the physical process that causes SET [1]. For example, the reverse migration of ions/vacancies during RESET may annihilate the conductive channels formed during SET. Since no such reversal exists in unipolar resistive switching, the RESET process in unipolar switching is often explained by thermal effects. The similarity between SET and RESET operation conditions in unipolar switching devices may cause competition between SET and RESET processes, which reflects as instability in the switching characteristics. This paper presents some experimental evidences supporting the hypothesis of power-induced thermal nature of RESET and analyzes the instability associated with the SET-RESET competition in unipolar resistive switching, using data measured on Cu_2O-based resistive switching devices [2, 3].

A common observation on many unipolar resistive switching devices is that the maximum RESET current (I_{RESET}) exceeds the maximum SET current (I_{SET}), as shown in the unipolar switching I-V characteristics in Fig. 1. It also indicates that RESET requires higher power (P_{RESET}) than the maximum SET power (P_{SET}), because $P = I^2 \times R_{on}$, where R_{on} is the LRS resistance. This is consistent with the power-driven nature of RESET; otherwise, devices at the end of SET may be switched again (i.e., RESET) instead of staying in LRS. Direct evidences of power-induced thermal effects in RESET come from the power-temperature equivalence in RESET. As shown in Fig. 2, the RESET power decreases with the increase of temperature, i.e., higher temperature compensates the power requirement for RESET. The bake test in Fig. 3 proves that RESET can also occur without electrical power at sufficiently high temperature. Devices switched off by high temperatures can be switched to LRS again and further cycled. Table 1 summarizes these experimental evidences that support the hypothesis that RESET in unipolar resistive switching is caused by the thermal effects induced by power consumption and heat dissipation.

In unipolar resistive switching, SET and RESET are operated under similar electrical conditions. Competition may exist between forces favoring SET and RESET, which could cause unstable switching behaviors [4, 5]. Fig. 4 illustrates a pulse RESET experiment where voltage pulses with increasing amplitude are applied on devices in LRS with a read verification after each pulse. Before the onset of RESET in some devices, a "partial SET" occurs and device resistance further decreases from R_{on}. This "partial SET" can be explained from the fact that RESET requires sufficiently high power, because the decrease of resistance may enable higher RESET power (i.e., $P = V^2/R$). It suggests that the actual RESET current and power may be much higher than the intuitive expectation of V_{RESET}/R_{on} and $(V_{RESET})^2/R_{on}$. This "partial SET" may also alternate with "partial RESET" and cause several rounds of resistance variation before a "clean RESET". At the end of RESET where devices are switched to HRS, unintentional SET may occur before the RESET voltage is removed. In Fig. 5, the RESET depth (calculated by the ratio of device conductance after and before RESET) degrades from below 10^{-2} (deep RESET) at optimal RESET voltage to above 1 (RESET failure) at higher voltage, due to increasing number of unintentional SET following RESET within the same voltage pulse. Consequently, RESET yield degrades significantly when RESET voltage is increased beyond the optimal values (Fig. 5, inset). The SET-RESET competition during unipolar switching can be directly observed in an AC RESET process with a 300 ns voltage pulse measured with an oscilloscope (Fig. 6). Resistance oscillation with ~ 35 ns period is clearly visible during RESET, as shown in the reading current response.

The dynamic competition between SET and RESET can be explained with Fig. 7-9. Fig. 7 illustrates the RESET where device resistance changes from LRS to HRS along a load line whose slope is determined by the serial resistance of selection devices. The increase of resistance may raise the voltage applied on the device and reduce device power, as shown in the calculation in Fig. 8 based on a one-transistor-one-resistor (1T1R) configuration. Higher voltage increases the chance of SET and lower power reduces the drive for RESET (Fig. 9). When the RESET force is overcome by the SET force, the RESET process may be reversed. As the SET process starts to reduce device resistance, device voltage decreases and power increases, which may reverse the trend that triggered SET and consequently RESET takes over again. This dynamic change between SET and RESET causes resistance oscillation during the switching process, as illustrated by the red arrows in Fig. 7.

In summary, RESET in unipolar resistive switching devices can be explained by power-induced thermal effects. The SET-RESET competition causes instability in unipolar switching behaviors, which may degrade yield, reduce switching speed, and increase switching power (table 2). Operation algorithms need to be carefully designed to accommodate the instability, but fundamental solution may require better understanding of the mechanisms and material improvements.

References: [1] M.J. Rozenberg, *et al*, Phys. Rev. B **81**, 115101 (2010); [2] A. Chen *et al.*, *IEDM Tech. Dig.*, p. 765 (2005); [3] T. Fang *et al.*, *IEDM Tech. Dig.*, p. 789 (2006); [4] D. Ielmini, IEEE Electron. Dev. Lett. **31**, 552 (2010); [5] D.K. Kim, et al, IEEE Electron. Dev. Lett. **31**, 600 (2010).

Table 1. Evidences supporting the hypothesis of power-induced thermal effects in the RESET switching process of unipolar resistive switching devices.

Evidences		Measurements
Power requirement of RESET	$P_{RESET} > P_{SET}$ or $I_{RESET} > I_{SET}$	SET and RESET power can be obtained from the measured switching current and voltage
Power-temperature equivalence in RESET	RESET power decreases with increasing temperature	Measure RESET power at different temperature
	RESET by heat alone without electrical power	Bake devices at high temperature and observe device resistance change

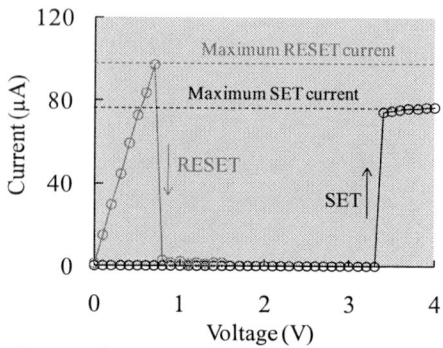

Fig. 1 SET and RESET switching I-V curve of Cu_2O devices.

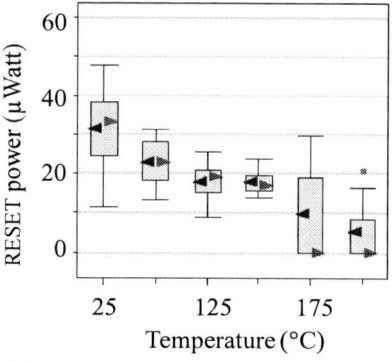

Fig. 2 RESET power versus temperatures.

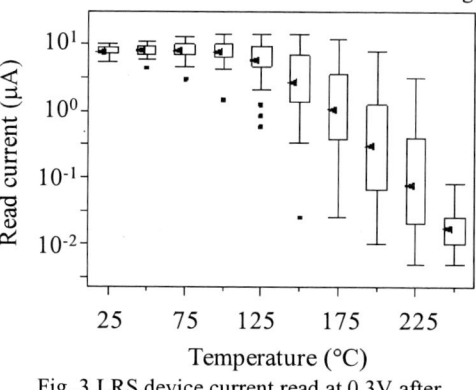

Fig. 3 LRS device current read at 0.3V after being baked from 50°C up to 250°C.

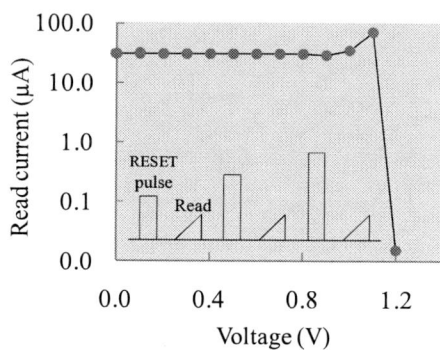

Fig. 4 AC RESET with a read verification after each pulse, showing "partial SET" before RESET.

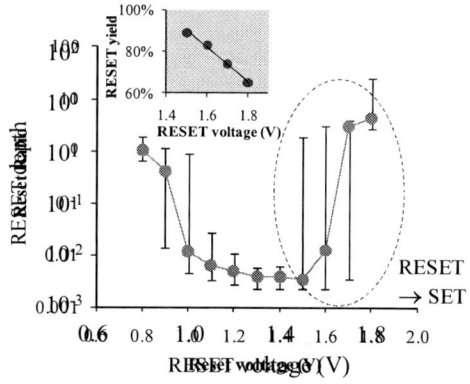

Fig. 5 RESET depth versus RESET voltage, showing unintentional SET following RESET. The inset plots RESET yield versus voltage.

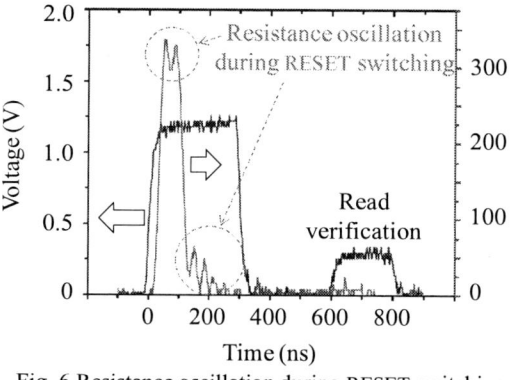

Fig. 6 Resistance oscillation during RESET switching process, showing instability induced by the dynamic competition between SET-RESET forces.

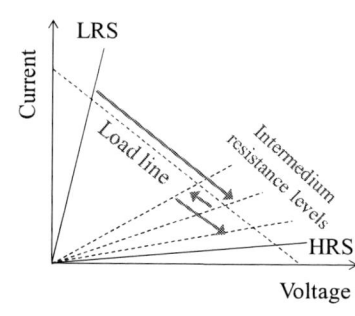

Fig. 7 Illustration of RESET along a load line of serial resistance, where the device resistance oscillates among intermedium levels between LRS and HRS.

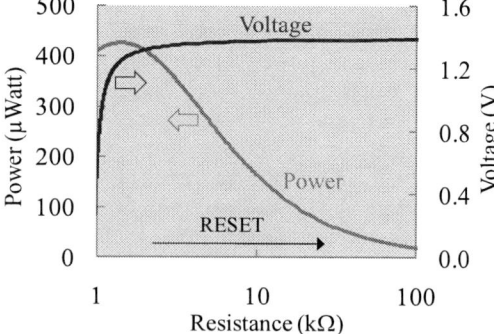

Fig. 8 Calculated power and voltage of a resistive switching device in an 1T1R configuration during RESET, with a fixed voltage applied on the 1T1R.

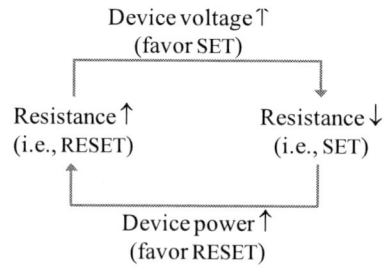

Fig. 9 Illustration of the competing forces: power (favoring SET) and voltage (favoring RESET), which may cause resistance oscillation during resistive switching process.

Table 2 Instability in unipolar resistive switching devices and its impact on memory performance

Instability in unipolar resistive switching	Impacts on memory performance
Resistance oscillation during switching	Longer time for the switching process to stabilize, i.e., slower switching speed
Unintentional SET following RESET in the same voltage pulses	RESET failure and yield degradation
Partial SET before the onset of RESET	Higher RESET current and power consumption

A Hybrid Ferroelectric and Charge Nonvolatile Memory

Shantanu R. Rajwade, Kshitij Auluck, Jonathan Shaw, Keith Lyon and Edwin C. Kan

School of Electrical and Computer Engineering, Cornell University, Ithaca NY 14853
Email: srr77@cornell.edu Phone: (607) 220 4344

Abstract

We introduce a new nonvolatile memory that incorporates ferroelectric (FE) switching layer and charge-storage floating node in a single gate stack. This hybrid FE-charge design reduces the depolarization field in the FE layer as well as increases the memory window over conventional FE-FET. The magnitude of the electric field in the tunnel dielectric is reduced at retention and enhanced at program/erase from the FE polarization. This paper discusses the working principle, gate stack design, fabrication and experimental results of this hybrid design compared to FE-FET and charge-trap Flash.

Keywords: hybrid memory, ferroelectric, gate injected charge

Introduction

In spite of low-voltage operation, ferroelectric (FE) FET memory is known to suffer from issues of poor retention due to the depolarization field setup up in the FE layer [1], charge leakage through defect sites and large FE film thicknesses (> 100 nm) for realistic memory windows. We propose a new highly scalable hybrid nonvolatile memory designed with FE PVDF film and HfO$_2$ as charge-trap layer that takes advantage of low-voltage operation from FE-FET as well as reduces depolarization field by gate-injection of charge.

Memory Design and Working Principle

Fig. 1 shows an example of experimental design for the gate stack and program/erase (P/E) operation of the hybrid cell. A floating gate (FG) (discrete/continuous) is included above the FE layer. A thin top oxide serves as a tunnel barrier for gate injection and removal of electrons [2]. Program condition (V_{PROG} < 0) orients FE polarization with positive surface charge facing the gate and hence increases the V_{TH} of the cell. The injected electrons into the FG add to the V_{TH} shift. These electrons also stabilize the positive dipole charge on FE during retention and thus decrease the depolarization field. Erase operation (V_{ERASE} > 0) removes the electrons from FG and reverses the polarization of FE layer.

Fabrication and Experimental Results

Fabrication procedure is outlined in Table I. 0.5 % solution of 70:30 P(VDF-TrFE) in MEK was spin coated and annealed at 140 °C for 10 min [3] to obtain a film thickness of 35 nm. Subsequent processing of the hybrid device was limited below 110 °C. Control FE-FET and gate inject (GI) Flash devices with the same corresponding FE layer and tunnel oxide thicknesses and identical EOTs were also fabricated for comparison. Fig. 2 shows hysteresis measurements for separate 80 nm thick PVDF MFM capacitors with inset (a) confirming the XRD peak in FE β-phase.

Hybrid devices showed larger ΔV_{FB} against FE-FET for V_{PROG} >10 V due to injected-electron contribution as seen in Fig. 3(a). Fig. 3(b) estimates the injected electron density and contribution of FE switching to ΔV_{FB} in hybrid devices. No channel injection of electrons was observed until V_{PROG}=13 V in all devices. GI Flash showed poor ΔV_{FB} due to lesser tunnel oxide fields compared to hybrid devices. Reduced depolarization field improves retention in hybrid devices over

FE-FET as seen in Fig. 4(a). GI Flash also demonstrated limited retention due to inferior quality of top oxide constrained by present process integration. Hybrid device showed > 10^4 s retention and can present significantly better results with improved quality tunnel barrier. Fig. 4(b) shows retention characteristics in a single hybrid device for identical ΔV_{FB} but different program condition. When programmed with stronger pulse for shorter time, most of the ΔV_{FB} results from FE switching since dipole polarization responds significantly faster than charge injection. Therefore it shows poor retention on account of uncompensated depolarization field against programming with smaller pulse for a longer duration. During program, large amount of charge is trapped in the polycrystalline PVDF film (I-V of MFM devices in Fig. 2 inset (b)). This charge gradually leaks out to the gate with minimal positive bias in FE-FET thereby smearing C-V characteristics as seen in Fig. 5. The existence of the trapping layer in hybrid design provides an ideal sink to this leak path. Trapped charge in the FE layer is emitted to the trap layer assisted by the depolarization field. Fig. 6 shows calculated band diagram before and after charge relaxation and subsequent reduction in depolarization field which prevents C-V smearing in the hybrid cell. Fig. 7 illustrates pulsed program measurements indicating over 1 s saturation time. Presence of tunnel oxide reduces the fluence of charge in the FE film during P/E cycling which results in superior endurance of hybrid design as depicted in Fig. 8.

Simulation Results

Electrostatic simulations incorporating FE hysteresis [4] were performed on the above gate stacks. Fig. 9 shows the decrease in the depolarization field with injected electrons. The stored electrons in the hybrid cell experience lower tunnel oxide fields during retention due to the counterbalancing positive surface charge on the FE compared to GI Flash as seen in Fig. 10. Due to the huge electric displacement in the FE layer at small voltages, electric field is enhanced in the tunnel oxide (Fig. 11) during the P/E conditions that enables low-voltage operation. The charge storage increases the bottom oxide electric field during retention as displayed in Fig. 12, but renders sufficient design room for memory operation.

Scaling Implementation

For the present geometry, Fig. 13 proposes a 40:60 division of ΔV_{FB}=2.5 V into FE and stored charge showing reduced FE hysteresis of hybrid device over FE-FET. Seen in Table II, this benefits into 50 % reduced operating voltage, 70 % reduced depolarization field and significantly superior P/E and retention fields in top oxide over GI Flash. Alternatively, shown in Fig. 14, to obtain realistic memory windows with inorganic FE films like BSTO or SBT [5] and 10 nm bottom oxide, hybrid design is more suited for aggressive scaling (in spite of increased EOT due to 3 nm tunnel oxide) than FE-FET due to lower FE thickness and longer retention for given P/E voltages.

978-1-61284-243-1/11 $26.00 © 2011 IEEE

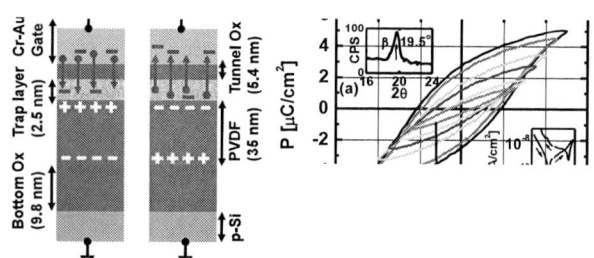

Figure 1 Design and operation of FE-charge hybrid memory

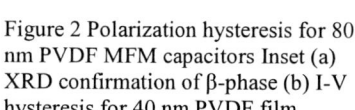

Figure 2 Polarization hysteresis for 80 nm PVDF MFM capacitors Inset (a) XRD confirmation of β-phase (b) I-V hysteresis for 40 nm PVDF film

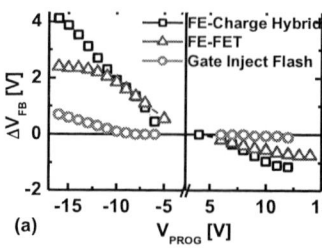

Figure 3 (a) Memory window of FE-charge hybrid shows significant improvement over FE-FET and GI charge-trap Flash at the same *EOT*; (b) Estimated contribution of FE and trapped charge density to memory window in FE-charge hybrid as a function of V_{PROG}.

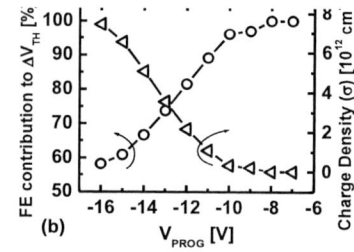

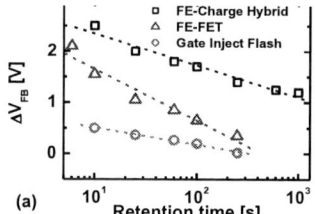

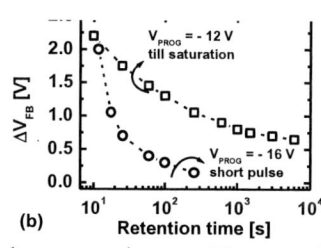

Figure 4 (a) Hybrid memory shows longer retention over FE-FET and GI Flash after program at -12 V for 5 s;(b) Longer t_{PROG} helps increased injection of electrons that stabilize polarization and improve retention.

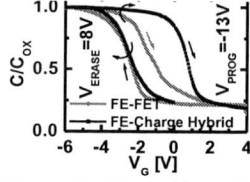

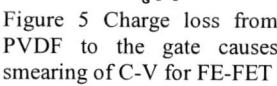

Figure 5 Charge loss from PVDF to the gate causes smearing of C-V for FE-FET

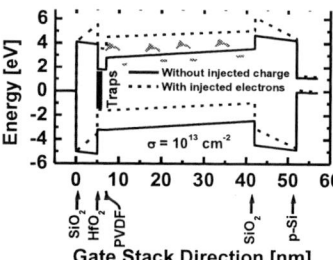

Figure 6 Band diagram at retention showing injected charge relaxing to the trap layer and thereby reducing the depolarization field.

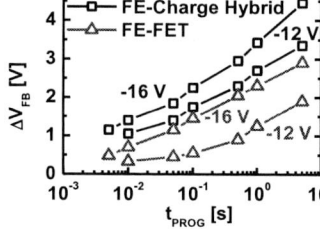

Figure 7 ΔV_{FB} against t_{PROG} for hybrid and FE-FET devices.

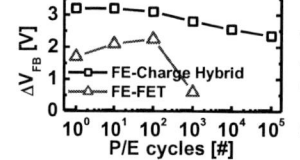

Figure 8 Reduced fluence of charge due to top oxide in hybrid memory results in improved endurance over conventional FE-FET.

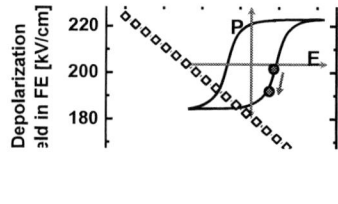

Figure 9 Simulation of decrease in depolarization field with gate-injected charge and illustrative inset.

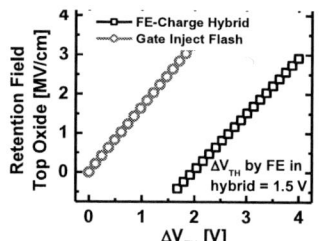

Figure 10 Hybrid design yields much reduced retention field in tunnel oxide over GI Flash.

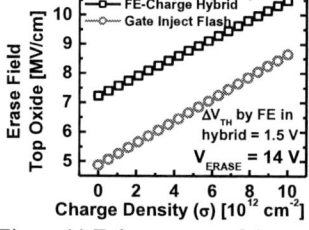

Figure 11 Enhancement of the erase field in the tunnel oxide of hybrid memory due to large electric displacement in the FE material.

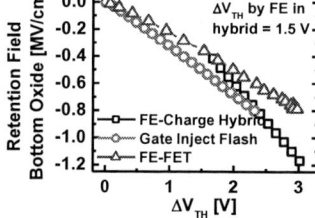

Figure 12 Retention field in the bottom oxide of hybrid memory follows FE-FET characteristics for small ΔV_{TH} and gate-inject Flash for higher ΔV_{TH}.

Figure 14 Hybrid cell with 40% FE contribution to ΔV_{TH} yields 50% reduction in required SBT thickness [5] even with additional 3 nm top oxide for 10 V P/E operation.

Table I Fabrication Procedure

Step	Hybrid	FE-FET	GI
1.Thermal bottom oxide	98 Å	170 Å	230 Å
2. PVDF	350 Å	350 Å	N/A
3. Evap. SiO₂	10 Å	N/A	10 Å
4. 110 °C ALD HfO₂	25 Å	N/A	25 Å
5. 110 °C ALD SiO₂	54 Å	N/A	54 Å
6. Cr-Au	1000 Å		

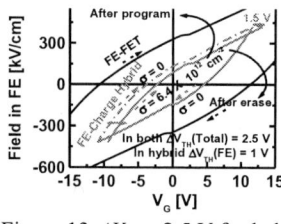

Figure 13 ΔV_{TH}= 2.5 V for hybrid memory has been split into 1 V through FE switching and 1.5 V by electron injection from the gate. FE in hybrid memory thus follows a smaller P-E loop against 2.5 V FE-FET. This provides electrostatic advantages over conventional devices as listed in Table II.

Table II Advantages of FE-charge hybrid design

Parameter	Hybrid	FE-FET	GI
Operating Voltage [V]	12	23	>20
Retention field top oxide [MV/cm]	1.89	-	4
Retention field bottom oxide [MV/cm]	-1.03	-0.64	-0.82
Program field top oxide [MV/cm] (-12 V)	-6	-	-4
Erase field top oxide [MV/cm] (12 V)	7.6	-	7.8
Depolarization Field [kV/cm]	88.5	329	-

References

[1] T. P. Ma and J.-P. Han IEEE Elec. Dev. Lett. 23 386 (2002)

[2] D. Wu et al Semi. Sci. Tech. 23, 7, 075035 (2008)

[3] G. A. Salvatore et al ESSDERC 2008 pp. 162-165

[4] H.-T. Lue, et al IEEE Tran. Elec. Dev 49 1790 (2002)

[5] S. Sakai et al IMW 2008 pp. 978

Spin-torque switchable perpendicular magnetic junctions for solid-state memory

J. Z. Sun[1,2], R. P. Robertazzi[1], J. J. Nowak[1], P. L. Trouilloud[1], G. Hu[1], M. C. Gaidis[1], S. L. Brown[1], D. W. Abraham[1], E. J. O'Sullivan[1], W. J. Gallagher[1], D. C. Worledge[1], and A. D. Kent[2]

[1] IBM-MagIC MRAM Development Alliance, IBM T. J. Watson Research Center, P. O. Box 218, Yorktown Heights, NY 10598, USA; [2] Dept of Physics, New York University, NY 10003, USA.

Introduction: Spin-torque switchable junctions are the leading candidates for future generations of magnetic random-access memory (MRAM) technologies[1]. Earlier, spin-valves with perpendicular magnetic anisotropy (PMA) were shown to lower the switching threshold current[2], consistent with spin-torque switching dynamics[3]. Recently, MgO-based magnetic tunnel junctions (MTJs) were developed to have fully PMA states[4-7], improving switching speed and margin. Here we review the essential device properties of some such PMA junctions.

Device Structure: MTJ devices are built using the IBM Yorktown MRL 200mm wafer Fab, with integrated CMOS front-end circuits underneath[8,9]. Magnetron sputtering was employed to prepare the magnetic layer stack structures that form the MTJ [7]. Wafers were post annealed at 240C. Optical lithography was used for MTJ-definition, followed by pattern transfer etching. A representative device stack reads: ||Substrate ||5 RuCoFe|2 Ta|0.8 CFB|0.9 MgO|0.5 Fe|0.8 CFB|0.3 Ta|0.25 Co|0.8 Pt|[0.25 Co|0.8 Pd]×4|0.3 Co|0.9 Ru|[0.25 Co|0.8 Pd]×14|20 Ru||, where numbers are thicknesses in nm, and CFB=$Co_{60}Fe_{20}B_{20}$ [7]. The magnetic free-layer is below MgO. Junction sizes range from about 60 nm to above 120 nm, with a resistance-area product around 5 to 10 $\Omega \, \mu m^2$. Both isolated and CMOS-integrated devices were studied. Fig.1 shows a transmission electron microscope (TEM) cross-section image of a resulting MTJ. Fig.2 is a die photo of a test chip. It shows areas where 4kbit memory arrays reside, complete with CMOS read and write circuits. To explore even faster switching, e-beam lithography-fabricated all-metal spin-valve PMA devices based on Co|Ni were also examined. These demonstrate sub-nanosecond spin-torque switching, and give quantitative comparison with physics models.

Results and Discussion: Quasi-static electrical results from a representative individual MTJ are summarized in Figs. 3(a)-(f). The resistance-magnetic field (RH) loop gives a magneto-resistance (MR) ~ 50 %. The RH loop center is controlled by the dipolar field-balancing layer. The current-voltage (IV) curves were measured using a triangular wave-form on the junction voltage bias (±0.5V) with sweep frequencies varying from 0.025 to 25Hz, 25-100 trace averaged (Fig.3b). Voltage thresholds for spin-torque switching show sweep-frequency dependence (Fig.3c) which was used to estimate the thermal activation barrier height E_b, giving E_b ~ 50 $k_B T$ with $T = 300K$. The quasi-static magnetic switching phase boundary is shown in Fig.3d. Dynamic switching probability is shown in Fig.3e-f at RH loop center, from which switching speed *vs* drive voltage amplitude is derived (Fig.4). Several traces with different easy-axis magnetic field bias are shown. RH loop center field corresponds to the highlighted trace. A steeper switching speed increase *vs* pulse height is seen for these PMA junctions, due to its magnetically thinner free-layer as compared to in-plane-magnetized (IMA) junctions of similar E_b (gray points). This is consistent with angular momentum conservation during spin-torque switching[8]. On a different sample with integrated CMOS circuits, one digitally counts a single junction's write bit error rate (BER) which is measured as a function of pulse height for widths between 2ns and 3.2μs (Fig.5a). The BER appears nearly Gaussian (i.e. linear) on this scale, giving a narrower threshold voltage distribution than macrospin-model would predict. The resulting threshold voltage at 50% (0 σ) is shown in Fig.5b as a function of pulse width. Switching times well below 10ns are accessible in these MTJs. Further test has demonstrated <10^{-11} BER at 10ns, limited by test time[11]. To explore even faster spin-torque switching for 3-terminal spin-torque devices for example[12], PMA two-terminal spin-valve model systems were also examined[3]. Fig.6 is a chip photo of one such two-terminal spin-valve chip. The junction stack is: ||Substrate|3 Pt|20 Cu|5 Ta|[0.25 Co|0.52 Pt]×4|0.25 Co|[0.6 Ni|0.1 Co]×2|4 Cu|[0.1 Co|0.6 Ni]×2|0.2 Co|5 Ta|30 Cu|3 Pt||. For 100×100 nm^2 junctions, the switching speed *vs* drive amplitude is shown in Fig.7. The long-time tail gives a thermal activation barrier height of about 63$k_B T$, much smaller than a volume-based estimate (with an H_c ~ 965 Oe) of the barrier ~ 195 $k_B T$, indicating sub-volume thermal activation, which has also been observed in PMA MTJs. The sub-volume fraction appears to approach junction size when the later is reduced to below about 50nm.

Conclusion: PMA spin-torque switchable junctions have been demonstrated with lower switching current and faster switching speed compared to IMA devices. They are promising for further technology exploration in solid-state memory applications.

978-1-61284-243-1/11 $26.00 © 2011 IEEE

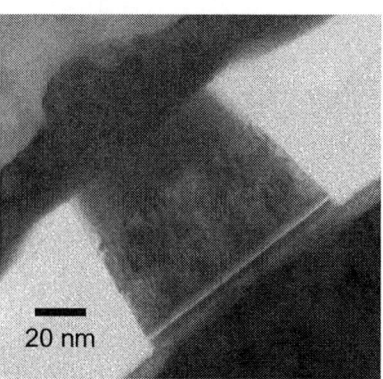

Fig. 1. A cross-section TEM view of an MTJ device, courtesy Dr. Yu Zhu, IBM Research.

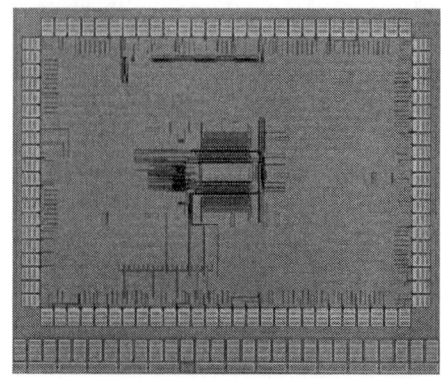

Fig. 2. An optical die-photo of the resulting chip containing 4,096 bits of spin-torque switched MTJs.

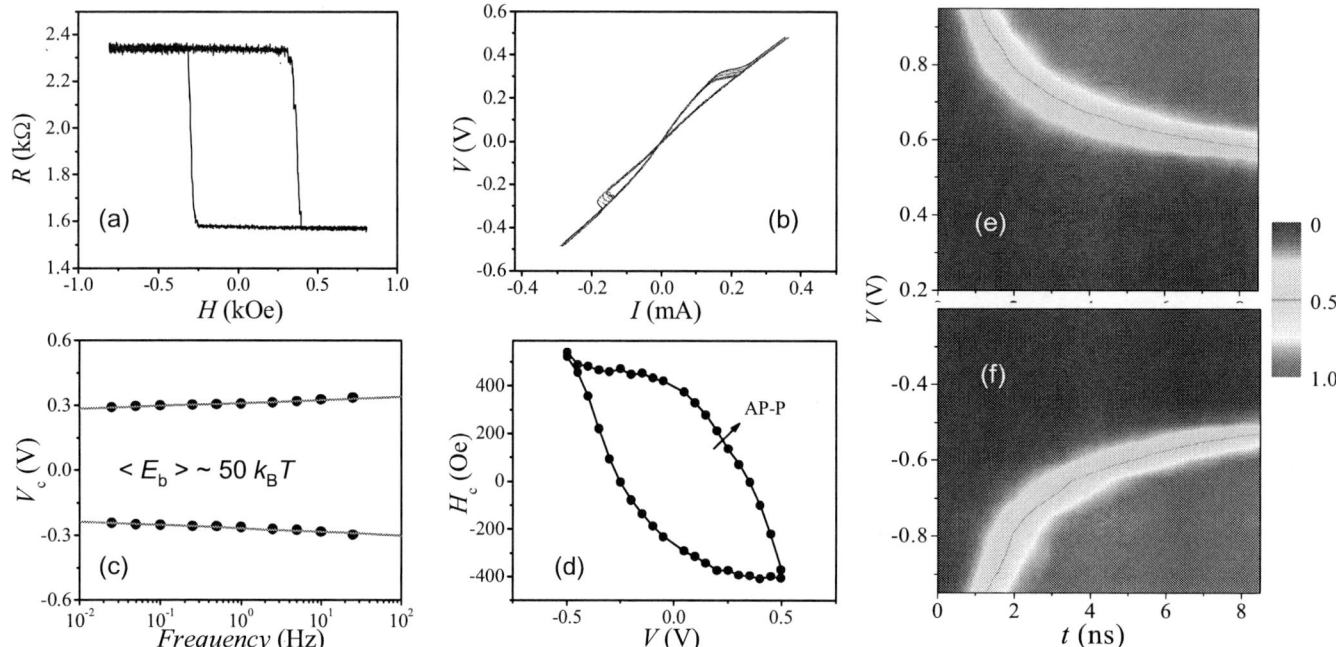

Fig. 3.(a) R-H loop of an MTJ, 120nm in diameter. Field ⊥ film surface. (b) Sweep-rate dependent spin-torque switching IV curves, 0.025 to 25 Hz, ±0.5V. (c) Threshold voltage *vs* sweep-rate. (d) Switching boundary in (H, V) space measured by static V-bias and 0.2 Hz triangular wave H-sweep. Arrow indicates direction for AP to P switching. (e-f) Dynamic switching probability *vs* drive voltage amplitude. Color gradient indicate switching probability from 0 to 1; (e) is for AP-P transition, (f) P-AP. A dc magnetic field bias of 50 Oe along the easy-axis was applied during pulses.

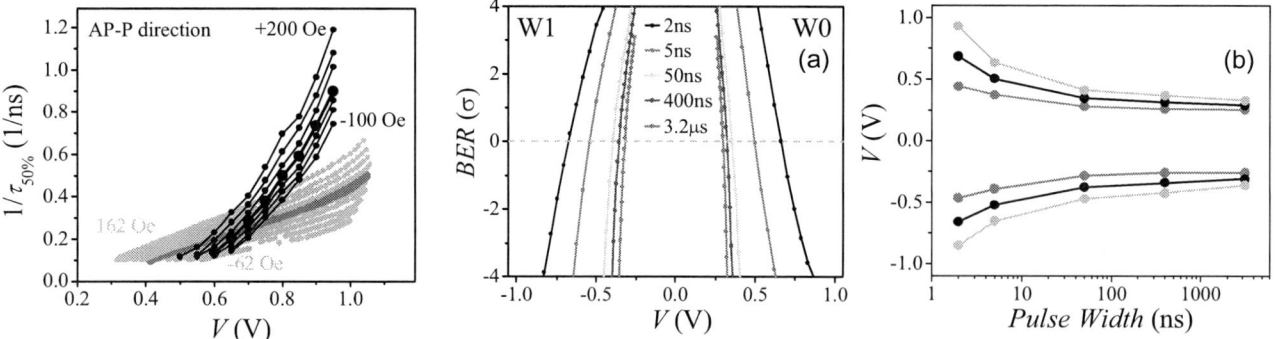

Fig. 4. Switching speed of the PMA junction. Gray data are a corresponding IMA junction of similar E_b and resistance.

Fig. 5. (a) Switching bit error rate (BER) as a function of drive voltage, at pulse widths from 2ns to 3.2μs. (b) Switching thresholds at +4σ, 0σ (i.e. 50%), and -4σ BER as a function of pulse width. +V corresponds to write-0 (AP-P).

978-1-61284-243-1/11 $26.00 © 2011 IEEE   172

Fig.6 A chip photo of the Co|Ni multilayer-based spin-valve junctions. This chip was provided by Jordan Katine of HGST[3].

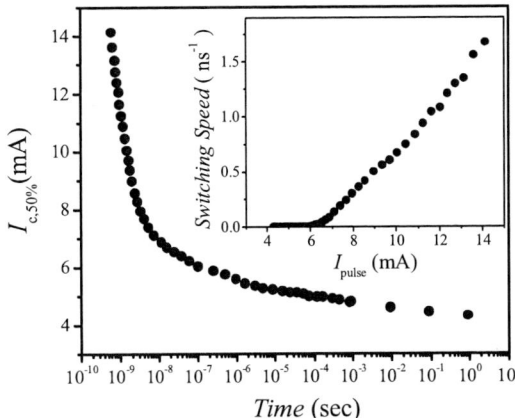

Fig.7 The switching speed (at 50% switching probability) vs drive current amplitude for a Co|Ni-based PMA spin-valve. From the long-time tail: $U_0/k_BT \sim 63$. Inset shows a 50% switching time of about 0.6ns at 14mA pulse height.

References:

[1] M. Hosomi *et. al,* IEDM (2005) Tech. Digest. IEEE International, 459-462 DOI: 10.1109 /IEDM.2005.1609379.

[2] S. Mangin *et. al,* Appl. Phys. Lett. **94**, 012502 (2009).

[3] D. Bedau *et al.* Appl. Phys. Lett. **97**, 262502 (2010).

[4] T. Kishi *et al, ,* IEDM (2008) Tech. Digest. IEEE International, DOI: 10.1109 /IEDM.2008.4796680.

[5] S. Ikeda *et al.* Nature Materials **9**, 721 (2010).

[6] D. C. Worledge *et al.* IEDM (2010) Tech. Digest. IEEE International, DOI: 10.1109 /IEDM.2010.5703349.

[7] D. C. Worledge *et al.* Appl. Phys. Lett. **98**, 022501 (2011).

[8] R. Beach *et al.* IEDM (2008) Tech. Digest. IEEE International, DOI: 10.1109 /IEDM.2008.4796679.

[9] T. Min *et al.* IEEE Trans. Magn. **46**, 2322 (2010).

[10] J. Z. Sun, Phys. Rev. B **62**, 570 (2000); J. Z. Sun *et al.* SPIE **5359**, 445 (2004).

[11] J. J. Nowak *et al.* unpublished (2011).

[12] J. Z. Sun, *et al.* Appl. Phys. Lett. **95**, 083506 (2009).

978-1-61284-243-1/11 $26.00 © 2011 IEEE   174

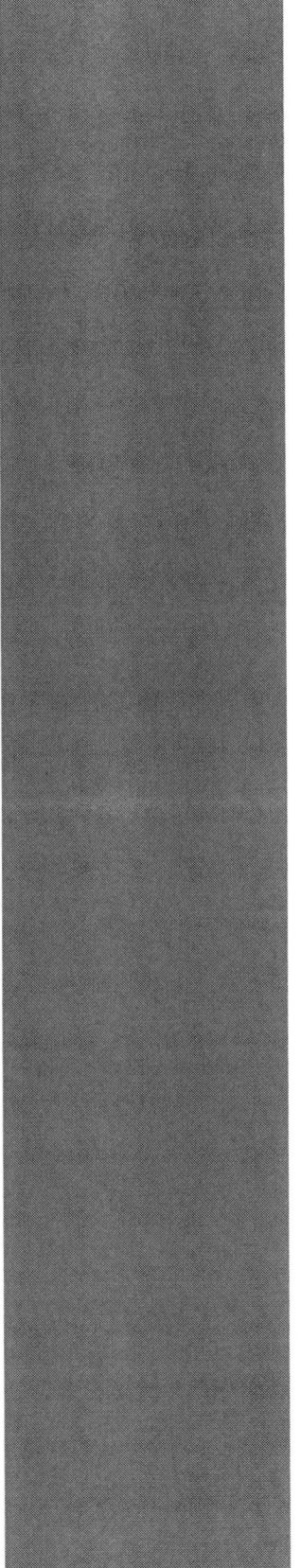

Session IV.B (Corwin Pavilion West)

Alternative Transistor Concepts

Tuesday AM, June 21st, 2011

Session Chair: TBA

8:20 AM IV.B-1 Student Paper
Combinational and Sequential Logic with Transistors based on Individual Carbon Nanotubes
H. Ryu[1], D. Kälblein[1], U. Zschieschang[1], O. G. Schmidt[2], and H. Klauk[1], [1]Max Planck Institute for Solid State Research, Stuttgart, GERMANY and [2]Faculty of Electrical Engineering and Information Technology, Chemnitz University of Technology, GERMANY

8:40 AM IV.B-2 Student Paper
Comparative Study of Fabricated Junctionless and Inversion-mode Nanowire FETs
C.-H. Park[1], M.-D. Ko[1], K.-H. Kim[1], C.-W. Sohn[1], C. K. Baek[1], Y.-H. Jeong[1,2], and J.-S. Lee[1,2], [1]Dept. of Electronic and Electrical Engineering, and [2]Division of IT-Convergence Engineering, POSTECH, Pohang, Gyeongbuk, KOREA

9:00 AM IV.B-3
Fabrication of Vertical InAs-Si Heterojunction Tunnel Field Effect Transistors H. Schmid, K. E. Moselund, M. T. Björk, M. Richter, H. Ghoneim, C. D. Bessire and H. Riel, IBM Research – Zurich, Rüschlikon, SWITZERLAND

9:20 AM IV.B-4 Invited Paper
Challenges for Post-CMOS Devices & Architecture
J. Welser[1,2] and K. Bernstein[2], [1]SRC-NRI,Durham, North Carolina, USA and [2]IBM Research, Yorktown Heights, New York & San Jose, California, USA

10:00 AM Break

10:20 AM IV.B-5 Invited Paper
Correlated Oxide Phase Transition Switch: A Paradigm in Electron Devices
Z. Yang, C. Ko, V. Balakrishnan, and S. Ramanathan Harvard School of Engineering and Applied Sciences, Harvard University, Cambridge, Massachusetts, USA

11:00 AM IV.B-6 Student Paper
Lateral Gate Suspended-Body Carbon Nanotube Field-Effect-Transistors with Sub-100nm Air Gap by Precise Positioning Method
J. Cao and A. M. Ionescu, Nanoelectronic Devices Laboratory (Nanolab), Ecole Polytechnique Fédérale de Lausanne, Lausanne, SWITZERLAND

11:20 AM IV.B-7
Late News

11:40 AM IV.B-8
Late News

978-1-61284-243-1/11 $26.00 © 2011 IEEE

978-1-61284-243-1/11 $26.00 © 2011 IEEE   176

Combinational and Sequential Logic with Transistors based on Individual Carbon Nanotubes

Hyeyeon Ryu,[a] Daniel Kälblein,[a] Ute Zschieschang,[a] Oliver G. Schmidt,[b] Hagen Klauk[a]

a) Max Planck Institute for Solid State Research, Heisenbergstr. 1, 70569 Stuttgart, Germany
(email: H.Ryu@fkf.mpg.de, phone: +49 711 6891317)
b) Faculty of Electrical Engineering and Information Technology, Chemnitz University of Technology, Germany

Field-effect transistors (FETs) that utilize an individual semiconducting carbon nanotube (CNT) as the channel are potentially useful for the realization of logic circuits with high integration densities that can be fabricated on transparent, large-area substrates, such as glass or flexible plastics. While FETs based on individual CNTs have already demonstrated excellent static characteristics [1,2], the realization of logic circuits with good static and dynamic performance based on individual-CNT FETs remains a challenge. Bachtold et al. realized 2-input NOR gates by connecting p-channel CNT FETs to external load resistors using coaxial cables and reported a signal delay of 30 msec per stage for a 3-stage unipolar ring oscillator [3]. Javey et al. realized complementary 2-input NAND, AND, NOR and OR gates by connecting p- and n-channel CNT FETs using coaxial cables and measured a delay of 750 µsec for a 3-stage ring oscillator [4]. The only monolithically integrated circuit based transistors utilizing individual carbon nanotubes was reported by Chen et al. who measured a signal delay of 1.9 nsec per stage in a complementary ring oscillator realized on a very long (19 µm) carbon nanotube [5].

While Sun et al. recently reported a flip-flop with FETs based on random carbon-nanotube networks [6], there are no reports of sequential circuits fabricated using FETs based on individual CNTs. Here we report on the fabrication of FETs based on individual CNTs with switching frequencies above 1 MHz and load resistors based on vacuum-deposited carbon films, and their integration into combinational and sequential logic circuits.

First, an array of probe pads was created by electron-beam lithography, metal evaporation, and lift-off. Next, gate electrodes were defined by e-beam lithography and deposition of 30 nm of aluminum. The Al gates were exposed to an O_2 plasma to create a 3.6 nm thick AlO_x layer, and a monolayer of octadecylphosphonic acid (2.1 nm thick) was allowed to self-assemble on the Al/AlO_x gates from solution. Thus, the AlO_x/SAM gate dielectric is 5.7 nm thick. CNTs produced by an arc discharge were deposited from a liquid suspension. Using scanning electron microscopy, an individual nanotube was located on each gate, and a pair of AuPd source/drain contacts was defined by e-beam lithography for each device. The channel length is 300 to 400 nm. **Fig 1.** shows the cross section, an SEM image, and the current-voltage characteristics of a completed FET, having a transconductance of 6 µS, an ON/OFF ratio of 10^7, and a subthreshold slope of 100 mV/decade.

To realize integrated logic circuits, load resistors are fabricated by depositing a thin layer of carbon by thermal evaporation in vacuum and lithographic patterning. Depending on the geometry and the carbon-film thickness, resistances between 10^5 and 10^8 Ω are obtained, and the resistors have excellent linearity (see **Fig. 2a**). Circuits are completed by connecting FETs and resistors with a dedicated metal interconnect layer or by wire bonding. Inverters composed of a carbon-nanotube FET and a carbon load resistor have full output swing and large gain (see **Fig. 2c**), and they show good switching response at frequencies up to 2 MHz (see **Fig. 3**). This frequency is limited by the load resistance (1.2 MΩ), so to estimate the dynamic performance of the FETs we have analyzed the inverter output signal and obtained a time constant of 12 nsec for the transition when the FET switches from the OFF-state to the ON-state (see **Fig. 3c**), suggesting a maximum frequency above 10 MHz.

Logic gates (NAND, NOR, AND, OR) comprised of two or three CNT FETs and one or two load resistors all show the correct logic function for supply voltages of -1 V (see **Fig. 4**). However, due to the positive threshold voltage of the FETs, the input and output signals of the circuits do not match (outputs produce only negative signals, while positive input signals are required to turn the drive FET completely OFF). This situation is fixed by integrating a level-shift stage comprised of two carbon resistors into each logic gate, as shown in **Fig. 5a**.

Fig. 5b and **5c** show the schematic and the correct electrical response of a NAND latch when operated with SET and RESET pulses having a pulse width of 20 msec. To our knowledge this is the first report of a sequential logic circuit realized using transistors based on individual carbon nanotubes.

[1] A. Javey et al., *Nano Lett.* **2005**, *5*, 345
[2] J. Chen et al., *Appl. Phys. Lett.* **2005**, *86*, 123108
[3] A. Bachtold et al., *Science* **2001**, *294*, 1317

[4] A. Javey et al., *Nano Lett.* **2002**, *2*, 929
[5] Z. H. Chen et al., *Science* **2006**, *311*, 1735
[6] D. M. Sun et al., *Nature Nanotechnology* **2011**, 6, 156

Fig. 1.
a) Schematic cross section and SEM image of an FET based on an individual carbon nanotube (CNT). **b)** Transfer and **c)** output characteristics of the transistor, which has a transconductance of 6 μS.

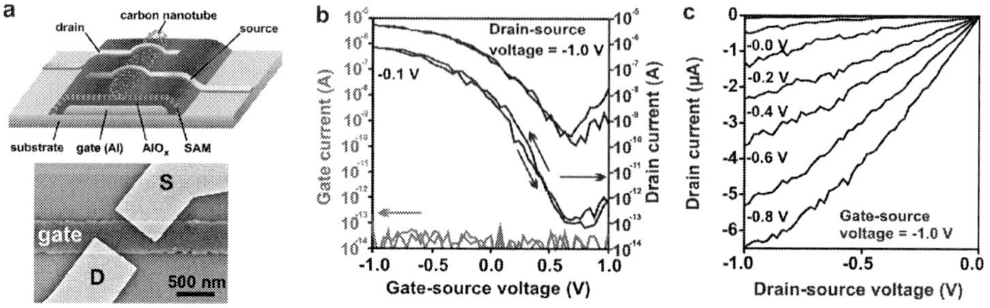

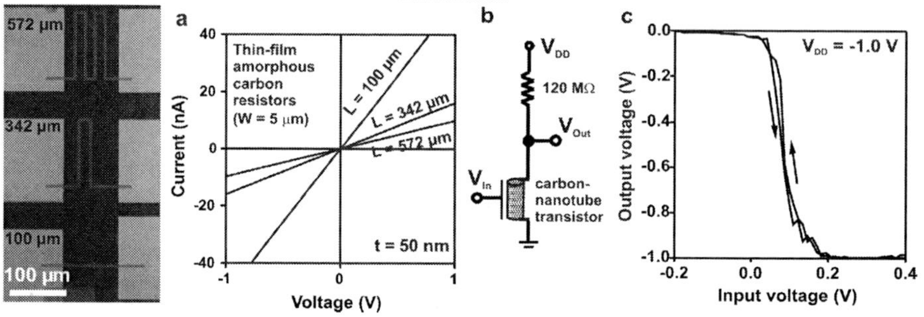

Fig. 2.
a) Photograph and current-voltage characteristics of three vacuum-evaporated carbon resistors with various lengths, having resistances between 20 MΩ and 100 MΩ. **b)** Schematic and **c)** transfer characteristics of a CNT-FET inverter with resistive load and a small-signal gain of ~10.

Fig. 3.
a) Schematic of the setup for the dynamic measurements. **b)** Input and output signals of the inverter for a signal frequency of 2 MHz. **c)** Exponential fits to the output signal, yielding a time constant of 12 nsec for the transition limited by the transistor.

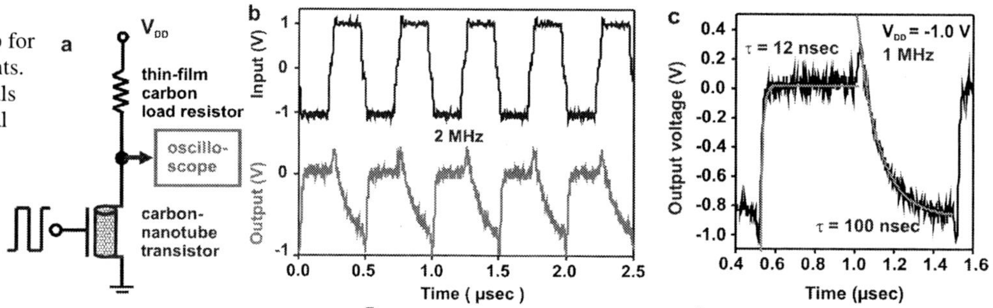

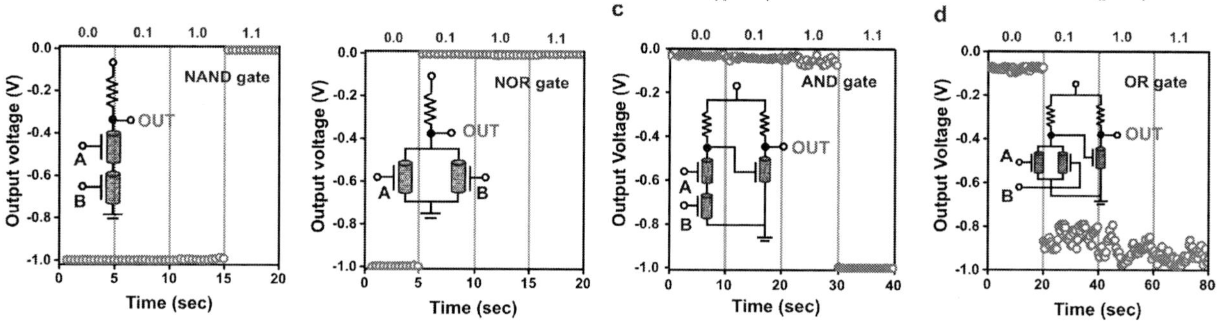

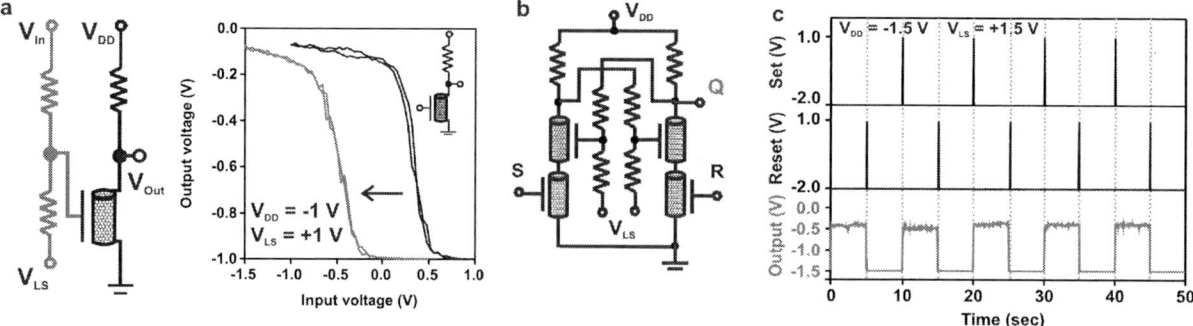

Fig. 4. Logic gates (NAND, NOR, AND, OR) consisting of two or three CNT FETs and one or two thin-film carbon load resistors.

Fig. 5. a) Circuit schematic and transfer characteristics of inverters with (red) and without (black) level-shift stage.
b) Circuit schematic and **c)** output response of a NAND latch (with level shifting). The S and R pulses have a width of 20 ms.

978-1-61284-243-1/11 $26.00 © 2011 IEEE

Comparative Study of Fabricated Junctionless and Inversion-mode Nanowire FETs.

Chan-Hoon Park, Myung-Dong Ko, Ki-Hyun Kim, Chang-Woo Sohn, Chang Ki Baek, Yoon-Ha Jeong[1,2], and Jeong-Soo Lee[1,2]

[1]Dept. of Electronic and Electrical Engineering, [2]Division of IT-Convergence Engineering, POSTECH, Pohang, Gyeongbuk, 790-784, Korea. Email : chpark82@postech.ac.kr Phone : 82-54-279-2897/ Fax : 82-54-279-2903

For the higher degree of integration and better performance of a device, the feature size of conventional MOSFET is expected to go down under 20 nm within a few years [1] and the nanowire FET (NWFET) is the most conspicuous candidate for the future device application. However, in the case of conventional inversion mode NWFETs (cINT), the formation of an abrupt junction for the source/drain (SD) is one of the technical obstacles [2]. Recently, junctionless NWFETs (JNT) where the channel and SD region are doped with the same dopant type has been suggested [3]. In this work, the n-type JNTs and cINT are fabricated with the gate length (L_G) of 20 ~ 250 nm and compared their electrical DC characteristics and low-frequency noise characteristics.

The fabrication of n-type JNT started with the SOI wafer having the thickness of p-type top silicon layer (T_{si}) with 100 nm and buried oxide (B_{OX}) with 200 nm. To form the channel and SD region, the wafer was doped with arsenic (dose: 4×10^{14} cm^{-2} with 65 keV) and then annealed at 1000 °C for 5 minutes (Fig. 1 - (a)). For the patterning of the active channel region and external electrode for SD pads, e-beam lithography and mask aligner are used, respectively, followed by dry etch in a ICP etcher (Fig. 1 - (b), (c)). The sacrificial oxide with thickness of 5 nm was thermally grown and removed by wet etch process to minimize the dry etch damage, and the gate oxide was grown by rapid thermal oxidation (Fig. 1 - (d)). Then, the gate electrode was formed by deposition of TiN with thickness of 40 nm (Fig. 1 - (e)). Finally, a gate line was formed by e-beam lithography and ICP etcher. For the comparative purposes, cINTs were also fabricated with the same structure as that of the JNT but the annealing condition is 1000 °C for 10sec.

Fig. 2 shows the TEM and SEM image of a NWFET with 20 nm of L_G, 15 nm of the bottom width, 20 nm of the height and 5 nm of the gate oxide thickness. It clearly showed that the TiN film is surrounding left, right and top sides of the channel. The doping density is measured as high as ~10^{19} cm^{-3} from the SIMS analysis (Fig. 3). The DC characteristics were measured using Keithly 4200 I-V meter and the low frequency noise characteristics are measured using BTA9812 noise analyzer and Cadence NoisePro software.

Both devices show relatively high on-state current ($I_{on} \approx$ 1 mA/μm) and high on-off ratio (I_{on}/I_{off}) > 10^6 with L_G = 20 nm as depicted in the Fig. 4 and 5. To investigate the short channel effect (SCE) in the NWFETs, the subthreshold slope (SS) is measured as low as 75 mV/dec for JNT and 92 mV/dec for cINT, respectively. The drain induced barrier lowering (DIBL) is extracted as low as 10 mV/V for JNT and 78 mV/V for cINT. As L_G decreases in conventional FETs, the potion of depletion region in the total channel region becomes larger, which degrades the SS and DIBL characteristics. Thus JNT shows better SS and DIBL characteristics than cINT because JNTs have no depletion region between SD region and channel region. The saturation current of JNT and cINT as a function of L_G are shown in Fig. 6 and 7, respectively.

Fig.8 shows low frequency noise characteristics of JNT, cINT and planar MOSFET, respectively. The measurement was performed with several samples of L_G = 100 nm at V_G - V_{TH} = 0.5 V and V_D = 0.05 V. For the comparisons of the noise profile between planar MOSFET and NWFET, a paper [4] reported that whereas MOSFETs show a uniform noise profile, NWFETs show large variations in their noise behaviors depending on the process configurations and three dimensional shape of the interface. As shown in Fig. 8, the JNT exhibited higher power spectral density compared with other devices, which suggest that JNTs have probably more traps at the interface. Hence, an additional annealing treatment is necessitated in order to improve the interface quality in JNTs.

In conclusion, we have demonstrated junctionless nanowire transistors for next generation device applications. The JNT with a gate length of 20 nm showed excellent electrical characteristics with high I_{on}/I_{off} ratio, good subthreshold slope, and low DIBL. The simpler fabrication process without junction formation makes the JNTs a promising candidate for future logic devices.

This research was supported by WCU (World Class University) program through the National Research Foundation of Korea funded by the Ministry of Education, Science and Technology (R31-2010-000-10100-0). Also, this work was supported by the BK21 program and National Center for Nanomaterials Technology (NCNT), Korea.

[1] http://www.itrs.net [2] T. Ito et al, *Jpn. J. Appl. Phys.*, vol. 41, no. 4B, pp. 2394, 2002. [3] J. P. Colinge et al, *Nat. Nanotechnol.*, vol. 5, pp. 225, Feb. 2010. [4] R. –H. Baek et al, published online, *IEEE Trans. Nanotechnol.*, 2010

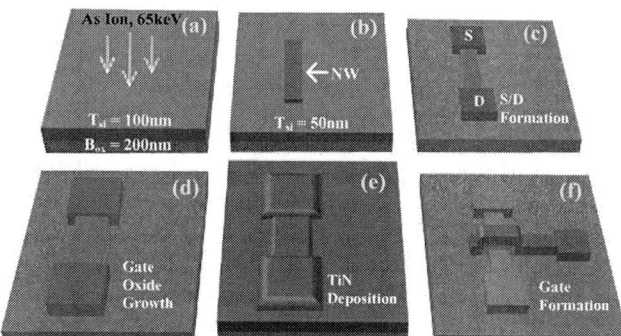

Fig.1. Schematics of the process flow for JNT.

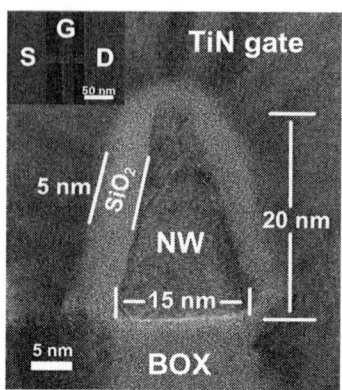

Fig. 2. The TEM image of JNT. The inset shows a SEM image(top-view) of a JNT.

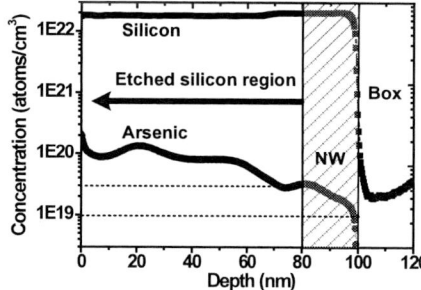

Fig.3. The doping profile of JNTs measured by SIMS.

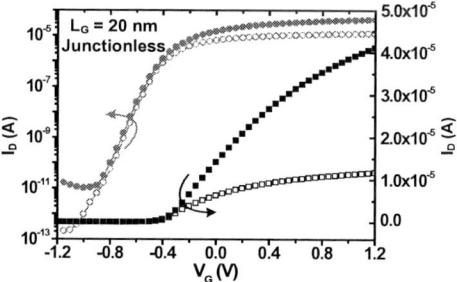

Fig.4. I_D-V_G characteristics of JNT with L_G=20nm (Open: V_D=0.05V, Solid: V_D=1V).

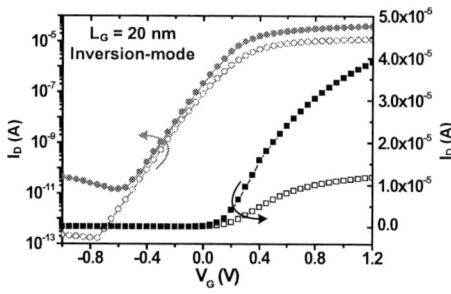

Fig.5. I_D-V_G characteristics of cINT with L_G=20nm.(Open: V_D=0.05V, Solid: V_D=1V)

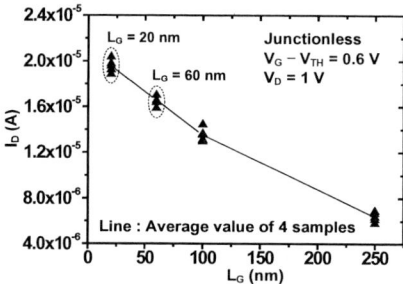

Fig.6. I_{ON} of JNT in the saturation regime with different L_G.

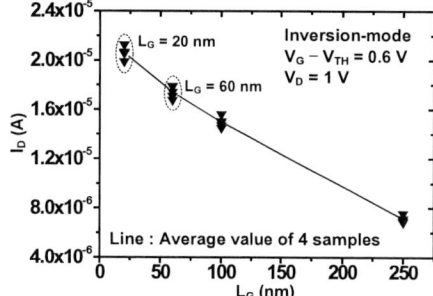

Fig.7. I_{ON} of cINT in the saturation regime with different L_G.

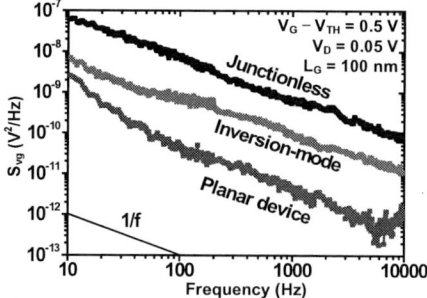

Fig.8. The 1/f noise characteristics of JNT, cINT and planar MOSFET devices.

978-1-61284-243-1/11 $26.00 © 2011 IEEE

Fabrication of Vertical InAs-Si Heterojunction Tunnel Field Effect Transistors

H. Schmid, K. E. Moselund, M. T. Björk, M. Richter, H. Ghoneim, C. D. Bessire and H. Riel

IBM Research – Zurich, Säumerstrasse 4, CH-8803 Rüschlikon, Switzerland.
Phone: +41 44 724 83 81, fax: +41 44 724 89 58, email: bjm@zurich.ibm.com

Gated *p-i-n* diodes operating as tunnel field effect transistors (TFETs) [1] are recently attracting much attention because of potential benefits over conventional MOSFETs. They are expected to have lower off-current, and operate at lower supply voltage compared to MOSFETs. Unfortunately, these promises are very difficult to realize using materials like Si, Ge and its alloys. However, encouraging experimental results were recently obtained using lower bandgap III-V (InGaAs) material systems [2, 3] offering higher tunneling probabilities. Here we report first results on the fabrication and electrical characterization of III-V / Si heterojunction TFETs with InAs as low bandgap source. This material combination maintains the advantages of Si as channel, drain and substrate material as proposed in [4].

Si-InAs heterojunctions cannot be grown by traditional methods due to the 11% lattice mismatch. However, growth of InAs nanowires (NWs) on Si was previously demonstrated using selective area epitaxy [5]. We grew InAs NWs having a diameter of 100 nm within e-beam patterned SiOx openings (Fig.1, inset) using MOCVD. The InAs NWs are *n*-type as grown and act as source. The Si substrate was p-type doped to 1×10^{19} cm^{-3} and an additional 150-nm-thick lowly doped (1×10^{15} cm^{-3}) Si layer was grown on top using MBE. The InAs NWs then serve as etch mask during RIE into the Si substrate (Fig. 1) forming a vertical and self-aligned channel and drain wire geometry (Fig. 2). A tungsten wrap gate was separated by a 7.5-nm-thick Al_2O_3 gate dielectric from the channel. BCB was used as dielectric spacer layers between drain, gate and source. The electrical contacts to source (Ti/Al) and drain (Ti/Au) were deposited by electron-beam evaporation.

Fig. 3 compares the diode characteristics of an InAs-Si-Si *n-i-p* device with only two contacts and minimal processing and a fully processed three-terminal device. The similar ideality factors of 1.4 to 1.7 indicate only minor process-induced defects, while the significant spread in forward current density (negative bias) points to large variation in contact resistances at the InAs NW side. The transfer- and output-characteristics of a single NW TFET are plotted in Fig. 4 and 5. The absence of a saturation current in the transfer curve is an expected behavior of a TFET while the high threshold voltage and threshold shift measured indicate a non-ideal (source) contact. This non-ideal contact likely contributes to the inverse subthreshold slope (SS) of ~220 mV/dec and in particular limits the drive current to ~ 0.4 µA/µm of the device. The same trend is visible in Fig. 6 displaying several transfer curves of individual devices showing a large spread in scaled current density and threshold voltage. Temperature-dependent measurements were performed to further elucidate the device operation and limitations. The transfer characteristics strongly depends on temperature (Fig. 7) showing a strong decrease in current and shift to lower voltages including a slight decrease of the inverse subtreshold slope. This is not expected for a TFET and can be explained by a barrier at the source contact. In Fig. 8 the I_{on} is plotted versus 1/T, which allowed extraction of an activation energy of 80meV which is attributed to the remaining barrier height due to the source contact metallization.

In summary, we showed that the combination of III-V NW growth on epilayers is a versatile fabrication approach that can provide well-defined vertical device dimensions and adaptable junction metallurgy. We fabricated first working *n-i-p* InAs-Si-Si gate all around NW TFETs with 100 nm channel diameter and 150 nm gate lengths. Enhanced device performance, in particular drive current, is anticipated by introducing *n*-type doping in the InAs wire [6], EOT scaling, and improved processing of the metal contacts.

This work was supported in part by the European Union FP7 Program STEEPER.

[1] T. Baba, Jpn. J. Appl. Phys, Vol. 31, 455-457, **1992**
[2] S. Mookerjea., D. Mohata, R. Krishnan, J. Sing, A. Vallett, A. Ali, T. Mayer, V. Narayanan, D. Schlom, A. Liu, S. Datta, IEDM, Tech. Dig. **2009**, p. 949
[3] H. Zhao, Y. Chen, Y. Wang, F. Zhou, F. Xue, J. Lee, IEEE EDL, Vol. 31, 1392, **2010**
[4] A. Verhulst, W.G. Vanderberghe, K. Maex, S. D. Gendt, M.M. Heyns, G. Groeseneken, EDL, Vol. 29, 1398, **2008**
[5] K. Tomioka, J. Motohisa, S. Hara, and T. Fukui, Nano Lett. Vol. 8, 3475, **2008**
[6] M. T. Björk, H. Schmid, C. D. Bessire, K. E. Moselund, H. Ghoneim, S. Karg, E. Lörtscher, H. Riel Appl. Phys. Lett. 97, 163501, **2010**

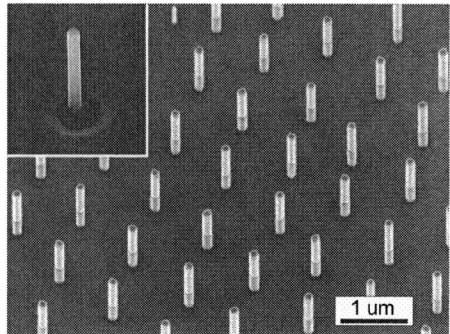

Fig. 1. InAs nanowires on Si <111>. The inset shows an InAs NW grown on Si <111> within the SiOx mask. The InAs NWs (bright part) serve as etch masks during RIE into the Si epilayer substrate as shown here.

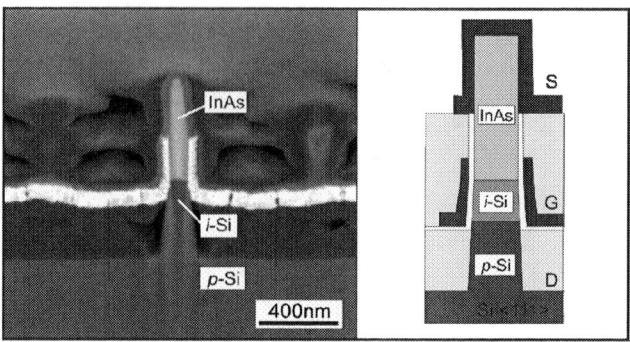

Fig. 2. SEM image showing a cross section of a test device without source contact and the corresponding schematic of the actual device geometry. The InAs NW has a diameter of 100nm, the undoped Si channel on the p-type substrate is 150nm long.

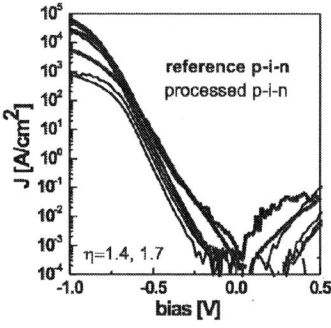

Fig. 3. Comparison of the diode characteristics of a simple two-terminal *p-i-n* structure and of a fully processed three-terminal *p-i-n* device.

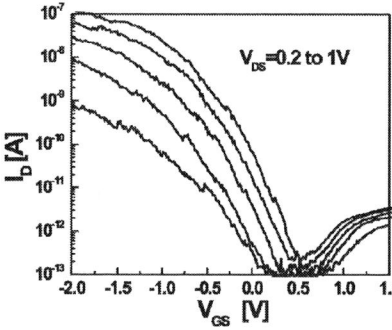

Fig. 4. Transfer characteristics of a single NW InAs-Si TFET.

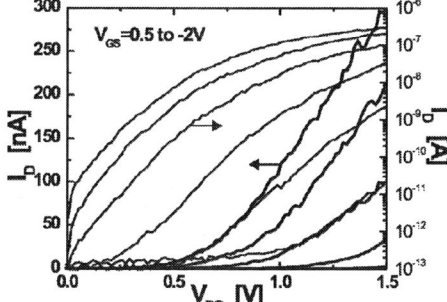

Fig. 5. Output characteristics of the TFET showing superlinear I_D vs. V_{DS}.

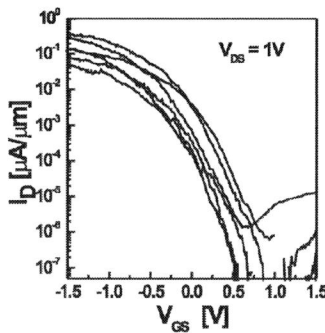

Fig. 6. Spread in the transfer characteristics of NW TFET devices illustrating the device variability.

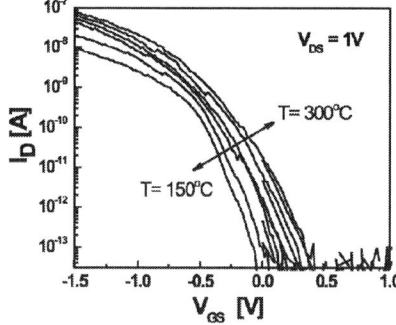

Fig. 7. Temperature dependence of the transfer characteristics showing variation of I_{on} and SS.

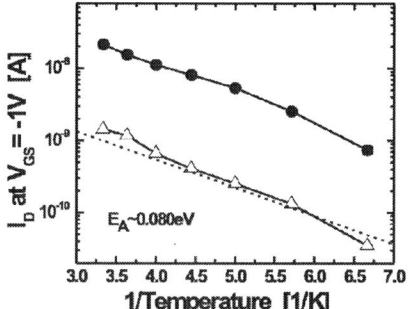

Fig. 8. Arrhenius plot of I_D at V_{DS} of 1 V. The slope corresponds to an activation energy of 80 meV.

978-1-61284-243-1/11 $26.00 © 2011 IEEE   182

Challenges for Post-CMOS Devices & Architectures

J. Welser[1,2] and K. Bernstein[2]
[1]SRC-NRI, 1101 Slater Rd., Suite 120, Durham, NC 27703; [2]IBM Research, Yorktown Heights, NY & San Jose, CA
Email: jeff.welser@src.org / Phone: 408-927-1017

Since 2006, the Nanoelectronics Research Initiative (NRI) has been actively funding work at universities across the U.S. with one specific mission: Demonstrate novel computing devices capable of replacing the CMOS FET as a logic switch in the 2020 timeframe. These devices must show significant advantage over FETs in power, performance, density, and/or cost to enable the semiconductor industry to extend the historical cost and performance trends for information technology. NRI seeks to find not just a one generation improvement on the FET, but rather a new extended scaling path. This is crucial to justify the expense of making any major change in the current technology infrastructure (both at the device and design level) – and the larger the change, the more benefit and longevity the new technology must offer.

NRI was conceived as a "goal-oriented, basic-science research program," and the research has focused on exploring many different materials, physics phenomena, and new device concepts, largely through simulation and more limited experimental work. The goal has been to find some promising avenues, before committing larger efforts to pursue them. In the past year, a large effort was made to benchmark the various concepts, both against each other and against CMOS [1]. It looked at all aspects of the devices, not only in isolation but also in small circuits, such as inverters and adders, which better judge the devices' potential for doing computation. A good summary of these results is captured in Fig. 1, showing how various devices compare on energy and delay when implementing a two-input NAND. Note these comparisons should be considered a "snapshot in time" of any given device's potential, as the research on all of them is at a very early stage and hence the data is evolving. In fact, that was one of the goals of the benchmarking work: by showing where various devices have advantages and challenges, it helps focus the research on the most pressing issues for each.

Overall, the results indicated that many of the devices showed advantages in power, but most also had challenges in speed as compared to modern CMOS. This is not entirely unexpected, since power is the primary challenge that is currently limiting the usability of scaled CMOS – which is what motivated NRI to look for devices beyond FETs in general –and power and speed tend to go together. It has also highlighted the need to consider circuit architecture hand-in-hand with the device research, to better understand the potential of any technology. If the individual devices are slow, can an architecture take advantage of some other aspect of their behavior to achieve high computational throughput? That is, maybe building Boolean logic gates is not the best utilization of the devices, but if some other approach is pursued, it needs to still meet the daunting needs of future chip design.

Post-CMOS Architectural Considerations

While it is true that the successor to the CMOS switch is anything but clear at this point, it can also be said that we know a great deal about the boundary conditions which will govern the selection of such a post-CMOS follow-on. In the era when these new switches will begin emerging, the compute problems encountered will be profoundly more challenging than the ones presently before us. Most likely, systems in this coming era will need to accommodate so-called "Big Data" arising in multiple fields. Most of the tasks confronting future computing engines will stream substantially larger source databases, as well as increasingly unstructured data, into the processors for transaction execution. These tasks include, but are not limited to: cyber-security, genomics, weather prediction, image recognition, augmented reality, and interpreting human communication nuance in interfaces. Effective solutions for Big Data, therefore, will require specific characteristics to be present in whatever future architecture eventually hosts these next switches. A few of these are listed below; while NRI is not directly researching these topics, they serve as an important backdrop for considering new architectural directions.

978-1-61284-243-1/11 $26.00 © 2011 IEEE

Computational Efficiency: The metrics of compute efficiency are typically captured in the amount of power required to retire a number of instructions per unit of time, e.g. millions of instructions per second per Watt (MIPS/W) or the number of transactions completed per CPU cycle (TpCC). However a figure of merit which ties fundamental technology to resulting logic transaction is most appropriate for device work. "Logical Effort" is a construct first offered by Ivan Sutherland at Sun Microsystems [2] which quantitatively measures the investment in hardware and energy needed to complete a fixed amount of logic. To this end, while microprocessors have continued to evolve, the fundamental MOSFET in principal has not changed in 50 years. MOSFETs still support at best a fan-out of four, for example. New architectures will need to extract an improved logical effort associated with any replacement switch.

Complexity Management: The number of machine states (i.e. registers) times the total amount of branching per block (i.e. fan-out) provides some sense of the complexity within an arbitrary microprocessor logic block. Increasing levels of functional integration in multi-core processors has steadily increased microprocessor complexity, challenging EDA tools as well as designers to efficiently accommodate an increasing number of nets. An unintended byproduct is that these increasing net counts spin off emergent behavior in the larger design; this behavior is difficult to anticipate, but critical to address. Against this backdrop, future architectures will need to more effectively deal with the burgeoning complexity that many more lower power devices will likely enable.

Self-Organized Reliability and Serviceability: It makes no sense to introduce a new technology and architecture unless it is more reliable and manufacturable than what it replaces. Given the microprocessor's increasing number of machine states mentioned above, future architectures will need to participate even more in their own verification and validation. Wafer, chip and module test currently is the most expensive portion of component fabrication, and insuring 100% test coverage has become increasingly untenable. In addition, the use of new switches on chip only increases the probability that previously undiscovered material defects and responses will present themselves.

Intrinsic Cyber-Security: A serious requirement of future architectures is that they can be made to be secure. With the advent of the coming cloud computing paradigm, the need to maintain thread separation becomes essential in large multi-threaded systems. Further, as the state machine size for each of the microprocessors in a system increases, it becomes increasingly easy to hide malicious additional content in a design, and increasingly harder to find it. In addition, it is possible to attack the reliability of a processor in a manner that does not appear during test, but only after it has been fielded. Anti-tamper technologies, digital design watermarking, encryption/decryption, and anti-counterfeit devices will likely become ubiquitous, expected, standard features.

NRI Device Focus

Based on the initial benchmarking work, as well as the increased recognition of the need for including circuit level analysis as part of the device research, each of the NRI university centers chose 2-3 devices to focus on for the next two years. They built teams to look at all aspects, from materials to circuits:

- Novel materials development, growth, and characterization
- Specific work on self-assembly or novel fabrication techniques, if required
- Physics theory and experimental verification of key phenomena, including work on phonon / heat flow for novel cooling and/or non-equilibrium behavior
- Device modeling, design, fabrication and characterization
- Novel circuit & architecture work to take advantage of unique device properties (e.g. non-volatility / built-in memory) for low-energy or energy-recovery during operation

A selection of the devices being researched is shown in Fig. 2, and more details on specific devices is available at the NRI website or on the individual center websites (see http://nri.src.org). Note that even though NRI is currently focusing on these device areas, there is no expectation that any of them are necessarily the "next switch". NRI continues to balance focusing on the current most-promising options at the centers, with continuing to explore other avenues through individual grants and projects. At a high level, the devices being pursued at the centers now fall into three categories:

Spintronics and Nanomagnetics: These devices all utilize spin state – either on individual carriers, or in collective states, such as spin waves or nanomagnetic particles – to represent digital data. The potential advantage is that it may be possible to move and manipulate spin at very small energies compared to moving charge, but there are continued challenges in finding low-energy clocking mechanisms and launching and maintaining a strong spin signal. The movement of spin is also quite slow compared to moving charge, so new architectures must be considered to achieve high throughput. An example is the use of multiple frequency spin waves that could run simultaneously on the same wires and gates, similar to multiple light wave frequencies traveling down a single fiber, to enable parallel processing.

Graphene: Rather than using graphene as a substrate for improved FETs, the NRI centers are trying to exploit other novel physics in the material to create devices. One device, the Bilayer pSeudospintronic FET (BiSFET) relies on the formation of a room temperature exciton condensate between graphene bilayers. The unique I-V characteristic of this device requires a different family of circuits to support the basic logic functions, and extensive simulation work indicates that it has the potential to run at speeds comparable to CMOS while consuming orders of magnitude less power. Another approach is looking at using pn junctions in graphene to steer carrier flow for an architecture based on both programmable interconnects and logic functions, taking advantage of graphene's bandstructure which allows carriers to move like massless particles, similar to photons, subject to lensing and reflection by gates.

Tunneling FETs: These devices seek to take advantage of tunneling between the source and channel of an FET, to achieve a faster turn-on (< 60 mV/dec) of the device and hence operate at lower voltage and power. These devices have the advantage of being almost a "drop-in" replacement for FETs, with little architectural change needed, but have suffered from a low on current due to the tunnel barrier resistance. The NRI group is looking at using extreme heterojunctions (staggered or broken gap) as well as reduced dimensionality (in nanowires and in graphene nanoribbons) to improve the on-off ratio substantially.

The goal of these efforts is to produce sufficient data on each device to see if it could pose a viable option for extending scaling beyond CMOS. While it is unlikely that any could entirely replace CMOS, several due seem to offer advantages, such as ultra-low power or non-volatility, which could be utilized to augment CMOS or to enable better performance in specific application spaces. The latter is an area of particular interest, given the move to multi-core chips. While most are homogeneous today, if scaling slows in delivering the historically expected performance improvements in future generations, heterogeneous multi-core chips may be a more attractive option, with specific, custom-designed cores dedicated to accelerate high-value functions. These cores could utilize different devices and architectures, with particular affinity to the function being implemented. While integrating dissimilar technologies and materials is a big challenge, advances in packaging and 3D integration may make this more feasible over time, but the performance improvement would need to be large to balance this effort.

Of course, this is the key test for any new technology: does it present enough of an advantage over the existing technology to justify the expense and complexity of implementing it. While it is too soon to answer that question for any post-CMOS technology – and CMOS is a high bar – NRI's goal is to expand the search for alternative devices, to insure many options are available for consideration.

Acknowledgements: The NRI is funded by a consortium of Semiconductor Industry Association (SIA) companies, as well as NIST, NSF, and several state governments, all under the management of the Semiconductor Research Corporation (SRC), (http://nri.src.org).

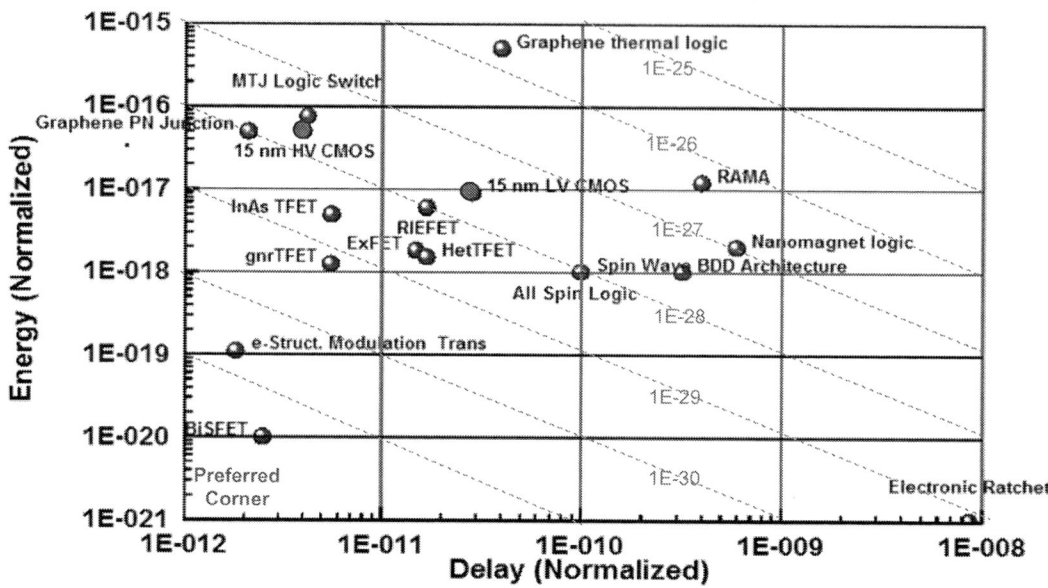

Figure 1: Sample data from the NRI benchmarking work [1], showing energy versus delay for various post-CMOS technologies. Values are compared to a projection for both high-performance and low-power 15nm CMOS as a convenient reference. Note all values are a snapshot in time, and continue to evolve with future research.

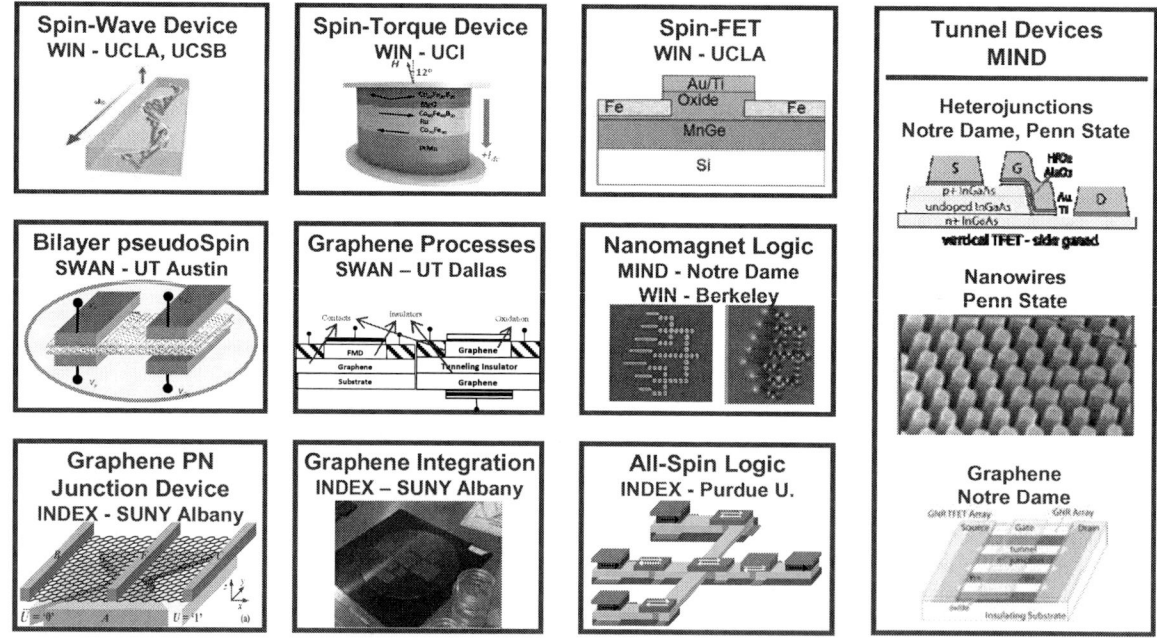

Figure 2: A selection of devices currently being researched in the four NRI centers: Western Institute for Nanoelectronics (WIN); Institute for Nanoelectronics Discovery and Exploration (INDEX); Midwest Institute for Nanoelectronics Discovery (MIND); Southwest Academy for Nanoelectronics (SWAN). (http://nri.src.org)

[1] "Device and Architecture Outlook for Beyond CMOS Switches," K. Bernstein, et al., Proceedings of the IEEE Special Issue - Nanoelectronics Research: Beyond CMOS Information Processing, Vol. 98, No. 12, December 2010.
[2] *Logical Effort: Designing Fast CMOS Circuits*, Morgan Kaufmann, ISBN-13: 978-1558605572, Feb 16, 1999.

CORRELATED OXIDE PHASE TRANSITION SWITCH: A PARADIGM IN ELECTRON DEVICES

Zheng Yang, Changhyun Ko, Viswanath Balakrishnan, and Shriram Ramanathan

Harvard School of Engineering and Applied Sciences, Harvard University, Cambridge, MA 02138, USA

(Emails: zyang@seas.harvard.edu & shriram@seas.harvard.edu)

Alternate information processing state variables and computing architectures are an active area of research given the rapidly approaching physical limitations of conventional CMOS scaling. In this context, it may be worth exploring the idea of ultra fast phase transitions in correlated oxides as a basic switching element that can be used in CMOS-like or neural circuits or other high-frequency circuit components in photonics. The basic paradigm is to trigger an insulator-to-metal phase transition in a correlated oxide with an external perturbation in form of thermal, electrical, optical, or magnetic excitations, while switching off upon the removal of the applied external excitation. Vanadium dioxide (VO_2) is a correlated oxide that undergoes a sharp insulator-to-metal transition (IMT) near room temperature (~340K for single crystals) with several orders of magnitude of resistance change. In this presentation, we discuss results on electrical triggering of the phase transition in VO_2 devices fabricated on semiconductor platforms such as Si and Ge along with studies on elementary semiconductor physics associated with such materials as a device component.

Figure 1(a) shows the schematic of a gated capacitor device, composed of a stack structure of HfO_2/VO_2/HfO_2/n-Si. The HfO_2 and VO_2 layers were grown using atomic layer deposition and magnetron sputtering, respectively. Figure 1(b) shows the temperature dependence of the normalized resistance $[R(T)/R(20\ °C)]$ of the VO_2 thin film sandwiched in the stack capacitor device. The measurement was done prior to top HfO_2 deposition. Nearly three orders of magnitude of resistance change are observed from 20 to 100 °C, indicating good quality of the VO_2 film. The insets in Fig. 1(b) show the phase transition temperatures of VO_2 at 63°C and 57 °C during heating up and cooling down, respectively, determined by standard Gaussian fitting on the derivative logarithmic plot. Figure 2 and its inset show the cross-sectional transmission electron microscope (TEM) image of the device and high-resolution (HR) TEM image near the VO_2-HfO_2 interface. Figure 3(a) shows the C-V characteristics at 1 MHz measured during temperature ramping up from 20 °C to 100 °C, while Fig. 3(b) show the same plots but only for high temperature curves in a linear scale, to zoom in on the high temperature range. The accumulation capacitance (that represents the dielectric stack contribution) of the device shows about one order of magnitude enhancement when temperature is increased from 20 to 100 °C (~251 pF at 20 °C vs. ~2628 pF at 100 °C). The accumulation capacitance of a metal/HfO_2/n-Si reference MOS capacitor device is nearly temperature independent within the range of 20 to 100 °C, indicating that the strong temperature dependent capacitance of the metal/HfO_2/VO_2/HfO_2/n-Si/metal capacitor device arises from the sandwiched VO_2 layer. This is to be expected due to the significant change of the dielectric properties of VO_2 with temperature. Figure 4 shows the temperature dependence of the relative dielectric constant of VO_2 based on the C-V data shown in Fig. 3 using equivalent circuit impedance transformation. The VO_2 permittivity increases significantly with increasing temperature. Figure 5 shows a VO_2 phase transition electronic switch fabricated on Ge. Figure 5(b) and (c) show the schematic of the Au/VO_2/Ge MIT switch and the cross-sectional TEM image near the VO_2/Ge interface area. Figure 5(a) shows the I-V characteristics of device, measured between Au and n^+-Ge contacts at three different temperatures. When the environment temperature is below the phase transition temperature of VO_2, clear switching behavior of the device is observed, which is due to the electrically-triggered phase transition in the sandwiched VO_2 layer, i.e. the VO_2 layer changes from high-resistance insulating state to low-resistance metallic state. On-going studies on three-terminal device fabrication on silicon will further be discussed in the presentation.

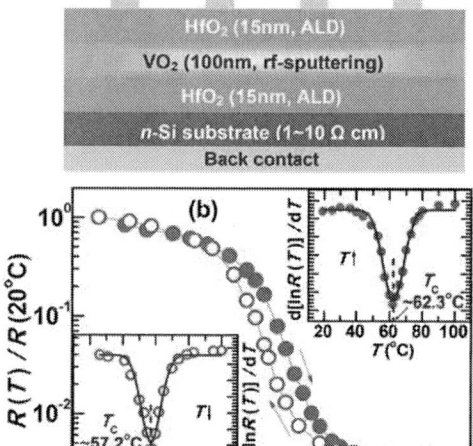

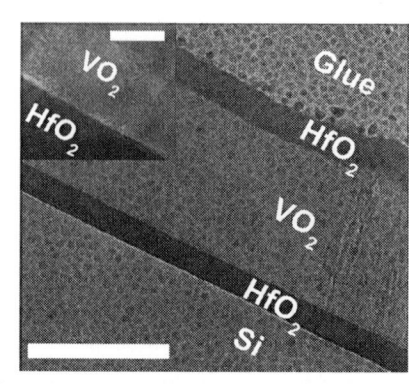

FIG. 2 Cross-sectional TEM image of the device and HRTEM image (inset) of the VO₂-HfO₂ interface. The scale bar of the TEM and HRTEM are 100 and 10 nm.

FIG. 1 (a) Schematic of a thermal capacitor device with a structure based on HfO₂/VO₂/HfO₂/Si. (b) Temperature dependence of the resistance (normalized) of the VO₂ thin film between two HfO₂ dielectric layers (measured before putting the top HfO₂ layer). The inset shows the determined phase transition temperatures.

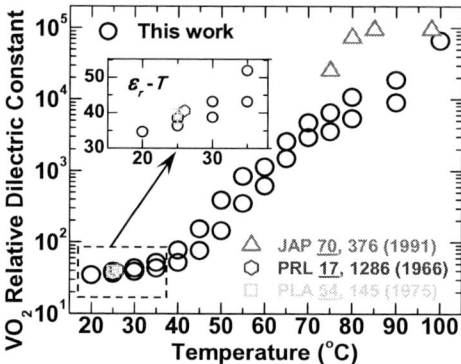

FIG. 4 Estimated relative dielectric constant of VO₂ at different temperatures based on equivalent circuit impedance transformation.

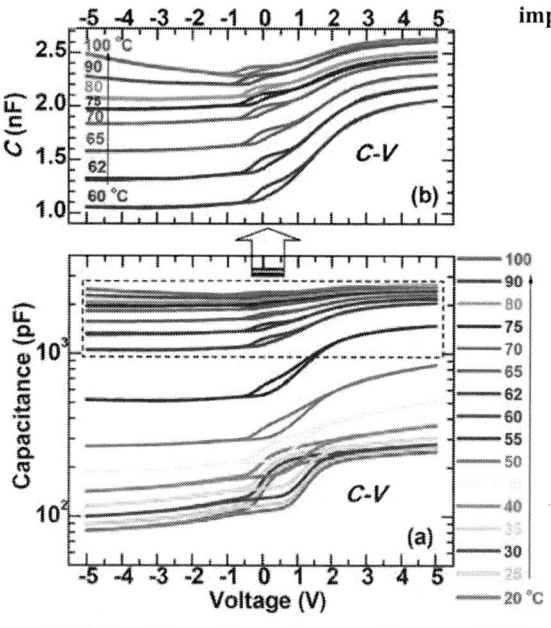

FIG. 3 (a) Capacitance-voltage (C-V) characteristics of the HfO₂/VO₂/HfO₂/Si stack device during temperature ramping up from 20 to 100 °C measured at 1MHz. (b) The same C-V plot as (a) but only for high temperature curves in linear scale, to magnify the high temperature range.

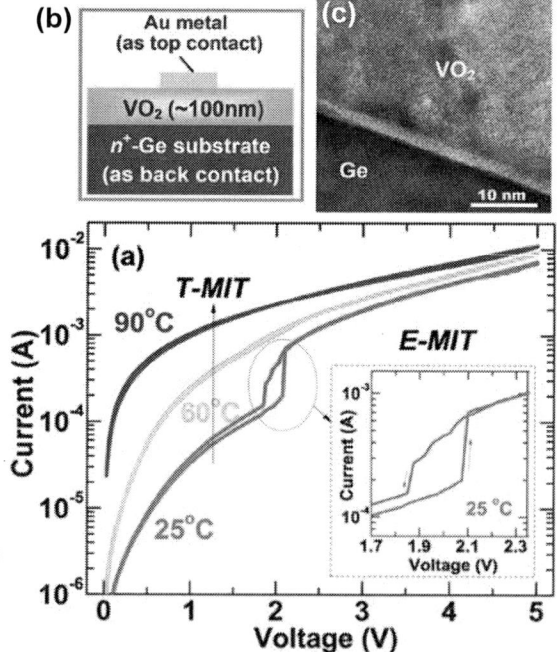

FIG. 5 (a) I-V characteristics of an Au/VO₂/Ge IMT switch. (b) The schematic of the Au/VO₂/Ge IMT switch. (c) TEM image of the VO₂/Ge interface.

978-1-61284-243-1/11 $26.00 © 2011 IEEE   188

Lateral Gate Suspended-Body Carbon Nanotube Field-Effect-Transistors with Sub-100nm Air Gap by Precise Positioning Method

Ji Cao, Adrian M. Ionescu

Nanoelectronic Devices Laboratory (Nanolab), Ecole Polytechnique Fédérale de Lausanne,
CH-1015 Lausanne, Switzerland; Tel: +41 21 693 7856; Email: ji.cao@epfl.ch.

Carbon nanotubes (CNTs) have been intensively studied for nanoelectromechanical systems (NEMS) applications owing to their remarkable electrical and mechanical properties. Efforts have been made in single-walled CNT field-effect transistor (SWCNTFET) based ultrasensitive mass detection, radio-frequency (RF) signal processing, etc [1]. However, current techniques of manipulating CNTs (including: in-situ CNT growth and post-synthesis fabrication) often precludes bottom-up integration with pre-existing complementary metal-oxide-semiconductor (CMOS) circuits [2], due to: high process temperature, lack of self-alignment accuracy, etc.

Here, we report, for the first time, a self-aligned lateral gate suspended-body CNTFET with sub-100nm air gap fabricated by an improved precise positioning method based on our previous work [3]. The superior I-V characteristics of the lateral gate CNT FET are experimentally studied. The proposed suspended-body CNTFETs hold promise for bottom-up fabrication of resonant NEMS devices for sensing and RF applications.

The fabrication process is depicted in Fig. 1. First, Ti/Pd electrodes were patterned on a SiO_2 (500 nm)/Si substrate (Fig. 1a), which is coated with 100 nm LOR/50 nm PMMA (Fig. 1b). Ultra-narrow trenches (50 nm) were transferred to the PMMA layer by e-beam lithography (EBL) (Fig. 1c). SWCNTs in CNT suspension were attracted into the trenches by dielectrophoresis applied between opposite electrodes (Fig. 1d and 1e). Compared to the previous method, only source/drain contacts were defined on a second PMMA layer by EBL and deposited by metal evaporation (Fig. 1f). Moreover, PMMA layers were stripped to remove the misaligned CNTs and impurities, keeping the LOR layer (Fig. 1g). Then, the lateral gate and source/drain contacts were defined by EBL on a new PMMA layer on LOR and deposited by metal evaporation (Fig. 1h and 1i). Finally, CNTFETs were released by stripping the resists and being dried in CPD (Fig. 1j).

The SEM image of a representative lateral gate suspended-body CNTFET fabricated by the self-assembly method is presented in Fig. 2a. No misaligned CNTs among electrodes and suspended body are observed. The 850 nm wide gate was self-aligned 90 nm (g) away from the suspended CNT body (2 nm thin, 1.5 μm long). The suspension height is determined by LOR thickness. As shown in the CNTFET layout (Fig. 2c), by the proposed technique, the lateral gate could be brought arbitrarily close to the CNT body within the EBL resolution, gaining strong gate-channel coupling. However, by using the previous method, misaligned CNTs often got clamped by the lateral gate and bridged the channel, as shown in Fig. 2b, resulting in the gate failure.

I-V characterizations were carried out in vacuum at room temperature. Fig. 3 displays the output characteristics of suspended body CNTFET with various lateral gate biases. Strong dependence on the lateral gate is observed.

Fig. 4 shows the transfer characteristics (I_{ds}-V_{gs}) of the same CNTFET at various V_{ds}. The device operates as a laterally gated p-FET with ultra high $I_{on}/I_{off} \sim 10^7$, ultra small off-current ($\sim10^{-14}$ A) and a small inverse subthreshold slope SS=$(d\log_{10}I_{ds}/dV_{gs})^{-1}$= 132 mV/dec. The excellent transfer characteristics suggest that the suspended CNT channel is effectively controlled by the lateral gate over the sub-100nm air gap and the tube quality was not degraded by the processing steps. As far as we know, these characteristics of our suspended body CNT FETs are comparable with or even better than those previously reported [4].

Fig. 6 shows the transconductance g_m versus V_{gs}. g_m reaches 0.16 μS at V_{ds}= 0.3V. Quasi-periodic peaks of g_m versus V_{gs} were observed for different small values of V_{ds}, which is possibly due to the Coulomb blockade at room temperature in the lateral gate suspended-body structure, as previously reported by Y. Ohno et al. [5]. However this effect seems dominated by the FET effect, as clearly demonstrated by the I_{ds}-V_{gs} characteristics.

Assuming that CNT FETs act as diffusive FETs with effective channel length L_g (Fig. 2b), we estimate the mobility of the holes in the linear region of $I_{ds}/g_m^{0.5}$-V_{gs} by Ghibaudo's method [6]: $d(I_{ds}/g_m^{0.5})/dV_{gs}$ = $(C_g\mu_hV_{ds})^{0.5}/L_g$, as in Fig. 7. In one dimension FET, the gate capacitance Cg is given by $Cg \approx 2\pi\varepsilon\varepsilon_0L_g/\ln(2g/r)$ and the quantum capacitance C_q of the CNT can be ignored. We obtain a hole mobility of $\mu_h \sim 920$ cm^2/Vs. However, this value is significantly underestimated, because the effective V_{ds} in the gating region is much smaller due to the voltage loss on the resistance of CNT segments out of the gating region and on the source/drain contacts. Moreover, the proposed suspended-body FET shows a slight gating hysteresis even in air, as shown in Fig. 8, which reflects a limited charge trapping.

In conclusion, the excellent I-V characteristics of self-aligned lateral gate suspended–body CNTFETs with sub-100nm gate/SWCNT body distance have been demonstrated by the improved positioning method. These unique CNT devices enable the bottom-up fabrication of resonant NEMS devices for sensing and RF applications.

Acknowledgement: This work was supported by the Swiss Nanotera project CABTURES (SNF Number: 20NAN0_123614), and by the FP6 IST-028158 project NANORF (Hybrid Carbon Nanotube – CMOS RF Microsystems).

[1] V. Sazonova, et al., *Nature*, 431, 284, 2004.
[2] B. Mahar, et al., *IEEE Sensors Journal*, 7, 2, 266, 2007.
[3] J. Cao, et al., *16th Transducers '11 Conference Digest*, 2011.

[4] R. Seidel, et al., *Nano Lett.*, 5, 1, 147, 2005.
[5] Y. Ohno, et a, *Jpn. J. Appl. Phys.*, 42, 4116, 2003.
[6] G. Ghibaudo, *Electronics Letters*, 24, 9, 543545, 1988.

978-1-61284-243-1/11 $26.00 © 2011 IEEE

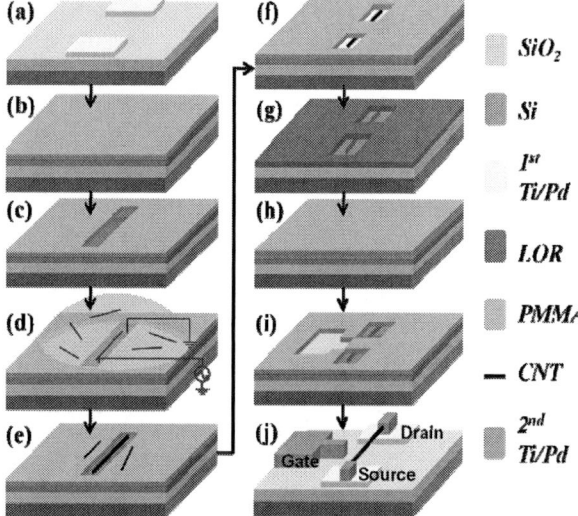

Figure 1. Bottom-up integration scheme used to fabricate the self-aligned lateral gate suspended-body CNTFETs by the improved precise positioning method.

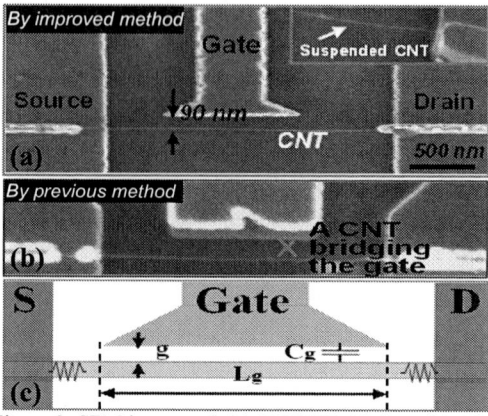

Figure 2. SEM images of: (a) CNTFET fabricated by the improved positioning method, with an accurately aligned lateral gate (850 nm long) with sub-100 nm air gap to the 2 nm thin, 1.5 μm long suspended CNT body (as shown in Inset); (b) Typical misalignment in the CNTFET fabricated by the previous method. A CNT is bridging the lateral gate and the channel. (c) Layout of the lateral gate CNTFET, marked with critical parameters.

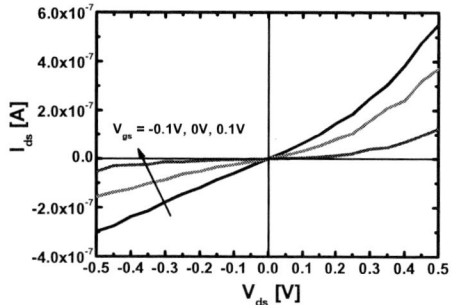

Figure 3. Output characteristics of the suspended-body CNTFET. Dependence on the lateral gate potential is shown.

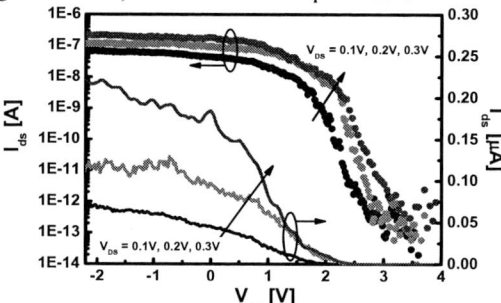

Figure 4. Transfer characteristics of the CNTFET at various V_{ds} in vacuum.

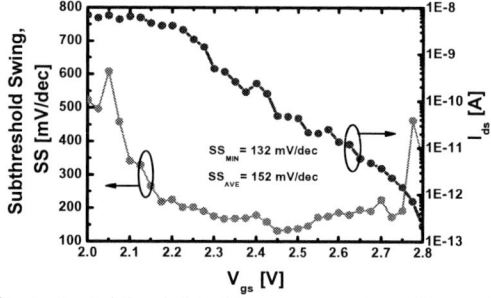

Figure 5. Subthreshold slope versus gate voltage of the CNTFETs.

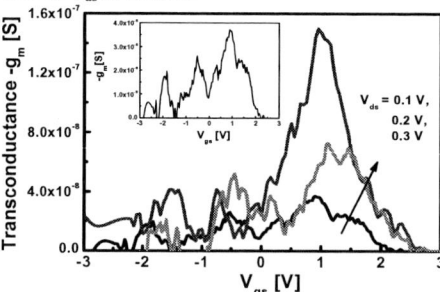

Figure 6. Transconductance versus gate voltage of the CNTFET at various V_{ds} in vaccum. Inset: Magnified g_m-V_{gs} at V_{ds} =0.1 V, showing quasi-periodic peaks.

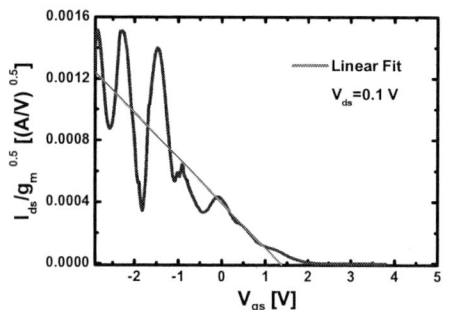

Figure 7. Illustration of Ghibaudo's method for mobility extraction for the fabricated CNTFET at V_{ds} = 0.1 V.

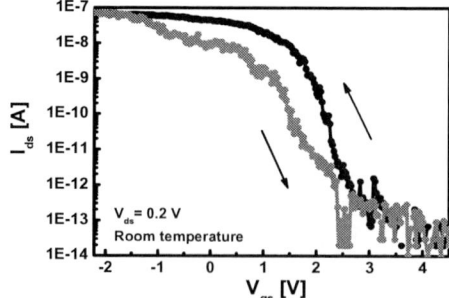

Figure 8. Transfer characteristics of the suspended-body CNTFET in forward and backward sweeps with 0.2 V source drain bias in air. Slight hysteresis is observed.

978-1-61284-243-1/11 $26.00 © 2011 IEEE

Session V.A (Corwin Pavilion East)

Tunnel FETs

Tuesday PM, June 21st, 2011

Session Chair: Heinz Schmid, IBM Zurich and William Frensley, University of Texas, Dallas

1:30 PM V.A-1 Invited Paper
Si-based Tunnel Field-Effect Transistors for Low-Power Nano-Electronics
A. S. Verhulst[1], W. G. Vandenbergh[1,2], D. Leonelli[1,2], R. Rooyackers[1], A. Vandooren[1], J. Zhuge[4], K-H. Kao, B. Sorée[1,5], W. Magnus[1,5], M. V. Fischetti[6], G. Pourtois[1], C. Huyghebaert[1], R. Huang[4], Y. Wang[4], K. De Meyer[1], W. Dehaene[1,2], M. M. Heyns[1,3], and G. Groeseneken[1,2], [1]imec, Leuven, BELGIUM, [2]Department of Electrical Engineering, [3]Department of Metallurgy and Materials Engineering, K.U.Leuven, Leuven, BELGIUM, [4]Institute of Microelectronics, Peking University, Beijing, CHINA, [5]Department of Physics, Universiteit Antwerpen, , Wilrijk, BELGIUM; [6]Department of Materials Science and Engineering, University of Texas Dallas, Richardson, Texas, USA

2:10 PM V.A-2
Compact Model and Performance Estimation for Tunneling Nanowire FET
P. M. Solomon, D. J. Frank, and S.O. Koswatta, IBM, SRDC, T.J. Watson Research. Center, Yorktown Heights, New York, USA

2:30 PM V.A-3 Student Paper
Using Dimensionality to Achieve a Sharp Tunneling FET (TFET) Turn-On
S. Agarwal and E. Yablonovitch, University of California, Berkeley, California, USA

2:50 PM V.A-4
Investigation on Superlattice Heterostructures for Steep-Slope Nanowire FETs
E. Gnani, P. Maiorano, S. Reggiani, A. Gnudi and G. Baccarani ARCES and DEIS, University of Bologna, Bologna, ITALY

3:10 PM Break

3:30 PM V.A-5 Student Paper
Self-aligned Gate NanoPillar $In_{0.53}Ga_{0.47}As$ Vertical Tunnel Transistor
D. K. Mohata, R. Bijesh, V. Saripalli, T. Mayer and S. Datta, The Pennsylvania State University, University Park, Pennsylvania, USA

3:50 PM V.A-6 Student Paper
Self-aligned $InAs/Al_{0.45}Ga_{0.55}Sb$ vertical tunnel FETs
G. Zhou[1], Y. Lu[1], R. Li[1], Q. Zhang[1], W. Hwang[1], Q. Liu[1], T. Vasen[1], H. Zhu[2], J. Kuo[2], S. Koswatta[3], T. Kosel[1], M. Wistey[1], P. Fay[1], A. Seabaugh[1], and H. G. Xing[1], [1]Department of Electrical Engineering, University of Notre Dame, Notre Dame, Indiana, USA, [2]IntelliEPI, Richardson, Texas, USA, and [3]IBM T. J. Watson Research Center, Yorktown Heights, New York, USA

4:10 PM V.A-7 Student Paper
P-type Tunneling FET on Si (110) Substrate with Anisotropic Effect
M. H. Lee[1], C.-Y. Kao[1], C.-L. Yang[1], and C.-H. Lee[2], [1]Institute of Electro-Optical Science and Technology, National Taiwan Normal University, Taipei, TAIWAN and [2]Graduate Institute of Electronics Engineering (GIEE) and Department of Electrical Engineering, National Taiwan University, Taipei, TAIWAN

4:30 PM V.A-8
Late News

4:50 PM V.A-9
Late News

978-1-61284-243-1/11 $26.00 © 2011 IEEE

978-1-61284-243-1/11 $26.00 © 2011 IEEE

Si-based Tunnel Field-Effect Transistors for Low-Power Nano-Electronics

A.S. Verhulst[a], W.G. Vandenberghe[a,b], D. Leonelli[a,b], R. Rooyackers[a], A. Vandooren[a], J. Zhuge[d], K-H. Kao, B. Sorée[a,e], W. Magnus[a,e], M.V. Fischetti[f], G. Pourtois[a], C. Huyghebaert[a], R. Huang[d], Y. Wang[d], K. De Meyer[a], W. Dehaene[a,b], M.M. Heyns[a,c], and G. Groeseneken[a,b]

[a]imec, Kapeldreef 75, 3001 Leuven, Belgium; [b]Department of Electrical Engineering, [c]Department of Metallurgy and Materials Engineering, K.U.Leuven, 3001 Leuven, Belgium; [d]Institute of Microelectronics, Peking University, Beijing 100871, China, [e]Department of Physics, Universiteit Antwerpen, Groenenborgerlaan 171, 2020, Wilrijk, Belgium; [f]Department of Materials Science and Engineering, University of Texas Dallas, Richardson, Texas 75080, USA
phone: +32 16 28 19 65, email: anne.verhulst@imec.be

Unlike MOSFETs, tunnel-FETs (TFETs) are not limited by a 60 mV/dec subthreshold swing and therefore scaling the supply voltage beyond the MOSFET's 1 V plateau becomes feasible. Supply voltage scaling is a necessary condition for reducing the power consumption per transistor, which enables further size scaling of the FETs. Designing a successful FET is however challenging, because it is not sufficient for the TFET to realize a sub-60 mV/dec subthreshold swing at one particular voltage. The next-generation FET must realize an average sub-60 mV/dec subthreshold swing over the whole supply voltage window, such that on-currents of about I_{on} = 100 µA/µm are achieved with I_{on}/I_{off} ratios of about 10^6 for sub 0.5 V supply voltages. Silicon-based TFETs are the most attractive because they allow for a full re-use of the existing expertise in fabricating silicon MOSFETs. However, the large bandgap of silicon (Si) results in low on-currents for the all-Si TFET and both input and output characteristics are inferior to the ones of all-Si MOSFETs.

To improve the TFET's on-current while maintaining a Si channel, the incorporation of heterostructures has been proposed [1-3]. In particular, a germanium(Ge)-source Si-channel TFET (Fig. 1), which acts as an n-TFET, maintains the low off-current of the all-Si TFET, while increasing the on-current with over an order of magnitude at 1 V supply voltage. Having a complementary p-TFET is a key requirement to perform logic with manageable static power consumption. Hence an indium(gallium)arsenide-source Si-channel TFET has been proposed as complementary p-TFET (Fig. 2) [4]. The on-current of the proposed heterostructure TFETs, in particular the on-current of the Ge-source Si-TFET at 1 V supply voltage is still 1 order of magnitude smaller than the on-current of the all-Si MOSFET. Additional current boosters are required.

Design improvements are considered first. The impact of the gate configuration is investigated. A short-gate configuration does not affect the on-current but improves the overall circuit performance by a reduced Miller capacitance [5]. Improvements of up to a factor of 2 in either power consumption for a given performance or in performance for a given power consumption are predicted (Fig. 3). This is beneficial but not sufficient to bridge the gap with the MOSFET. A tighter gate control, by changing the single-gate design to a double-gate or gate-all-around design, further improves the characteristics (Fig. 4) [6]. The effective improvement in on-current is expected to be close to a factor of 10, with more pronounced improvements for one-dimensional nanowires, like extremely small diameter (< ~5 nm) semiconducting nanowires or carbon nanotubes (CNTs). The former choice will be subjected to size-induced quantum confinement, with processing variability imposing a serious limitation to the circuit performance. The change from a Si-based FET to CNT-based FETs is still premature today due to the difficulty to reproducibly fabricate CNTs of a given diameter and chirality. Except for the gate control, also the impact of the gate oxide thickness is investigated. Scaling the gate oxide results in larger improvements of the TFET input characteristics than scaling the body thickness. More importantly, scaling the gate oxide increases the TFET on-current superlinearly while it increases the MOSFET on-current only linearly [5-7]. Advances in gate-dielectric design, especially the fabrication of gate oxide with higher k value, will therefore significantly impact the future of TFETs. Performance improvements are also expected from the shape of the source region (Fig. 5). Decreasing the angle of the source-channel junction with the gate oxide to angles smaller than 90 degrees improves the overall subthreshold swing with up to 20% [8]. Overall, design improvements are likely to result in similar performance at 1 V supply voltage for the all-Si MOSFET and the Ge-source Si-channel TFET, provided the gate configuration, body thickness, oxide thickness and source are properly optimized.

978-1-61284-243-1/11 $26.00 © 2011 IEEE

To beat the MOSFET performance and to allow lower supply voltages, additional improvements are required. We have theoretically analyzed the impact of heterostructure strain in a nanowire-based Ge-source Si-channel TFET. Improvements in on-current of about 2 orders of magnitude are predicted, together with improvements in subthreshold swing, allowing successful TFET operation at an ultra-low supply voltage < 0.5 V (see Fig. 6). The impact of body thickness on TFET performance in the strained nanowire-based heterostructure TFET is much stronger than would be expected from improved gate control only. This is due to the fact that the body thickness not only changes the electrostatic profile, but also the strain profile in the device.

Not only the input characteristics are important, also the output characteristics are crucial for proper operation of logic circuits. In the TFET output characteristics, an unwanted superlinear onset is often observed, yet poorly understood. We discuss an analytical model which explains this unwanted onset and offers design improvements to decrease it. At the same time, the added insight in the output characteristics offers a powerful new tool to interpret experimental data. The latter will help to improve the link between theory and experiment.

Theory and experiment are not in perfect agreement for TFETs, yet agreement is crucial to come to credible predictions. On the one hand, a lot of the modeling and simulation efforts are incomplete because semi-classical approximations are made to allow for predictions which can be calculated today. On the other hand, the processing optimizations, like for example source/drain doping schemes, are all based on MOSFET optimizations, which are typically not resulting in optimal TFET performance. To tackle the first issue, we are setting up a quantum mechanical treatment of the TFET. Efforts have been focused on the impact of a non-uniform electric field [9], and a proper treatment of phonon-assisted transitions in indirect bandgap materials. An important effect we have identified is field-induced quantum confinement (see Fig. 7) [10]. An analysis is made for the line-tunneling TFET, where the gate stack is positioned on top of the source region. The quantum-mechanical treatment reveals that the strong band bending near the gate dielectric, required for creating short tunnel paths, results in quantization of the energy band (conduction band for n-TFET, valence band for p-TFET). Compared to semi-classical models, a shift of the onset of 750 mV and a 20% increase in subthreshold swing are predicted for an n-channel Si TFET with a source doping of 10^{20} at/cm^3 and effective oxide thickness of 0.5 nm. The degraded subthreshold swing is mainly due to the distinct onset of tunneling to the two electron valleys. This effect, which has so far not been taken into account in theoretical predictions, has a significant impact on all TFET configurations, with the strongest impact for the configurations with high source doping and small effective masses.

In an effort to further improve the link between theory and experiment, the performance of all-Si FinFET-based TFET is studied in detail and compared with predictions (see Fig. 8) [11-12]. We discuss the observed dependence of the on-current on fin width and oxide thickness. We report our largest achieved on-currents for all-Si TFETs and compare with literature.

In conclusion, for the Si-based TFET to beat the MOSFET performance and allow ultra-low voltage operation with re-use of a lot of the existing processing expertise, critical device optimization is needed whereby a combination of several performance boosters must be implemented. Heterostructures and an appropriate stress profile are necessary requirements. The largest design impact is expected from scaling the effective oxide thickness and the body thickness. Field-induced quantum confinement affects most theoretical predictions today and needs to be addressed in the design optimization. Overall, there are still significant challenges both in modeling, processing and characterization of the device. Progress in all three areas is required to uncover the full potential of the TFET.

W. Vandenberghe acknowledges the support of a Ph.D. stipend from the Institute for the Promotion of Innovation through Science and Technology in Flanders (IWT-Vlaanderen). This work was also supported by imec's Industrial Affiliation Program.

[1] K. Bhuwalka et al., Trans. Electron Dev. 52, 1541 (2005), [2] E-H. Toh et al., Appl. Phys. Lett. 91, 243505 (2007), [3] A.S. Verhulst et al., J. Appl. Phys. 104, 064514 (2008), [4] A.S. Verhulst et al., Electron Dev. Lett. 29, 1398 (2008), [5] J. Zhuge et al., revisions submitted to Semicond. Science and Technol., [6] A.S. Verhulst et al., J. Appl. Phys. 107, 024518 (2010), [7] J. Knoch et al., 63rd DRC IEEE Digest, Vol. 1, p. 153 (2005), [8] K-H. Kao et al., Proc. of ULIS (2011), [9] W.G. Vandenberghe et al., J. Appl. Phys. 107, 054520 (2010). [10] W.G. Vandenberghe et al., accepted to Appl. Phys. Lett., [11] D. Leonelli et al., Jap. J. Appl. Phys. 49, 04DC10 (2010). [12] D. Leonelli et al., Proc. of ESSDERC, p. 107 (2010).

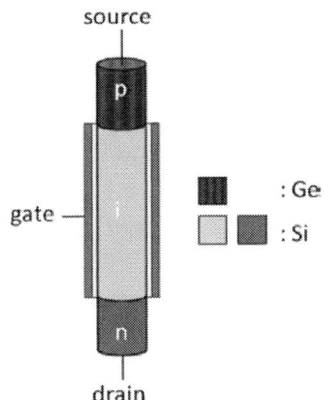

Fig. 1. Schematic of nanowire-based Ge-source Si-channel TFET.

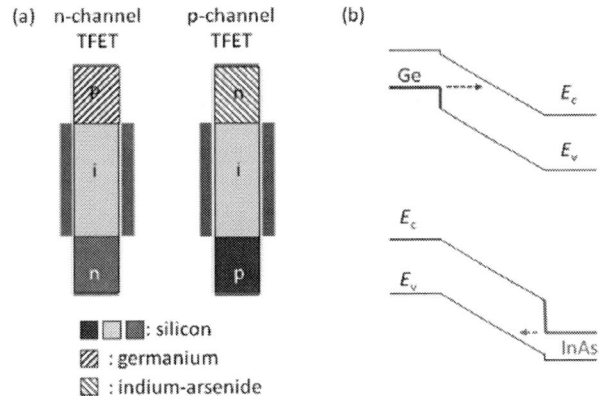

Fig. 2. (a) Schematic of complementary heterostructure Si-channel TFETs. (b) Schematic of the positions of the band edges (left-to- right corresponding to a top-to-bottom cut-line from source to drain of the configurations in (a)). The p-doped section is at the left and the n-doped at the right for both drawings.

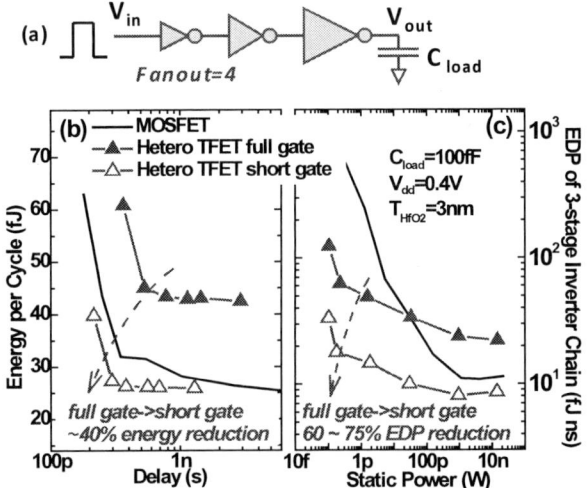

Fig. 3. (a) Schematic of 3-stage inverter chain with a fanout of 4 and a load capacitance. (b) Energy per cycle of 3-stage inverter chain as a function of delay at V_{dd} of 0.4V. (c) Energy delay product (EDP) of 3-stage inverter chain as a function of static power consumption at V_{dd} of 0.4V.

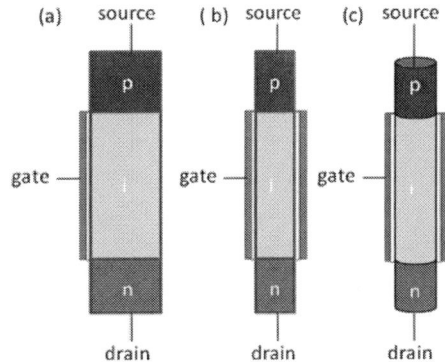

Fig. 4. Schematic of (a) single-gate, (b) double-gate and (c) gate-all-around TFET.

978-1-61284-243-1/11 $26.00 © 2011 IEEE   195

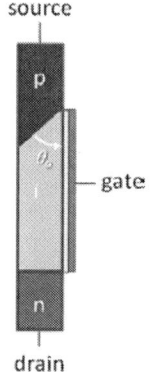

Fig. 5. Schematic of TFET with a non-orthogonal junction angle θ_s.

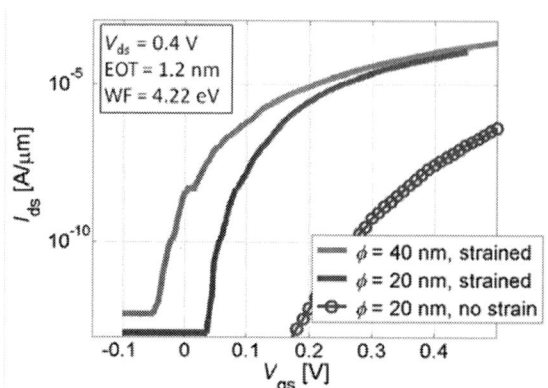

Fig. 6. Input characteristics corresponding to the TFET of Fig. 1, whereby the impact of strain on the electronic band structure is included under the assumption of a dislocation-free heterostructure interface. A self-consistent calculation is performed with a non-local direct-bandgap band-to-band tunneling model (SentaurusDevice, v. 2009.06, Synopsys).

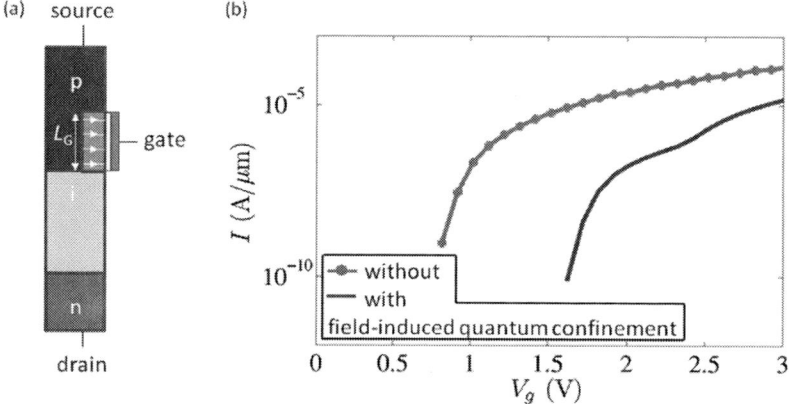

Fig. 7. (a) Schematic of the line-tunneling TFET used to study the impact of field-induced quantum confinement. (b) Input characteristics for the configuration of (a) with L_G = 30 nm, source doping of 10^{20} at/cm³, effective oxide thickness of 0.5 nm and work function of the gate metal of 4 eV.

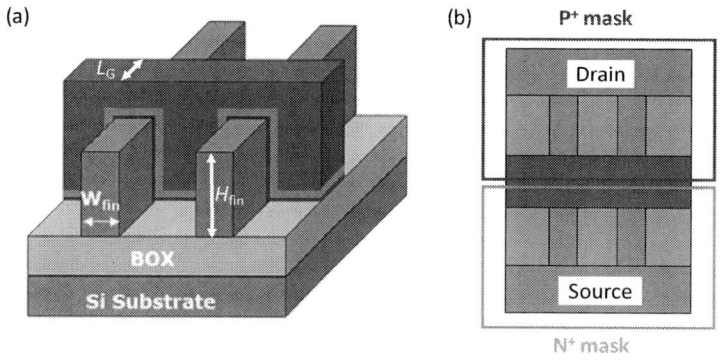

Fig. 8. Schematic of the all-Si FinFET-based TFETs, fabricated on 300 mm wafers, of which the experimental performance is analyzed: (a) three-dimensional view, (b) top view with applied implantation masks. The fin width is W_{fin} = 10 to 250 nm, the fin height is H_{fin} = 65 nm and the gate length is L_G = 150 nm.

978-1-61284-243-1/11 $26.00 © 2011 IEEE   196

Compact Model and Performance Estimation for Tunneling Nanowire FET

P. M. Solomon, D. J. Frank, and S.O. Koswatta

IBM, SRDC, T.J. Watson Research. Center, Yorktown Heights, NY 10598, e-mail: solomonp@us.ibm.com

Tel: (914) 945-2841, Fax: (914) 945-2141, Email: solomonp@us.ibm.com

Introduction: In his work we present a compact model for a heterojunction tunnel FET (TFET) which captures the essential features peculiar to this device [1-3] in a way that faithfully preserves their dependency on the physical device parameters. In addition to the principal tunneling mechanism, effects of source degeneracy, back-injection from the drain, and direct source-drain tunneling are captured. This model therefore fulfils a need for design and optimization of TFET devices and circuits and enables the use of sophisticated performance estimation tools for this technology. We show smooth and continuous I-V curves which are compared with device level simulations and demonstrate the robustness of the model by running it in a SPICE-like transient simulator and in the demanding environment of a system-level optimizer which tests the model over an extensive sample of parameter space. The optimization was applied to the realistic case of (In,Al)As/(Ga,Al)Sb.

Model Assumptions and Simplifications: Our approach is based on the idealized geometry and band diagram of Fig. 1, where the source potential is assumed constant and the gate induces an exponential shaped tunneling barrier of height φ_b extending over a range V_{cl}, where V_{cl} is proportional to the gate voltage. The exponential decay length, $\lambda_s = (r + t_i)/2$ [4], assuming the same permittivity in the channel and the gate insulator. Analytical scaleable functions of energy and channel length $G(u, L_r)$ were derived for the WKB integrals. These were integrated numerically to obtain the current as

$$I = n_v g_0 \int_{u_v}^{\infty} \exp\{-\tfrac{kT}{e}\ln F(u, u_s, u_d) - \tfrac{2\lambda_s}{h}\sqrt{2emV_{cl}}\,G(u, L_r)\}\,du$$

where F is the difference of source and drain Fermi functions.

The channel potential is calculated self-consistently taking the back-injected drain charge into account The transport charge is then added to the total. but This reduces the channel potential, which is simulated by adding a resistor in series with the source. Capacitances are derived through numerical differentiation. Table I gives the input parameter description.

Results and Discussion: All characteristics expected from a TFET are exhibited by the model. Log I_d vs. V_g characteristics (see Fig. 2a) for a 20nm channel length show the steep subthreshold slope and a degradation due to direct S/D tunneling. The I_d vs., V_d characteristics (Fig. 2b) show smooth saturating behavior at positive V_d and the expected negative resistance at negative V_d for a degenerate source. This disappears for a nondegenerate source (Fig. 2c) replaced by strongly non-linear turn-on. CV curves (Fig. 3a) show the strong influence of drain back-injection with only a minor component due to the transport charge [3].

Our model obtains characteristics qualitatively similar to more rigorous NEGF simulations of carbon nanotubes. To obtain more quantitative comparison we inputted our potential profiles into the CNT Hamiltonian, with results shown in Fig.

4. The main adjustments to bring the two into agreement was that the effective mass used is about ½ the band edge value for the CNT, partly due to the parabolic dispersion we use compared to the full 2-band dispersion for the NEGF simulations, and the decay length is reduced by ~25% to account for the higher Fourier terms c.f.[4] in the potential near the source.

AC switching waveforms for an inverter chain are shown in Fig. 3b. A potential problem for the TFET is shown where the large feedback capacitance and high dynamic resistance at zero V_d result in poor input-output isolation and large voltage overshoots [3]. Table II compares the stage delay with a simple effective current/effective charge calculation. Agreement is good except at the lowest V_{dd} where the overshoots cause an increased delay.

The TFET model was run in an optimizer program [5], which adjusts device design parameters to achieve optimal chip-level performance under constraints such as power or power density. Fig 5(a) shows the minimized logic delay at different power levels when simultaneously optimizing over V_{dd}, L, V_t, V_e, widths, V_e and φ. Local and global device-level variations are included in L, R, t_{ox}, V_b, V_e and φ, in addition to power supply and coupling noise [5]. The length tolerances are consistent with 11nm technology. Arrays of nanowires were considered with appropriate parasitic capacitances added. This figure compares TFET technology (with 4nm diameter nanowires and m^*=0.02) with FinFETs. The results (Fig. 5a and b) show that the principal advantage of the TFET is in achieving >10x smaller switching energies at moderate performance levels. We also include in Fig. 5a more realistic InAs/GaSb case, where the variation and variability of φ and m with wire radius [6], due to quantum confinement, is taken into account (Fig. 6). This results in a fairly large optimum radius, an increased λ_s, and hence some performance loss.

Conclusions: A compact model is presented which realistically reproduces TFET characteristics and allows complex circuit simulation and parameter optimization studies. The model has been applied to circuit simulations which reveal anomalous switching behavior, and to a multi-parameter optimization study which quantifies the power-performance advantage of the TFET over conventional MOSFETs.

References: [1] J. Knoch and J. Appenzeller, *Phys. Stat. Sol. (a)*, v. 205, p. 679, 2008. [2] S. O. Koswatta, M. S. Lundstrom, and D. E. Nikonov, *IEEE TED*, v. 56, p. 456, 2009. [3] S. Mookerjea, R. Krishnan, S. Datta, and V. Narayanan, *IEEE TED*, v. 56, p. 2092, 2009. [4] B. Yu, , L. Wang, Y. Yuan, P. M. Asbeck, and Y. Taur, IEEE Trans. Electron. Dev., vol. 55, p. 2846, 2008. S.O. Koswatta and M.S. Lundstrom and D.E. Nikonov, Appl Phys Lett., vol. 92, 043125, 2008. [5] D. J. Frank, et al., IBM J. Res. Dev., 50, pp.419-431, 2006. [6] S.E. Laux, unpublished.

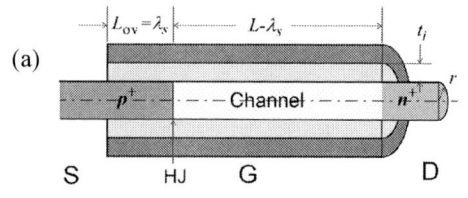

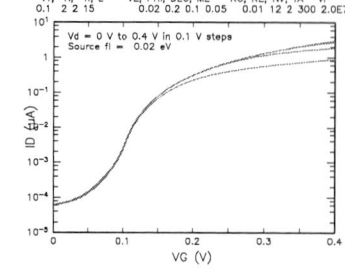

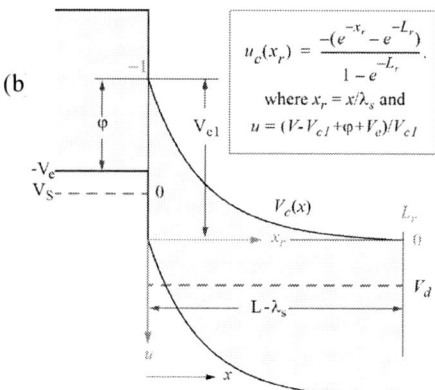

$$u_c(x_r) = \frac{-(e^{-x_r} - e^{-L_r})}{1 - e^{-L_r}}.$$

where $x_r = x/\lambda_s$ and
$u = (V - V_{c1} + \varphi + V_e)/V_{c1}$

Fig. 1: (a) Cross-section of TFET (b) Definition of potentials (V) and scaled potentials (u) used in the TFET model.

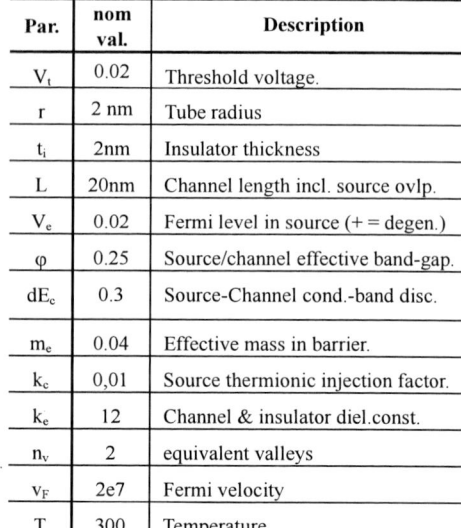

Fig.2: (a) Id vs Vg ad (b) Id vs Vd with (a) and (b) for positive and (c) for negative source Fermi level. Model parameters are shown above each figure.

Table 1. Model Parameters

Par.	nom val.	Description
V_t	0.02	Threshold voltage.
r	2 nm	Tube radius
t_i	2nm	Insulator thickness
L	20nm	Channel length incl. source ovlp.
V_e	0.02	Fermi level in source (+ = degen.)
φ	0.25	Source/channel effective band-gap.
dE_c	0.3	Source-Channel cond.-band disc.
m_e	0.04	Effective mass in barrier.
k_c	0,01	Source thermionic injection factor.
k_e	12	Channel & insulator diel.const.
n_v	2	equivalent valleys
v_F	2e7	Fermi velocity
T	300	Temperature

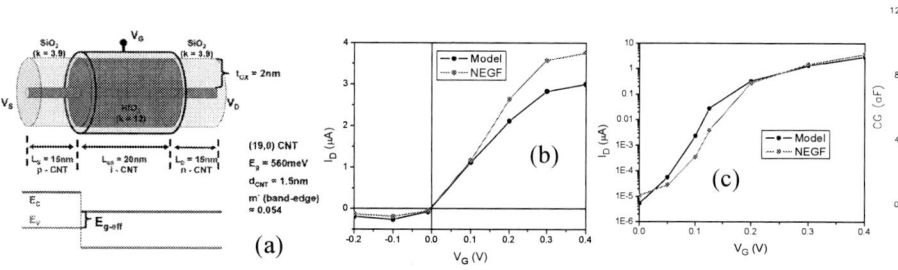

Fig.4: (a): Diagram of CNT TFET used in NEGF simulations. (b) and (c): Comparison of compact model to NEGF simulations of CNT in (a). The model parameters were c.f. Table I: 0.1, 0.75, 2, 17.75, 0.01, 0.35, 0.4, .04, 0,12, 2,1E8, 300.

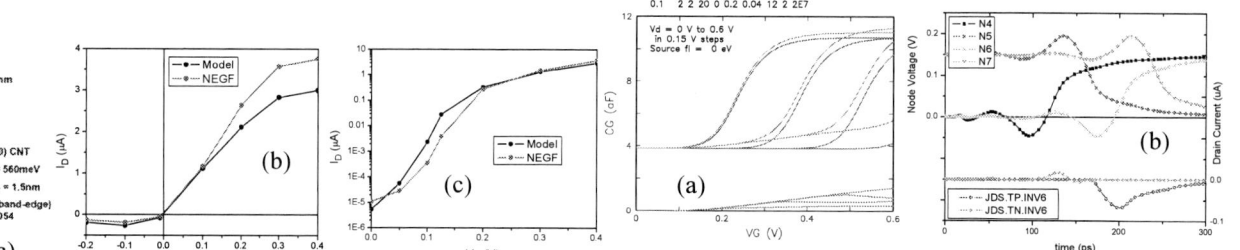

Fig.3: (a) C_{gs} vs. V_g, showing components due to drain back injection (blue) transport charge (red) and net capacitance (green). (b) Waveforms in stages 5,6,7 of a 10 stage inverter chain using n-and-p channel TFETS, and simulated using ASX (SPICE-like) program.

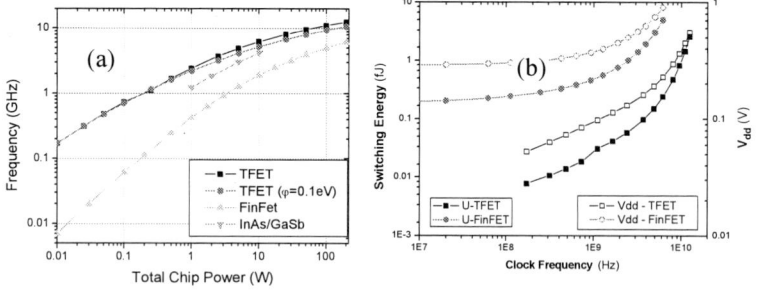

Fig. 5: Performance optimizer results showing (a) clock frequency vs. total chip power and (b) switching energy and power supply voltage vs. clock frequency. 16 cores of 1.5×10^6 circuits each are assumed. The total chip power was constrained and V_{dd}, L, V_t, V_e, widths, V_e and φ were varied to maximize the clock frequency.

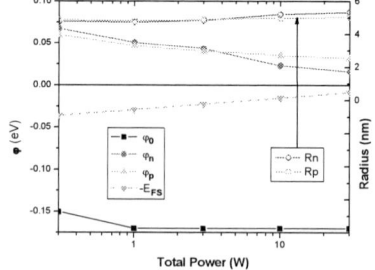

Fig. 6: Optimized values for effective band offsets (φ_p and φ_n) source Fermi energy ($-E_{fs}$), and nanotube radii (R_p and R_n)e vs. power for a (Ga,Al)Sb/(In,Al)As TFET. The composition dependent bulk band offset φ_0 is varied along with the tube radii and the other variables (see text) to obtain the minimum delay for a constrained total power.

978-1-61284-243-1/11 $26.00 © 2011 IEEE

Using Dimensionality to Achieve a Sharp Tunneling FET (TFET) Turn-On

Sapan Agarwal and Eli Yablonovitch

University of California, Berkeley, 262M Cory Hall, Berkeley, Ca 94704
Email: sapan@berkeley.edu, Phone: 510-642-1023

In order to achieve significantly reduced power consumption, the transistor operating voltage needs to be reduced. To do this, a tunneling based transistor needs to rely on the density of states turn-on as shown in Fig 1 [1]. Current can only flow when the conduction and valence bands overlap. If the band edges are ideal, one might expect an infinitely sharp turn on when the band edges overlap. Surprisingly, in a typical 3d bulk TFET, the nature of the turn on is actually quadratic in the gate voltage. Nevertheless, it is possible improve this if dimensionality is reduced. Consequently, we explored the nature of the band overlap for the various dimensionalities shown in Fig 2. We find that a 2d-2d pn junction, as shown in Fig. 2(i) brings us significantly closer to an ideal step function. Confining each side of the pn junction will also significantly increase the on state conductivity at low voltages.

First we consider a simple nanowire tunneling junction as shown in Fig 2(a). To model the tunneling current we look at the overlap voltage, V_{OL} between the conduction and valence band as shown in Fig 3b. We consider small voltages on the order of k_BT where the tunneling probability is roughly constant and the current is dominated by the density of states. We also initially assume a full valence band and empty conduction band. The current in a nanowire is simply given by a quantum of conductance times a voltage. In a metallic nanowire the voltage would be given by the difference in the Fermi levels as shown in Fig 3(a). However, in a tunneling junction the current is cut off by the band edges as shown in Fig 3(b) and so the relevant voltage is V_{OL}. Thus the maximum current is: $I_{1d} = 2q^2/h \times V_{OL} \times T(\vec{E})$. We can account for partially full bands by multiplying by $qV_{SD}/4K_BT$ at small biases. This allows us to express a conductivity with respect to V_{SD}. If we expand the junction along the other dimensions such as in Fig 2(d) or 2(g) we just need to multiply the current by number of transverse states such that $\partial I = N_{\perp states} \times 2q^2/h \times T \times \partial E$. This gives:

$$I_{3d} = \frac{1}{2}\left(\frac{Am^*}{2\pi\hbar^2} \times \frac{qV_{OL}}{2}\right) \times \frac{2q^2}{h} V_{OL} \text{ and } I_{2d} = \frac{2}{3}\left(\frac{L\sqrt{m^*}}{\pi\hbar} \times \sqrt{qV_{OL}}\right) \times \left(\frac{2q^2}{h} \times V_{OL} \times T(\vec{E})\right).$$

Now we consider what happens when one quantum well is overlapping another as in Fig 2(i). During the tunneling process both the transverse momentum and the energy have to be conserved. This results in a step function turn on. Fig 4(a) shows the energy-momentum conservation. The lower paraboloid represents all of the available states on the left side of the junction and the upper paraboloid represents the available states on the right side of the junction. In order for current to flow the initial and final energy must be the same and so the paraboloids must overlap. However, as seen in the right part of the figure, they can only overlap at a single energy. Furthermore, the number of states at a given transverse energy is given by the 2d density of states which is a constant regardless of energy as seen in Fig 4(b). Actually finding the current requires finding the transition rate for each state that tunnels and summing over all of the states that tunnel through a clever application of fermi's golden rule [2]. This gives the following current: $I_{2d-2d} = \left(qmA/\pi^2\hbar^3\right) \times E_{Z,i} \times E_{Z,f} \times T(\vec{E})$ where $E_{Z,i}$ and $E_{Z,f}$ are the confinement energies on the initial and final sides on the junction respectively.

A comparison of the different tunneling junctions is shown in Fig 5. As seen from the figure, confining each side of the junction results in a sharper turn on and a higher initial conductivity. Overlapping 1d nanowires and 2d quantum wells show very interesting behavior as they reflect the 1d and 2d density of states respectively. Of course the 1d divergence is cut off by the series resistance of a nanowire. Even after considering various non-idealities, using a 2d-2d pn junction could bring us significantly closer to realizing a step function turn on.

---

[1] J. Knoch, et al., *Solid-State Electronics*, vol. 51, pp. 572-578, Apr 2007.

[2] W. A. Harrison, *Physical Review*, vol. 123, pp. 85-89, 1 July 1961.

978-1-61284-243-1/11 $26.00 © 2011 IEEE

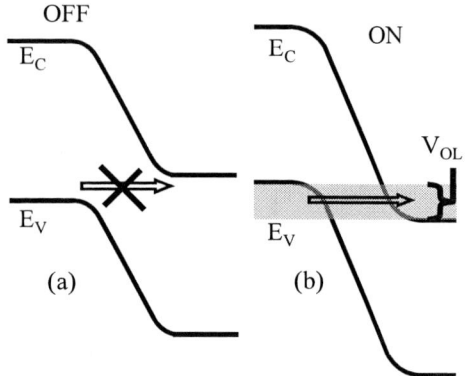

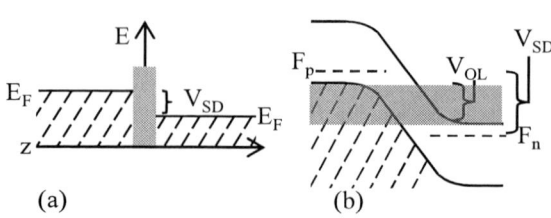

Figure 3: (a) In a metallic nanowire the current is given by the quantum of conductance and the separation of the fermi levels (b) In a tunneling junction the band edges limit the current

Figure 1: When the bands overlap current can flow

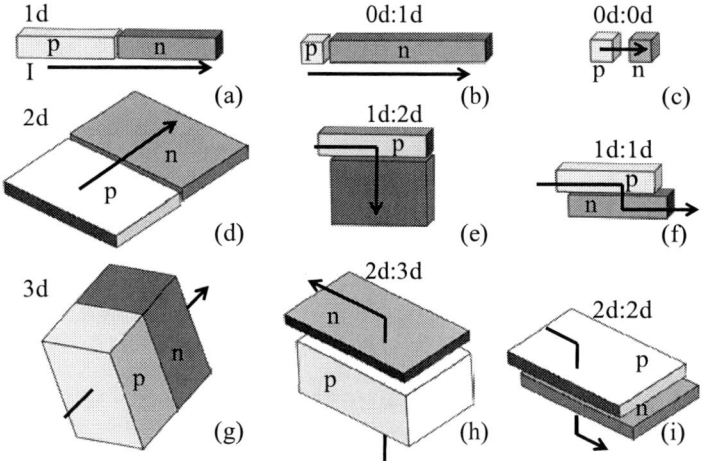

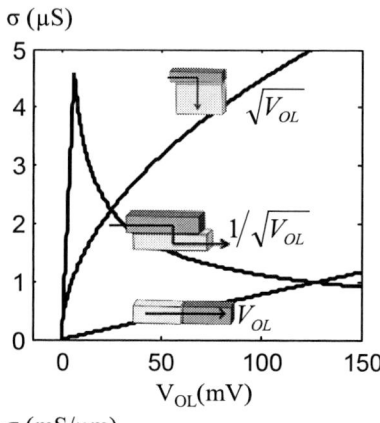

Figure 2: The different tunneling junction dimensionalities shown have different turn on characteristics.

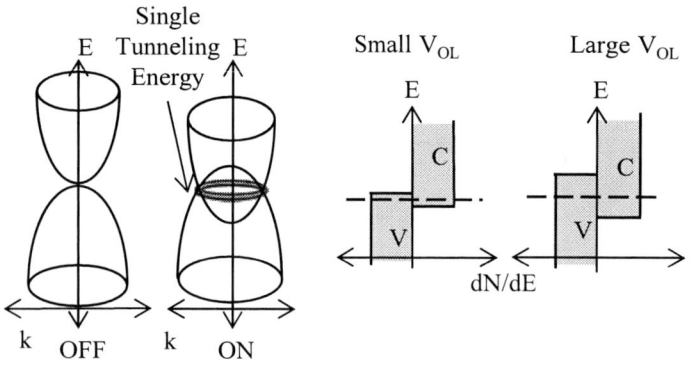

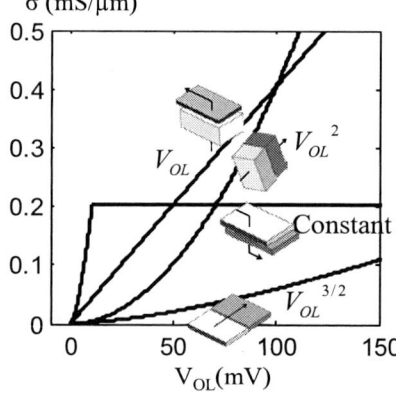

Figure 5: The source drain conductivity as a function of the overlap voltage for various junctions is plotted. The functional dependence of the current on V_{OL} is indicated next to each curve. The confinement energies are taken to be 25 meV, the effective mass is 0.1, all overlap lengths are 100 nm and the tunneling probability is 1%

Figure 4: (a) In a 2d-2d junction the conservation of energy and momentum results in a single tunneling energy (b) The density of states is a constant regardless of the tunneling energy and so the tunneling current is a constant

978-1-61284-243-1/11 $26.00 © 2011 IEEE   200

Investigation on Superlattice Heterostructures for Steep-Slope Nanowire FETs

E. Gnani, P. Maiorano, S. Reggiani, A. Gnudi and G. Baccarani

ARCES and DEIS, University of Bologna, Viale Risorgimento 2, 40136 Bologna, Italy

Phone: +39 051 209 3773, Fax: +39 051 209 3779, E-mail: egnani@arces.unibo.it

Abstract

In this work we investigate the feasibility of a steep-slope nanowire FET based on the filtering of the high-energy electrons via a superlattice heterostructure in the source extension. Several material pairs are investigated for the superlattice, with the aim to identify the most promising ones with respect to the typical FET evaluation metrics. We found that the GaN-AlGaN pair provides excellent results, which led us to optimize its device structure. We obtain a peak SS ≈ 15 mV/dec and an ON-current approaching 1mA/μm.

1. Introduction

The objective of a current turn-on rate much steeper than 60 mV/dec has been pursued by several approaches, some of which are mentioned in the ITRS document devoted to the emerging devices [1]. One of the most promising approaches is based on the filtering of the high-energy electrons injected into the channel. The typical example of this approach is the tunnel FET (T-FET) [2], where the filtering function is entrusted to the band-to-band tunneling (BTBT) mechanism. However, BTBT injection suffers from severe limitations of the on-state current, and of the sustained switching slope. In addition, the upward curvature of the output characteristics and the related small drain conductance at zero V_{DS} limits the T-FET performance below that of standard CMOS FETs at the same supply voltages [3].

In order overcome the above limitations, a new device concept was devised by using a superlattice (SL) interposed between the source and channel regions of a NW-FET as the filtering structure [4]. In this work numerical investigations show the effectiveness of the SL to filter out the high-energy electrons. A number of well-known III-V materials pairs, commonly used for the fabrication of HEMTs and lasers, were investigated. Next, we carried out an optimization of the GaN-AlGaN SL structure, which appears to be one of the most promising SL pairs for our purposes.

2. Device concept and physical model

Fig. 1 sketches the cylindrical nanowire (NW) geometry of the investigated device. The SL is realized with multiple barrier (b) and well (w) layers interposed between the source and the channel. The quantum-mechanical treatment is carried out by decoupling the Schrödinger equation into the transverse and longitudinal problems. Current transport is assumed to be ballistic. Each region is characterized by its specific transport mass, dielectric constant and electron affinity. An energy-adaptive mesh is used to accurately describe the resonant states generated within the superlattice.

Fig. 2 compares the turn-on characteristics of different SL structures, simulated with b = w = 2 nm at V_{DS} = 100 mV. Among them, the GaN-AlGaN appears to be one of the most promising SL pairs despite its relatively small subband offset, as it provides the steepest switching slope with a relatively high on-current (see Fig. 3). This material pair can provide a very small SS thanks to the large effective masses which reduce source-to-drain tunneling. The next most promising pair appears to be the InGaAs-InAlAs, which provides a higher on-current at the expense of a larger inverse slope. At the opposite site, the InAs-GaSb pair does not provide any serious improvement with respect to the traditional SS = 60 mV/dec.

3. GaN-AlGaN steep-slope NW-FET

We carried out the optimization of the SL structure for the GaN-AlGaN material pair. First we found that b = 0.9567 nm (3 atomic layers) and w = 1.5945 nm (5 atomic layers) give the best trade-off between SS and on-state current. The thickness values are fairly challenging, but are not beyond the possibilities offered by an advanced MBE machine.

Next, we investigated the effect on the device properties of doping in the SL well regions (see Fig. 4). By increasing the doping density N_W we observe an increase of the off-state current and of the switching voltage. The reason for this trend is clarified in Fig. 5, showing the local density of states of the first subband, where two minibands are clearly visible in the SL region. For a higher doping the potential energy in the SL region decreases and the second miniband comes closer to the Fermi level, thus providing a higher off-state current. Moreover, electrons entering the channel from the first miniband experience a higher potential barrier, which shifts the switching voltage to the right. Fig. 6 summarizes the obtained results for the various doping densities: on the left the on/off current ratio (top) and the on-state current (bottom) are plotted as function of the off-current. On/off current ratios larger than 10^4 can be obtained for off-currents smaller that 10^{-8} A/μm with an on-state current of about 0.7 mA/μm at V_{DD} = 0.4V. Fig. 6 (right) shows that a peak SS of 15 mV/dec can be obtained with a sustained SS always lower than 30 mV/dec. The peak and average SS are stable for gate voltage ranges of about 25 mV and more than 100 mV, respectively (bottom).

As can be seen in Fig. 7, the number of barriers can be as small as 7 without impairing the SL filtering effectiveness and the FET on-state current, with a shorter device length.

Finally, we studied the effect of the Al molar fraction (x) on the device performance. Fig. 8 (left) shows the drain current as x ranges from 0.15 to 0.25. With x = 0.15 an on-state current of about 1mA/μm at V_{DS} = 0.1 V can be obtained at the expense of some SS degradation, as indicated in Fig. 8 (right). The choice of $N_w = 3 \times 10^{19}$ cm^{-3} and x = 0.25 ensures the simultaneous fulfilment of the specs on the on/off current ratio and off-state current for HP, LOP and LSBP at V_{DD} = 0.4V for the 22 nm technology node [1]. On the other and, the choice of x = 0.15 would provide the best performance for HP applications.

4. Conclusion

In this work, we investigated the feasibility of a steep-slope NW-FET based on a SL in the source extension. The most promising material pair for the SL turned out to be the GaN-AlGaN system, which provides an SS ≈ 15 mV/dec and a sustained SS of about 28 mV/dec over a V_{GS} range of ≈ 0.1V. An on/off current ratio in excess of 10^4 can be achieved with an off-state current of 10^{-9}A/μm, i.e. two orders of magnitude smaller than the ITRS spec of 0.1 μA/μm [1] at V_{DD} = 0.4 V.

Acknowledgments

This work has been supported by the EU Grant No. 257267 (STEEPER) via the IUNET consortium.

References:
[1] http://public.itrs.net
[2] K. Boucart et al., *ESSDERC*-2004, pp. 383-386.
[3] T. Krishnamohan, *IEDM*-2008, pp. 947-949.
[4] E. Gnani et al., *ESSDERC 2010*, p. 380, 2010.
[5] S. Adachi et al., Properties of Group-IV, III-V and II-VI Semiconductors. Wiley, 1950.

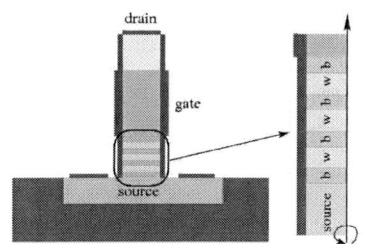

Fig 1: Pictorial view of the nanowire cross section. Different materials can be used for the source, super-lattice, channel and drain regions.

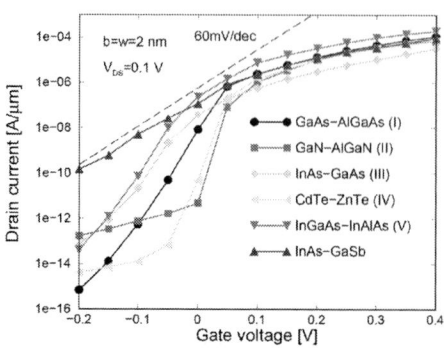

Fig 2: Comparison of the turn-on characteristics for different SL materials. The barrier and well lengths b = w = 2 nm and V_{DS} = 0.1V. Due to the very low effective mass, the InAs/GaSb pair exhibits poorer subthreshold properties due to the high tunneling probability across the super-lattice barrier, which prevents the filtering action.

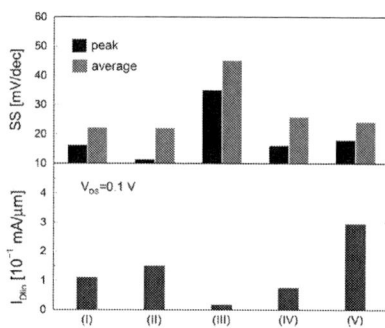

Fig. 3: Top: peak and average inverse SS for the various SL materials. The figure shows that the lowest inverse SS ≈ 10 mV/dec for the GaN-AlGaN SL-FET. Bottom: Normalized I_{Dlin} with respect to the device diameter, computed at V_{DS} = 0.1 V and V_{GS} = 0.4V.

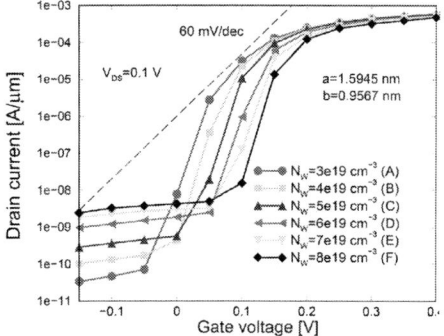

Fig. 4: Drain current at V_{DS} = 0.1V for different well doping densities N_W. The device parameters are: D = 5 nm, t_{ox} = 1 nm and N_D = 1x10^{20} cm^{-3} in the source and drain regions. The barrier and well lengths b and w are multiple of the AlGaN and GaN lattice unit cells, respectively.

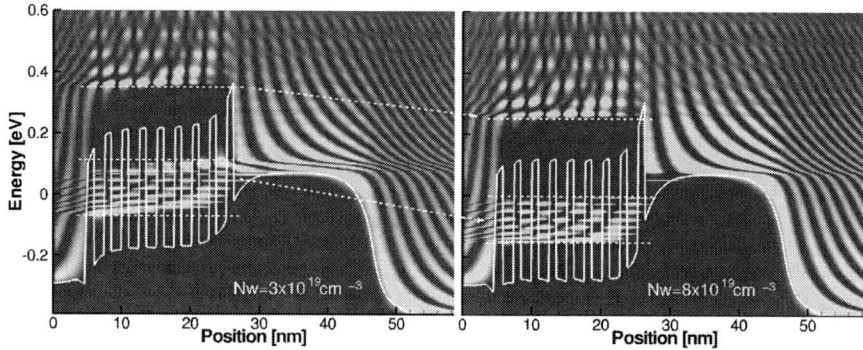

Fig. 5: Local density of states of the first subband at V_{DS} = 0.1V and V_{GS} = 0.2V for the GaN/Al$_{0.25}$Ga$_{0.75}$N nanowire with N_W = 3x10^{19} cm^{-3} (left) and N_W = 8x10^{19} cm^{-3} (right). For a higher doping the potential energy in the SL region decreases, and the second miniband comes closer to the Fermi level, thus giving a higher off-state current (see Fig. 4). For a fixed V_{GS} electrons entering the channel from the first miniband experience a higher potential barrier, which provides a shift of the switching voltage to higher values (see Fig. 4).

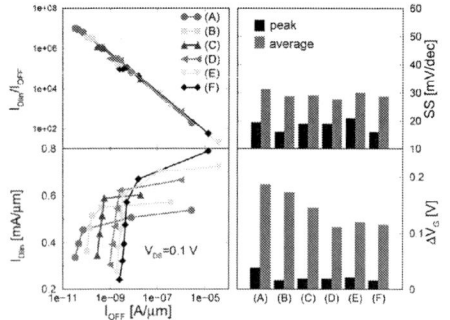

Fig. 6: Left: on/off current ratio (top) and on-current (bottom) vs. off-state current for various well doping levels. The on-state current is computed at V_{DS} = 0.1V and V_{GS} = 0.4 V. Right: peak and average SS (top) and gate voltage ranges over which the peak and average SS are stable (bottom) for various well doping levels.

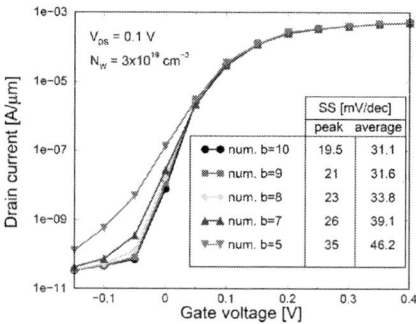

Fig. 7: Drain current at V_{DS} = 0.1V for different number of barriers (num. b). The decrease of the number of barriers does not affect the on-state current while causing some degradation of the peak and average SS which, however, is very moderate and quite acceptable down to 7 barriers.

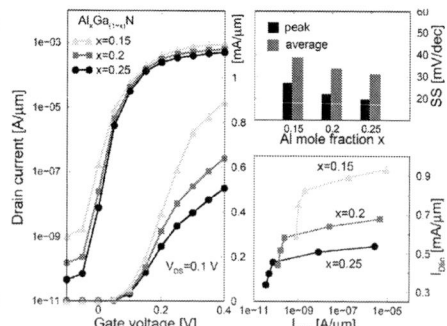

Fig. 8: Left: Drain current at V_{DS} = 0.1V for different Al mole fractions. With x = 0.15 an I_{on} > 1mA/μm can be achieved at the expense of some SS degradation (see right-top). Right-bottom: on-state current vs. off-current for various x values. With x = 0.25 the 22 nm node specs are fulfilled for either HP, LOP and LSP.

978-1-61284-243-1/11 $26.00 © 2011 IEEE

Self-aligned Gate NanoPillar $In_{0.53}Ga_{0.47}As$ Vertical Tunnel Transistor

D. K. Mohata, R.Bijesh, V.Saripalli, T. Mayer and S.Datta[1]

[1]Pennsylvania State University-PA-U.S.A

Phone: (814) 753 0465, Fax: (814) 865 7065, e-mail: dkm154@psu.edu

Introduction: Tunnel field effect transistors (TFET) have gained interest recently owing to their potential in achieving sub-kT/q steep switching slope, thus promising low Vcc operation[1-5]. Steep switching slope has already been demonstrated in Silicon TFET [2]. However, it has been theoretically shown and experimentally proved that Si or Si_xGe_{1-x} based homo-junction or hetero-junction TFETs would not meet the drive current requirement for future low power high performance logic applications [3]. III-V based hetero-junction TFETs have shown promise to provide MOSFET like high drive currents at low operating Vcc while providing the sub-kT/q steep switching slope[1,4,5]. However, the device design demands extremely scaled EOT and ultra-thin body double-gate geometry in order to achieve the desired transistor performance [4]. In this paper, we discuss a vertical TFET fabrication process with self-aligned gate [6] which can ultimately lead to the ultra-thin double-gate device geometry in order to achieve the desired TFET performance.

Nanopillar Tunnel Transistor Fabrication: Fig. 1 shows the cross-section schematics of the fabricated TFET following key process steps. Also shown is a summary of the entire fabrication process flow. $In_{0.53}Ga_{0.47}As$ layers were epitaxially grown on semi-insulating InP substrate using solid state MBE. The top P+ layer is 260nm thick and C doped at $5x10^{19}/cm^3$. This was followed by 100nm thick intrinsic channel and 260nm thick N+ layer, doped at $10^{19}/cm^3$ with Si. 250nm thick Molybdenum(Mo) was blanket deposited on P+ InGaAs using ebeam evaporation. Cr/Ti dry etch masks with minimum width of 250nm were created on Mo using e –beam Lithography, ebeam evaporation and lift-off techniques. Mo and InGaAs were dry etched using Cl_2 and SF_6 based chemistry. Dry etch of InGaAs was carried to a depth of 300nm. To remove sidewall damage and produce undercut, wet etch was performed using H_3PO_4, H_2O_2 and DI water (1:2:40) for 30 secs. An undercut of 55nm was obtained and the net InGaAs MESA height became 380nm. This undercut is important for the formation of self aligned gate. After the wet etch, 5nm Al_2O_3 (high-k) was deposited as gate dielectric using plasma-enhanced atomic layer deposition technique. 20nm Palladium (Pd) gate was vertically deposited using ebeam evaporation. The entire structure was then planarized with BCB and cured at 250^oC for 60mins in nitrogen ambient. After curing, BCB was etched back to expose Pd on top of Mo. Pd and Al_2O_3 were then bombarded off using Cl_2 and Ar based dry etch recipe. Lithography was followed to open large contact pads for source, drain and the gate. Ti/Pd/Au probing contacts were then deposited and lifted off.

Device Characterization and Modeling: Figures 2 (a) and (b) show the transmission electron microscopy (TEM) image of a 250nm drawn mesa width fabricated TFET device. Figure 3 shows the two terminal PIN current densities plotted for different MESA widths. Clearly, the Zener (reverse) side of the characteristics overlap on top of each other confirming the success in achieving the isolation between the top source and the side wall gate (self aligned). This also indicates that the PIN leakage floor in the transfer characteristics can be scaled by scaling the mesa width with this planarization approach. Figure 4(a) shows the simulated [7] transfer characteristics for different mesa width devices. Nearly 2 orders of improvement in Ion/Ioff is expected by scaling the mesa area from 20x20 μm^2 to $0.25x5\mu m^2$. Improvement in switching slope (Figure 4(b)) is also expected with MESA scaling due to reduction in leakage floor. Figure 4(c) shows the expected output characteristics for the 250nm width mesa device with an EOT of 2.25nm. The short channel effects observed in the simulation can be reduced in future by scaling the EOT. Figure 5(a) shows the measured Id-Vg characteristics for different mesa width TFET devices. Clearly, the reduction in the leakage floor is observed experimentally due to mesa scaling. Figure 5(b) shows the variation in the switching slope with the mesa width. With mesa scaling, leakage floor reduces and the steeper part of the SS is revealed. Present devices exhibit >kT/q switching slope because of trap assisted tunneling [8]. Better oxide-semiconductor interface and switching slope can be demonstrated in the future and is independent of the current process flow. In order to achieve SS smaller than 60mV/dec slope, not only the leakage floor needs to be reduced, the Tsi needs to be reduced below 20nm [4]. The current process flow has the capability to achieve the desired Tsi width, which will require an optimization of different layer thicknesses and MESA undercut. Figure 5(c) shows output characteristics for the smallest mesa device and shows existence of short channel effects as expected from the simulations. Figure 5(d) shows increasing peak-to-valley ratio with reducing mesa dimensions in the NDR part of the output characteristics. If TFET is expected to operate as a memory element using NDR, reduced mesa dimension and increasing gate control are desired.

[1] S. Mookerjea et al., *IEEE IEDM Tech. Dig.,* Dec. 2009 .   [2]R. Gandhi et al., accepted *IEEE Electron Device. Lett.* (2011)
[3] S. Koester et al., ECS Trans., vol. 33, no. 6, October 2010.   [4]Y. Liu et al., *68th Device Research Conf. Dig.,* 2010, p. 17.
[5] H. Zhao et al., *IEEE Electron Device Lett.* 31 (2010) 1392.   [6] H. Saito et al., *Applied Phys. Exp.* 3 (2010) 084101.
[7] Sentaurus Users Guide, Ver. C-2009.06 .   [8] S. Mookerjea et al., *IEEE Electron Device Lett.* 31 (2010) 564.

Cross-section Schematic

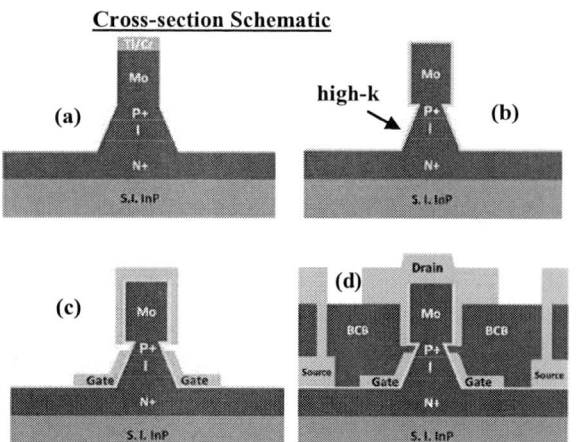

Process Flow

(a)
1. Blanket deposit Mo on InGaAs
2. Define Ti/Cr etch mask
3. Dry etch Mo and InGaAs
4. Strip off Ti/Cr etch mask

(b)
5. Wet etch undercut InGaAs
6. Deposit High-k

(c)
7. Lift-off self aligned Pd gate
8. Device Isolation

(d)
9. Planarization with BCB and etch back
10. Remove Pd and high-k on top of Mo
11. Lift-off Ti/Pd/Au Source, Drain and Gate contact Pads.

Fig. 1 Cross section schematics of the device follwoing key process steps. Fabrication process flow is shown to the right.

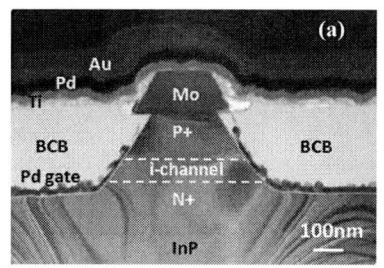
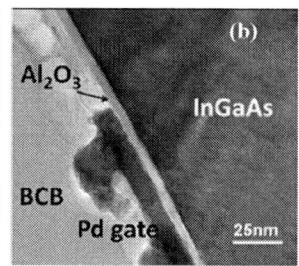
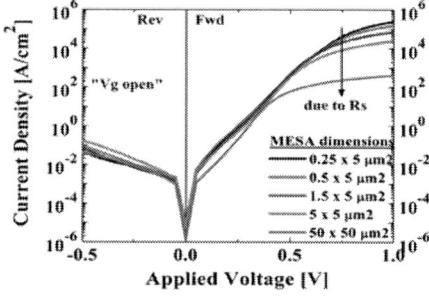

Fig. 2 (a) Cross-section Transmission Electron Microscopy image of the fabricated device structure. (b) Zoomed in image of the left sidewalll showing 20nm self-aligned Pd gate on Al_2O_3 gate dielectric.

Fig. 3 Two terminal PIN current densities for different MESA area devices. Zener (Rev) BTBT current density scales with MESA area and implies isolation of gate pad from the top source contact

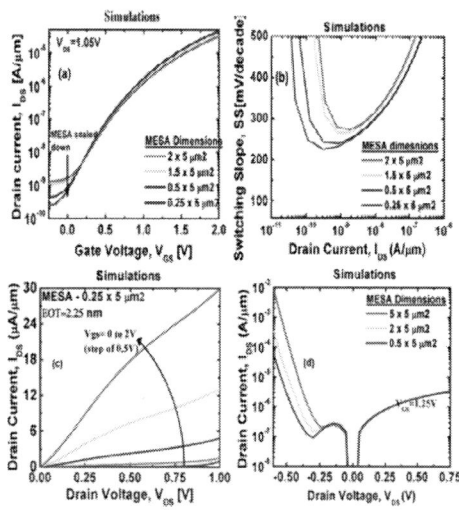
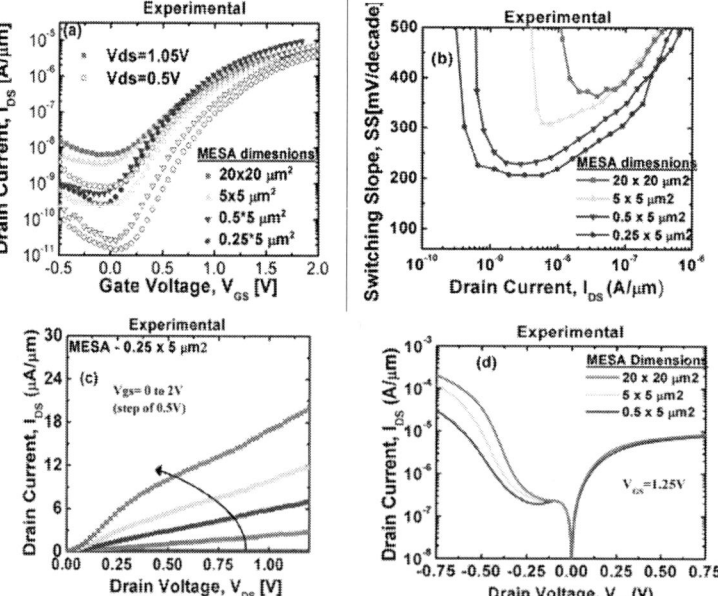

Fig. 4 (a) Simulated Id-Vg curves showing improved on-off ratio due to MESA scaling. (b) Point switching slope improves with MESA scaling (c) Simulated Id-Vd curves for 0.25 x 5 μm^2 MESA (d) NDR characteristics show increasing peak to valley ratio with MESA scaling.

Fig. 5 (a) Measured Id-Vg curves showing improved on-off ratio with MESA scaling. (b) Point switching slope improves with MESA scaling due to reduction in leakage floor. (c) Measured Id-Vd characteristics for the smallest MESA (d) Higher peak to valley ratio in NDR characteristics with MESA scaling indicates reduced thermionic/recombination currents for a given gated current.

978-1-61284-243-1/11 $26.00 © 2011 IEEE

Self-aligned InAs/Al$_{0.45}$Ga$_{0.55}$Sb vertical tunnel FETs

Guangle Zhou[1], Y. Lu[1], R. Li[1], Q. Zhang[1], W. Hwang[1], Q. Liu[1], T. Vasen[1], H. Zhu[2], J. Kuo[2], S. Koswatta[3], T. Kosel[1], M. Wistey[1], P. Fay[1], A. Seabaugh[1], and Huili (Grace) Xing[1]

(1) Department of Electrical Engineering, University of Notre Dame, Notre Dame, IN 46637 (2) IntelliEPI, Richardson, TX 75081 (3) IBM T. J. Watson Research Center, Yorktown Heights, NY 10598
Email: hxing@nd.edu, Phone: +1 574 631-9108

Tunnel field-effect transistors (TFETs) are under intense investigation for low-power applications because of their potential for extremely low subthreshold swing (SS) and low off-state leakage [1]. III-V semiconductors with small effective mass and near broken band alignment are considered to be ideal for TFETs in that they promise high on-current and I_{ON}/I_{OFF} ratios [2-3]. In this paper, we report the first demonstration of an InAs/Al$_{0.45}$Ga$_{0.55}$Sb heterojunction TFETs fabricated using an optical-lithography-only, self-aligned process and also investigate the effects limiting the InAs/Al$_{0.45}$Ga$_{0.55}$Sb TFET performance.

Fig. 1(a) shows a cross section of the n-channel InAs/Al$_{0.45}$Ga$_{0.55}$Sb TFET in a new tunneling geometry with the tunnel transport directed normal to the gate.. The TFETs were grown by molecular beam epitaxy (MBE) on a GaSb substrate. The epitaxial structure, starting from the substrate, consists of: 200 nmAlSb/AlAs superlattice buffer layer, 300 nm of n+InAs$_{0.91}$Sb$_{0.09}$,10 nm of n-InAs (Si-doped, 1 x 10^{17} cm^{-3}), 110 nm of p+GaSb, and 30 nm of p+Al$_x$Ga$_{1-x}$Sb (Be-doped, 4 x 10^{18} cm^{-3}), with the Al composition x increased in three steps from 0 to 0.45, and concluding with a top 30 nm n-InAs layer (Si-doped, 1 x 10^{17} cm^{-3}). Three samples were processed; for one sample TFETs were fabricated on the heterostructures as grown, while in the other two the top InAs layer was thinned using Citric acid:H$_2$O$_2$ (1:1) to 22 nm and 15 nm thickness, respectively. A 7 nm thick Al$_2$O$_3$ gate dielectric was deposited by atomic layer deposition (ALD) immediately after cleaning in 1HCl:1H$_2$O for 30 s. A Ti/W/SiN$_x$ gate stack was blanket-deposited, then patterned using optical lithography, and reactive-ion etched (RIE). Plasma-enhanced chemical vapor deposition (PECVD) SiN$_x$ sidewalls were then formed around the gate, followed by removal of Al$_2$O$_3$ gate dielectric using AZ 400K developer. After drain metallization and lift-off (Ti/Au), InAs was selectively etched in 1citric acid:1H$_2$O$_2$, followed by a selective AlGaSb etch using tartaric acid:H$_2$O$_2$:HCl:H$_2$O (3.75 g : 4 ml : 40 ml : 400 ml) until the AlGaSb under the drain and the SiN$_x$ spacer was removed, forming the undercut mesa structure. Fig. 1(b) shows the cross sectional image of a fabricated InAs/AlGaSb vertical TFET, taken after cross sectioning in a focused-ion beam and imaging by scanning electron microscopy (FIB/SEM). The SEM images clearly indicate that the InAs/AlGaSb tunnel junctions were fully overlapped by the gate electrode.

Shown in Fig. 2 (a), (b) and (c) are the measured $I_D - V_{DS}$ characteristics of a TFET with a 30 nm, 22 nm and 15 nm top InAs thickness at 300 K, respectively. The on-current is about 1200, 275 and 1 µA/µm at V_{DS}= 0.5 V, respectively, while the gate leakage is smaller than the drain current. The low on-current of the 15 nm InAs TFET is due to the overetching of the AlGaSb under the gate and consequent higher access resistance. Shown in Fig. 3(a) and (b) are the $I_D - V_{GS}$ characteristics of TFETs with 22 nm and 15 nm of InAs at 300 K, respectively. While the drain current on/off ratio of a 30 nm InAs TFET is about 2, it increases to 100 and 2000 as the top InAs is thinned to 22 nm and 15 nm, respectively. The measured SS is approximately 580 mV/dec and 170 mV/dec on the 22 nm and 15 nm thickness InAs TFETs, respectively. The stretch-out of the SS likely results from interface states at the Al$_2$O$_3$/InAs interface.

Simulation using Synopsis Sentaurus 2010.03 in Fig. 3(c) show that, including the impact of the calculated drain and source resistance (R_S = 250 Ω µm, R_D = 1000 Ω µm), the extrinsic on-current is expected to be one order magnitude lower than the intrinsic value for these devices. Larger on-current is expected with reduced parasitic resistances. The off current (I_{OFF}), however, appears to be limited by trap-assisted tunneling in the interface between the InAs and Al$_{0.45}$Ga$_{0.55}$Sb. The I-V characteristics of vertical tunnel diodes fabricated from the same epitaxy are shown in Fig. 4(a) and confirm that the observed gate-modulated drain current stems from tunneling since negative differential resistance (NDR) can be clearly observed under negative bias. Fig. 4(b) shows the simulated I_D-V_{GS} characteristics at V_{DS} = 0.3 V with different InAs thicknesses, but assuming a doping of 1 x 10^{18} cm^{-3}. These simulations show the reduction of on current and improvement of SS with decreasing InAs thickness, which is consistent with the measurements.

In conclusion, a new self-aligned tunnel FET geometry is described and demonstrated for the first time. The process features staggered InAs/Al$_{0.45}$Ga$_{0.55}$Sb heterojunctions, and self-alignment using Si3N4 sidewall spacers. This work is supported by the Semiconductor Research Corporation's Nanoelectronics Research Initiative and the National Institute of Standards and Technology through the Midwest Institute for Nanoelectronics Discovery (MIND).

[1] A. Seabaugh and Q. Zhang, *Proc. IEEE*, vol. 98, no. 12, p. 2095, 2010.   [2] J. Knoch and J. Appenzeller , *IEEE EDL.*, vol. 31, no. 4, p. 305, 2010.   [3] S. O. Koswatta et al., *IEDM*, pp. 909, 2009.   [4] S. Mookerjea et al. *IEDM*, pp. 949, 2009.   [5] D. Mohata, *Appl. Phys. Express*, 4 (2011) 024105.   [6] H. Zhao et al, *IEEE EDL*, vol. 31, no. 12, p. 1392, 2010.

978-1-61284-243-1/11 $26.00 © 2011 IEEE

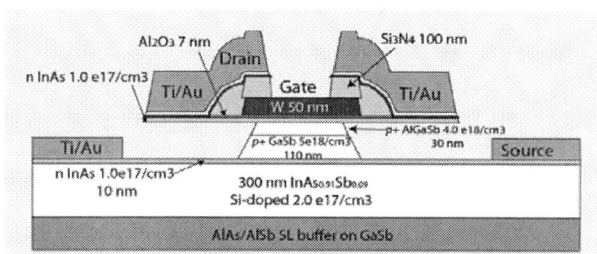 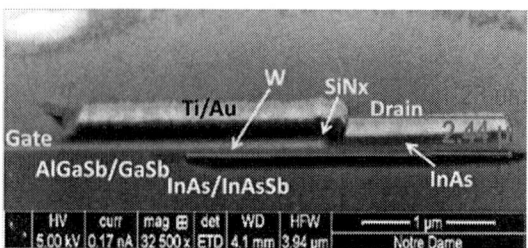

Fig. 1: (a) Scale cross section of an InAs/Al$_{0.45}$Ga$_{0.55}$Sb TFET fabricated using a gate first self-aligned process. **(b)** FIB/SEM cross-sectional image of a fabricated InAs/Al$_{0.45}$Ga$_{0.55}$Sb TFET featuring self-aligned drain contact and AlGaSb/GaSb undercut etch.

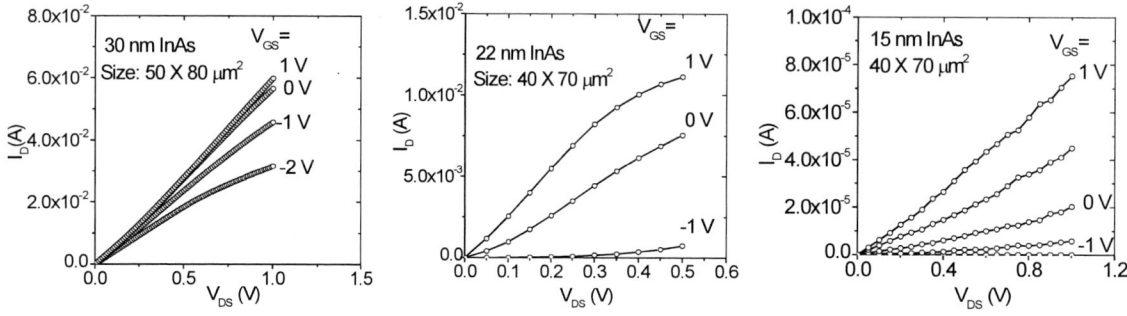

Fig. 2: (a) I_D-V_{DS} for a 30-nm-thick-InAs/Al$_{0.45}$Ga$_{0.55}$Sb TFET at 300 K. **(b)** I_D-V_{DS} for a 22-nm thick-InAs/Al$_{0.45}$Ga$_{0.55}$Sb TFET at 300 K. **(c)** I_D-V_{DS} for a 15-nm-thick-InAs/Al$_{0.45}$Ga$_{0.55}$Sb TFET at 300 K.

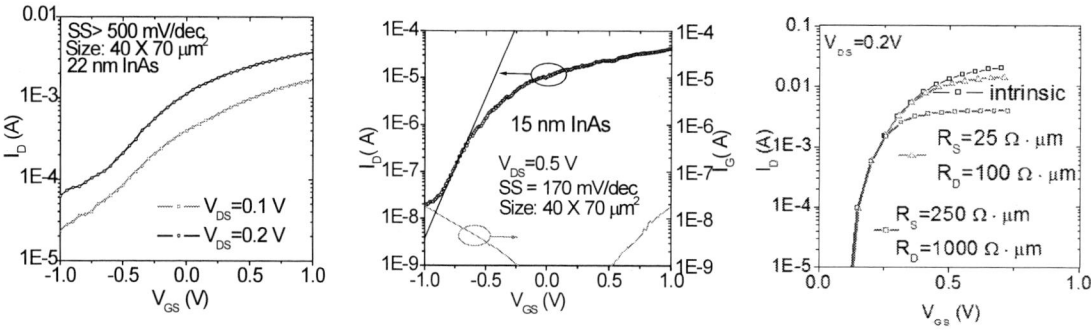

Fig. 3: (a) I_D-V_{GS} for a 22-nm-thick-InAs/Al$_{0.45}$Ga$_{0.55}$Sb TFET at 300 K. **(b)** I_D-V_{GS} for a 15-nm-thick-InAs/ Al$_{0.45}$Ga$_{0.55}$Sb TFET at 300 K. **(c)** Comparisons of simulated I_D-V_{GS} for 22-nm–thick-InAs/Al$_{0.45}$Ga$_{0.55}$Sb TFETs with different parasitic resistances for a 40 x 70 µm^2 device.

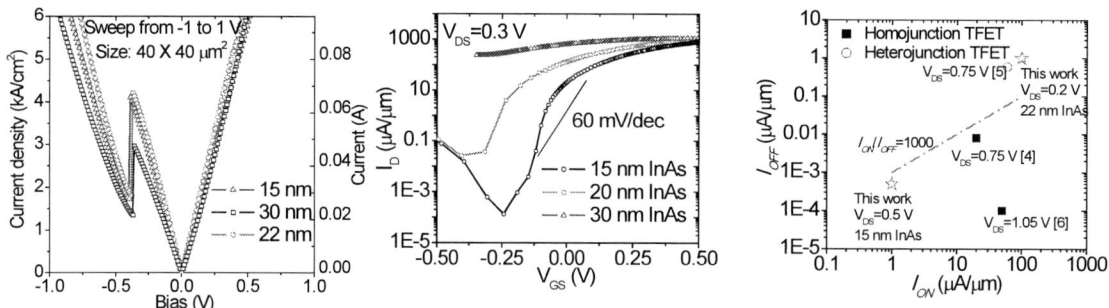

Fig. 4: (a) I-V characteristics of InAs/AlGaSb tunnel diodes with different InAs thickness. **(b)** Simulated I_D-V_{GS} for InAs/AlGaSb TFETs with different InAs thicknesses at V_{DS} = 0.3 V using Synopsys.Sentaurus 2010.03. **(c)** Comparisons of I_{ON} and I_{OFF} vs. published III-V TFETs.

978-1-61284-243-1/11 $26.00 © 2011 IEEE

P-type Tunneling FET on Si (110) Substrate with Anisotropic Effect

M. H. Lee[1,*], C.-Y. Kao[1], C.-L. Yang[1], and C.-H. Lee[2]

[1] Institute of Electro-Optical Science and Technology, National Taiwan Normal University, Taipei, Taiwan
[2] Graduate Institute of Electronics Engineering (GIEE) and Department of Electrical Engineering, National Taiwan University, Taipei, Taiwan
[*]Tel: 886-2-77346747 / Fax: 886-2-86631954   e-mail: mhlee@ntnu.edu.tw

The promising potential of tunneling FETs (TFETs) for steep switch behavior with gate controlled band-to-band tunneling (BTBT) mechanism has attracted much attention for supply voltage (V_{DD}) scaling and power consumption next generation CMOS [1, 2]. However, the challenge for TFETs is lower drive currents as compare with MOSFET due to a high conductance resistance while reverse bias. Tunneling FETs (TFETs) operates with band-to-band tunneling current that change with the channel potential more abruptly than thermionic emission current. In order to obtain high I_{ON} without sacrificing I_{OFF}, and the high-k dielectric and metal gate are integrated as gate stack. To obtain high quality and avoid crystallizing of high-K layer, the gate last process was performed in this work. For N-TFET, much works have been reported on the SS improvement [4, 5]. For P-TFET, Bhuwalka et al. reported the ambipolar working of vertical TFET with negative gate bias, which obtain SS < 60mV/dec [6, 7]. In this work, we will demonstrate HK/MG (high-K/metal gate) P-TFET with the gate last process, and discuss the anisotropic effect on (110) substrate.

Standard 6 inch MOS based line and gate last process are employed for this study. First, the p^+ and n^+ regions for drain and source were defined and implanted on p type Si substrate (100) and (110), and were implanted by BF_2 (60keV, 4×10^{15} cm^{-2}) and P (90keV, 4×10^{15} cm^{-2}), respectively. The annealing process for dopant activation was performed by RTA in an N_2 ambient with 2 steps. The step 1 is 600 °C for 100 sec and step 2 is 650 °C for 30 sec. A physical thickness ~13 nm HfSiO$_x$ as gate dielectric by Metal Organic Chemical Vapor Deposition (MOCVD) and 200 nm TiN as metal gate by Physical Vapor Deposition (PVD) were deposited (Fig. 1). In order to improve the interface layer between Si substrate and HfSiO$_x$, the annealing with 600 °C 30 sec before PVD was performed. The TiN and HfSiO$_x$ were defined by dry etching and dipping in diluted HF solution to accomplish metal gate at last process (Fig. 2).

The C-V characteristics of MOSCAP on (100) and (110) orientation of Si substrate shows a significant shoulder, and indicates the trap formation as combination center for (110) TFET (Fig. 3). The EOT (equivalent oxide thickness) is ~ 7.2 nm by C_{max}, and the dielectric constant of HfSiO$_2$ is about 10. The high dielectric constant is due to gate last process without thermal annealing of source/drain activation for gate last process. The extracted D_{it} (interface trap density) by high-low frequency C-V method shows the value at mid-gap is 5×10^{11} cm^{-2}eV^{-1} for (100) and 1×10^{12} cm^{-2}eV^{-1} for (110), respectively (Fig. 4). The channel direction is defined for band to ban tunneling (BTBT) on (100)&(110) orientation (Fig. 5). The transfer characteristics I_{DS}-V_{GS} of P-TFET on (100) and (110) substrate shows > 10^5 ON/OFF ratio (Fig. 6). The threshold voltage (V_T) of the TFET has been extracted based on the constant current method with $I_{DS}=10^{-9}$ A/μm for V_{DS} = -1 V. The output characteristics I_{DS}-V_{DS} of P-TFET on (100) and (110) substrate shows 79% saturation current enhancement of <112> as compare with that of <111> on (110) orientation (Fig. 7). Note that it is forward bias for p/i/n diode at positive V_{DS}, and the BTBT would be occurred at negative V_{DS} for reverse bias. One of the characteristics for TFET is driving current independent on L_g due to the mechanism of BTBT current [8] (Fig. 8). The I_{ON} of <112> has 90% higher than that of <111> for bias in linear region on (110) substrate (Fig. 9). However, there is no significant difference for I_{OFF} (Fig. 10). The effective mass of electron and hole for different tunneling direction and wafer orientation may explain the results. The devices uniformity is improved for (100) by observing the slope of the cumulative probability (Fig. 9 & 10). There is no significant different for SS, which are as low as ~ 100 mV/dec for <112>/(110) and <111>/(110) (Fig. 11). For the I_{DS} of (110) as compare with that of (100) wafer, the <112>/(110) exhibits enhancement and the degradation for <111>/(110) (Fig. 12).

We have successful demonstrated the anisotropic effect for (110) P-type TFET with gate last process. The channel direction of <112>/(110) has highest I_{ON} with similar I_{OFF} as compare with <111>/(110) and <110>/(100). The effective mass of electron and hole for different tunneling direction and wafer orientation may explain the results. Finally, the authors are grateful for funding supporting by National Science Council (NSC 98-2221-E-003-020-MY3), National Nano Device Laboratories (NDL), and Nano Facility Center (NFC), Taiwan.

References: [1] W. Y. Choi, et al., IEEE EDL, 28, p. 743, 2007. [2] D. Leonelli, et al., SSDM, p. 767, 2009. [3] T. Krishnamohan, et al., IEDM Tech. Dig., p. 947, 2008. [4] S. Mookerjea, et al., DRC, p. 47, 2008. [5] P.-F. Guo, et al., IEEE EDL, 30, p. 981, 2009. [6] K. Bhuwalka, et al., JJAP, 45, p. 106, 2006.   [7] H. Virani, et al., SSDM, p. 384, 2009. [8] S. H. Kim, et al., in VLSI Symp. Tech. Dig., p. 178, 2009.

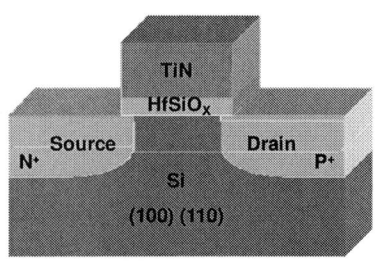

Fig. 1. The structure of TFET with TiN/HfSiO$_x$ gate stack on Si (100) and (110) substrate. A physical thickness ~13 nm HfSiO$_x$ as gate dielectric and 200 nm TiN as metal gate.

PROCESS FLOW

- (100) (110) substrates preparation
- S/D implantation
- RTA activation
- MOCVD HfSiO$_2$ deposited
- PVD TiN deposited
- Gate stack definition

Fig. 2. The process flow of the gate last process for TFET with HKMG gate stack.

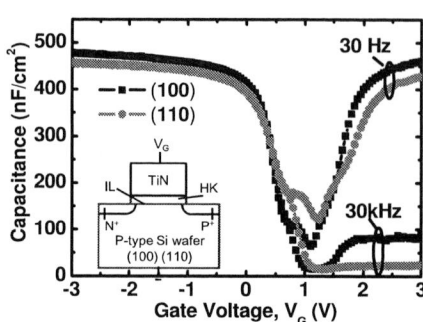

Fig. 3. The C-V characteristics of MOSCAP on (100) and (110) orientation of Si substrate. The inset show the schematic diagram of the measurement setup.

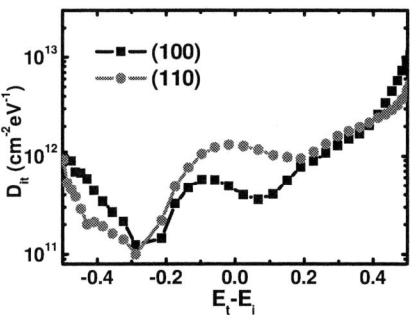

Fig. 4. The extracted D$_{it}$ (interface trap density) by high-low frequency C-V method. The higher D$_{it}$ was obtained for (110) TFET.

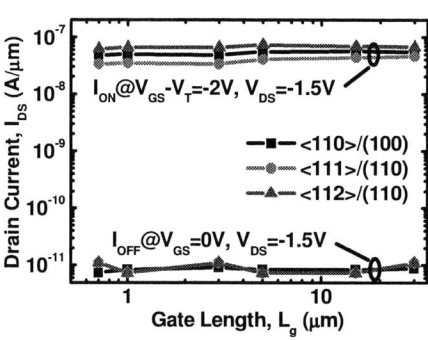

(a)   (b)

Fig. 5. The schematic diagram of the channel direction definition for BTBT on (a) Si (100) wafer and (b) Si (110) wafer.

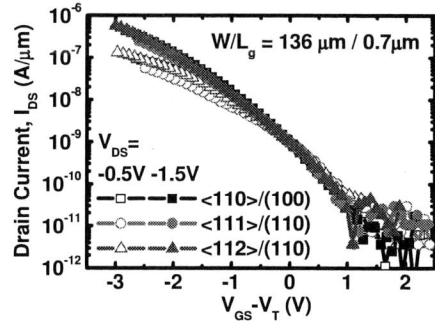

Fig. 6. The transfer characteristics I$_{DS}$-V$_{GS}$ of P-TFET on (100) and (110) substrate. It shows > 10^5 ON/OFF ratio.

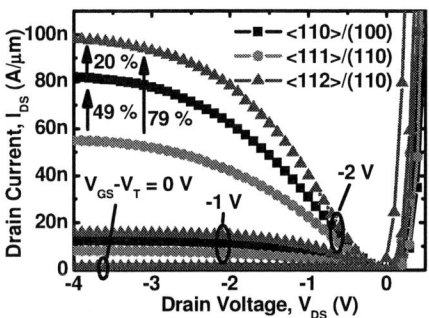

Fig. 7. The output characteristics I$_{DS}$-V$_{DS}$ of P-TFET on (100) and (110) substrate. It shows 79% saturation current enhancement of <112>/(110) as compare with that of <111>/(110).

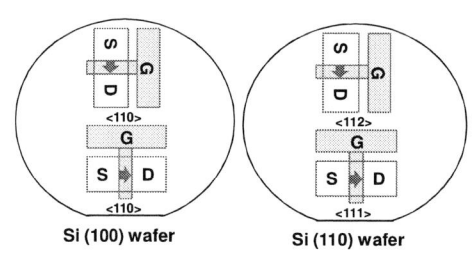

Fig. 8. Measured I$_{ON}$ and I$_{OFF}$ of TFET vs. L$_g$. The characteristics for TFET is driving current independent on gate length due to the mechanism of BTBT current.

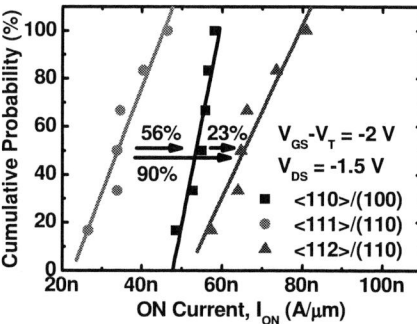

Fig. 9. The I$_{ON}$ cumulative probability for (100) and (110) orientation. The channel direction of <112>/(110) has highest I$_{ON}$. Note that the slope indicates the uniformity.

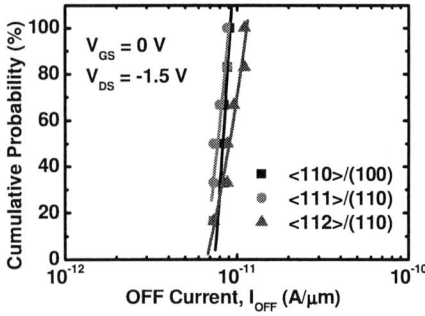

Fig. 10. The I$_{OFF}$ cumulative probability for (100) and (110) orientation. There are no significant difference I$_{OFF}$ for the three channel direction.

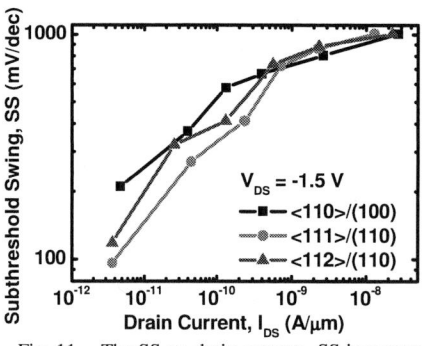

Fig. 11. The SS vs. drain current. SS increases as I$_{DS}$ increases, as expected for a tunnel transistor.

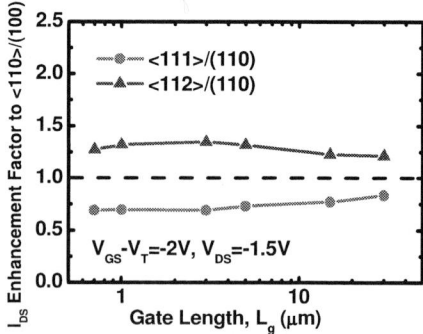

Fig. 12. The I$_{DS}$ enhancement factor to <110>/(100) vs. L$_g$. The <112>/(110) exhibits enhancement and the degradation for <111>/(110).

978-1-61284-243-1/11 $26.00 © 2011 IEEE   208

Session V.B (Corwin Pavilion West)

Wide Bandgap

Tuesday PM, June 21st, 2011

Session Chairs: Siddarth Rajan, Ohio State University

1:30 PM V.B-1
Anomalous output conductance in N-polar GaN-based MIS-HEMTs
M. H. Wong[1], U. Singisetti[1], J. Lu[1], J. S. Speck[2], and U. K. Mishra[1], [1]Electrical and Computer Engineering and [2]Materials Departments University of California, Santa Barbara, Californa, USA

1:50 PM V.B-2
Normally-off Gate-Recessed AlGaN/GaN-on-Si Hybrid MOS-HFET with Al_2O_3 Gate Dielectric
A. L. Corrion, M. Chen, R. Chu, S. D. Burnham, S. Khalil, D. Zehnder, B. Hughes, and K. Boutros, HRL Laboratories LLC, Malibu, California, USA

2:10 PM V.B-3 Student Paper
N-Polar AlGaN/GaN MIS-HEMTs on SiC with a 16.7 W/mm Power Density at 10 GHz Using an Al_2O_3 Based Etch Stop Technology for the Gate Recess
S. Kolluri, S. Keller, S. P. DenBaars and U. K. Mishra, Department of ECE, University of California, Santa Barbara, California, USA

2:30 PM V.B-4 Invited Paper
Total GaN Solution for Electircal Power Conversion
Y.-F. Wu, R. Coffie, N. Fichtenbaum, Y. Dora, C.S. Suh, L. Shen, P. Parikh and U.K. Mishra, Transphorm Inc., Goleta, California, USA

3:10 PM Break

3:30 PM V.B-5
First AlN/GaN HEMTs power measurement at 18 GHz on Silicon substrate
F. Medjdoub, M. Zegaoui, D. Ducatteau, N. Rolland and P.A. Rolland, IEMN, Villeneuve d'Ascq, FRANCE

3:50 PM V.B-6
Enhanced mobility for MOCVD grown AlGaN/GaN HEMTs on Si substrate
S. L. Selvaraj, A. Watanabe and T. Egawa, Research Center for Nano-Device and System, Nagoya Institute of Technology, Gokiso-cho, Showa-ku, Nagoya, JAPAN

4:10 PM V.B-7
High Performance GaN-on-Si Power Switch: Role of Substrate Bias in Device Characteristics
R. Chu, D. Zehnder, B. Hughes, and K. Boutros, HRL Laboratories LLC, Malibu, California, USA

4:30 PM V.B-8
Late News

4:50 PM V.B-9
Late News

978-1-61284-243-1/11 $26.00 © 2011 IEEE

978-1-61284-243-1/11 $26.00 © 2011 IEEE

Anomalous output conductance in N-polar GaN-based MIS-HEMTs

Man Hoi Wong[1], Uttam Singisetti[1], Jing Lu[1], James S. Speck[2], and Umesh K. Mishra[1]

[1]Electrical and Computer Engineering and [2]Materials Departments
University of California, Santa Barbara, CA 93106, USA
Email: mhwong@ece.ucsb.edu / Phone: +1-805-893-8594/ Fax: +1-805-893-8714

Introduction: N-polar $(000\bar{1})$ GaN-based high electron mobility transistors (HEMTs) have been proposed for the next generation millimeter-wave GaN electronics [1]. One major advantage anticipated of N-polar HEMTs is their inherent AlGaN back-barrier for enhanced electron confinement and improved output resistance in highly-scaled GaN HEMTs [2]. Contrary to expectation, the design of N-polar HEMTs seems to show increased DC output conductance (G_{DS}), leading to large threshold voltage shifts (ΔV_T) [3,4]. This study aims at providing physical understanding and insight into the anomalous G_{DS} in N-polar HEMTs, as well as its impact on device performance and design.

Experiments: We begin by noting that high G_{DS} has been observed in N-polar HEMTs with a long gate length (L_G) and a high aspect ratio where short channel effects were insignificant, independent of the growth technique employed [3, 4]. In all cases, the off-state leakage was low and could not account for the increase in drain current (I_D). To rule out impact ionization as a probable cause, we observe that gate-pulsed I-V measurements (to minimize device heating) with a 200-ns pulse width on a well-passivated device reported in [3] showed poor I_D saturation over a wide range of drain bias (V_D) from the knee voltage (V_{knee}) $\leq$ 5 V till destructive breakdown at V_{br} = 60 V (Fig. 1). Such high V_{br} would have been unreasonable had impact ionization been responsible for the onset of high G_{DS} at a much lower drain bias close to V_{knee}.

The device structures adopted in this study were inspired by earlier experimental observations of a deep donor-like hole trap state near the valence band at the net negative polarization interfaces in both Ga- and N-polar heterostructures [5-7]. In a prototypical N-polar GaN HEMT comprising AlGaN (cap)/GaN (channel)/AlGaN (back-barrier)/GaN (buffer) where the 2-D electron gas (2DEG) was induced at the GaN (channel)/AlGaN (back-barrier) interface (Fig. 2), the ionized traps at the net negative AlGaN (back-barrier)/GaN (buffer) interface provided compensating positive charges for the interface and imaged electrons in the 2DEG. Alternatively, a Si δ-doping sheet could be inserted below the AlGaN barrier to supply electrons to the 2DEG channel while the ionized Si donors provided the positive charges. Two devices with a 30-nm $Al_{0.3}Ga_{0.7}N$ back-barrier, one of which was doped with Si (1×10^{13} cm^{-2}) below the back-barrier but were otherwise identical to each other in epitaxial structure, were designed for the same 2DEG density (Fig. 2). The $Al_{0.1}Ga_{0.9}N$ cap (5 nm) protected the GaN channel layer (25 nm) during the deposition of a 5-nm Si_xN_y gate dielectric by high temperature chemical vapor deposition (CVD) at 1020°C prior to fabrication.

The devices were grown by plasma-assisted molecular beam epitaxy (PAMBE) on C-face 6H-SiC substrates. An alloyed Ti/Al/Ni/Au metal stack was used as the ohmic contact. Mesas were formed with BCl_3/Cl_2 reactive ion etch. A 160-nm Si_xN_y layer was deposited by plasma-enhanced CVD (PECVD) for surface passivation, followed by a CF_4-based trench gate process with self-aligned Ni/Au/Ni deposition [8,9]. The devices were 2×75 µm wide with L_{GS} = 0.5 µm, L_{GD} = 2 µm, and a nominal L_G of 0.7 µm that corresponded to a high aspect ratio of ~1:18.

Results and Discussion: The common-source output characteristics of the undoped device showed high G_{DS} with a large ΔV_T = -0.45V when V_D was increased from 10V to 30V, while the Si-doped device exhibited satisfactory I_D saturation with a much smaller ΔV_T = -0.1V across the same range of drain bias (Fig. 2). Both devices showed low off-state leakage (< 0.1 mA/mm) at V_D = 30 V. The increase in I_D in the undoped device was reflected in the source current.

Small-signal characterization was performed on an Agilent E8361A network analyzer. First, an off-wafer probe tip calibration on a standard substrate was carried out. Next, on-wafer open-short calibration standards were used to de-embed the pad parasitics and to extract the intrinsic RF output conductance (g_{ds}) at peak f_{max} at V_D = 20 V [10,11]. The extracted RF-g_{ds} of the two devices were very similar at ~1200 Ω despite their large DC-G_{DS} ratio of 17500 Ω/5500 Ω (Fig. 3). Comparable RF-g_{ds} values of 1100 – 1500Ω were measured in an MBE-grown Ga-polar $Al_{0.3}Ga_{0.7}N$/GaN/SiC HEMT of similar dimensions. These results suggested that traps, which were unresponsive to RF, could be responsible for the high DC-G_{DS}. Moreover, the f_{max} of the undoped device showed no apparent degradation compared to that of the Si-doped device (Fig. 4), which further indicated that the output conductance was a low-frequency phenomenon.

We propose a simple physical mechanism behind the anomalous G_{DS} based on these results. The absence of Si-doping below the AlGaN back-barrier in an N-polar HEMT brings the equilibrium Fermi energy in close proximity to the donor-like trap state as suggested by the band diagram (Fig. 2). As the V_D is increased beyond the V_{knee}, a high-field depletion region forms on the drain side of the gate where the traps ionize and emit electrons to the 2DEG channel. A higher V_D is therefore needed to re-establish I_D saturation, at which point more traps will ionize and the process repeats itself. Si-doping suppresses this process by separating the trap level from the Fermi energy. However, our experiments could not rule out the possibility that Si-doping could result in higher formation energy of the donor-like traps.

To further examine the plausibility of our proposed model, another set of N-polar MIS-HEMTs were fabricated. The control device had an $Al_xGa_{1-x}N$ back-barrier linearly-graded in x ($0.05 \leq x \leq 0.3$) and uniformly-doped with Si. It was a variation of the doped structure described above by distributing the negative polarization sheet charge and δ-doping, and would therefore exhibit high DC output resistance [12]. The test device had in addition a 2-nm AlN interlayer that was commonly adopted to improve 2DEG mobility (Fig. 5). The device with an AlN interlayer clearly exhibited higher DC-G_{DS} (Fig. 5), which could be attributed to the action of donor-like traps at the negative AlN/$Al_{0.3}Ga_{0.7}N$ interface.

Summary: We propose that the anomalous output conductance in N-polar GaN MIS-HEMTs was caused by ionization of donor-like traps from a net negative polarization interface. It is a low-frequency phenomenon that changes the V_T of the device with V_D, while no evidence of increased output conductance or related device performance degradation was found under RF conditions. Appropriate back-barrier designs are needed to mitigate the DC-G_{DS} in N-polar GaN MIS-HEMTs.

978-1-61284-243-1/11 $26.00 © 2011 IEEE

[1] U. Mishra *et al.*, IEEE Int. Microw. Symp. 2010, pp. 1130-1133. [2] P. S. Park *et al.*, IEEE Trans. Electron Devices **58**, 704 (2011).

[3] M. H. Wong *et al.*, IEEE Electron Device Lett. **29**, 1101 (2008). [4] D. F. Brown *et al.*, Appl. Phys. Lett. **94**, 153506 (2009).

[5] A. Chini *et al.*, 32nd International Symposium on Compound Semiconductors (ISCS), 2005, Rust, Germany.

[6] J. Simon *et al.*, Science **327**, 60 (2010). [7] M. Grundmann, Ph.D. Thesis, UCSB (2007)

[8] Y. Dora *et al.*, IEEE Electron Device Lett. **27**, 713 (2006). [9] S. Kolluri *et al.*, Electronic Materials Conference 2011.

[10] M. C. A. M. Koolen *et al.*, Proc. Bipolar Circuits and Technology Meeting, 1991, pp. 188-191.

[11] G. Dambrine *et al.*, IEEE Trans. Microw. Theory Tech. **36**, 1151 (1988). [12] S. Rajan *et al.*, J. Appl. Phys. **102**, 044501 (2007).

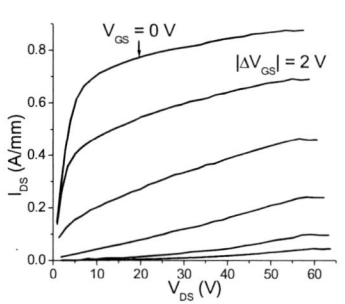

Fig. 1. 200-ns gate-pulsed *I-V* measurement of a well-passivated N-polar GaN MIS-HEMT with an AlN back-barrier [3], showing poor I_D saturation over a wide range of V_D.

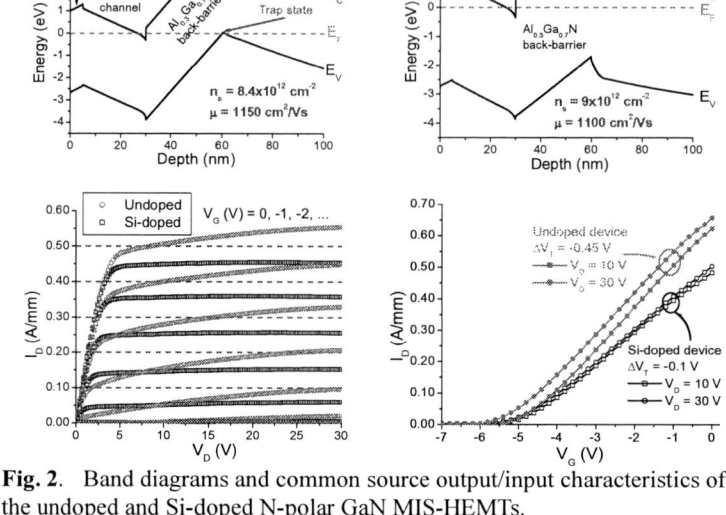

Fig. 2. Band diagrams and common source output/input characteristics of the undoped and Si-doped N-polar GaN MIS-HEMTs.

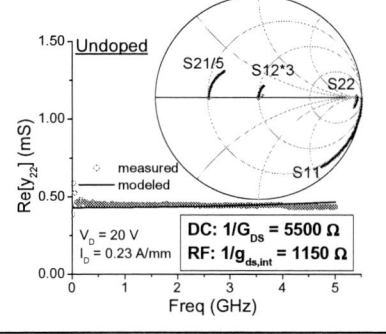

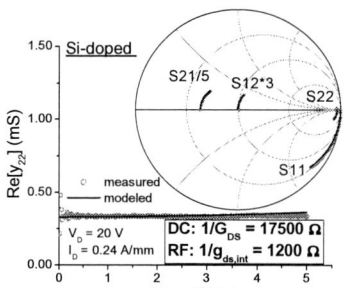

Fig. 3. The measured (circles) and modeled (line) s-parameters and Re[y_{22}] of the undoped (left) and Si-doped (right) devices from 10 MHz to 5 GHz after de-embedding the pad parasitics, showing a good match. The intrinsic output conductance ($r_{ds,int} = 1/\mathrm{Re}[y_{22,int}]$) was obtained by further de-embedding the series source/drain resistances.

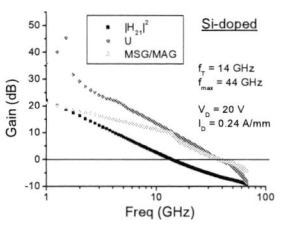

Fig. 4. Small signal performance of the undoped (top) and Si-doped (bottom) N-polar GaN MIS-HEMTs. The higher f_{max} for the undoped device could be accounted for by its higher intrinsic transconductance and lower source resistance (due to process variation).

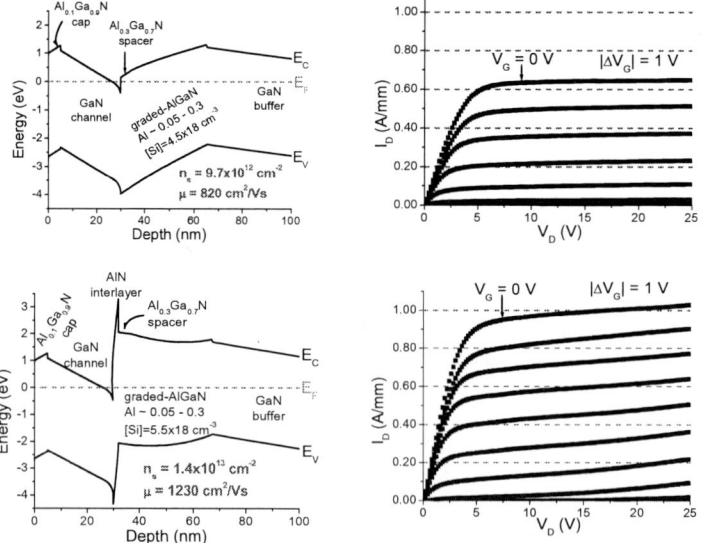

Fig. 5. Band diagrams and *DC-IV* of the graded-AlGaN back-barrier N-polar GaN MIS-HEMTs without (top) and with (bottm) an AlN interlayer.

978-1-61284-243-1/11 $26.00 © 2011 IEEE

Normally-off Gate-Recessed AlGaN/GaN-on-Si Hybrid MOS-HFET with Al$_2$O$_3$ Gate Dielectric

A. L. Corrion, M. Chen, R. Chu, S. D. Burnham, S. Khalil, D. Zehnder, B. Hughes, and K. Boutros

HRL Laboratories LLC, 3011 Malibu Canyon Road, Malibu, CA 90265, U.S.A.

Phone: 310-317-5254, Email: alcorrion@hrl.com

GaN-based HFETs offer a combination of high breakdown field, high current densities, and low on-resistance, making them well-suited for power-switching applications. Normally-off FETs are preferred in power switching applications for circuit simplicity and safety. Recently, a new type of normally-off GaN device has been reported: the hybrid metal-oxide-semiconductor (MOS)- or metal-insulator-semiconductor (MIS)-HFET, consisting of an MOS-type structure under the gate for normally-off operation and an HFET-like structure in the access regions for low on-resistance [1-6]. Optimization of the insulator-epi interface and insulator quality is critical for this type of device, since the electrons under gate electrode are in direct contact with the gate insulator. Previous reports of hybrid MOS-HFETs used SiO$_2$ or SiN gate dielectrics deposited by plasma-enhanced chemical vapor deposition (PECVD). However, alternative deposition methods such as atomic layer deposition (ALD) have been shown to result in superior thickness control, uniformity, conformality, and film quality, while ALD high-k gate dielectrics such as Al$_2$O$_3$ have generated significant interest for GaN HFETs due to excellent GaN interface quality. In this work, we fabricated normally-off AlGaN/GaN hybrid MOS-HFETs on (111) Si substrates using gate recess etching combined with an ALD Al$_2$O$_3$ gate dielectric for low gate leakage, low on-resistance, and high breakdown voltage. The gate fabrication process was optimized to reduce the trap density associated with the dielectric and eliminate threshold voltage hysteresis, which can result from slow traps in the dielectric or at the dielectric-epi interface [7]. A three-terminal breakdown voltage (V_B) of 1370V was measured at a gate bias of 0 V on a device with a 20 mm gate periphery and a low specific on-resistance (R_{on}) of 9.0 mΩ-cm^2. The resulting V_B^2/R_{on} figure of merit of 208 MW/cm^2 is among the highest values reported to-date for normally-off GaN-on-Si HFETs.

Figure 1 shows a schematic cross-section of the hybrid MOS-HFET device. The epitaxial material was grown by MOCVD on 3-inch (111) Si substrates and had a sheet resistance of 510 Ω/sq. PECVD Si$_3$N$_4$ was deposited for surface passivation, and gate recess etching was performed by low-power reactive ion etching (RIE). The ohmic contacts consisted of alloyed Ti/Al/Ni/Au. The Al$_2$O$_3$ gate dielectric was deposited by thermal ALD. Devices with gate peripheries of both 200 μm and 20 mm were fabricated with varying gate-drain spacings (L_{gd}).

Figure 2 shows transfer curves from a hybrid MOS-HFET with a gate periphery of 200 μm. The gate bias was swept from -3 V to +4 V and back to investigate hysteresis associated with the gate dielectric. The hysteresis was negligible, reflecting a high-quality ALD Al$_2$O$_3$ film and interface. The threshold voltage (defined as the x-intercept of a line tangent to the drain current at the point of peak transconductance) was +1.1V, while the pinch-off voltage (defined as the gate bias at a drain current of 1 mA/mm) was +0.4V. The average and standard deviation of the threshold voltage measured across the wafer were +1.17 +/- 0.14 V. The gate current was very low – <1x10^{-8} A/mm up to a gate bias of +4V – due to suppression of gate leakage by the ALD gate dielectric. Figure 3 shows the subthreshold characteristics of the 200 μm device. A low subthreshold slope of 85 mV/dec and high I_{on}/I_{off} ratio of ~3 x 10^8 were measured. The average and standard deviation of the three-terminal off-state breakdown voltages of the 200 um devices measured at V_{gs} = 0 V across a 3" wafer are shown in Fig. 4. The breakdown voltage was defined at a drain current of 1 mA/mm and was 1049 +/- 60 V at a gate-drain spacing of 12 μm. The DC IV and breakdown characteristics of the 20 mm devices are shown in Fig. 5. The on-resistance measured at V_{gs} = +3 V was 0.85 Ω (16.9 Ω-mm), corresponding to a specific on-resistance of 9.0 mΩ-cm^2, while the three-terminal off-state breakdown voltage at V_{gs}=0 V was 1370 V. The specific on-resistance was calculated using the entire active device area including ohmic contact pads. Figure 6 shows the breakdown characteristics on a log scale; the off-state drain current was ~1 x 10^{-4} A (~5 μA/mm) and gate current was ~5 x 10^{-5} A (~2.5 μA/mm). The breakdown and on-resistance of the 20 mm gate periphery hybrid MOS-HFET is compared to previously-reported normally-off GaN-on-Si HFETs in Fig. 7. These results show the potential of hybrid AlGaN/GaN MOS-HFETs using an Al$_2$O$_3$ gate dielectric for normally-off operation with good uniformity, low dielectric and interface trap density, high breakdown voltage, and low on-resistance.

[1] Oka et al., IEEE Elec. Dev. Lett., *29* (7) p. 668 (2008) [2] Ikeda et al., Proc. of the IEEE, *98* (7) p. 1151 (2010) [3] Huang et al., ISPSD, p. 295 (2008) [4] Kambayashi et al., ISPSD, p. 21 (2009) [5] Kambayashi et al., Solid State Elec. *54* (6) p. 660 (2010) [6] Kambayashi et al, Solid State Elec. *56* (1) p. 163 (2011) [7] Meyer et al., Solid State Elec. *54* (10) p. 1098 (2010) [8] Medjdoub et al., Elec. Dev. Lett. *31* (2) p. 111 (2010) [9] Boutros et al., IEDM (2009) [10] Ohmaki et al., Jpn. J. Appl. Phys. *45* L1168 (2006) [11] Kaneko et al., ISPSD, p. 25 (2009) [12] Chen et al., Elec. Dev. Lett. *30* (5) p. 430 (2009) [13] Ota et. al., IEDM (2009) [14] Uemoto et al., IEEE Trans. Elec. Dev. *54* (12) p. 3393 (2007) [15] Ikeda et al., ISPSD, p. 369 (2004) [16] Derluyn et al., IEDM (2009)

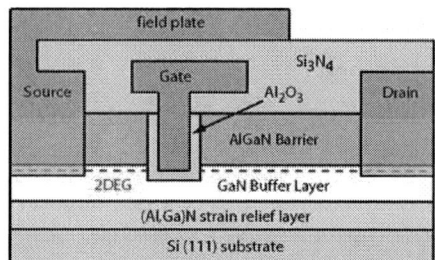

Fig. 1: Hybrid MOS-HFET schematic.

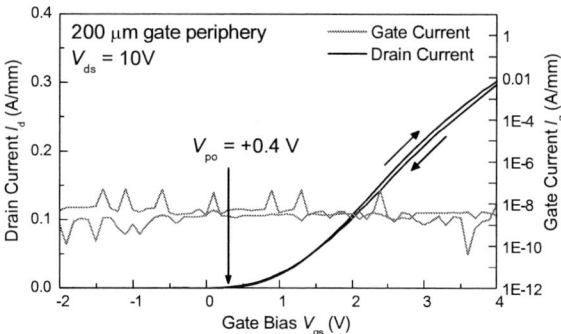

Fig. 2: Transfer characteristics of the hybrid MOS-HFET. (L_{gd} = 10 μm)

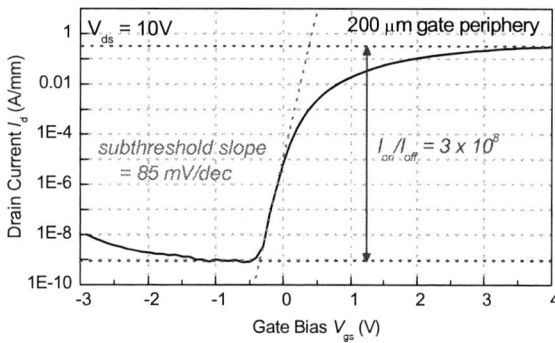

Fig. 3: Subthreshold characteristics of the hybrid MOS-HFET.

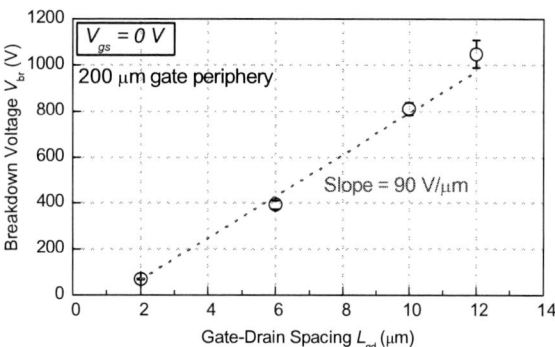

Fig. 4: Three-terminal off-state breakdown voltages of hybrid MOS-HFETs with a 200 μm gate periphery and varying gate-drain spacings measured at V_{gs}=0V (average +/- standard deviation across 3" wafer).

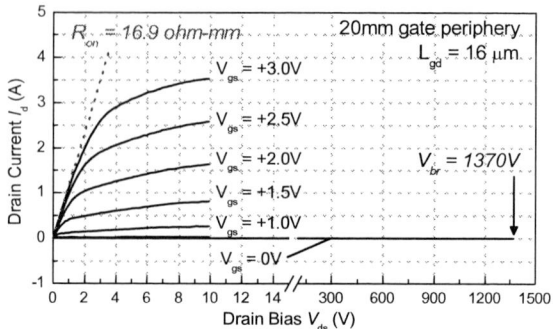

Fig. 5: DC IV and breakdown characteristics of the hybrid MOS-HFET with a 20 mm gate periphery.

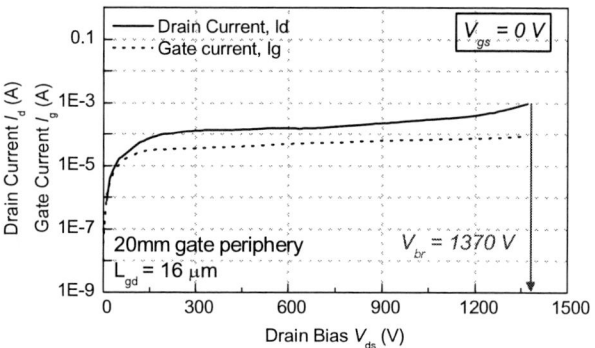

Fig. 6: Three-terminal off-state breakdown characteristics of the hybrid MOS-HFET with a 20 mm gate periphery measured at V_{gs}=0V.

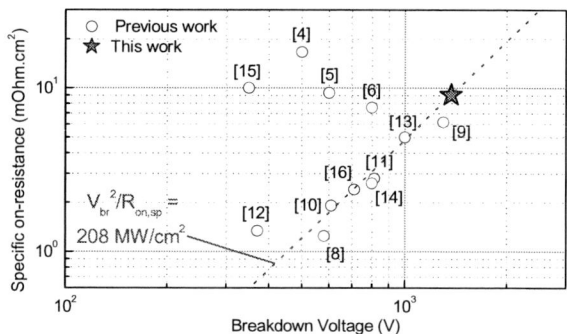

Figure 7: Comparison of this work (20 mm device) with previously-reported normally-off GaN-on-Si HFETs.

978-1-61284-243-1/11 $26.00 © 2011 IEEE   214

N-Polar AlGaN/GaN MIS-HEMTs on SiC with a 16.7 W/mm Power Density at 10 GHz Using an Al_2O_3 Based Etch Stop Technology for the Gate Recess

Seshadri Kolluri[*], Stacia Keller, Steven P. DenBaars and Umesh K. Mishra

Department of ECE, University of California, Santa Barbara, CA, USA

This paper presents the X-band and C-band power performance of MOCVD grown N-polar AlGaN/GaN MIS-HEMTs grown on semi-insulating SiC substrates. Additionally, an Al_2O_3 based etch stop technology was demonstrated for improving the manufacturability of N-polar GaN HEMTs with Si_xN_y passivation. The reported output power densities of 16.7 W/mm at 10 GHz and 20.7 W/mm at 4 GHz represent the highest reported values so far for an N-polar device, at both of these frequencies.

N-polar AlGaN/GaN HEMTs have demonstrated excellent small signal performance [1] and large signal performance comparable to Ga-polar devices in terms of power added efficiency (PAE) [2,3]. However, maximum output power obtained from the N-polar devices was limited by the low breakdown voltage of MBE grown devices [2], and high self-heating for devices grown on sapphire substrate by MOCVD [3]. N-polar HEMTs grown by MOCVD on n-type SiC by Brown et al. [4] suffered from very high parasitic capacitances between the pads and n-type SiC substrate, rendering them unsuitable for power measurements. In this study, N-polar HEMTs grown by MOCVD, on semi-insulating 4H-SiC substrates mis-oriented 4° toward the m-plane, were fabricated to enable the devices for high-frequency power applications.

Fig.1 shows the epitaxial structure of the fabricated device. Growth conditions were similar to the device reported in [4], with an additional AlGaN nucleation layer, to improve the structural properties of the GaN buffer layer. The GaN channel was capped with a 2 nm thick $Al_{0.6}Ga_{0.4}N$ layer and a 5 nm thick high temperature CVD (HTCVD) deposited Si_xN_y gate insulator to reduce the gate leakage. Ohmics were made of annealed Ti/Al/Ni/Au multilayer stack, and devices were isolated by mesa isolation with a BCl_3/Cl_2 reactive ion etch. The surface was subsequently passivated with a 1.5 nm thick Al_2O_3 deposited by atomic layer deposition (ALD), followed by a 160 nm thick Si_xN_y by PECVD. A high Si_xN_y to Al_2O_3 etch selectivity of 86 was observed for a dry etch using a CF_4/O_2 RIE plasma, as shown in Fig. 2. The Al_2O_3 layer acted as an effective etch stop layer for the gate recess-etch through the PECVD Si_xN_y, without affecting the HTCVD Si_xN_y layer which was used as the gate insulator. After the gate recesses etch, Al_2O_3 under the gate was removed by wet etching in AZ 726 MIF developer, followed by the deposition of Ni/Au/Ni (30/350/50 nm) gate metallization into the trench. All the reported measurements, except the 4 GHz power measurement, were carried out on a device 2 x 50 μm wide, with a nominal gate length of 0.7 μm, and a source-drain spacing of 2 μm (Device A). The 4 GHz power measurements were carried out on a 2 x 75 μm wide device with a source drain spacing of 2.3 μm (Device B).

Pulsed I-V and transfer characteristics of Device A are shown in Fig. 3 and Fig. 4 respectively. Though there was no significant current collapse in the device under pulsed conditions, some kinks were observed in the DC currents. The cause for these kinks is yet to be understood. However, these kinks did not seem to have a significant negative impact on device performance. Peak transconductance of the device was 230 mS/mm (Fig. 4). Small signal measurements using Agilent E8361A network analyzer yielded an extrinsic f_T and f_{Max} of 16 GHz and 58 GHz respectively (Fig. 5), at a drain bias of 20 V and gate bias of -5 V.

Large signal measurements were performed at 10 GHz on a Maury loadpull system. As shown in Fig. 6, the output power increased with increasing drain bias, without much degradation in PAE, indicating good dispersion control in the device. At a drain bias of 58V, an output power density of 16.7 W/mm, with an associated PAE of 44 % was observed (Fig. 7). This is the highest reported X-band power density for an N-polar device and, it compares well with Ga-polar devices reported by Pei *et al.*, [5] (20.9 W/mm output power density, with a 40% associated PAE at a 83V drain bias). 4 GHz power measurements were performed on a similar device with a width of 2 x 75 μm and source drain spacing of 2.3 μm (Device B). At a drain bias of 70 V, Device B yielded an output power density of 20.7 W/mm with an associated PAE of 60% at 4 GHz, which is the best reported power performance for an N-polar device at 4 GHz (Fig. 8).

In conclusion, high power N-polar GaN MIS-HEMTs grown by MOCVD on a semi-insulating SiC substrate, for X-band and C-band power applications have been presented. Future directions include scaling of the devices to deep-submicron gate lengths, to achieve Ka-band power performance.

References: [1] Singisetti *et al.*, Appl. Phys. Express 4 (2011). [2] Wong *et al.*, Appl. Phys. Lett., 94, 182103 (2009). [3] Kolluri *et al.*, IEEE EDL, Vol. 30, No. 6 (2009). [4] Brown *et al.*, Appl. Phys. Lett., 94, 153506 (2009). [5] Pei *et al.*, Jap. Jrnl. Of Appl. Phys., Vol. 46, No. 45, 2007.

* Corresponding Author: seshadri@ece.ucsb.edu, Phone: +1-617-959-3585, Fax: +1-805-893-8714

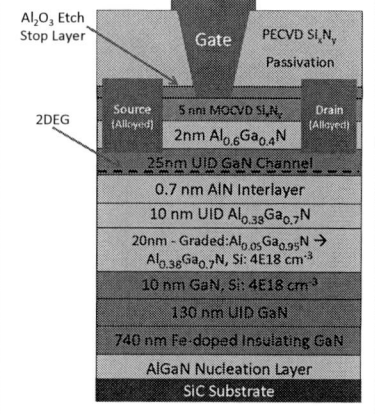

Fig. 1. Epitaxial layer structure and schematic cross-section of the device.

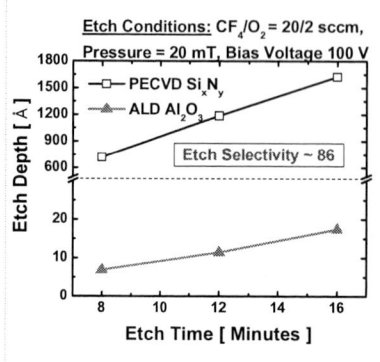

Fig. 2. Comparison of etch rates for PECVD Si_xN_y and ALD Al_2O_3 using a CF_4/O_2 etch chemistry.

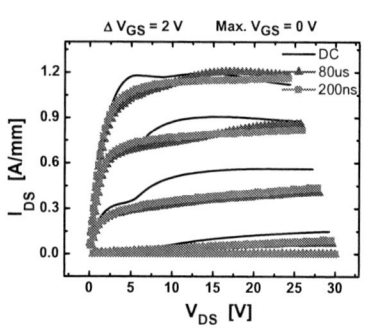

Fig. 3. Pulsed I-V characteristics of the device. (W = 100 μm, L_{SD} = 2 μm, L_G = 0.7 um)

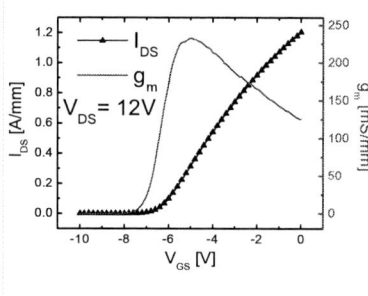

Fig. 4. Transfer characteristics and transconductance (g_m) of the device at a drain bias of 12 V. (W = 100 μm, L_{SD} = 2 μm, L_G = 0.7 um)

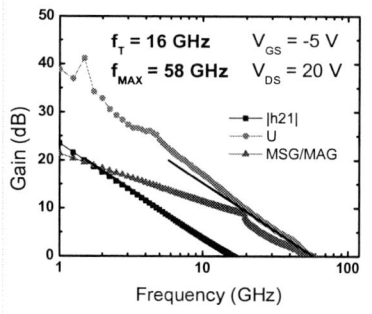

Fig. 5. Small signal performance of the device. (W = 100 μm, L_{SD} = 2 μm, L_G = 0.7 um)

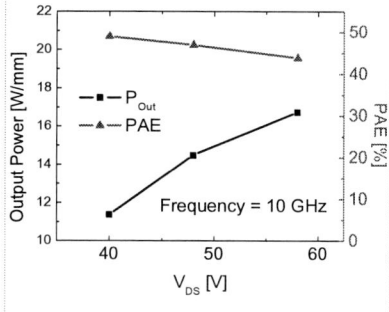

Fig. 6. Power Measurements at 10 GHz: Dependence of P_{out} and PAE on drain bias. (W = 100 μm, L_{SD} = 2 μm, L_G = 0.7 um)

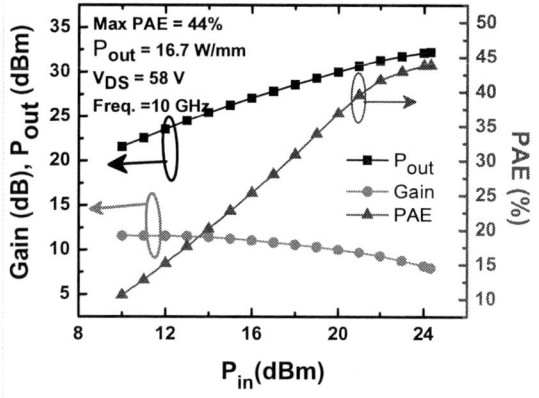

Fig. 7. Large signal performance of the device at 10 GHz. (V_{DS} = 58 V, W = 150 μm, L_{SD} = 2.3 μm, L_G = 0.7 um)

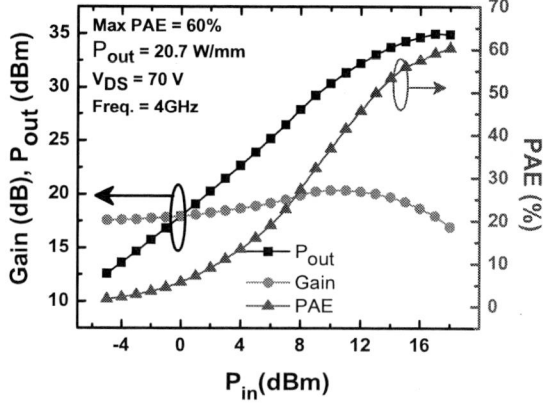

Fig. 8. Large signal performance of the device at 4 GHz. (V_{DS} = 70V, W = 150 μm, L_{SD} = 2.3 μm, L_G = 0.7 um)

978-1-61284-243-1/11 $26.00 © 2011 IEEE   216

Total GaN Solution to Electrical Power Conversion

Y.-F. Wu, R. Coffie, N. Fichtenbaum, Y. Dora, C.S. Suh, L. Shen, P. Parikh and U.K. Mishra

Transphorm Inc., 115 Castilian Drive, Goleta, CA 93117

Email: ywu@transphormusa.com

ABSTRACT: *We present the first 600V-class, total GaN solution for electrical power conversion applications. A 220V-400V boost converter using a GaN transistor and a GaN diode with fast & clean hard-switched waveforms has been demonstrated. The conversion efficiency was >99.1% at 100 kHz and >98.2% at 800 kHz.*

INTRODUCTION: Since the advent of GaN high electron mobility transistor (HEMT) in 1993 and the demonstration of its RF power capability in 1996, dramatic progress has been made in the last 15 years and commercial GaN products are now available in RF transmission applications with unmatched bandwidth and efficiency. The 2nd wave of GaN Electronics is currently emerging fueled by the need for energy saving in electricity conversion. Although research for high voltage GaN devices has resulted in numerous publications, no GaN devices were shown to outperform the matured Si counterparts at 400V and above. Here we announce the first 600V-class, total GaN solution for power conversion applications.

DEVICEs & SPECIFICATIONS: The devices in this presentation include a GaN HEMT and a GaN diode produced by Transphorm Inc., both in TO-220 packages. The GaN HEMT features the Quiet-Tab[TM] package scheme, with Gate-Source-Drain (GSD) pin-out arrangement and a package base (or tab) connected to the source terminal.

The plastic-packaged GaN HEMT has a gate threshold of +2.1V typical at 1mA drain current and the drain leakage is $10\mu A$ typical at Vgs=0V & Vd=600V. The on-resistance is 0.25Ω typical and 0.31Ω maximum. The pulsed drain current is 40A at a Vgs=8V & Vds=10V. The rated continuous (CW) drain current is 8.5A at a case temperature (Tc) of 25°C. Compared with similarly-rated state-of-the-art Si super-junction MOSFETs on the market, this 1st generation GaN HEMT has about 28% reduction in on-resistance, 80% increase in pulsed current and 40% reduction in output charging energy.

The GaN diode has a Schottky barrier with a forward voltage of 1.3V typical at 2A & 25°C. Pulsed current is 20A at 25 °C and CW current is rated as 2.5A at Tc=125°C. Reverse leakage is 10uA typical and 100uA maximum at 600V & 25°C.

ELECTRICITY CONVERSION EXPERIMENT: To observe the operation characteristics and performance advantages of the GaN devices over their Si counterparts, two continuous-mode 220V-400V dc-to-dc boost converters have been constructed with identical layout and passive components. One uses state-of-the-art Si MOSFET and Si ultra-fast diode and another uses Transphorm GaN HEMT and GaN Schottky diode. In addition to the transistor and diode, each converter includes a gate driver set to 0V off &12V on, a film capacitor for input energy storage, a high Q inductor for boost action, a surface-mount ceramic capacitor at the diode cathode for transient reduction and another storage capacitor for stable dc output. The converter design strives for text-book simplicity with a tight layout, without any gate drive resistor and snubbing components. The Quiet-Tab[TM] package allows the GaN HEMT to be mounted on a grounded heat-sink without an insulating spacer, which facilitates heat dissipation and minimizes source inductance.

The gate and drain waveforms (Vgs and Vds) were first examined at the rated power of 760W. During the transistor turn-on process, the Si converter showed violent Vgs spikes right at the Vds transition from blocking to conducting. There are attributed to the diode recovery charge, the MOSFET feedback capacitance and the interaction between gate & drain currents at the single source lead. In comparison, the GaN converter exhibited a fast yet smooth transient with small ripples, confirming the advantages of the GaN devices including free of minority charges, low capacitances and the low-inductance package configuration. During the turn off process, the Si converter spent 25 ns for the Vds to ramp up while the GaN converter took only 7 ns, indicating a reduction of 3.5 times in output capacitance.

Conversion performance was then characterized at 100 kHz, a popular PWM frequency in today's power factor correction circuits. The GaN converter achieved >99% efficiency from 200 W to the rated power of 760 W with a peak value of 99.16% at 490W. It outperformed the Si converter by 0.4-0.5%, corresponding to a loss reduction of 35% at mid power. Performance was also tested at a much higher frequency of 800 kHz. The GaN converter maintained an efficiency >98% from 320W to rated power, with a peak value >98.2% at 500-600W. In contrast, the Si converter suffered from a drastic increase in power loss, resulting in efficiencies below 93.5%. At 410W output level the power loss of the Si converter reached 29W, pushing the MOSFET close to its maximum junction temperature. In comparison, the GaN converter dissipated only 7.5W at the same power level, translating to a loss reduction of 75%. The ability of the GaN devices to maintain high efficiency at high frequencies opens doors to future high-density power circuits.

CONCLUSION: 600V-class GaN HEMTs and GaN diodes were presented as Total GaN[TM] solution to electricity conversion. A 220V-400V boost converter using the GaN-only devices outperformed that of Si-only devices by loss reduction of 35% at 100 kHz and 75% at 800 kHz. These GaN devices represent the initial thrust of the III-V wide-gap semiconductor into the Power Electronics industry.

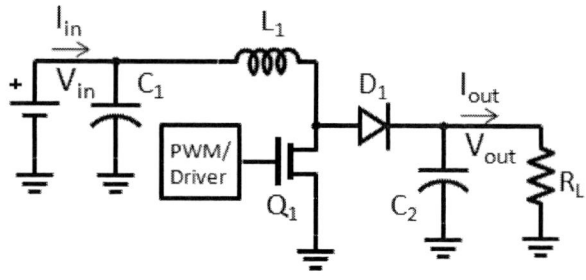

Fig.1. Schematics of the dc boost converter. With GaN Transistor and diode, no gate resistor, no snuber and no insulating shim between transistor & heatsink were necessary.

Fig.2. Photo of the dc converter before mounting inductor L1.

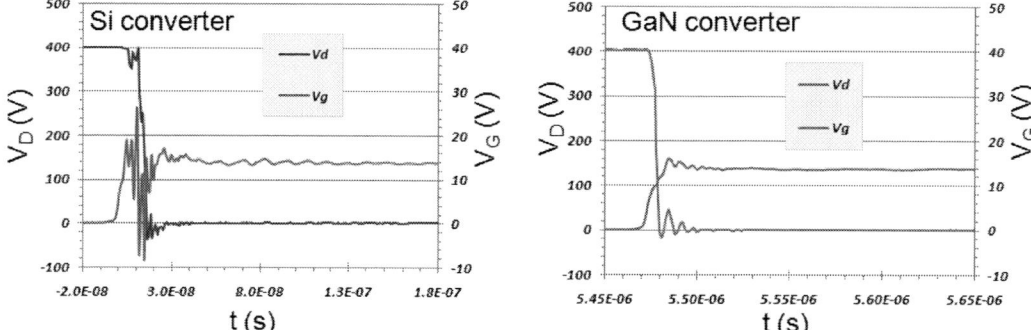

Fig.3. Gate and drain turn-on waveforms of the Si-only and total GaN dc converters.

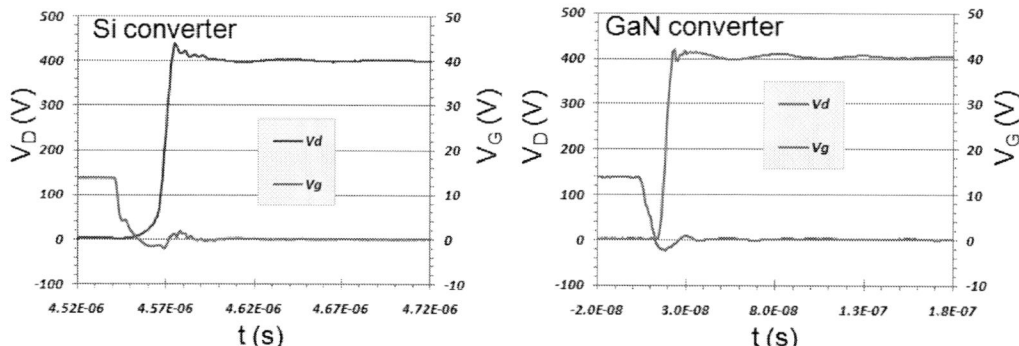

Fig.4. Gate and drain turn-off waveforms of the Si-only and total GaN dc converters.

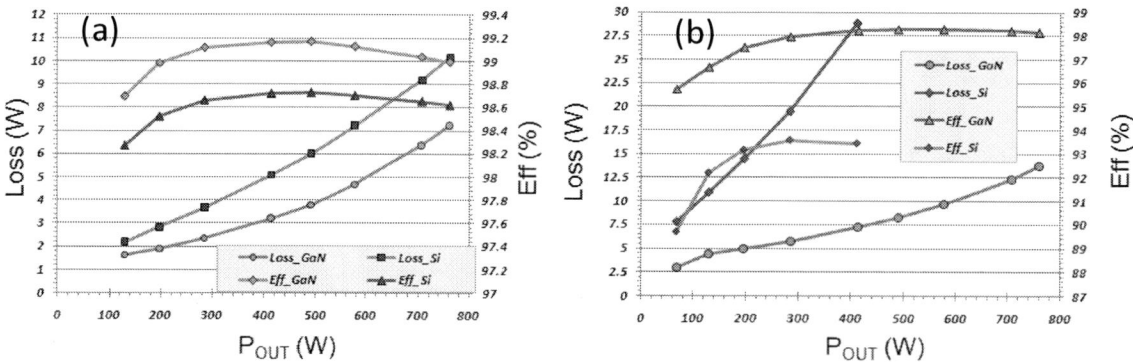

Fig.5. Conversion efficiency & loss as a function of output power at (a) 100kHz and (b) 800kHz.

978-1-61284-243-1/11 $26.00 © 2011 IEEE   218

First AlN/GaN HEMTs power measurement at 18 GHz on Silicon substrate

F. Medjdoub, M. Zegaoui, D. Ducatteau, N. Rolland and P.A. Rolland

IEMN, Avenue Poincaré, 59652 Villeneuve d'Ascq, France.
Email: farid.medjdoub@iemn.univ-lille1.fr

AlN/GaN heterostructure is an ideal candidate to push the limits of microwave GaN-based devices owing to the maximum theoretical spontaneous and piezoelectric difference between the epitaxial AlN barrier and the underlying GaN layer. If the tricky growth conditions of this binary can be controlled, AlN/GaN HEMTs promise breakthrough performances, superior to any other III-V nitride-based heterostructure [1]. In particular, this structure should allow the extension of the GaN-based frequency operation due to the possibility to significantly reduce the gate length while maintaining an appropriate gate-to-channel aspect ratio to mitigate short channel effects. However, gate leakage current remains a serious issue with such ultrathin barrier heterostructure and gate dielectrics that often leads to device instability are generally used to overcome this problem. Furthermore, there is an increasing interest in the growth of GaN-on-Si substrates because of its low cost, large size, good thermal conductivity and the potential for integration with Si-based devices. In this work, we developed a novel AlN/GaN HEMT technology on Si substrate. The highest GaN-on-Si drain current density as well as a record transconductance together with excellent RF performance have been achieved. Additionally, AlN/GaN HEMT power measurements at 18 GHz have been performed for the first time. These results show the outstanding potential of this structure to extend GaN-on-Si performances to millimeter wave applications.

In this study, AlN/GaN heterostructures having various barrier thicknesses with a 5 nm thick in-situ Si_3N_4 cap layer on 4-inch Si (111) substrate have been grown by MOCVD (Fig. 1). The use of an in-situ SiN cap layer enables to prevent 2DEG depletion from the surface, to control the strain relaxation and allows for strengthening the surface robustness [2]. Gate field plates have been fabricated through a stack of 50 nm plasma enhanced CVD Si_3N_4 and 5 nm in-situ Si_3N_4. Devices have been passivated with 200 nm PECVD Si_3N_4. A benchmarking of the 2DEG carrier density (measured by Hall Effect) as a function of the barrier thickness of 30% Al content AlGaN/GaN, lattice matched InAlN/GaN and AlN/GaN HEMT structures has been performed (Fig. 2). These measurements clearly confirm the superiority of the AlN/GaN 2DEG properties, especially when using ultrathin barrier thickness below 10 nm. For instance, carrier density higher than 2×10^{13} cm^{-2} are observed with an AlN thickness as low as 4 nm, unreachable with any other nitride heterostructure for such a thin barrier.

Fig. 3 shows the DC output as well as the transfer characteristics of 0.2×50 μm^2 6 nm barrier thickness AlN/GaN-on-Si HEMT. The maximum DC output current density at $V_{gs} = +2$ V exceeds 2 A/mm (2.03 A/mm for the presented device), which represents to the best of our knowledge the highest drain current density of any GaN-based HEMT structure grown on Si substrate. This reflects the high sheet carrier density in this novel heterostructure. The same DC characteristics for a 3 nm ultrathin barrier thickness AlN/GaN-on-Si HEMT are shown in Fig. 4. Maximum drain current density drops to 1.2 A/mm due to the lower 2DEG carrier concentration. On the other hand, a record GaN-on-Si HEMT extrinsic transconductance as high as 470 mS/mm is achieved resulting from the combination of high 2DEG density and short gate to channel distance. It has to be noted that low leakage current without the use of any gate dielectrics are observed in both cases, which is remarkable and quite unique regarding the ultrathin barrier thicknesses. This is attributed to the high AlN growth quality as well as the surface robustness enhanced by the in-situ SiN cap layer.

Despite the relatively high access resistances of 0.65 Ω.mm and residual losses about 0.7 dB/mm at 50 GHz (Fig. 5), excellent RF parameters have been extracted for a 0.2 μm gate length resulting from the high transconductance. Current gain cut-off frequency and maximum oscillation frequency $f_T = 59$ GHz and $f_{max} = 102$ GHz (shown in Fig. 6) extrapolated from the current gain H_{21} and the power gain (U) at the optimum bias conditions, respectively. This shows that downscaling this technology should allow very high frequency operation considering the high aspect ratio achievable with sub-100 nm gate lengths.

The pulsed IV and DC characteristics at open channel ($V_{gs} = +2$ V) and at pinch-off ($V_{gs} = -4$ V) of a 3 nm barrier thickness AlN/GaN-on-Si HEMT for different quiescent bias points (cold point: $V_{DS0} = 0$ V and $V_{GS0} = 0$ V, gate lag: $V_{DS0} = 0$ V and $V_{GS0} = -5$ V, drain lag: $V_{DS0} = 15$ V and $V_{GS0} = -5$ V) using 500 ns pulses and 1% duty cycle are shown in Fig. 7. The dynamic load-line at $V_{DS} = 20$ V (Fig. 7) acquired with a Large Signal Network Analyzer (LSNA) confirms the reduced RF dispersion at this bias. Fig. 8 depicts the first on-wafer CW power sweep at 18 GHz performed on a 0.2×200 μm^2 3 nm barrier thickness AlN/GaN-on-Si HEMT. At $V_{DS} = 20$ V, output power density P_{OUT} close to 3 W/mm is achieved, as expected from the pulsed and load line measurements. The power added efficiency PAE peaks above 20% and the linear gain is about 11 dB. Power measurements on the 6 nm barrier AlN/GaN HEMTs at 18 GHz that delivers much higher current density are under way.

Acknowledgement: The authors would like to thank the company EpiGaN for providing the epi-material.

[1] M. Higashiwaki *et al.*, *IEEE Electron Dev. Lett.*, 27, p. 719, 2006
[2] J. Derluyn *et al*, *J. Appl. Phys.*, vol. 98, 054501, 2005

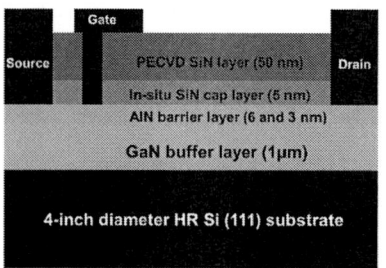

Fig. 1: Cross section of the fabricated AlN/GaN-on-Si HEMTs

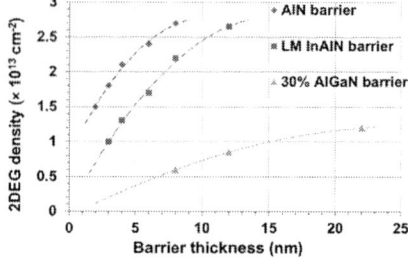

Fig. 2: HEMT 2DEG density vs. various barrier thickness material

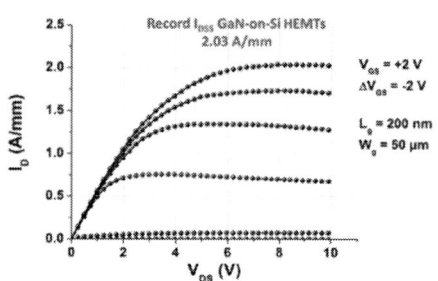

Fig. 3 : Record DC output (left) and transfer (right) characteristics at V_{DS} = 5 V of a 0.2×50 μm^2 **6 nm thick AlN barrier** GaN-on-Si HEMT

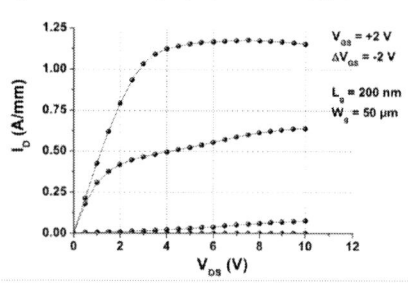

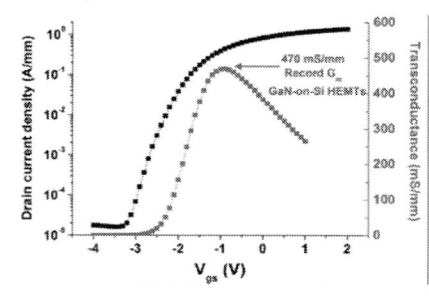

Fig. 4 : DC output (left) and transfer (right) characteristics at V_{DS} = 5 V of a 0.2×50 μm^2 **3 nm thick AlN barrier** GaN-on-Si HEMT

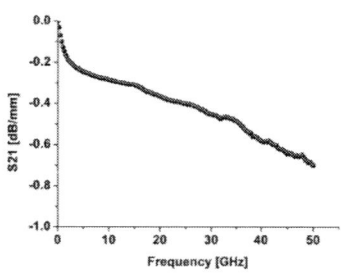

Fig. 5: Standard CPW measurement of AlN/GaN on Si up to 50 GHz

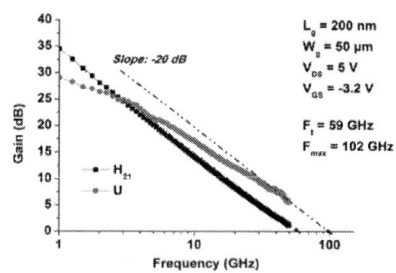

Fig. 6: RF performance of a 0.2×50 μm^2 **6.0 nm AlN barrier thickness** GaN-on-Si HEMT

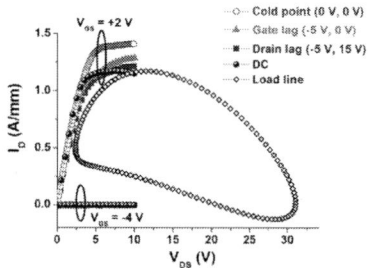

Fig. 7: Pulsed IV for different quiescent bias points and DC characteristics at open channel and at pinch-off as well as the dynamic load-line at V_{DS} = 20 V of a 3 nm barrier thickness AlN/GaN-on-Si HEMT

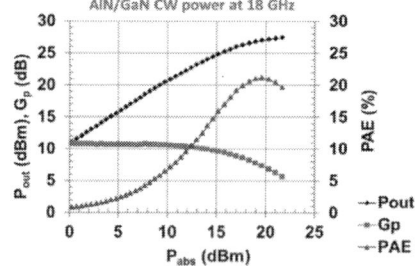

Fig. 8: CW power sweep at 18 GHz of a 200 μm AlN/GaN-on-Si HEMT (L_g = 0.2 μm) at V_{DS} = 20 V, showing almost 3 W/mm P_{OUT}

978-1-61284-243-1/11 $26.00 © 2011 IEEE

Enhanced mobility for MOCVD grown AlGaN/GaN HEMTs on Si substrate

S. Lawrence Selvaraj, Arata Watanabe and Takashi Egawa

Research Center for Nano-Device and System, Nagoya Institute of Technology, Gokiso-cho, Showa-ku, Nagoya 466-8555, Japan, Fax: +81-52-7355546, email: *lawrence@nitech.ac.jp*

Growth and optimization of *AlGaN/GaN* transistors on *Si* substrate is an important subject of investigation to surpass the cost effective substrates like *GaN*, *SiC* and sapphire. In the ongoing study of *GaN* devices on *Si*, we have achieved record high room temperature mobility (μ_{RT}) of 3215 cm^2/Vs for *AlGaN/GaN* HEMTs grown by *MOCVD*. Our approach to increase the mobility involves (i) reducing dislocation density by using thick buffer on *Si* and (ii) using 1.5 nm *AlN* spacer. This is the highest μ_{RT} so far reported for *AlGaN/GaN* grown on *GaN*, *SiC* and sapphire substrates. The growth and device characteristics of these HEMTs which have high mobility are presented in this report.

The highest μ_{RT} of 2200 cm^2/Vs was reported recently for *AlGaN/GaN* HEMTs grown on *SiC* [1]. However, for *AlGaN/GaN* heterostructure grown on *Si*, we reported the highest μ_{RT} as 1863 cm^2/Vs in 2007 [2]. A large lattice mismatch (over 17%) between *Si* substrate and *GaN* leads to high density of dislocations at the interface which affects the *2DEG* mobility. The effect of dislocations on *2DEG* mobility becomes more prominent when the dislocation densities are above 10^8 - 10^{10} cm^{-2} [3]. By reducing the dislocation density as low as 10^{10}cm^{-3} it is possible to achieve a high mobility. Hence reducing the dislocations for *AlGaN/GaN* HEMTs on *Si* continues to be the priority of our study. To improve the quality of *GaN* grown on 4-inch *p-Si*, thick buffers were used which lowered the screw dislocation density as low as 1×10^9 cm^{-2} [4]. The hetero-structure of our HEMTs shown in Fig. 1 have 1.5 nm *AlN* spacer layer for mobility enhancement. In order to compare the mobility and other device performance, the device structures were grown using thin (0.5 μm) and thick buffer (5.0 μm). The *AFM* picture of *AlGaN/GaN* grown on thick buffer (Fig. 2) has a *RMS* roughness of 0.14 nm and *P-V* distance of 1.9 nm signifying a high quality growth with uniform surface morphology similar to the growth achieved on *SiC* substrates. The μ_{RT} measured for 1.5 μm and 7 μm total thickness (T_{Tot}) samples are respectively 1863 cm^2/Vs and 3215 cm^2/Vs. A lowest sheet resistance (R_{sh}) of 171 $\Omega/\square$ and carrier density of 1.1×10^{13} cm^{-2} was observed for *HEMT* of 7 μm T_{Tot}. Such a high mobility is attributed to improved *AlGaN/GaN* hetero-interface quality arising out of low dislocations due to growth on thick buffer [5]. Fig. 3 summarizes the *Hall* properties of the two HEMTs having different T_{Tot}. In Fig. 4, our mobility values are compared with theoretical μ vs *2DEG density* for *AlGaN/GaN* grown on *SiC* substrates.

To study the implications of the high mobility on the device properties, *SiO₂* passivated HEMTs (W_g/L_g = 15/1.5 μm) were processed. A contact resistance (R_c) and sheet resistance (R_{sh}) of 2.54 Ω.mm and 320 $\Omega/\square$ was observed respectively. Though there was decrease in the R_c and R_{sh}, the observed R_{sh} from *TLM* was higher than R_{sh} measured from *Hall* measurement. Because of the presence of 1.5 nm insulating *AlN* spacer layer, the flow of electrons in the *2DEG* channel across source and drain comes under additional resistance. So we recess-etched about 60 nm at the source/drain region prior to *Ohmic* contact and obtained low resistance values (R_c = 0.9 Ω.mm; R_{sh} = 270 $\Omega/\square$) as summarized in Fig. 5. The I_{DS}-V_{DS} characteristics of the HEMTs with high mobility as shown in Fig. 6 yielded a high I_{DSmax} of 900 mA/mm, an I_{DSmax} equivalent to that obtained for *AlGaN/GaN* grown on *SiC* [6] or *GaN* [7]. The high μ of the grown HEMTs give a low drain resistance (R_d) of 3.1 Ω.mm and *specific on-resistance* of 2.9×10^{-4} Ω.cm^2 calculated in the linear I_{DS}-V_{DS} region of the Fig. 6. The summary of R_d and R_{d-ON} for devices with low and high μ are shown in Figures 7 and 8. A g_{mmax} of 183 mS/mm observed for HEMTs with L_g of 1.5 μm. In Fig. 9, a three order decrease in the drain leakage through *i-GaN* for HEMTs grown on thick buffer signifying a high quality *i-GaN* with less dislocations. The *Depth vs 2DEG* profile from the *C-V* measurements (Fig. 10) also confirm the depletion layer spreading into the *Si* substrate has relatively low carrier concentration compared to the HEMTs grown on thin buffer. This low carrier concentration at the buffer/*Si* interface reduces the buffer and substrate leakage to offer a high *three terminal-OFF breakdown* of 486 V as shown in Fig. 11. Thus the figure of merit (*FOM* = BV^2/R_{d-ON}) scales to a record high 8.0×10^8 V$^2\Omega^{-1}$cm^{-2} for our high mobility devices.

As a summary, a high quality *AlGaN/GaN* transistors equivalent to the growth quality on *GaN* or *SiC* substrates was achieved on *Si* substrate with a high μ_{RT} of 3215 cm^2/Vs by using thick buffer layers.

References: [1] J. W. Chung et. al., *IEEE Electron. Dev. Lett.*, vol. **31**, 195 (2010); [2] S. L. Selvaraj et. al., *Appl. Phys. Lett.*, vol. **90**, 173506 (2007); [3] D. Jena et. al., *Appl. Phys. Lett.*, vol. **76**, 1707 (2000); [4] S. L. Selvaraj et. al., *IEEE Electron. Dev. Lett.*, vol. **30**, 587 (2009); [5] L. F. Eastman et. al., *IEEE Trans. Electron. Dev.*, vol. **48**, 479 (2001); [6] L. Shen et. al., *IEEE Electron. Dev. Lett.*, vol. **22**, 457 (2001); [7] D. F. Storm et. al., *Electron. Lett.*, vol. **40**, 1226 (2004).

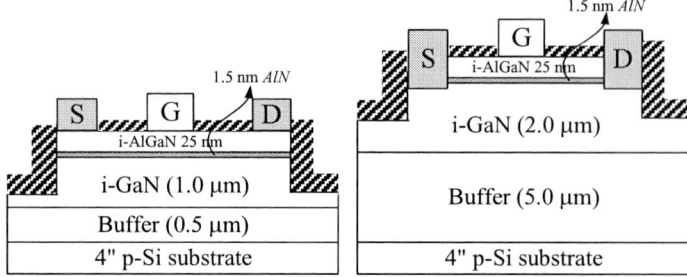

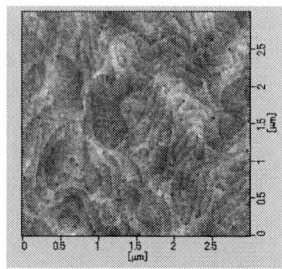

Fig.1. *MOCVD* grown heterostructures with (a) thin buffer of $\mu = 1863$ cm²/Vs and (b) thick buffer of $\mu = 3215$ cm²/Vs. The use of thick buffer to reduce dislocations and 1.5 nm *AlN* spacer attributes to the high mobility.

Fig. 2. *AFM* image of the *MOCVD* grown *AlGaN* surface with *RMS* of 0.14 nm and *P-V* of 1.9 nm.

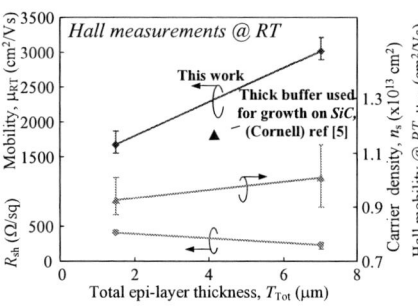

Fig. 3. Summary of *Hall* measurements at room temperature for *MOCVD* grown *HEMTs* with high mobility.

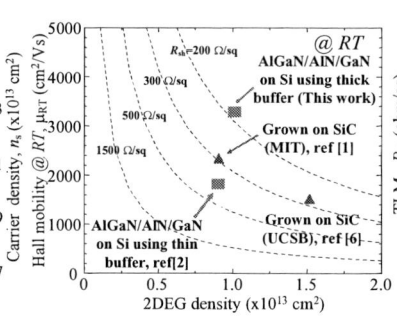

Fig. 4. Our mobility values for *AlGaN/GaN* grown on *Si* compared with theoretical and literature reports for growth on *SiC*.

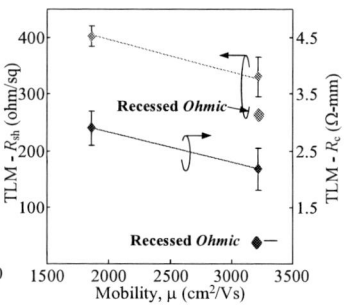

Fig. 5. Sheet and contact resistance for the *HEMTs* measured using *TLM* (Un-recessed and recessed).

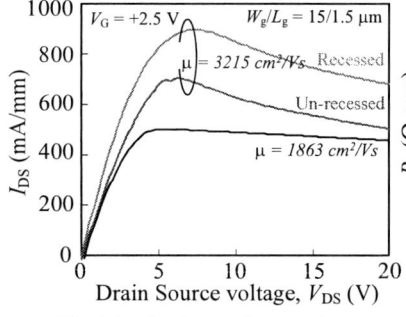

Fig. 6. I_{DS}-V_{DS} characteristics on the devices and contribution of the mobility towards *turn-ON* voltage.

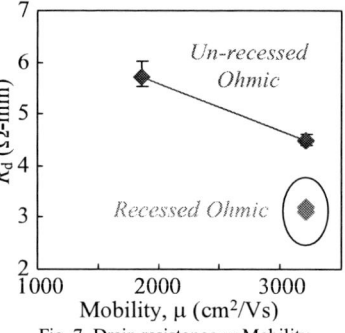

Fig. 7. Drain resistance *vs* Mobility for *HEMTs* with un-recessed and recessed source/drain contact.

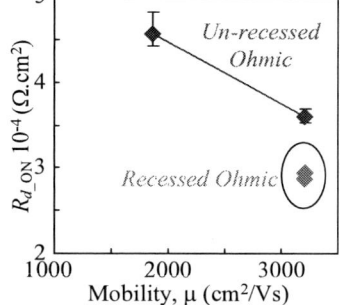

Fig. 8. *Specific-ON* resistance *vs* Mobility for *HEMTs* with un-recessed and recessed source/drain contact.

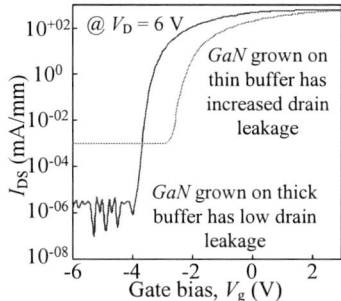

Fig. 9. I_{DS} vs V_g shows a low drain leakage for *i-GaN* grown on thick buffer owing to low dislocations.

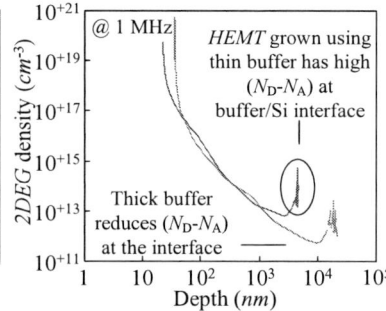

Fig. 10. *2DEG* density *vs* Depth profile from *C-V* measurements confirm a low doping concentration at the Si interface.

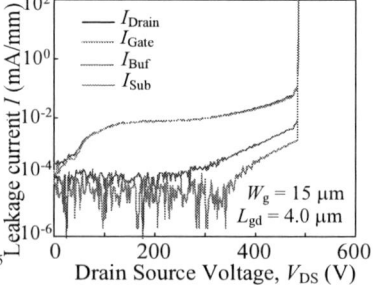

Fig. 11. Three terminal-OFF breakdown for HEMTs. A high breakdown of 486 V was observed.

978-1-61284-243-1/11 $26.00 © 2011 IEEE

High Performance GaN-on-Si Power Switch: Role of Substrate Bias in Device Characteristics

Rongming Chu, Daniel Zehnder, Brian Hughes, and Karim Boutros
HRL Laboratories LLC, 3011 Malibu Canyon Road, Malibu, CA 90265 (e-mail: rchu@hrl.com)

Field-effect transistors based on the low-cost GaN-on-Si platform are promising candidates for high-efficiency power switching at high frequencies. We have reported a normally-off GaN-on-Si switch with a blocking voltage of 1200V, and a very low dynamic on-resistance [1]. For future improvement of the GaN-on-Si switching technology, it is important to understand the role of the non-insulating Si-substrate in device characteristics. In this paper, we discuss the static (DC) and dynamic (switching) characteristics of the GaN-on-Si device, focusing on the impact of bias conditions applied on the Si substrate. It was found that state-of-the-art dynamic on-resistance characteristics of the GaN-on-Si switch can be achieved by properly terminating the Si substrate potential.

Device structure and top view photograph is shown in Fig. 1. A float-zone, high resistivity, p-type Si (111) wafer was used as the substrate. More information about the fabrication process can be found in Ref. 1. After thinning down the substrate to about 100 μm, the backside of the substrate was coated with metal and can be independently biased. Fig. 2 shows the output IV characteristics of a device with its substrate grounded. The static on-resistance was 0.5 Ω; the maximum output current was 5 A; and the blocking voltage was 1200 V with a drain leakage of 3 mA when the gate was biased at 0 V. Fig. 3 shows the transfer characteristics of the device under different substrate bias conditions. A negative bias applied on the substrate significantly shifted the threshold voltage toward positive, leading to increased on-resistance and decreased output current. The amount of threshold voltage shift caused by a positive substrate bias was much less; because the metal/p-Si junction can block a positive (rather than negative) substrate bias from modulating the 2DEG.

Dynamic on-resistance measured at 5 μs after the device was switched from off-state to on-state is plotted in Fig. 4. When the substrate was connected to a fixed bias, there was little degradation of dynamic on-resistance up to 600 V bias. Significant degradation of dynamic on-resistance can be observed when the substrate was floated. A model was developed to explain this observation. When the substrate-floated device was subjected to a positive drain bias, electrons and holes were generated in the Si substrate toward the drain and the source side respectively, as depicted in Fig. 5. Immediately after the device was switched to the on-state, electrons on the drain side did not have enough time to recombine with holes on the source side. The excessive electrons in the Si substrate depleted the 2DEG in the channel, causing dynamic on-resistance degradation. When the substrate was terminated to a fixed bias, those excessive electrons were able to be quickly removed through the substrate terminal; hence the dynamic on-resistance was not affected. Fig. 6 shows the waveform of the substrate terminal potential during an off-to-on switching cycle of a substrate-floated GaN-on-Si device. During the off-state, the positive drain bias led to a positive potential on the substrate. When the device was switched from off to on, sudden drop in the drain bias pulled the substrate bias down to a negative value. After that, the substrate bias gradually changed toward zero as the electrons were recombining with the holes in the Si substrate. The time constant associated with the electron-hole recombination process was in the order of milliseconds, indicating that the substrate bias is important for switching characteristics at frequencies greater than 1 KHz.

Reference:
1. R.M.Chu, A. Corrion, M. Chen, R. Li, D. Wong, D. Zehnder, B. Hughes, and K. Boutros, "1200 V normally-off GaN-on-Si field-effect transistors with low dynamic on-resistance," accepted by IEEE Electron Dev. Lett..

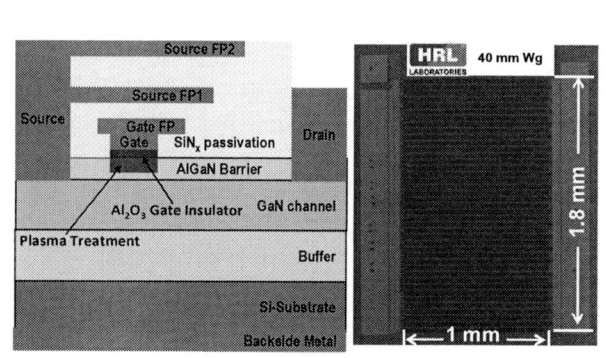

Fig. 1 Cross-sectional schematic and microscopy top view photograph of the GaN-on-Si power switching device. The active area of the chip is 1.8 mm².

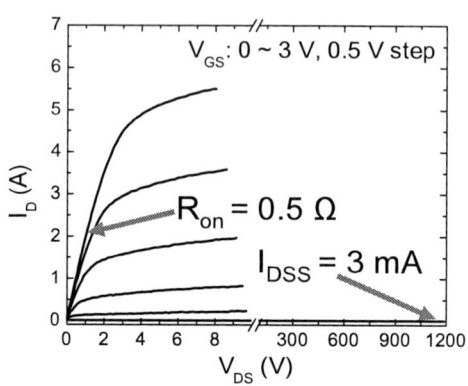

Fig. 2 Output IV characteristics of a device with the substrate grounded. R_{on} is 0.5 Ω; V_{br} is 1200 V; I_{DSS} is 3 mA at V_{GS} = 0 V and V_{DS} = 1200 V.

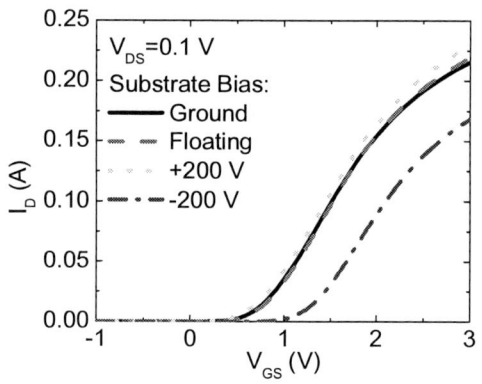

Fig. 3 Transfer IV characteristics of a device with the substrate biased at different conditions.

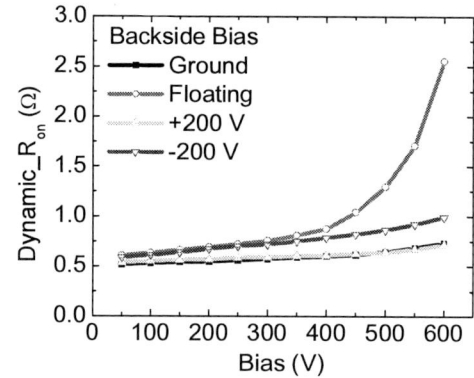

Fig. 4 Dynamic on-resistance as a function of output voltage at different substrate bias conditions.

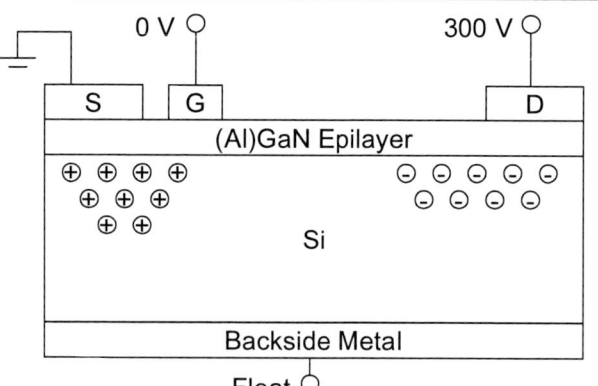

Fig. 5 Schematic showing the generation of electrons and holes in the Si substrate when a substrate-floated GaN-on-Si FET is blocking a positive drain.

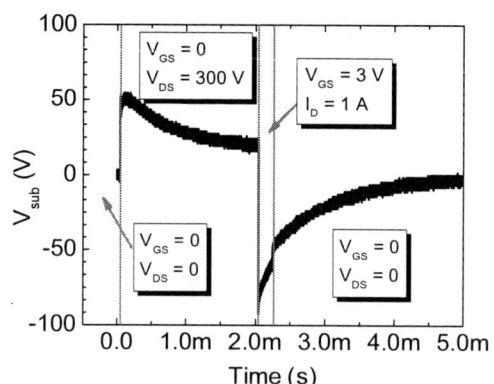

Fig. 6 Substrate potential as a function of time during an off-to-on switching cycle for a device with floating substrate.

978-1-61284-243-1/11 $26.00 © 2011 IEEE   224

Corwin Pavilion, MultiCultural Center Theater

Rump Sessions

Tuesday PM, June 22nd, 2010

8:30 PM (Corwin East)
Large Area OFETs: Organic or Oxide?

Session Organizers:
Ioannis Kymissis, Columbia University
Yongtaek Hong, Seoul National University

Panelists:
Hagen Klauk, Max Planck Institute for Solid State Research
John Wager: Oregon State University
Michael Chabinyc: University of California, Santa Barbara
Carl Taussig: HP Labs

Significant work has been invested in developing the next generation thin film semiconductors for large area display, sensor, and actuator systems. Work has focused on developing transistors with higher performance, reduced thermal budgets, increased process flexibility, and expanded substrate compatibility compared with amorphous silicon. Organic semiconductor-based field effect transistors have made incredible progress in performance over the past decade, and now surpass amorphous silicon in nearly all figures of merit while retaining a high processability and low thermal budget. In parallel, zinc oxide and related systems have begin to show carrier mobilities superior to polysilicon at processing temperatures approaching those of organic systems. In addition to providing significantly higher mobilities, oxide FET devices offer the potential stability of inorganic semiconductor systems, which is the Achilles heel of organic devices. In this rump session, "OFETs: Organic or Oxide?", we bring together a distinguished panel of experts from academia and industry to battle it out in predicting the next generation thin film transistors.

8:30 PM (Corwin West)
What is the Ultimate Low Power Device?

Session Organizers:
Kirsten Moselund, IBM Corporation
Siyuranga Koswatta, IBM Corporation
Erik Lind, Lund University

Panelists:
Jeff Bokor, University of California, Berkeley
Paul Solomon, IBM Corporation
Rafael Rios, Intel Corporation
Dejan Markovic, University of California, Los Angeles

The increasing power consumption, both static and dynamic, is limiting the performance of integrated circuits today, and looking forward, it's only getting worse. Multi-core architectures have maintained system performance and allowed Moore's law to continue. But, from the power perspective, silicon CMOS is hamstrung by the 60mV/dec limit on the subthreshold swing.

A number of potential candidates have been proposed to overcome these limitations. III-V devices offer the promise of a better power-performance trade-off due to lower voltage operation, even though they are still constrained by the thermionic limit. So-called steep slope devices are based on different transport mechanisms, and are not limited by the 60mV/dec limit. Among these, the tunnel FETs promise low off-currents, but achieving high Ion seems to be a daunting challenge. NEMS devices demonstrate inherently abrupt transitions and low Ioff, but scalability and reliability are main concerns.

When reaching beyond the electronic domain, ultra-low power may be offered by alternative computation methods, such as nanomagnetic logic and relay logic, but

perhaps at a heavy price on performance. Which device will win the race for low power? – Or, is silicon CMOS "too big to fail"? In this rump session, we bring together a distinguished panel of experts from academia and industry to battle it out

8:30 PM (MultiCultural Center Theater)
Graphene--What is it good for?

Session Organizers:
Eric Pop, University of Illinois Urbana-Champaign
Dimitris Pavlidis, Technical University Darmstadt
Jeong Moon, HRL

Panelists:
James Hone, Columbia University
Debdeep Jena, University of Notre Dame
Alex Balandin, UC Riverside
Rashid Bashir, University of Illinois Urbana-Champaign

Graphene is a material with remarkable physical and electrical properties. These properties and its discovery led to the award of the 2010 Nobel Prize in Physics. For the past number of years, Graphene electron devices have had a large presence at the DRC through contributed papers and past rump sessions. For the most part, Graphene discussions at the DRC have focused on its application as a field effect transistor for high performance RF or digital circuit applications. Graphene's usefulness for these applications has been vigorously debated due to some inherent limitations (lack of a bandgap).

Given Graphene's novel properties and in honor of this year's conference location, this rump session wishes to discuss Graphene in the context Kroemer's Lemma of New Technology[1] which states:

The principal applications of any sufficiently new and innovative technology have always been—and will continue to be—applications created by that technology.

We ask a panel of experts with diverse backgrounds to give us their view of the applications and opportunities that are being created from Graphene's unique properties and predict what the future may hold.

[1] H. Kroemer, Rev Mod. Phys., 73 (2001) 783-793.

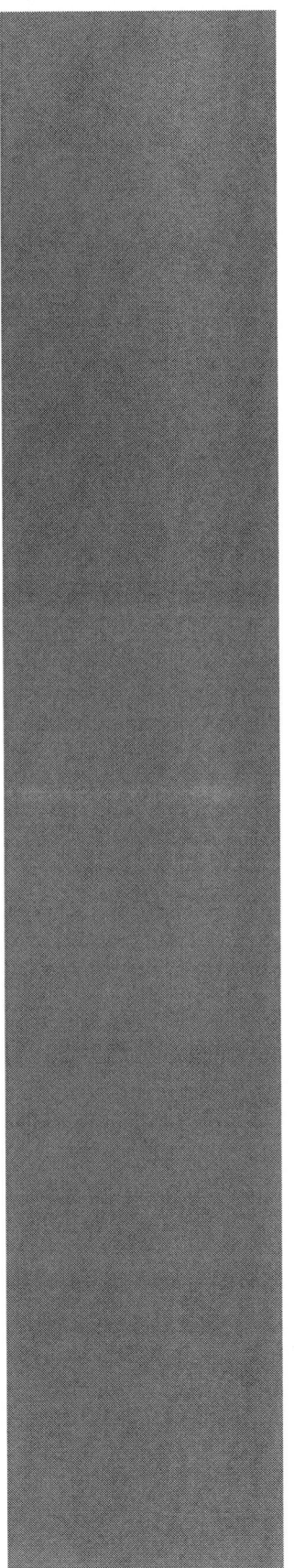

(Corwin Pavillion)

Plenary Session

Wednesday AM, June 22nd, 2011

Joint DRC/EMC Plenary Session

8:20 AM Awards Ceremony

8:30 AM Plenary Paper
New Concepts and Materials for Solar Power Conversion
Wladyslaw Walukiewicz, Lawrence Berkeley National Laboratory

9:20 AM Break

978-1-61284-243-1/11 $26.00 © 2011 IEEE

Session VI.A (Corwin Pavilion East)

pFETs

Wednesday AM, June 22nd, 2011

Session Chair: TBA

10:00 AM VI.A-1 Invited Paper
High Mobility Strained P-Channel Germanium Quantum Well Field Effect Transistor for Low Power (V_{cc} = 0.5 V) III-V CMOS Applications
R. Pillarisetty, Intel Corporation, Components Research, Technology and Manufacturing Group, Hillsboro, Oregon, USA

10:40 AM VI.A-2 Student Paper
Performance enhancement of GaAs UTB pFETs by strain, orientation and body thickness engineering
A. Paul[1], S. Mehrotra[1], G. Klimeck[1] and Mark Rodwell[2], [1]ECE Department and NCN, Purdue University, West Lafayette, Indiana, USA and [2]ECE Department, University of California, Santa Barbara, California, USA

11:00 AM VI.A-3 Student Paper
Highly-Strained SGOI p-Channel MOSFETs Fabricated by Applying Ge Condensation Technique to Strained-SOI Substrates
J. Suh, R. Nakane, N. Taoka, M. Takenaka and S. Takagi, School of Engineering, The University of Tokyo, Tokyo, JAPAN

11:20 AM VI.A-4 Student Paper
Hole Mobility Enhancement in Uniaxially Strained SiGe FINFETs: Analysis and Prospects
R. Bijesh[1], I. Ok[2], M. Baykan[2], C. Hobbs[2], P. Majhi[2], R. Jammy[2], and S. Datta[1], [1]The Pennsylvania State University, University Park, Pennsylvania, USA and [2]Sematech, Austin, Texas, USA

11:40 AM VI.A-5
Late News

978-1-61284-243-1/11 $26.00 © 2011 IEEE

978-1-61284-243-1/11 $26.00 © 2011 IEEE

High Mobility Strained P-Channel Germanium Quantum Well Field Effect Transistor for Low Power (Vcc = 0.5 V) III-V CMOS Applications

Ravi Pillarisetty

Intel Corporation, Components Research, Technology and Manufacturing Group, Hillsboro, OR 97124, USA

Abstract

In this talk, we review recent research results[1] investigating the germanium quantum well field effect transistor (QWFET) for use as the p-channel device option for future low power (Vcc = 0.5V) III-V CMOS architecture. We demonstrate a high mobility Ge p-channel QWFET, with scaled TOXE = 14.5Å and mobility of 770 cm^2/V*s at n_s =5×10^{12} cm^{-2}. For TOXE < 40 Å, this represents the highest hole mobility reported for any Ge device and is 4x higher than state-of-the-art strained silicon[2]. Furthermore, at Vcc = 0.5V, the Ge QWFET exhibits 2x higher drive current at fixed Ioff than the best III-V [3] and germanium devices [4] reported to date. These results suggest the Ge QWFET is a viable p-channel option for non-silicon CMOS.

[1] R. Pillarisetty et al., *IEDM Tech. Dig.,* pp. 150-153 (2010).
[2] P. Packan et al., *IEDM Tech. Dig.,* pp. 3.4.1 (2008).
[3] M. Radosavljevic et al., *IEDM Tech. Dig.,* pp. 30.3.1 (2008).
[4] J. Mitard et al., *IEDM Tech. Dig.,* pp. 873-876 (2008).

978-1-61284-243-1/11 $26.00 © 2011 IEEE

Performance enhancement of GaAs UTB pFETs by strain, orientation and body thickness engineering

Abhijeet Paul[1*], Saumitra Mehrotra[1*], Gerhard Klimeck[1] and Mark Rodwell[2]

[1]ECE Department and NCN, Purdue University, West Lafayette, IN 47906.
Phone: 765-404-3589, Email: abhijeet.rama@gmail.com. *equally contributed to the work.
[2] ECE Department, University of California, Santa Barbara, CA 93106.

III-V semiconductors can provide a viable option for continuous scaling of future CMOS technology [1-3]. We report a significant enhancement in the ON-current (I_{ON}) of ultra-thin body (UTB) GaAs intrinsic channel p-MOSFETs using biaxial compressive strain. Our theoretical investigation shows that valence bands (VB) become hyperbolic under compressive strain in GaAs rendering effective mass approximation (EMA) invalid. The ballistic I_{ON} (~Q_{inv} (hole density) x V_{inj} (injection velocity)) is governed mainly by the asymptotic group velocity ($V_{grp} \sim \alpha.V_{inj}$) of the hyperbolic VBs, These bands can be engineered using GaAs body thickness (T_{ch}) scaling, compressive strain value and wafer orientation. V_{inj} is primarily controlled by strain and T_{ch} whereas, Q_{inv} is governed mainly by the gate electrostatics, thus providing two separate design parameters to control I_{ON}. Isotropic strain enhances V_{inj} which gives a maximum improvement in I_{ON} of ~23-40% for [100]/(100) and [110]/(111) pMOSFETs for 5 nm body thickness at 4% compressive biaxial strain. Scaling body thickness from 5nm to 2nm improves I_{ON} by ~2X for all the device orientations considered in this study.

The analysis of I_{ON} involves the following steps. VB E(k) is calculated using an atomistic sp3d5s* tight-binding model with spin orbit (SO) coupling [4] under the action of biaxial compressive strain (0 to 4%) for the following [transport]/(wafer) orientations: [100]/(100), [110]/(110) and [110]/(111) (Fig.1a), for T_{ch} varying from 2 nm to 5 nm. DOS and modes (M(E)) are calculated numerically using the VB E(k) which are eventually used to calculate the inversion hole density (Q_{inv}) and drain current (I_{ds}), respectively. V_{inj} is calculated using I_{ds}/Q_{inv}. Gate electrostatics is accounted for by considering a gate oxide of thickness T_{ox}. Quantum hole charge density correction (T_{inv}) is obtained using SCHRED [5]. The entire procedure and the equation to calculate gate over drive (V_{gt}) is similar to the method in [3] (Fig. 1).

With III-V n-FETs, the density of states (DOS) is low, the bands are approximately parabolic, and electron velocity varies as the square root of kinetic energy; the channel effective mass, m^*, is selected for highest drive current by balancing its opposing effects on charge density and on injection velocity [2-3]. In marked contrast, over the range of Fermi energies expected in p-FET operation, the computed VB E (k) fits closely to a hyperboloid, with carrier group velocity approaching an asymptotic maximum with increased kinetic energy. The calculated VB E(k) for a 4nm thick GaAs UTB under 4% strain fits very well to hyperbolic bands (Fig 2a), and the injection velocity shows little variation with energy (Fig. 2b). Further, because the state density is high, highest current is obtained by designing the channel for highest group velocity, and by selecting a thin body and dielectric for high charge density. Approximating the hyperbolic bands as $E(k) = \hbar v_0 |k|$, the injection velocity is constant, $v_{inj} = (2/\pi)v_0$ and the sheet hole density, $p_s = k_f^2/\pi = \pi^{-1}(E_f/\hbar v_0)^2$ is large and varies as the inverse square of the asymptotic velocity (v_0); strained VB in GaAs can be well represented by hyperbolic bands where the velocity, rather than the effective mass, is constant with energy.

The valence bands are highly anisotropic and respond very differently to strain applied in different directions. Figure 3 shows the VB E(k) for 4nm thick GaAs UTB channel for -4%, -1% and 0% biaxial strain. Strain causes an isotropic compression of VB E(k) for [100]/(100) (Fig.3 a-c) and [110]/(111) (Fig. 3 g-i) channels. However, for [110]/(110) channel the strain causes anisotropic compression in VB E(k) (Fig.3 d-f).

The calculated I_{ON}, under strain at V_{gt} =0.3V and t_{ox} = 0.5nm (low-operating-power devices [1]) are shown for various GaAs body thickness. The I_{ON} improves ~2X for all the orientations with body scaling from 5nm to 2nm under 0% strain (Fig.4 a-c). The anisotropic strain effect on VB E(k) reveals itself in the computed I_{ON} vs. strain. [100]/(100) and [110]/(111) pFETs improve monotonically with strain (Fig.4 a&c) showing a maximum improvement of ~38% and 23%, respectively for 5nm thick GaAs channel under -4% biaxial strain. This is an outcome of the compression of the VB E(k) under strain which increases the V_{inj} (Fig.3). The highest I_{ON} is obtained in the [110]/(110) orientation, though strain has no benefit (Fig.4 b).

A crucial aspect for designing III-V pFETs is the action of strain and gate electrostatics on the drive current. As an example the V_{inj}, Q_{inv} and I_{ON} for [100]/(100) pFETs with T_{ch} for two different V_{gt} are shown in Fig.5. An important observation is that V_{inj} is primarily enhanced by strain (Fig. 5a) whereas Q_{inv} is dominated by the gate electrostatics. This behavior is observed in all the orientations. The independent control of Q_{inv} and V_{inj} can essentially allow us to design III-V pFETs with required I_{ON}. Thus, a proper choice of wafer/transport orientation and strain, providing a high hole group velocity, along with an optimal gate oxide thickness can lead to better III-V pFETs for the future CMOS technology.

Acknowledgements: MSD and the Nonclassical CMOS center under SRC, NRI under MIND and NSF for financial support. NCN and nanoHUB.org for computational support.

References: **[1]** ITRS Road map, http://www.itrs.net/, **[2]** M. Rodwell et. al, IEEE DRC 2010, pp 149-153.
[3] A. Nainani et. al. IEDM 2010, pp 138-141, **[4]** G. Klimeck et. al, CMES Vol.3, 5 pp 601-642 (2002),
[5] D. Vasileska et. al, "SCHRED", DOI: 10254/nanohub-r221.4.

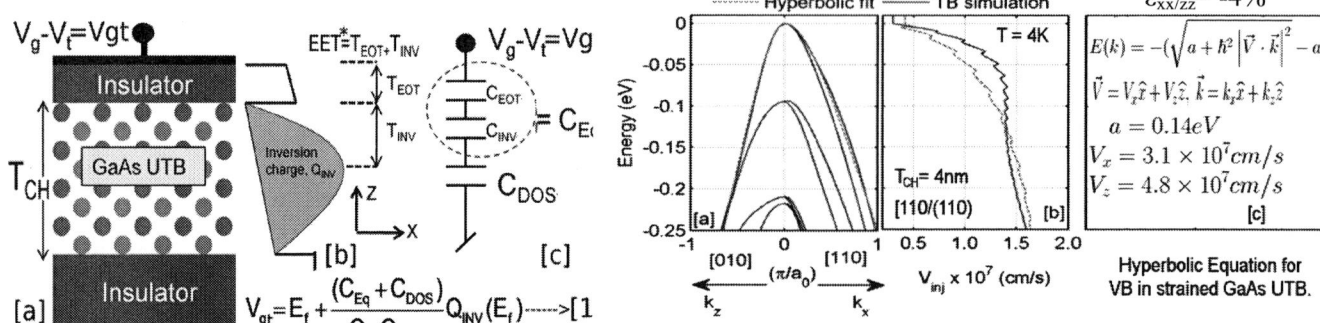

Figure 1: Schematic of an atomic GaAs UTB pFET channel along with inversion hole density. T_{inv} is calculated using SCHRED [5]. Gate electrostatics is taken into consideration using Eq. (1).

$$V_{gt}=E_f+\frac{(C_{Eq}+C_{DOS})}{C_{Eq}C_{DOS}}Q_{INV}(E_f)---->[1]$$

$EET^* = $ Equivalent electrostatic thickness

Figure 2: (a) TB calculated (grey) and hyperbolic fitted (brown) E(k) along kx (at kz=0) and kz (at kx=0). (b) Comparison of injection velocity obtained from hyperbolic E(k) to the simulated value at T = 4K. Constant Vinj vs. E_f is a signature of hyperbolic band. (c) Hyperbolic E(k) expression and the parameters values for the fitted TB E(k).

$$E(k)=-(\sqrt{a+\hbar^2\left|\vec{V}\cdot\vec{k}\right|^2}-a)$$
$$\vec{V}=V_x\hat{x}+V_z\hat{z},\ \vec{k}=k_x\hat{x}+k_z\hat{z}$$
$$a=0.14eV$$
$$V_x=3.1\times10^7cm/s$$
$$V_z=4.8\times10^7cm/s$$

Hyperbolic Equation for VB in strained GaAs UTB.

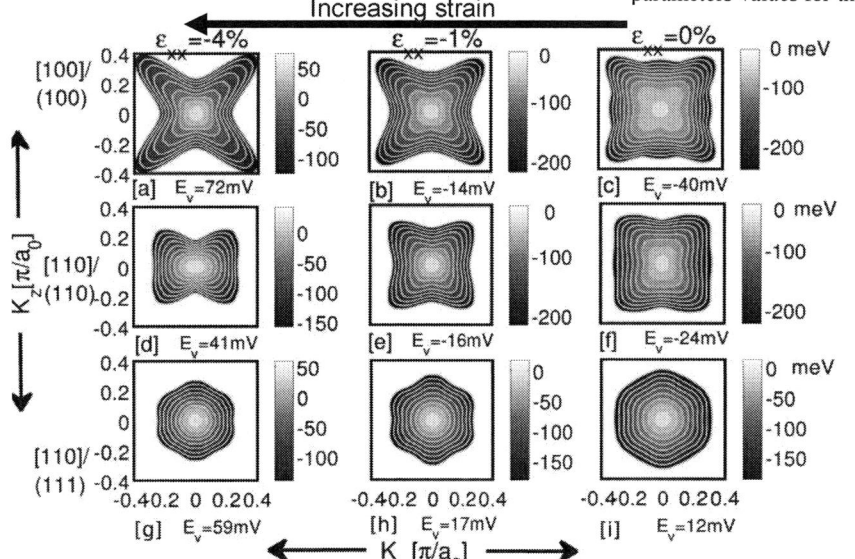

Figure 3: 2D E(k) of the highest VB in the 4nm thick strained GaAs UTB for (a-c) [100]/(100) **high V_grp** (d-f) [110]/(110) **low V_grp** and (g-i) [110]/(111). **high V_grp** Strain value is -4%, -1% and 0% for left, middle and right column, respectively. The VB max values (Ev) are shown Energy range is from Ev to 8KT below Ev. Kx is the transport direction.

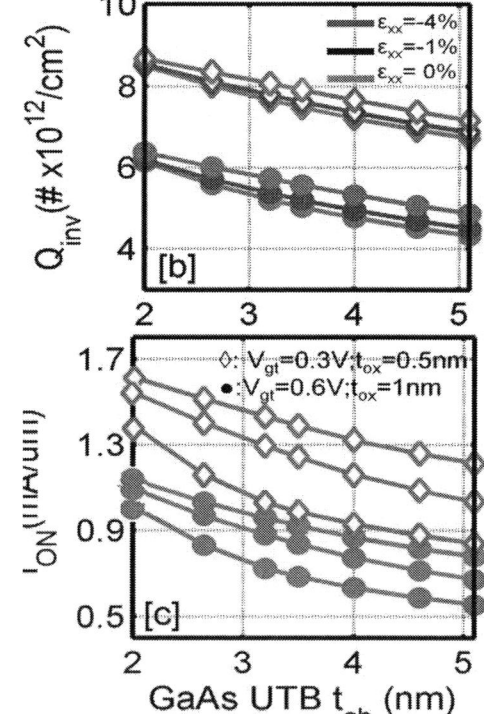

Figure 5: Variation in (a) Vinj, (b) Qinv and (c) I_{ON} with GaAs body thickness for 3 strain values (-4%, -1% and 0%) and two gate overdrive biases (Vgt). Filled (open) symbol represents Vgt =0.6V (0.3V) for Tox = 1nm (0.5nm). **Inversion charge is governed by the gate electrostatics while injection velocity is governed mainly by strain.**

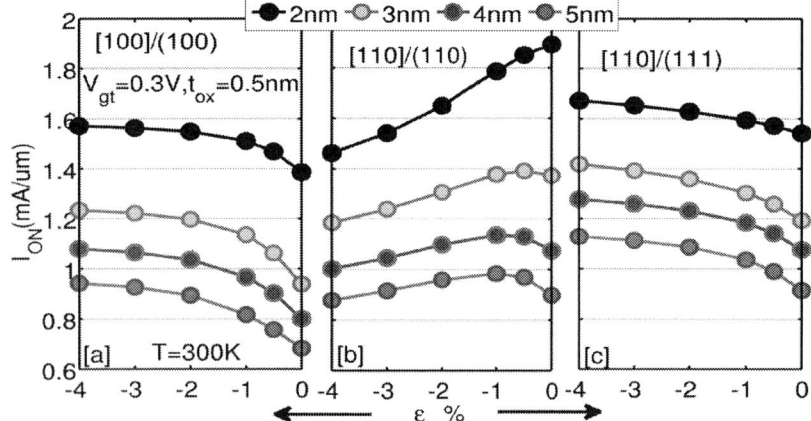

Figure 4: I_{ON} variation with strain for (a) [100]/(100), (b) [110]/(110) and (c) [110]/(111) oriented GaAs p-FETs for 4 different body thickness. [100]/(100) and [110]/(111) devices show improvement with strain and body thickness scaling, whereas [110]/(110) degrades with strain. **In [100]/(100) I_{ON} improves a maximum of 38% (5nm,-4%) with strain and ~2X with body scaling. [110]/(110) I_{ON} degrades by 29% (2nm,-4%) but improves by ~2X with body scaling. In [110]/(111) I_{ON} improves a maximum of 23%**

978-1-61284-243-1/11 $26.00 © 2011 IEEE   234

Student paper

Highly-Strained SGOI p-Channel MOSFETs Fabricated
by Applying Ge Condensation Technique to Strained-SOI Substrates

Junkyo Suh, Ryosho Nakane, Noriyuki Taoka, Mitsuru Takenaka and Shinichi Takagi

School of Engineering, The University of Tokyo, Bunkyo-ku, Tokyo, 113-8656, Japan

Phone & Fax: +81-3-5841-6733   Email : jksuh@mosfet.t.u-tokyo.ac.jp

Much attention has recently been paid to MOS channel materials with high mobility and resulting high injection velocity that can increase I_{ON} and reduce delay [1]. Among them, ultrathin body SiGe-On-Insulator (SGOI) structure with high compressive strain and high Ge content is a promising channel material for pMOSFETs under future technology nodes. Here, many theoretical studies [2-4] have reported that incorporation of a large amount of compressive strain into SiGe materials is a key technology for boosting the performance. Also, one of promising techniques for fabricating the SGOI structures is Ge condensation technique, composed of epitaxial growth of SiGe layers on SOI substrates and successive thermal oxidation [5, 6]. It is known, however, in Ge condensation using conventional unstrained SOI substrates [5, 7, 8] that strain relaxation occurs when Ge content becomes ~0.60 and strain significantly decreases with an increase in Ge content. This strain relaxation has been attributed to crystal defect generation during Ge condensation, induced by large strain in the SGOI due to the lattice mismatch between Si and Ge [8-10].

However, if high compressive strain is maintained even in Ge content higher than 0.85, very high performance SiGe pMOSFETs due to the combination of high strain and Ge-like band structure [2] are realized. Also, such SGOI layers are expected to have high crystal quality with low defect density, because of the mitigation in strain relaxation. Actually, Tezuka et al. [11] have recently shown that pseudomorphic sSGOI layers can be achieved up to ~0.75 by using strained-SOI substrates, because of decrease in the lattice mismatch between the initial substrates and Ge. However, devices on these sSGOI layers have not been realized yet and, thus, the electrical characteristics have not been studied yet. In this paper, we systematically examine the electrical properties of sSGOI on sSOI in comparison with those of sSGOI on conventional SOI and demonstrate much superior device performance and characteristics.

The process flow of Ge condensation is shown in Fig. 1, where there are two groups of sSGOI devices. One is named as sSGOI on sSOI and the other is named as sSGOI on SOI. The only difference in the two groups is starting substrates, sSOI with ~0.8 % tensile strain and unstrained SOI. Ge content and compressive strain of SGOI during Ge condensation were evaluated by Raman spectroscopy [12] (Fig. 2). Strain relaxation in sSGOI on sSOI occurs in much higher Ge content region than that on SOI. As a result, the amount of strain on sSOI is higher in Ge content over 0.60 than on SOI. These results suggest that SGOI fabricated on sSOI can enjoy the higher amount of strain as well as better crystal quality, because of the small degree of relaxation.

Back gate SGOI pMOSFETs were fabricated for examining the electrical properties. The fabrication process is summarized in Fig. 3. The I_{ON}/I_{OFF} ratio of ~10^5 is obtained (Fig. 4) except for Ge-rich sSGOI on SOI due to large amount of strain relaxation. The extracted V_{TH} and substhreshold swing (S. S.) are shown in Fig. 4. It is found that, while V_{TH} and S. S. of SGOI fabricated on sSOI do not change so much, those on SOI significantly increase in Ge content higher than ~0.80, indicating that the crystal defect generation due to strain relaxation causes undesirable hole generation and the degradation of interface quality. The mobility was evaluated by using Kelvin pattern (Fig. 6) in order to avoid the influence of series resistance. The effective mobilities on sSOI and SOI at given Ge contents of 0.58 (Fig. 7) and ~0.90 (Fig. 8) are plotted as a function of surface carrier density (N_s), together with the hole mobility in Si pMOSFETs [13]. It is found that the mobility of sSGOI on sSOI is higher than that on SOI at both Ge contents, while both the SGOI pMOSFETs provide much higher hole mobility than that of Si pMOSFETs. The mobility enhancement factor against the hole mobility in Si pMOSFETs is almost independent of N_s, as shown in Fig. 9. The mobility enhancement factor at N_s of 4×10^{12} cm^{-2} is plotted for SGOI on sSOI and SOI in Fig. 10. The enhancement factor in SGOI on sSOI is higher than that on SOI, particularly in higher Ge content, which is consistent with the larger strain relaxation in SGOI on SOI in this Ge content region. The mobility enhancement of as high as 8 is obtained for SGOI on sSOI. In order to quantitatively understand this mobility enhancement, the experimental mobility was compared with the theoretical ones [3] (Fig. 11). The good agreement for both SGOIs on sSOI and SOI means that the mobility enhancement can be quantitatively explained by the combinational effect of strain and Ge content on hole mobility and the higher enhancement in SGOI on sSOI is attributed to the higher compressive strain due to less strain relaxation by utilizing sSOI.

In summary, we have demonstrated the high mobility enhancement of SGOI pMOSFETs through the larger compressive strain by employing sSOI substrates in Ge condensation process. This highly-strained SGOI pMOSFET is quite promising for a component of future high performance CMOS.

References [1] M. Lundstrom, IEEE Electron Device. Lett., 18, 361 (1997). [2] M. V. Fischetti et al., J. Appl. Phys., 80, 2234, (1996). [3] F. M. Bufler et al., J. Appl. Phys. 84, 5597 (1998). [4] K. C. Saraswat et al., IEDM Tech. Dig., pp. 659-662 (2006). [5] T. Tezuka et al., Jpn. J. Appl. Phys., 40, 2866 (2001). [6] S. Nakaharai et al., Appl. Phys. Lett., 83, 3516 (2003) [7] T. Tezuka et al., IEEE Trans. Electron Devices, 50, 1328 (2003). [8] S. Nakaharai et al., Semicond. Sci. Technol., 22, S103 (2007). [9] S. W. Bedell et al., Appl. Phys. Lett., 85, 5869 (2004). [10] S. W. Bedell et al., Electrochem. Solid-State Lett., 7, G105 (2004). [11] T. Tezuka et al., Appl. Phys. Lett., 90, 181918 (2007). [12] F. Pezzoli et al, Mat. Sci. Semicond. Proc., 11, pp. 279-284 (2008). [13] S. Takagi et al., IEEE Trans. Electron. Devices, 41, 2357 (1994).

978-1-61284-243-1/11 $26.00 © 2011 IEEE

Student paper

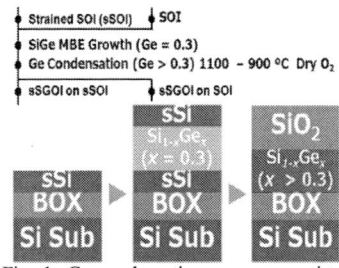

Fig. 1. Ge condensation process consisting of SiGe epitaxial layer growth on sSOI or SOI, followed by dry O_2 oxidation.

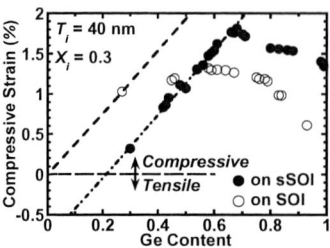

Fig. 2. Compressive strain during Ge condensation evaluated by Raman spectroscopy for sSGOI layers on sSOI (solid circle) and sSGOI layers on SOI (open circle).

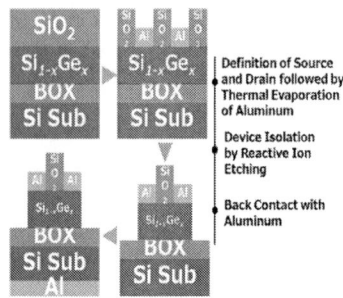

Fig. 3. Fabrication process flow of back gate SGOI pMOSFETs.

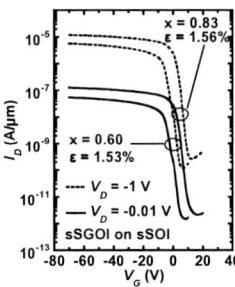

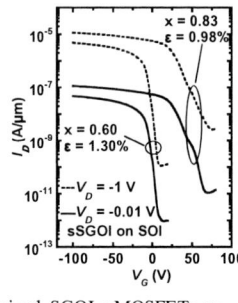

Fig. 4. I_D-V_G characteristics for strained SGOI pMOSFETs on sSOI and SOI. Channel length and width are 50 and 110 μm, respectively.

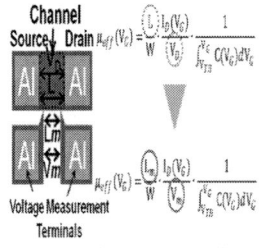

Fig. 6. Kelvin pattern used for effective mobility extraction in this work. The parameters in the equation are substituted for L_m and V_m.

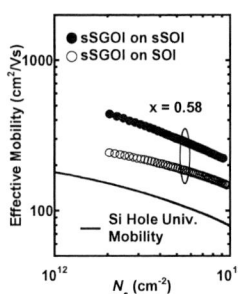

Fig. 7. Effective mobility curves as a function of N_s at Ge content, 0.58.

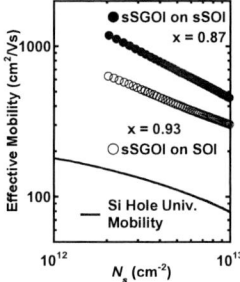

Fig. 8. Effective mobility curves as a function of N_s at Ge content, ~0.90.

Fig. 5. Extracted V_{TH} and subthreshold swing (S. S.) as a function of Ge content. A significant shift in V_{TH} and an increase in S. S. with an increase in Ge content are observed for sSGOI on SOI.

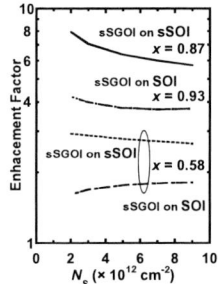

Fig. 9. Mobility enhancement against the universal mobility as a function of N_s for sSGOI pMOSFETs on SOI and sSOI.

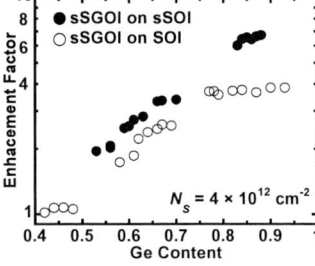

Fig. 10. Mobility enhancement factor against the universal mobility at $N_s = 4 \times 10^{12}$ cm^{-2} for sSGOI pMOSFETs fabricated on sSOI (solid circle) and SOI (open circle). This result means that sSGOI devices on sSOI show superior performance over those on SOI in the entire Ge content.

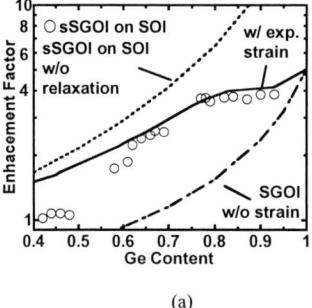

(a)

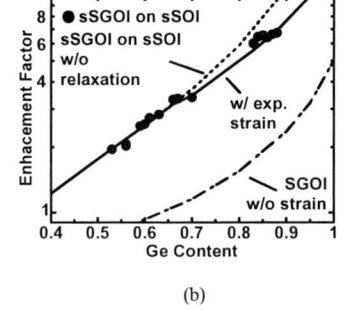

(b)

Fig. 11. Comparison of experimental results with theoretical ones [3] for sSGOI on (a) SOI and (b) sSOI. The dotted line indicates the enhancement factor without relaxation. The solid line means the calculated enhancement factor with the experimental strain on SOI in Fig. 2. The dashed line means the calculated enhancement factor without strain.

978-1-61284-243-1/11 $26.00 © 2011 IEEE

Hole Mobility Enhancement in Uniaxially Strained SiGe FINFETs: Analysis and Prospects

R.Bijesh[1], I.Ok[2], M.Baykan[2], C.Hobbs[2], P.Majhi[2], R.Jammy[2], S.Datta[1]

[1]The Pennsylvania State University, University Park, PA 16802, USA

[2]Sematech, Austin, TX ,USA

Email: bor5067@psu.edu

Introduction: Strain induced mobility enhancement in silicon (Si) FINFETs is attractive because of higher drive current and immunity to short channel effects. Uniaxial compressive strain combined with 110 channel orientation is found to give the best hole mobility enhancement in FINFETs [1]. However, further hole mobility enhancement in (110)/<110>Si pFINFETs is limited by strain relaxation issues and commonly observed limited impact of strain on (110) compared to (100) [2]. SiGe channel FINFET is an attractive replacement. Mobility enhancement with FINFET structure(SiGe channel with 25% Ge) that gives the best performance at low power has already been reported [3]. In this paper, we show for the first time that a 157% increase in hole mobility is possible in (110)/<110> SiGe FINFETs by increasing Ge content to 50% even in the presence of scattering due to enhanced interface charge density from increasing D_{it}, enhanced alloy scattering [4] and strain relaxation.

Process Technology: Biaxial compressively strained $Si_{0.75}Ge_{0.25}$ layer, 30nm thick, is epitaxially grown on 10nm thick (100) SOI wafers which, upon patterning, produced uniaxially strained 20nm wide, 40nm tall fins with {110}<110> channel orientation. Si fins were also fabricated for comparison. The FINFETs were formed by conventional process [3] and included a neutral/compressively strained contact etch stop layer (CESL) followed by interlayer dielectric (ILD) deposition, contact plug formation and copper metallization. Fig. 1 shows the TEM image of the fin cross section.

Results and Discussions: Finite element method (FEM) simulation in $SSGOI_{0.25}$ pFINFETs show an average sidewall stress of 1600 MPa (1% strain) which takes into account relaxation through amorphized S/D regions following the ion implantation. Higher performance of $SSGOI_{0.25}$ over (110)/<110>Si (SOI) pFINFETS is evident from Fig. 2. At 10n/um of Ioff, the $SSGOI_{0.25}$ pFINFETs with neutral and compressive CESL show 17% and 46% increase in Ion, respectively, over the silicon counterpart.. The subthreshold slope(SS) is slightly lower due to increased Dit in $SSGOI_{0.25}$ pFINFETs (Fig.3). A 57% net enhancement in hole mobility is seen at sheet carrier density (Ns) of $1x10^{13}cm^{-2}$ (Fig. 4) at 300K from long channel devices. However, the presence of alloy scattering in $SSGOI_{0.25}$ negates this enhancement at 77K (Fig 5). Hole mobility is extracted across temperatures and is then modeled by including the following scattering mechanisms: interface charge scattering (μ_{int}), surface roughness scattering (μ_{SR}), acoustic phonon(μ_{ac}) and optical phonon (μ_{opt}). In $SSGOI_{0.25}$ pFINFETs additional mechanism of alloy scattering (μ_{alloy}) is included the effect of which on $SSGOI_{0.25}$ mobility is seen clearly at T=77k (Fig. 5). Bulk coulomb scattering is ignored due to low doping levels in FINFET and remote high-k phonon scattering due to metal gate screening [4]. Excellent agreement between measured and modeled mobility values is seen (Fig 6). Mobility in $SSGOI_{0.25}$ and SOI pFINFETs at 300K and at Ns=$1x10^{13}cm^{-2}$ is dominated by phonon scattering whereas alloy scattering becomes important in $SSGOI_{0.25}$ at 77K. The different mobility components in $SSGOI_{0.25}$ are scaled appropriately (using linear extrapolation for the parameters listed in table1) to project the mobility for SiGe channel pFINFETs with higher Ge content(50%). A 157% enhancement is thus estimated at T=300K, Ns=$1x10^{13}cm^{-2}$ for $SSGOI_{0.50}$ over $SSGOI_{0.25}$ assuming 2x increase in the interface states(Fig. 7). At 300K hole mobility is still limited by phonon scattering but at 77K alloy scattering becomes the single dominant mechanism (Fig. 8). Decreasing fin length to 0.5um results in further strain relaxation (Fig. 9) and the net strain retained reduces to 1.5% which will degrade the mobility to 360 $cm^2V^{-1}s^{-1}$.

Conclusions: Experimental and theoretical hole mobility study in uniaxially strained (110)<110> $Si_{0.75}Ge_{0.25}$ pFINFETs shows that alloy scattering contributes only a small fraction of the overall mobility at 300K but plays a bigger role limiting 77K hole mobility. Increasing the Ge content to 50% increases the strain level. However, the extent of strain relaxation depends on the length of the fin. Fig. 10 shows the measured and projected hole mobility for SiGe FINFETs with 25% and 50% Ge mole fraction. Higher strain induced reduction of effective mass compensates for the increased interface charge density, Dit, in $SSGOI_{0.5}$ pFINFET and alloy disorder and results in 157% increase in the hole mobility observed at Ns=$1x10^{13}cm^{-2}$ and T=300K. Fig. 11 benchmarks the hole mobility in $SSGOI_{0.25}$ and $SSGOI_{0.5}$ pFINFETs as a function of electrical oxide thickness(TOXE) and shows its advantage over relaxed Ge channel MOSFETs. However strain relaxation for shorter length fins need to be addressed using careful layout techniques. High mobility combined with excellent short channel behavior make these devices a promising candidate for future technology node.

[1] T.Irisawa et al., IEDM 2005

[2] P.Packan et al., IEDM 2009

[3] I.Ok et al., IEDM 2010

[4] M.V. Fischetti et al., J.Appl.Phys ,vol.80, August(1996)

[5] R.Chau et al., IEEE EDL,vol.25, June (2004)

[6] Wei-Chin Wang et al., INEC, Jan(2010)

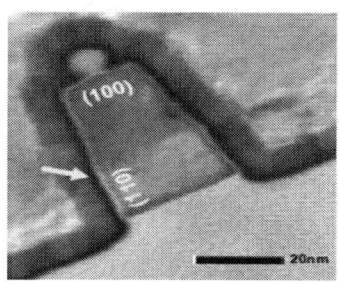

Fig 1. Cross section of SiGe/Si stack fin under gate with $Si_{0.75}Ge_{0.25}(30nm)/Si(10nm)$

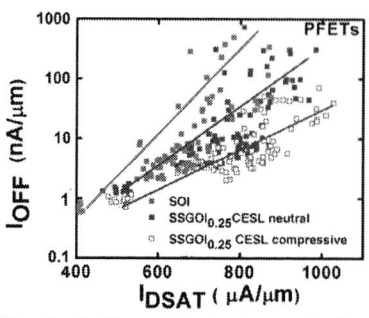

Fig 2. Ioff vs Idsat shows better performance of $SSGOI_{0.25}$ over SOI pFINFETs

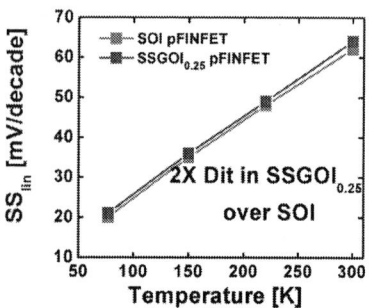

Fig 3. SS_{lin} vs T for $SSGOI_{0.25}$ and SOI pFINFETs

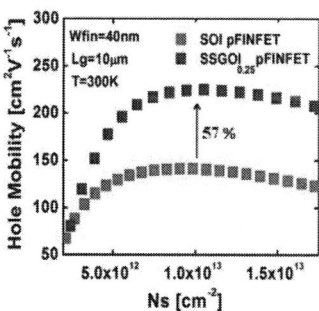

Fig 4. $SSGOI_{0.25}$ pFINFETs shows 57 % improvement in mobility over SOI pFINFETs

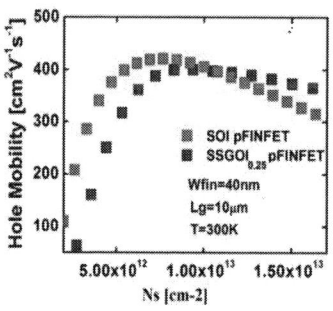

Fig 5. Hole mobility at T=77K in $SSGOI_{0.25}$ pFINFET flattening out at high Ns due to alloy scattering

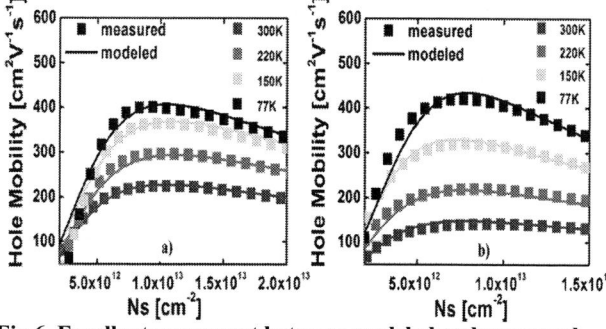

Fig 6. Excellent agreement between modeled and measured mobility values for a) SOI pFINFET b) $SSGOI_{0.25}$ pFINFET

$\mu_{int} \ \alpha \ (\varepsilon^{1.6} \ m_c^{-1} \ N_{it}^{-1})$
$\mu_{SR} \ \alpha \ (\varepsilon^2 \ m_c^{-1} \ m_{dos}^{-1})$
$\mu_{ac} \ \alpha \ (\rho\varepsilon^{1/3} m_c^{-1} m_{dos}^{-1} m_z^{-1/3})$
$\mu_{opt} \ \alpha \ (\rho \ m_c^{-1} \ m_{dos}^{-1})$
$\mu_{alloy} \ \alpha \ (x(1-x)\varepsilon^{1/3} m_c^{-2} \ m_z^{1/3})$

Table 1. Scattering parameters that change upon increasing Ge content to 50%. ε, m_c, m_{dos}, m_z, ρ and x are dielectric constant, conductivity mass, DOS mass, quantization mass ,density and Ge mole fraction respectively.

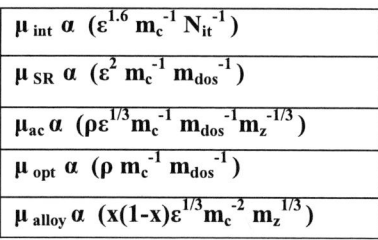

Fig 7. Measured and predicted hole mobilities in SiGe FINFETs

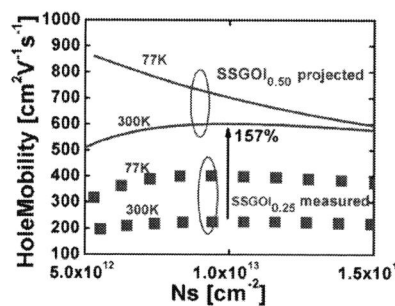

Fig 8. Alloy scattering is the single dominant mechanism at 77K in $SSGOI_{0.5}$ FINFET

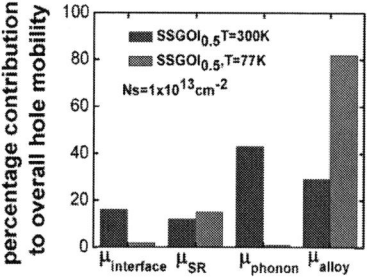

Fig 9. Strain relaxes at the fin edge due to the free surface after patterning

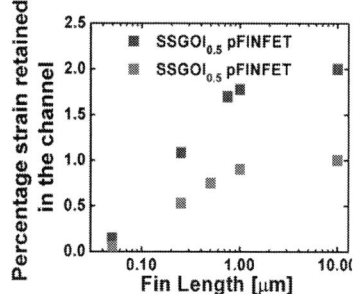

Fig 10. Measured and projected hole mobilities in SiGe FINFETs

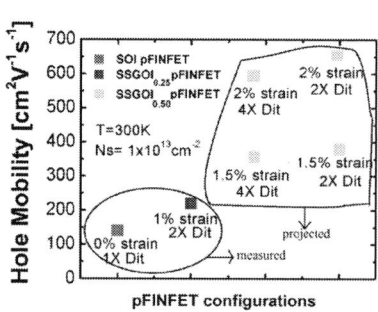

Fig 11. Hole mobility benchmarking as a function of TOXE

978-1-61284-243-1/11 $26.00 © 2011 IEEE

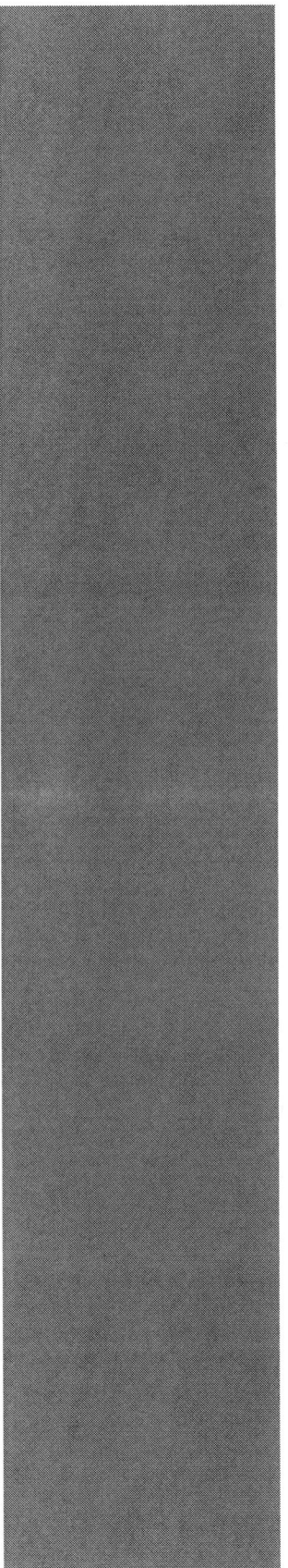

Session VI.B (Corwin Pavilion West)

Thin-Film Devices

Wednesday AM, June 22nd, 2011

Session Chair: Michael Chabinyc, University of California, Santa Barbara

10:00 AM VI.B-1 Invited Paper
Defect Analysis of Roll-to-Roll SAIL Manufactured Flexible Display Backplanes
C. Taussig[1], R. E. Elder[1], W. B. Jackson[1], A. Jeans[1], M. Jam, E. Holland[1,] H. Luo[1], J. Maltabes[1], C. Perlov[1], S. Trovinger[1], M. Almanza-Workman[2], R. A. Garcia[2], H. Kim[2], O. Kwon[2], and F. Jeffrey[2], [1]HP Labs, Palo Alto, California, USA and [2]Phicot Inc., Palo Alto, California, USA

10:40 AM VI.B-2
Indium-free Transparent Thin Film Transistors Based on Nanocrystalline ZnO
B. Bayraktaroglu[1], K. Leedy[1] and R. C. Scott[2], [1]Air Force Research Laboratory, Sensors Directorate, AFRL/RYDD, Wright Patterson AFB, Ohio, USA and [2]Arizona State University, Tempe, Arizona, USA

11:00 AM VI.B-3 Student Paper
Circuit applications based on solution-processed zinc-tin oxide TFTs
C.-G. Lee, T. Joshi, K. Divakar, and A. Dodabalapur, Microelectronics Research Center, University of Texas at Austin, Austin, Texas, USA

11:20 AM VI.B-4 Student Paper
Aluminum Top-Gate ZnO Nanowire Transistors with Improved Transconductance
D. Kälblein[1], B. Fenk[1], K. Hahn[2], U. Zschieschang[1], K. Kern[1,3], H. Klauk[1], [1]Max Planck Institute for Solid State Research, Stuttgart, GERMANY, [2]Max Planck Institute for Metals Research, GERMANY, and [3]Ecole Polytechnique Fédérale de Lausanne, Lausanne, SWITZERLAND

11:40 AM VI.B-5
Late News

978-1-61284-243-1/11 $26.00 © 2011 IEEE

Defect Analysis of Roll-to-Roll SAIL Manufactured Flexible Display Backplanes.

Carl Taussig, Richard E. Elder, Warren B. Jackson, Albert Jeans, Mehrban Jam,
Ed Holland, Hao Luo, John Maltabes, Craig Perlov, and Steven Trovinger
HP Labs, Palo Alto, California, USA, 650-857-4258, carl.taussig@hp.com

Marcia Almanza-Workman, Robert A. Garcia, HanJun Kim, Ohseung Kwon,
and Frank Jeffrey
Phicot Inc., Palo Alto, California, USA

Abstract

HP and Phicot have made the world's first roll-to-roll (R2R) manufactured active matrix displays. Currently we are developing a wrist-worn solar powered display for the U.S. Army. As we scale from research to preproduction on our 1/3 meter wide pilot line defect analysis and mitigation is our primary focus. In this presentation we will review the self-aligned imprint lithography (SAIL) process and discuss defects we observe, and the tools, and processes we have developed to detect and eliminate them

1. Introduction

HP and Powerfilm have together been developing methods for fabrication of electronics on flexible substrates using roll-to-roll (R2R) processes for over 10 years. We have created the world's first R2R active matrix display using the Self-Aligned Imprint Lithography (SAIL) process [1]. SAIL has been used to fabricate both electrophoretic and OLED displays and amorphous silicon and metal oxide based backplanes have been demonstrated with the process [2-4]. We are currently developing small quantities of wrist-worn solar powered displays for the Army Research Laboratory. The production of these qqVGA displays is the first step in scaling the SAIL process from the lab to volume manufacturing. To achieve necessary yields we have developed tools for inspection and electrical test of flexible substrates that are not bonded to a carrier as opposed to other efforts which utilize a bond - debond method for fabricating flexible backplanes on rigid carriers using conventional flat panel tools [5]. The leading sources of defectivity fall into two categories: electrical defects in the thin films of the TFT (thin film transistor) and defects affecting the imprint lithography process.

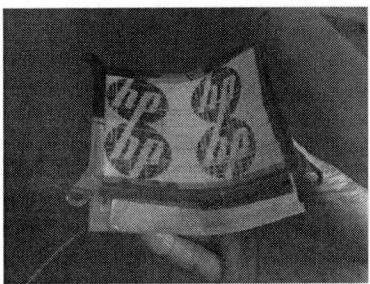

Figure 1: 160X120 pixel e-Ink based display with 500μm pixel pitch being developed for the Army

2. Self Aligned Imprint Lithography

The SAIL process was developed not only to make flexible displays but to enable low cost. We started with the single assumption that we would use R2R processes exclusively. The high level process flow for SAIL is shown on Figure 2. Unlike a conventional panel based process all of the deposition steps for the complete thin film transistor (TFT) stack are completed before any of the patterning steps. The multiple patterns required to create the backplane are encoded in the different heights of a 3D masking structure that is molded on top of the thin film stack once before any of the etching steps. By alternately etching the masking structure and the thin film stack the multiple patterns required for the backplane are transferred to the device layers. Because the mask distorts with the substrate perfect alignment is maintained regardless of process induced distortion.

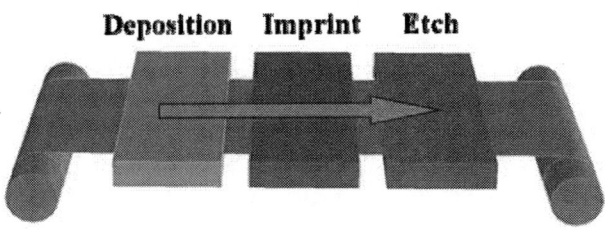

Figure 2: Schematic flow for SAIL

Figure 3 is an SEM image showing the 4 discrete heights in the mask used to produce bottom gate a-Si TFTs. Imprint lithography is ideally suited for R2R implementation because of its high resolution, compatibility with flexible substrates, high throughput and ability to reproduce complex 3-D structures. We have imprinted 40nm lines on 50μ thick polyimide and have developed materials that can maintain fidelity for thousands of impressions at throughputs of greater than 5m/min. Figure 4 shows transfer curves for 180 °C amorphous silicon TFTs produced with the R2R SAIL process. The curves are normalized by the W/L ratio of the devices to illustrate the scaling for channel lengths from 100μm to 1μm. At channel lengths below around 5μm the normalized on-current begins to drop due to the larger relative effect of resistance in the contacts. It is significant to note that while the on-current scales inversely with channel length, the speed of the TFT increases inversely with the square of channel length.

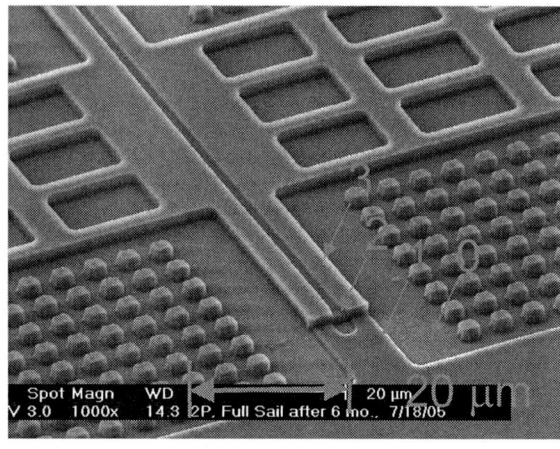

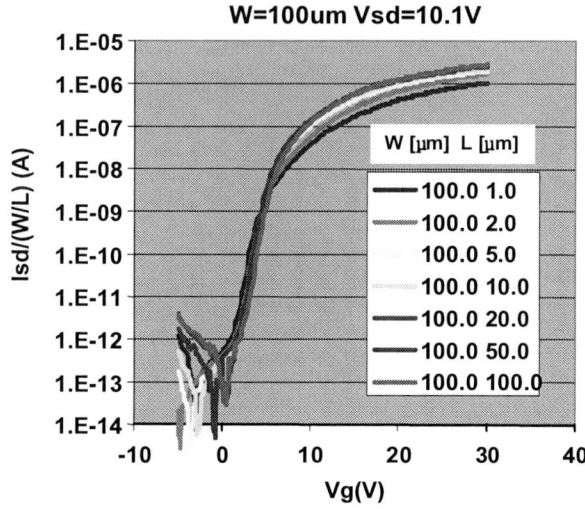

Figure 3: 4 level imprinted mask for active matrix backplane

Figure 4: Transfer curves normalized to the W/L ratio for R2R fabricated SAIL amorphous silicon TFTs

3.   Fabrication

Our backplanes are fabricated on 50 μm thick 1/3 m wide polyimide films. The web is wrapped on 6" diameter cores and moved from one machine to another during processing, each machine having its own unwind and rewind station. All equipment has been built in-house or externally to our specifications. The full TFT stack is deposited using vacuum deposition equipment at PowerFilm Inc. The imprint polymer is coated and embossed on a coater/imprinter at HP Labs. The subsequent etch process which transfers the imprint mask into the TFT geometries are performed by in a R2R tank etch and RIE (Reactive Ion Etch) at HP Labs. The backplane arrays are then singulated, tested, and some simple defects such as shorts are repaired before laminating them with an E Ink frontplane.

4.   Inspection and Electrical Test

We developed an electrical tester which uses bumped flex circuit contacts to connect to contact pads all sides of the backplane (Figure 5). The flex circuits are aligned to the backplane contacts through the use of alignment pins. Accurate alignment holes are cut in the substrate outside of the array near to the contact pads using a laser wafer dicer. A pressure ring applies even pressure around the array to ensure contact on all pads. Fixtures are included to facilitate loading and unloading the backplane under test without damage and to apply contact pressure repeatably. Compact electronics were built to obtain rapid test results. The system measures each data line and gate line for opens or shorts to common in less than 10 seconds.

It is necessary to perform R2R optical inspection both to evaluate the quality of the incoming substrate prior to processing and secondly to be able to study the evolution of defects during SAIL processing. Cutting samples from the web and imaging them with conventional inspection tools is not an option for two reasons. First, splices or holes in the web can interfere with further processing and second, once a sample is removed from the web the affects of the subsequent process steps cannot be studied. Figure 6 is a photograph of a R2R inspection tool that we have developed for inspection of the web at intermediate points in the process. The web is imaged by a microscope on the crown of a precision roller. The microscope can be scanned in the cross web direction.

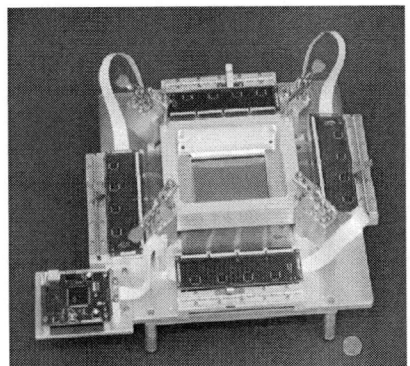  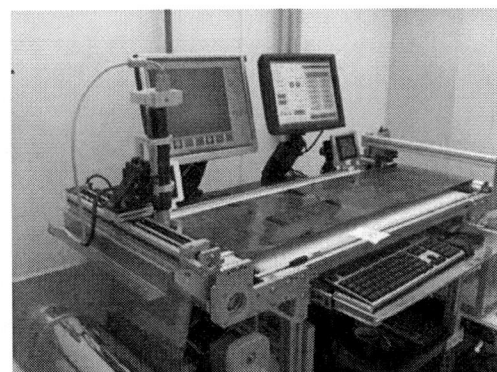

Figure 5: Flexible backplane electrical tester   **Figure 6: R2R optical inspection system**

5.   Defects

Currently two of the most common defect types in the SAIL process are shunts and surface particles. A shunt is a low resistance path between the top and bottom metal in the TFT stack. In the SAIL process the TFT stack is used for all of the components of the pixel circuit including data lines, gate lines, and the hold capacitor in addition to the TFT. A shunt in any of these components can result at minimum in the loss of a pixel or possibly a row or column loss. Shunts can be caused by asperities in the substrate, irregularities in the sputter or CVD processes, or voids resulting from mechanical damage. Figure 7 is a TEM showing embedded particles in TFT stack that can lead to electrical failure.

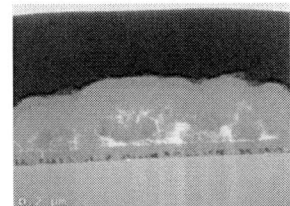

Figure 7: TEM of TFT stack with particle defect

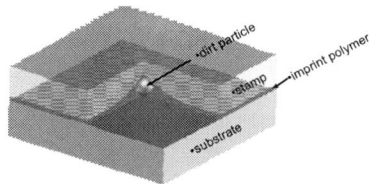

Figure 8: 'tent-pole' imprint defect

A second common defect occurs during the imprint process when the elastomeric imprint stamp is impinged onto the liquid photopolymer. This class of defects are generally referred to as 'tenting defects' because they produce a locally raised surface that deflects the imprint stamp (like a tent pole) creating an excess of photopolymer as shown on Figure 8. like a tent the area covered depends on the height of the pole so a relatively small particle can disrupt the pattern over a much larger area. In the SAIL process device layers are mapped to different thicknesses of the photopolymer masking structure so a significant change to the thickness of the masking layer (~0.7µm) can result in an incorrect mapping of the device layers. Typically this results in a short. Sources of tenting defects may be particles on top of the TFT stack, asperities within the TFT stack, or irregularities on the surface of the substrate. If the tenting is formed by a particle that remains on the web then the particle can acerbate the problem by masking subsequent etch processes or in some cases become adhered to the backside of the web as it is wound up then tear loose from its original location resulting in local delamination of the stack. Figure 9 shows the evolution of tenting defect through the SAIL process

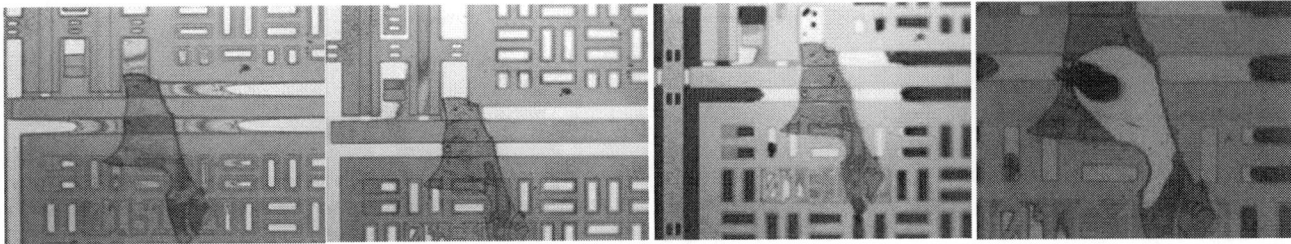

Figure 8: 'tent-pole' imprint defect results in tenting, micromasking and delamination. The evolution of the defect through the SAIL process starts at the left following the imprint, through the successive etch processes and finally adhesion and tearout on the right.

6. Acknowledgements

The authors gratefully acknowledge the support of their collaborators and sponsors including: the Army Research Laboratory (contract W911NF-08-2-0063), E Ink Corporation, and the ASU Flexible Display Center.

7. References

[1] Han-Jun Kim, Marcia Almanza-Workman, Bob Garcia, Ohseung Kwon, Frank Jeffrey, Steve Braymen, Jason Hauschildt, Kelly Junge, Don Larson, Dan Stieler, Alison Chaiken, Bob Cobene, Richard Elder, Warren Jackson, Mehrban Jam, Albert Jeans, Hao Luo, Ping Mei, Craig Perlov, and Carl Taussig, "Roll-to-roll manufacturing of electronics on flexible substrates using self-aligned imprint lithography (SAIL)," J. Soc. Inf. Display 17, 963 (2009)

[2] C. Taussig et al., "Architecture and Materials for R2R Manufactured OLED Displays using Self-Aligned Imprint Lithography (SAIL)", presented at 8th Annual Flexible Displays & Microelectronics Conf., Phoenix, USA, 2009.

[3] A. Jeans, M. Almanza-Workman, R. Cobene, R. Elder, R. Garcia, R. F. Gomez-Pancorbo, W. Jackson, M. Jam, H.-J. Him, O. Kwon, H. Luo, J. Maltabes, P. Mei, C. Perlov, M. Smith and C. Taussig, Proc. SPIE, 2010, 7637, 763719.

[4] H.-J. Kim, M. Almanza-Workman, B. Garcia, O. Kwon, F. Jeffrey, S. Braymen, J. Hauschildt, K. Junge, D. Larson, D. Stieler, A. Chaiken, B. Cobene, R. Elder, W. Jackson, M. Jam, A. Jeans, H. Luo, P. Mei, C. Perlov and C. Taussig, *J. Soc. Inf. Disp.*, 2009, **17**(11), 963.

[5] Haq, Jesmin; Ageno, Scott; Raupp, Gregory B.; Vogt, Bryan D.; Loy, Doug; , "Temporary bond-debond process for manufacture of flexible electronics: Impact of adhesive and carrier properties on performance," Journal of Applied Physics , vol.108, no.11, pp.114917-114917-7, Dec 2010

Indium-free Transparent Thin Film Transistors Based on Nanocrystalline ZnO

Burhan Bayraktaroglu, Kevin Leedy and Robin C. Scott[a]

Air Force Research Laboratory, Sensors Directorate, AFRL/RYDD

2241 Avionics Circle, Wright Patterson AFB, OH 45433

[a]Arizona State University, Tempe, AZ

E-mail: burhan.bayraktaroglu@wpafb.af.mil Tel: (937)528-8881 Fax: (937)255-8656

Wide bandgap semiconductors based on (Zn, In, Ga, Sn)-oxides are all good candidates for the channel material in transparent thin film transistors (TTFT) because of their simultaneous high electron mobility and optical transparency properties. The choice of contact layers are, however, more limited because not all metal oxides can be doped high enough to yield low resistivity layers. Historically, the most common contact layers are ternary compounds that include indium (e.g. indium-tin-oxide, indium-zinc-oxide etc). These indium-containing transparent conductive oxide (TCO) films find widespread applications in flat panel displays and touch-sensitive surfaces of many communication devices. Because of the rapidly expanding markets for such devices, and the limited availability of indium in the world markets, the increased demand-to-supply ratio has caused the cost of indium to increase very rapidly. There are concerns about the continuity of indium supply for future devices.[1]

Nanocrystalline ZnO (nc-ZnO) thin films have been used to fabricate high performance thin film transistors that exhibited field effect mobilities in excess of 100 cm^2/V.s[2] and microwave cutoff frequencies of f_{max}=10GHz[3]. When doped with Al- or Ga, nc-ZnO films can be highly conductive so that such films can be used as contact layers for indium-free transparent transistors. However, the use of the same material for both channel and contact layers can produce challenges in the device fabrication. The first challenge is in the fabrication of source and drain contacts on the undoped ZnO layer. Because the etch rate is nominally the same for both doped and undoped ZnO films, and the etch rate of nc-ZnO is very high in most etchants, it is difficult to etch the doped ZnO layer uniformly without destroying the underlying channel material. On the other hand, the use of lift-off techniques and metal shadow masks are not compatible with high performance devices that require shorter gate lengths. Another difficulty arises if the doped contact layer requires post-growth annealing in nitrogen, vacuum, or forming gas to increase film conductivity, since such annealing also increases the conductivity of the undoped channel layer and causes the device leakage current to increase.

We have developed film deposition and device fabrication techniques to overcome the difficulties described above and fabricated high performance indium-free TTFTs based on nc-ZnO. To overcome the contact layer etch problem, we used a thin layer of Ti (3nm) between the doped and the undoped layers, as shown in Fig.1. This layer acted as an etch-stop layer and allowed a uniform removal of the doped contact layer without etching the channel film. The etch-stop layer readily turns into TiO_2 during the process and does not introduce a leakage path between source and drain. To ensure that the contact layers do not require post-growth annealing for high conductivity, we have employed the recently developed Ga-doped ZnO films grown in Ar ambient in a Pulsed Layer Deposition (PLD) system.[4] These films can be grown at low temperatures and exhibit excellent conductivity and stability (see Fig. 2). As shown in Fig 3, the film stack has an average 90% optical transparency in the visible spectrum. The entire film stack was grown on glass or quartz substrates before device fabrication. PLD was used for doped and undoped nc-ZnO films, whereas Atomic Layer Deposition (ALD) was used for the gate insulator. The top Ga-doped ZnO (GZO) layer was selectively etched to produce source and drain contacts first. This was followed by mesa-type etching of undoped ZnO, HfO_2, and the gate GZO layers in sequence. Various gate length devices were fabricated in the range of 3-25μm. The process temperature budget was 200°C. The transistor I-V characteristics, transfer characteristics, electron mobility and transconductance characteristics are shown in Fig. 4 - 6. Drain current on/off ratio, subthreshold voltage swing, and the maximum field effect mobility were 5×10^{10}, < 200mV/decade, and 15 cm^2/V.s, respectively. To our knowledge, these are the highest values obtained with indium-free thin film transistors.

---

[1] M. Cook, Semiconductor Today, *Compounds & Advanced Silicon*, p.114, Vol. 5, (2010).

[2] B. Bayraktaroglu, K. Leedy and R. Neidhard, *IEEE Electron Dev. Lett.*, p. 1024, (2008).

[3] B. Bayraktaroglu, K. Leedy and R. Neidhard, *Proc. of SPIE*, Vol. 7679, 767904-1, (2010).

[4] R. C. Scott, *et al, Appl. Phys. Lett.*, Vol. 97, 072113, (2010).

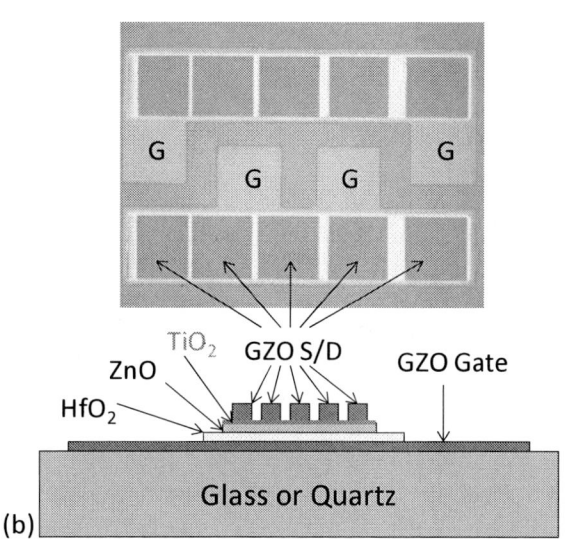

Fig 1: Cross sectional drawing and images of nc-ZnO TTFTs with various gate lengths.

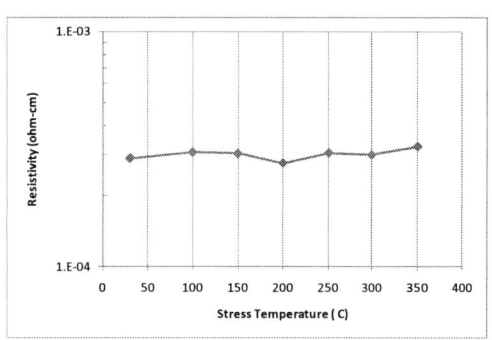

Fig 2: Resistivity stability of Ga-doped ZnO films grown at 200°C. Samples were stressed at each temperature for 1hr.

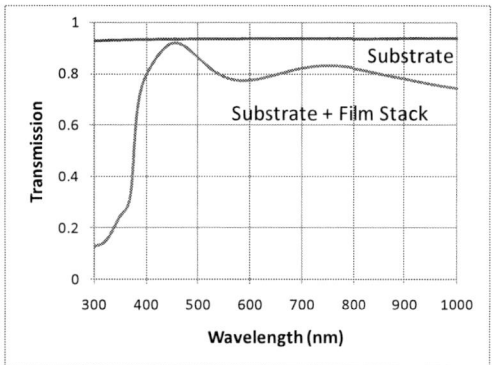

Fig 3: Optical transparency of the entire film stack (gate GZO, HfO$_2$, channel ZnO and S/D GZO) on quartz substrate.

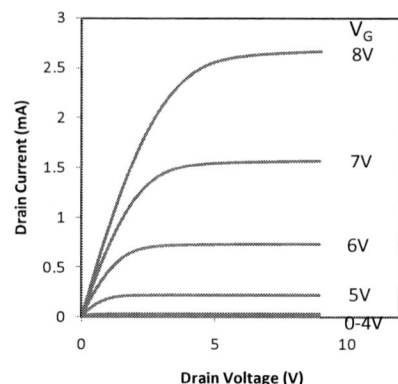

Fig 4: Common source I-V characteristics of nc-ZnO TTFT with L_G = 5um.

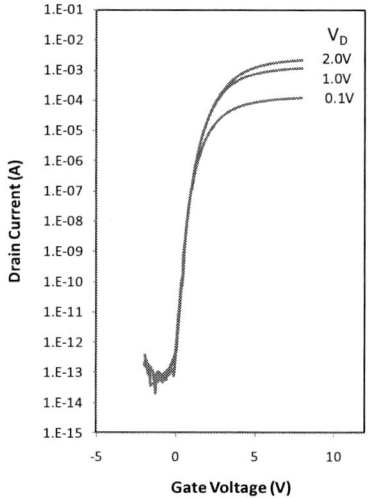

Fig 5: Transfer characteristics of nc-ZnO TTFT with L_G = 5um.

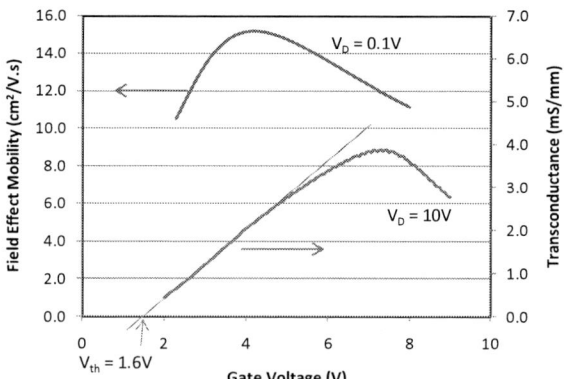

Fig 6: Field-effect mobility and transconductance dependence on gate voltage.

978-1-61284-243-1/11 $26.00 © 2011 IEEE   246

Circuit applications based on solution-processed zinc-tin oxide TFTs

Chen-Guan Lee, Tanvi Joshi, Kiran Divakar, and Ananth Dodabalapur
Microelectronics Research Center, University of Texas at Austin, TX 78758, USA
Email: chenguanlee@gmail.com, phone: (512) 983-1528

Amorphous oxide semiconductors (AOS) have been extensively studied for circuit applications, such as inverters [1], oscillators [2], and memory devices [3]. Most of the AOS-based TFTs used in the circuits are processed with high-vacuum systems even though solution-based processes have the advantages of easy processing, low fabrication cost and potential for large coverage area. In this paper, we demonstrate circuits based on solution-processed zinc-tin oxide (ZTO) TFTs, including inverters, ring oscillators and amplifiers. Performance uniformity and circuit functionality have been achieved with solution-based processing technique.

A patterned bottom-gate, top-contact device architecture was employed in all circuits, as shown in Fig. 1(a). The substrate was glass and the gate metal was deposited after photolithography. Prior to the gate metal deposition, an etching process with a reactive ion etcher (RIE) was employed to fabricate a recessed-gate structure. Detailed process of fabricating solution-processed ZrO_2 and ZTO films are reported in Ref. [4]. Subsequently, via area was patterned with photolithography and etched with RIE. Silver was deposited to form via plug and followed by liftoff. Then, the ZTO active layer was patterned by photolithography and etched with RIE to isolate individual devices. Finally, the source and drain (S/D) regions were patterned with photolithography. Al was deposited as S/D contacts and followed by liftoff. In Fig. 1(a), the left part of the figure shows the cross section of the transistor region while the right part shows the gate contact region. Fig. 1 (b) and (c) show the output characteristics and transfer curve of such a solution-processed ZTO TFT. The channel width and length are 80 µm and 4 µm, respectively. A mobility of 4 cm^2/V·s is observed.

Fig. 2 demonstrates an enhancement-load inverter circuit. The device dimensions are 1000 µm/18 µm (W/L) for the driver transistor and 20 µm/12 µm (W/L) for the load transistor. Voltage transfer characteristics with the inverter circuit diagram as inset are shown in Fig. 2 (a). With a V_{DD} of 10 V, the resulted minimum input voltage at the logic high state (V_{IH}) is 2.2 V and the maximum input voltage at the low logic state (V_{IL}) is 0.8V. The inverter has a minimum output high voltage (V_{OH}) of 8.7 V, a maximum output low voltage (V_{OL}) of 0.2 V, i.e. an output voltage swing of 8.5 V. Static gain of the inverter with its optical image as inset is illustrated in Fig. 2 (b). The maximum gain (dV_{OUT}/dV_{IN}) of this inverter is as high as -9.

Fig. 3 demonstrates a seven-stage ring oscillator circuit with all n-channel, solution-processed ZTO TFTs. Each inverter stage in this oscillator has a beta ratio of 20 with driver transistor dimensions of W_{driver} = 600 µm and L_{driver} = 4 µm, and load transistor dimensions of W_{load} = 30 µm and L_{load} = 4 µm. Fig. 3 (a) shows the output characteristics of the seven-stage oscillator circuit oscillating at 76.5 kHz with a supply voltage of 11 V. Fig. 3 (b) shows the oscillation frequency and propagation delay as a function of the supply voltage. The circuit operates at a V_{DD} as low as 3 V and oscillates at 106 kHz with a V_{DD} of 14 V, corresponding to a propagation delay of 0.67µs/stage.

Fig. 4 demonstrates amplifier circuits. Fig. 4 (a) shows a common-source amplifier of which the input is a small-amplitude ac signal (v_{in}) superimposed on a dc voltage (V_{IN}). The amplifier is biased such that the inverter operates with the highest static gain. By connecting the inverter circuit shown in Fig. 2 as a CS amplifier, the input voltage V_{IN} is set at 1.8 V and an ac signal with amplitude of 350 mV is applied as v_{in}. Fig. 4 (b) shows the input and amplified output voltage signal at a frequency of 1 kHz and a supply voltage V_{DD} of 10 V. The resulted ac gain is around 7. Fig. 4(c) demonstrates the circuit diagram of an operational amplifier. The operational amplifier is composed of three stages. In the first stage, a differential voltage input is amplified and subsequently converted to a single-ended output. Then, the signal is amplified by a cascode stage with a Miller-connected feedback capacitor. Finally, a third current-conveyor stage serves to perform further amplification and buffering. The design and characteristics will be discussed in the presentation.

Basic circuits using n-channel, solution-processed ZTO TFTs have been demonstrated, including inverters, ring oscillators and common-source amplifiers. With the achieved performance uniformity, it is possible to realize circuits with more complex configurations, such as operational amplifiers.

[1] D. P. Heineck *et al. IEEE Electron Dev. Lett.* **30**, 514 (2009).
[2] D. A. Mourey *et al. IEEE Electron Dev. Lett.* **31**, 326 (2010).
[3] H. X. Yin *et al. IEEE Trans. Electron Dev.* **55**, 2071 (2008).
[4] C. G. Lee *et al. Appl. Phys. Lett.* **96**, 243501, (2010).

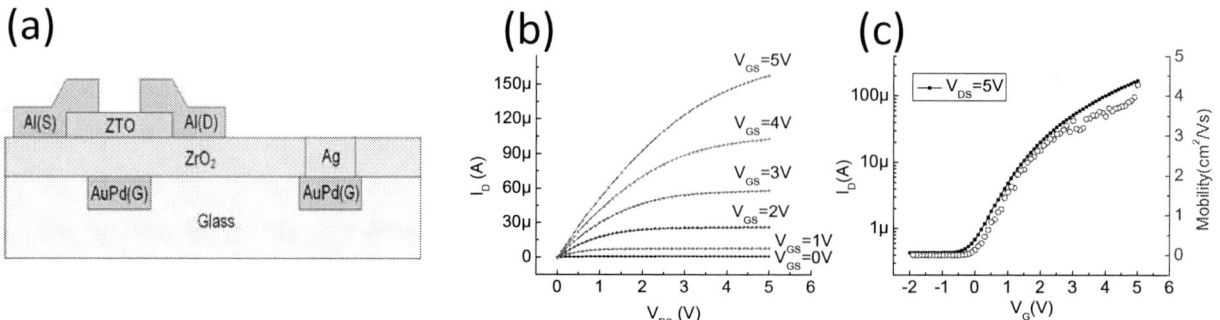

Fig. 1. (a) Cross section, (b) output characteristics, and (c) transfer curve of a solution-processed ZTO TFT. The channel width and length are 80 μm and 4 μm, respectively.

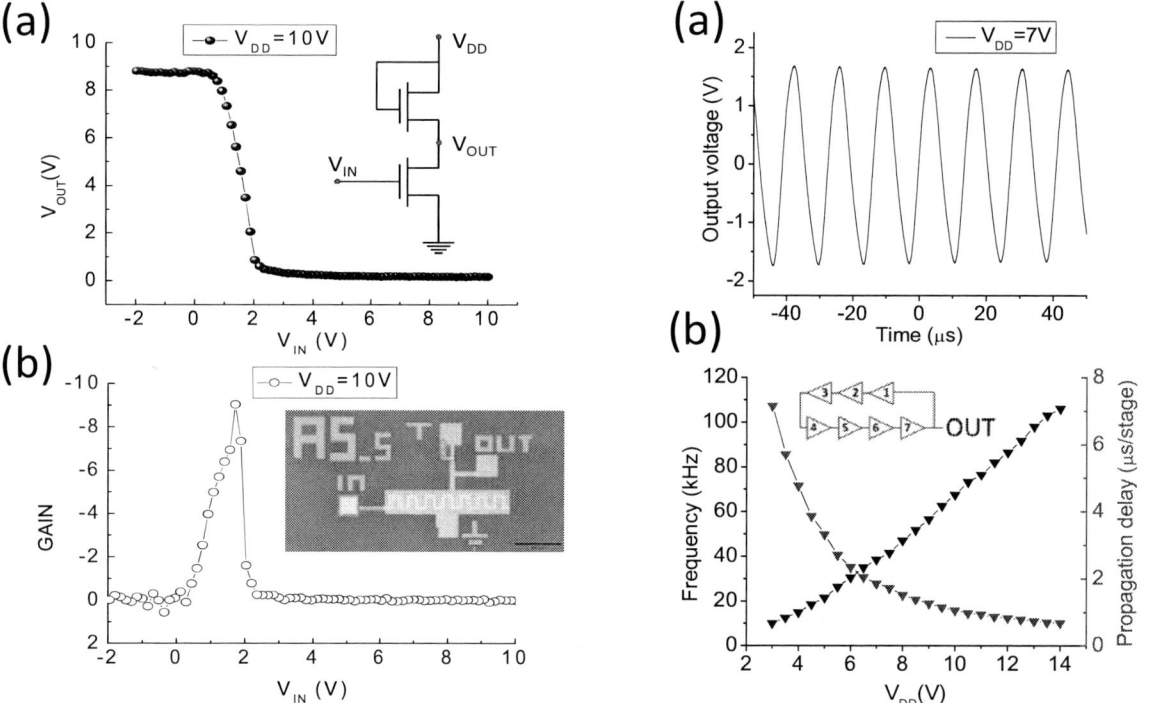

Fig. 2. Transistor-loaded inverter circuit. (a) Voltage transfer characteristics and circuit diagram; (b) Static gain and an optical image of the inverter.

Fig. 3. Seven-state ring oscillator circuit. (a) Oscillation at 76.5 kHz with a voltage supply of 11 V; (c) Frequency and propagation delay as a function of supply voltage.

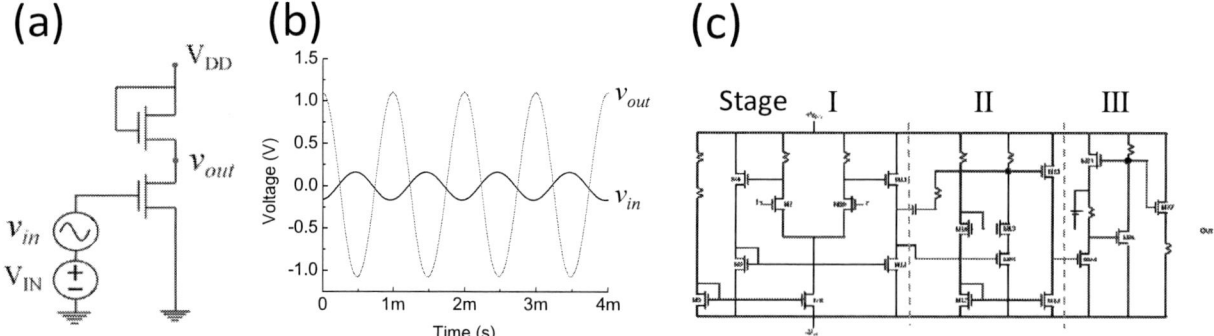

Fig. 4. Amplifier circuits. (a) Circuit diagram and (b) small signal input and output waveform of a common-source amplifier; (b) Circuit diagram of an operational amplifier.

978-1-61284-243-1/11 $26.00 © 2011 IEEE   248

Aluminum Top-Gate ZnO Nanowire Transistors with Improved Transconductance

Daniel Kälblein,[a] Bernhard Fenk,[a] Kersten Hahn,[b] Ute Zschieschang,[a] Klaus Kern,[a,c] Hagen Klauk[a]

a) Max Planck Institute for Solid State Research, Heisenbergstr. 1, 70569 Stuttgart, Germany
(phone: +49 711 689-1317; email: D.Kaelblein@fkf.mpg.de)
b) Max Planck Institute for Metals Research, Germany; c) Ecole Polytechnique Fédérale de Lausanne, Switzerland

Field-effect transistors (FETs) based on semiconducting nanowires are potentially useful to replace thin-film transistors (TFTs) in active-matrix displays, since the larger mobility and smaller footprint of nanowire FETs compared with a-Si:H and organic TFTs provide faster pixel charging and larger aperture ratio. Nanowire growth often requires high temperatures, but if the nanowires can be grown on a temperature-compatible substrate and then be transferred to the target substrate for FET fabrication, and if the temperature during FET fabrication is below ~150 °C, nanoscale FETs can be fabricated on polymeric substrates for flexible displays.

Previously we reported on the fabrication of FETs [1] and circuits [2] based on individual ZnO nanowires with patterned gold top-gate electrodes and a very thin gate dielectric consisting only of an alkylphosphonic acid self-assembled monolayer. These Au top-gate FETs show transconductance up to 1 μS and on/off ratio of 10^7, but because of the small dielectric thickness their gate current increases significantly when the gate-source voltage is increased above 0.5 V, making it difficult to integrate these FETs with organic LEDs for displays.

Here we show that by replacing gold with aluminum for the top gate, the thickness of the gate dielectric increases by a few nanometers due to the spontaneous formation of an interfacial AlO_x layer. As a result, the Al top-gate FETs operate with gate currents below 1 pA for voltages up to 3 V, making them fully compatible with organic LEDs. The ZnO nanowires were grown hydrothermally on a zinc foil [3]. Transmission electron microscopy (TEM) confirms that the nanowires grow along the c-axis and are single-crystalline. The as-grown nanowires are degenerately doped and require a post-growth anneal (600 °C) to make them suitable for FETs.

To fabricate Al top-gate FETs, the annealed nanowires are dispersed on a glass substrate, and source and drain contacts are defined by electron-beam lithography (EBL), evaporation of 80 nm thick Al, and lift-off (Fig. 1a). Immediately prior to the evaporation of the Al S/D contacts, the contact areas are exposed to an Ar plasma. After the lift-off, the substrates are exposed to an O_2 plasma and immersed in a fluoroalkylphosphonic acid solution (Fig. 1b,c). This leads to the formation of a dense hydrophobic monolayer on the ZnO nanowires and on the Al S/D contacts. Top gates are patterned again using EBL and evaporation of 80 nm thick Al (Fig. 1d).

Fig. 2 shows a photograph and the current-voltage characteristics of an Al top-gate ZnO nanowire FET with a channel length of 1 μm. To increase the drive current and transconductance, the S/D contacts were designed in a comb-like pattern with a total of eight contact fingers. The FET has a peak transconductance of 50 μS, an on/off ratio of 10^8, a subthreshold slope of 100 mV/dec, and a maximum gate current below 1 pA at a gate-source voltage of 3 V. To demonstrate that these FETs are capable of driving organic LEDs to sufficiently high brightness for display applications, we have connected this FET to a blue fluorescent OLED (Novaled AG, Dresden, Germany). The LED has an area of 0.067 cm^2 and an efficiency of 7 cd/A. By applying potentials of 5.5 V to the gate and 5 V to the drain of the ZnO nanowire FET, a current of 85 μA, which corresponds to a current density of 1.26 mA/cm^2, is driven through the LED, producing a brightness of 88 cd/m^2 (Fig. 3).

To characterize the dynamic properties of the Al top-gate ZnO nanowire FETs we fabricated inverters with an integrated level-shift stage (Fig. 4). The FETs have a channel length of 1.5 μm. For a frequency of 500 kHz, the inverter output follows the input signal and shows symmetric switching around 0 V.

To investigate the composition of the gate dielectric we performed TEM analyses of the nanowire cross section in the S/D contact region (Fig. 5a) and in the channel region (Fig. 5b). In the contact region the S/D metal (Al) is in intimate contact with the upper facets of the hexagonal nanowire, forming a low-resistance ohmic contact. In contrast, the TEM image of the channel region shows a bright shell around the nanowire which separates the Al gate from the nanowire surface. Also a thin AlO_x layer is visible at the bottom interface of the Al gate near the nanowire. The presence of this AlO_x indicates that atmospheric oxygen diffuses to the Al/SAM interface. A similar observation has recently been reported for Al top gates evaporated onto graphene [4]. The thickness of the bright shell around our SAM-coated nanowire is 2 to 6 nm, which is greater than the thickness of the SAM (2.1 nm). A plausible explanation for the bright shell around the nanowire in the channel region is the formation of a hollow region due to poor wetting of the Al on the hydrophobic SAM-coated nanowire.

[1] D. Kälblein et al., *DRC*, **2009**

[2] D. Kälblein et al., *IEDM*, **2009**

[3] C. Lu et al., *Chem. Comm.* **2006**, 3551

[4] S.L. Li et al., *Nano Lett.* **2009**, 10, 2357

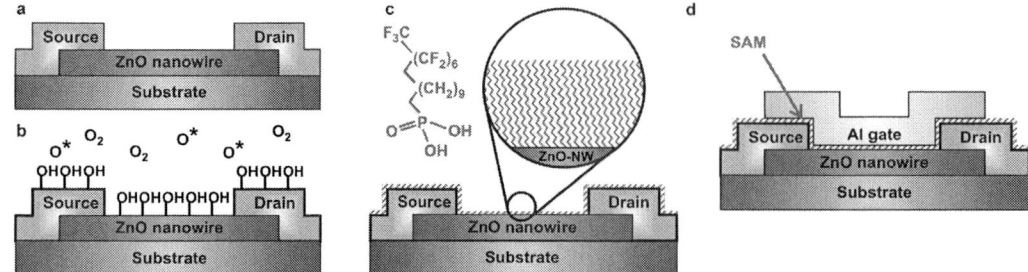

Fig. 1. Fabrication of Al top-gate ZnO nanowire FETs. (a) Al S/D contacts are patterned on the nanowires. (b) The substrate is exposed to an O_2 plasma to prepare for the adsorption of the self-assembled monolayer (SAM). (c) Formation of a fluoroalkylphosphonic acid SAM from solution on the ZnO nanowire and on the contacts. (d) Patterning of the Al top gate.

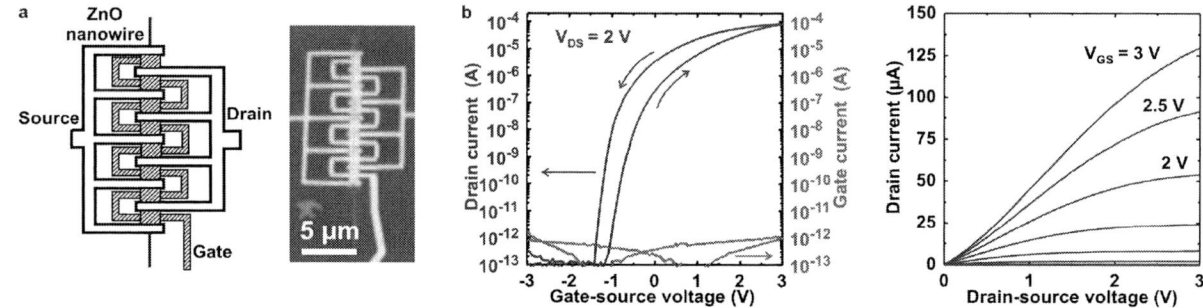

Fig.2. (a) Layout and photograph of an Al top-gate ZnO nanowire FET with a channel length of 1 μm, using a comb-like S/D pattern with eight contact fingers to maximize drive current and transconductance. (b) Transfer and output characteristics of the transistor. The gate current is below 1 pA for gate-source voltages up to 3 V. The transconductance is 50 μS.

Fig. 3. Large-area (0.067 cm²) organic LED driven by a nanowire FET with a drain potential $V_D = 5$ V and various gate potentials V_G. At a gate potential of 5.5 V, the OLED brightness reaches 88 cd/m².

Fig. 4. Circuit schematic and dynamic characteristics of an inverter with integrated level-shift stage. All four FETs were realized on the same nanowire. For an input frequency of 500 kHz, the output signal follows the input signal. The level-shift stage adjusts the output levels to make them compatible with the input levels, so the output signal is symmetric around 0 V.

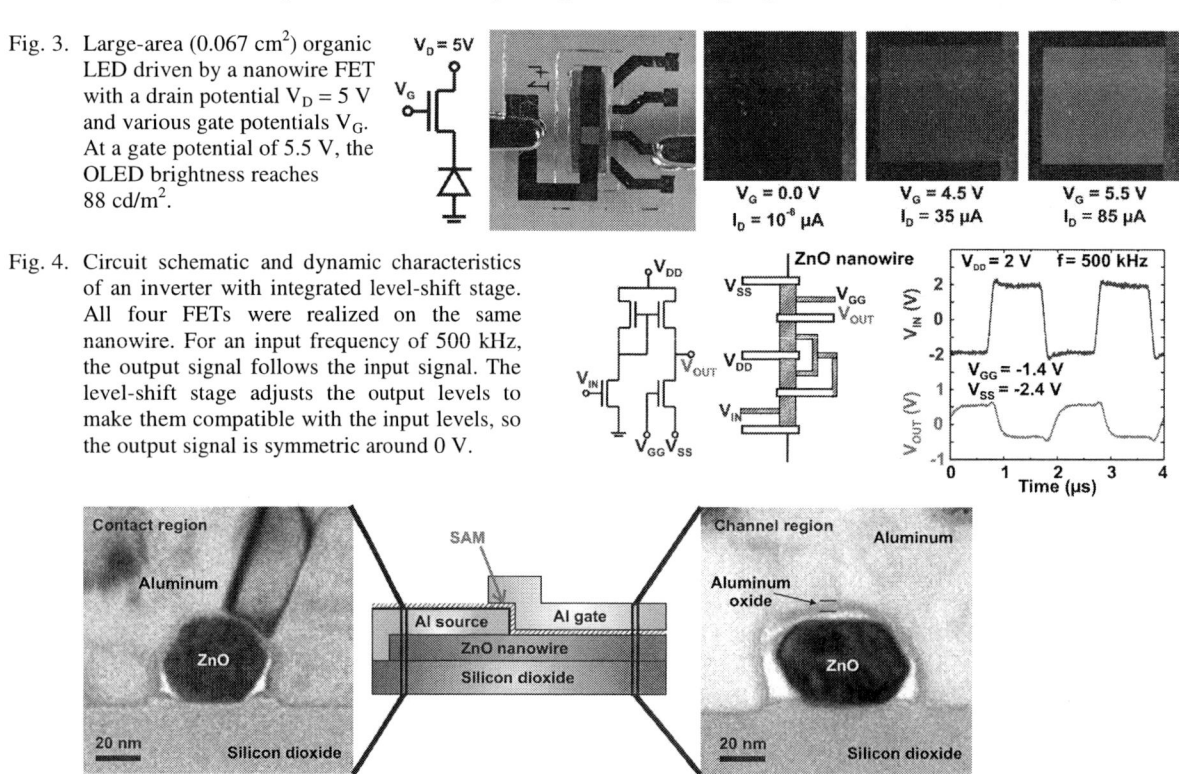

Fig. 5. TEM analysis of the cross section of an Al top-gate FET. In the S/D contact region (left), the Al contact is in intimate contact with the nanowire. In the channel region (right), the Al gate is separated from the SAM-coated nanowire by a bright shell, and an AlO_x layer is visible at the bottom interface of the Al gate. The bright shell is attributed to a hollow region caused by a poor wetting of the Al gate on the hydrophobic SAM.

978-1-61284-243-1/11 $26.00 © 2011 IEEE

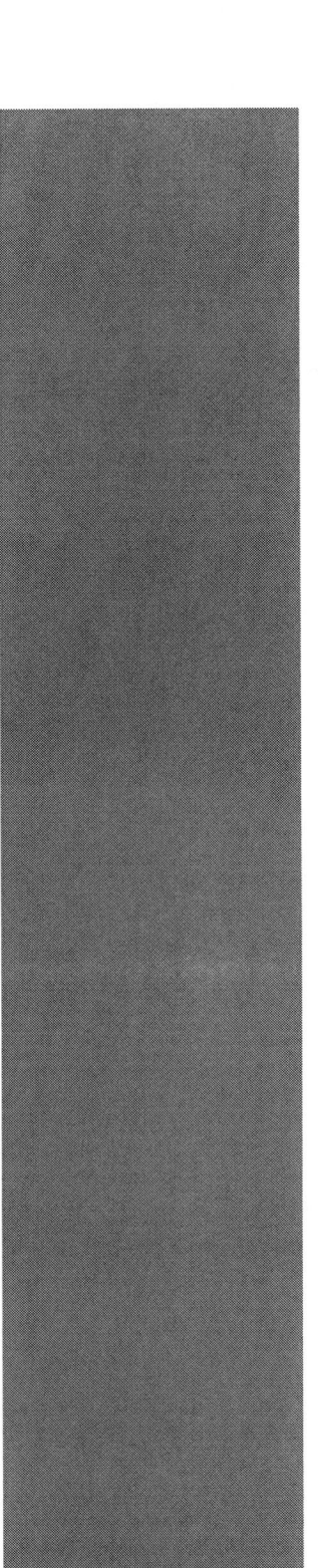

Session VII.A (Corwin Pavilion East)

Optoelectronic Devices

Wednesday PM, June 22nd, 2011

Session Chair: Mark Wistey, University of Notre Dame and Ganesh Balakrishnan, University of New Mexico

1:30 PM VII.A-1 Invited Paper
Monolithic Integration of CMOS and Nanophotonic Devices for Massively Parallel Optical Interconnects in Supercomputers
S. Assefa[1], W. M. J. Green[1], A. Rylyakov[1], C. Schow[1], F. Horst[2] and Y. A. Vlasov[1], [1]IBM Thomas J. Watson Research Center, Yorktown Heights, New York, USA and [2]IBM Zurich GMBH, Rueshlikon, SWITZERLAND

2:10 PM VII.A-2
Electrical pumped integrated III/V laser lattice-matched to a Silicon substrate
B. Kunert[1], S. Liebich[2], M. Zimprich[2], A. Beyer[2], S. Ziegler[1], K. Volz[2], W. Stolz[2], N. Hossain[3], S. R. Jin[3], and S. J. Sweeney[3], [1]NAsP III/V GmbH, Marburg, GERMANY and [2]Material Sciences Center and Faculty of Physics, Philipps-University Marburg, Marburg, GERMANY, and [3]Advanced Technology Institute and Department of Physics, University of Surrey, Guildford, Surrey, UK

2:30 PM VII.A-3 Student Paper
Lateral Carrier Injection with n-type Modulation-doped Quantum Wells in VCSELs
C.-H. Lin[1], Y. Zheng[1], M. Gross[3], M. J. W. Rodwell[1], and L. A. Coldren[1,2], [1]Department of Electrical and Computer Engineering, University of California, Santa Barbara, California, USA, [2]Department of Materials, University of California, Santa Barbara, California, USA, and [3]Ziva Corporation, San Diego, California, USA

2:50 PM VII.A-4 Student Paper
Near UV AlGaN-Cladding Free Nonpolar InGaN/GaN Laser Diodes
D. A. Haeger[1], C. Holder[1], R. M. Farrell[1], P. S. Hsu[1], K. M. Kelchner[2], K. Fujito[3], D. A. Cohen[2], S. P. DenBaars[1,2], J. S. Speck[1], and S. Nakamura[1,2], [1]Materials Department, University of California, Santa Barbara, California, USA, [2]Electrical and Computer Engineering Department, University of California, Santa Barbara, California, USA, and [3]Optoelectronic Laboratory, Mitsubishi Chemical Corporation, Ushiku, Ibaraki, JAPAN

3:10 PM Break

3:30 PM VII.A-5 Student Paper
RIE Lag Directional Coupler based Integrated InGaAsP/InP Ring Mode-locked Laser
J. S. Parker[1], P. R .A. Binetti[1], Y.-J. Hung[2], Erik J. Norberg[1], and L. A. Coldren[1], [1]Electrical and Computer Engineering Department, University of California, Santa Barbara, California, USA and [2]Dept. of Electronic Engineering, National Taiwan University of Science and Technology, Taipei, TAIWAN

3:50 PM VII.A-6 Student Paper
Integrated Non-III-Nitride/III-Nitride Tandem Solar Cell
N. G. Toledo[1], S. C. Cruz[2], C. J. Neufeld[1], M. A. Scarpulla[2], T. Buehl[2], A. C. Gossard[1,2], S. P. Denbaars[1,2], J. S. Speck[2] and U. K. Mishra[1], [1]Department of Electrical and Computer Engineering, University of California, Santa Barbara, and [2]Materials Department, University of California, Santa Barbara, California, USA

4:10 PM VII.A-7
Late News

4:30 PM VII.A-8
Late News

4:50 PM VII.A-9
Late News

978-1-61284-243-1/11 $26.00 © 2011 IEEE

978-1-61284-243-1/11 $26.00 © 2011 IEEE   252

Monolithic Integration of CMOS and Nanophotonic Devices for Massively Parallel Optical Interconnects in Supercomputers

Solomon Assefa, William M. J. Green, Alexander Rylyakov, Clint Schow, Folkert Horst* and Yurii A. Vlasov

IBM Thomas J. Watson Research Center, Yorktown Heights, NY 10598, USA; *IBM Zurich GMBH, Rueshlikon, Switzerland

High-performance computing (HPC) systems capable of delivering Exaflops performance are envisioned to become a reality by the end of this decade. In order to provide the enormous communication bandwidth that is necessary, hundreds of millions of optical interconnects will have to be deployed to connect together racks, modules and chips. To achieve such massive level of parallelism for HPC systems, monolithic integration of deeply scaled silicon optical circuits into the front-end of standard CMOS process have been demonstrated as a promising technology [1-3].

The monolithic integration of electronic and nanophotonic components was performed at the IBM using 200 mm SOI wafers (SOITEC) having a 220 nm silicon device layer on top of a 2 μm BOX [1-4]. Several processing modules have been added to a standard CMOS processing flow at the front-end of the line (Fig. 1). These modules require a minimal number of additional unique masks and processing steps, while sharing most mask levels and processing steps with the rest of CMOS. For example passive waveguides and electro-optical and thermo-optical modulators share the same silicon device layer with CMOS PFETS and NFETS. The Ge waveguide photodetectors were fabricated by utilizing a rapid melt growth (RMG) technique wherein the Ge was melted and crystallized during the source-drain anneal step [5-6]. As opposed to traditional approaches where Ge is typically grown by CVD after the source-drain anneal step, the RMG approach enables the sharing of many CMOS steps and mask levels, thus minimizing cost, while yielding a very thin defect-free Ge layer.

After completion of the monolithic integration including the metallization levels, the waveguides (200 nm x 500 nm cross-section) exhibit propagation loss below 3dB/cm and bending losses less than 0.02dB per 90°-bend with 6 μm radius. Also, a cascaded Mach-Zehnder (CMZ) 8-channel WDM filter provided 8 flat-top 2 nm-wide pass-bands with in-band ripple less than 1 dB and cross-talk less than -10 dB [7]. Furthermore, reverse-biased modulators were demonstrated to have flat RF response with a 3 dB cutoff exceeding 20GHz [8], while the Germanium photodetectors showed 3dB cutoff beyond 40 GHz at 1.5-2.0 V bias with some devices producing up to 10 dB avalanche gain at the same low bias conditions [9].

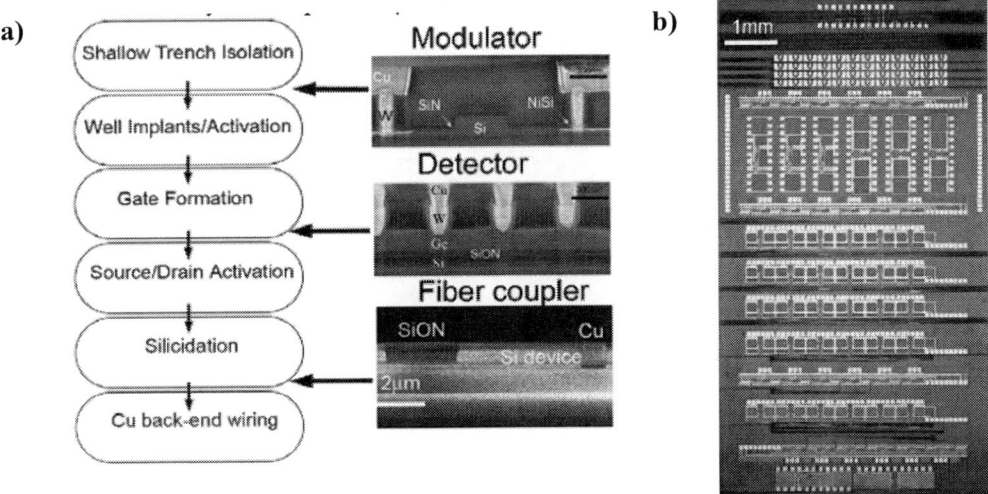

Fig. 1. a) Schematic flow diagram of processing flow in the front-end of the CMOS line, with addition of modulators, detectors, and fiber couplers into the CMOS circuitry, **b)** Microphotograph of a part of a CMOS die showing monolithically integrated CMOS and Nanophotonics circuitry

The CMOS ring oscillator (RO) testsite is taken from a standard IBM digital library and consists of a 65-stage ring oscillator, 10-stage ripple counter, and 4-stage divider (Fig. 2a)). After integration of all nanophotonics processing modules the RO exhibits 12 ps delay per stage at saturation, under a 1.5 V bias. This performance is on-target for 130 nm bulk technology.

The receiver amplifier (Fig. 2b)) consists of a DC-coupled common-gate transimpedance amplifier (TIA) occupying 170 μm x 40 μm, followed by a cascade of seven current-mode logic (CML) buffer limiting amplifier (LA) stages, followed by a single-ended open-drain output driver. The total area occupied by the multi-stage LA and output driver is 160 μm x 50 μm. The total circuit area is dominated by decoupling capacitors used to minimize the supply voltage noise. After incorporation of all nanophotonic processing modules, the receiver amplifier exhibits an open eye diagram at 5 Gbps with total power dissipation of about 28 mW (5.6 pJ/bit).

The modulator driver (Mod DRV) circuitry (Fig. 2c)) consists of the input pre-driver (105 μm x 175 μm) made of a 6-stage differential CML amplifier, and an output stage (105 μm x 70 μm) consisting of a cascoded differential driver with a dedicated power supply. The total circuit area is similarly dominated by the decoupling capacitors. High-speed operation of the output stage is enabled by a differential pair of low-threshold thin oxide NFETs. High-voltage, high-output swing capability of the output stage is enabled by a pair of long-channel thick oxide devices in a cascode configuration, protecting the thin oxide devices from the output voltages. After incorporation of all the nanophotonic processing modules, the transmitter amplifier exhibits an open eye diagram at 5 Gbps with total power dissipation of about 36 mW (7.2 pJ/bit).

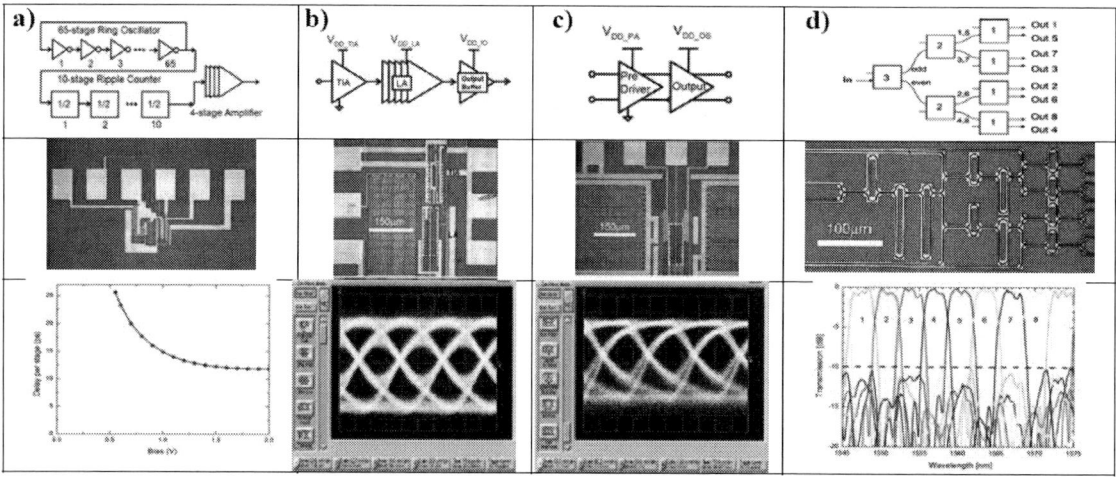

Fig. 2. Schematics, die photos, and performance measurements of CMOS digital circuits (ring oscillator in a)), CMOS analog circuits (transimpedance amplifier in b), modulator driver amplifier in c)), and nanophotonic circuits (cascaded Mach-Zehnder WDM in d)), taken after completion of the full integrated process flow.

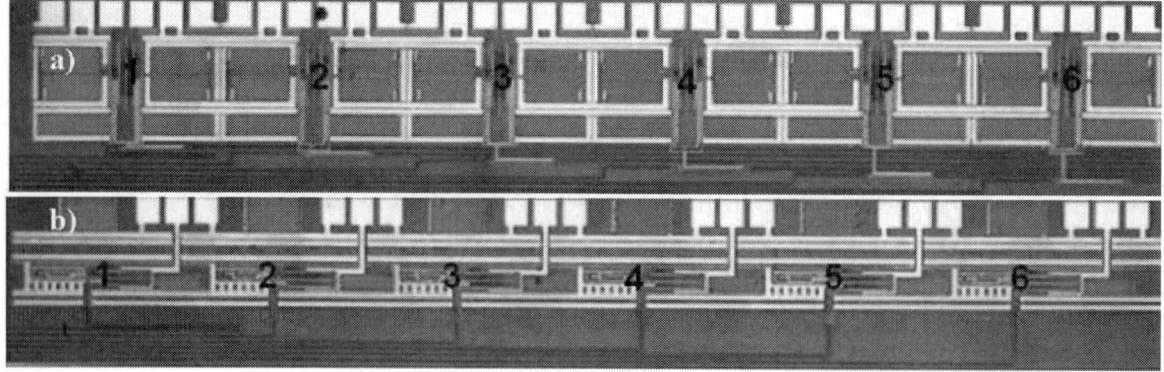

Fig. 3. Micrographs of an integrated silicon nanophotonic a) 6-channel transmitter and b) 6-channel receiver.

Figure 3 contains die photographs of a 6-channel transmitter (Fig. 3a)) and 6-channel receiver (Fig. 3b)). The chip footprint of the transmitter in Fig. 3a) (counted as shown by the dotted outlines in Fig. 1b)) is 3700 μm x 350 μm. The 0.21 mm2 area per channel is limited by the pad frame and decoupling capacitors. The area of the 6-channel receiver (Fig. 3b) counted the same way is 3700 μm x 500 μm, which totals to an area of 0.31 mm2 per channel. Correspondingly, the area per single transceiver channel can be estimated as 0.52 mm^2, which is approximately 10 times smaller than previous demonstrations [10].

To conclude, a proof-of-principle demonstration of dense monolithic integration of silicon nanophotonic components with analog and digital CMOS circuitry has been described. The integration density offered by the technology is at least 10 times higher than previous reports. The technology offers a scalable solution for massively parallel Terabit/sec-class optical transceivers, with 50 channels supporting serial lines rates of 20 Gbps each, occupying only 4 mm x 4 mm on a single CMOS die.

Acknowledgements
The contributions of many of our colleagues across various organizations at IBM Research are acknowledged. We are particularly grateful to Fengnian Xia, Leathen Shi, Jeffrey Sleight, Young-Hee Kim, Chris Jahnes, and the staff at the Microelectronics Research Laboratory.

References

1. W. Green, S. Assefa, A. Rylyakov, C. Schow, F. Horst, Y. Vlasov, SEMICON 2010.
2. S. Assefa, W. Green, A. Rylyakov, C. Schow, F. Horst, Y. Vlasov, OMM6, OFC 2010.
3. J. Van Campenhout, et al., Optics Letters **35**(7), 1013-1015 (2010).
4. J. Van Campenhout, et al., Optics Express **17**(26), 24020-24029 (2009).
5. S. Assefa, et al., IEEE Journal of Selected Topics in Quantum Electronics, **16**(5) 1376-1385 (2010).
6. S. Assefa, et al., Optics Express **18**(5), 4986-4999 (2010).
7. C.G.H. Roeloffzen et al., IEEE Photonics Technology Letters **12**, 1201-1203 (2000).
8. W. M. J. Green, et al., Optics Express, **15**(25), 17106-17113, (2007).
9. S. Assefa, et. al, Nature **464**, 80 (2010).
10. B. Analui et al., IEEE Journal of Solid-State Circuits **41**(12), 2945-2955 (2006).

978-1-61284-243-1/11 $26.00 © 2011 IEEE

Electrical pumped integrated III/V laser lattice-matched to a Silicon substrate

B. Kunert*, S. Liebich, M. Zimprich, A. Beyer, S. Ziegler*, K. Volz, W. Stolz

*) NAsP III/V GmbH, Am Knechtacker 19, 35041 Marburg, Germany
Material Sciences Center and Faculty of Physics, Philipps-University Marburg,
Hans-Meerwein-Strasse, 35032 Marburg, Germany

N. Hossain, S. R. Jin, S. J. Sweeney
Advanced Technology Institute and Department of Physics, University of Surrey, Guildford,
Surrey GU2 7XH, UK

The enormous development of Silicon (Si) based integrated circuits (ICs) and micro-electronics is based on the downscaling of semiconductor devices. This driving force, however, is approaching fundamental limitations and therefore new technologies are necessary to guarantee future progress in IC functionalities.
In particular the integration of III/V compound materials on Si based microelectronics and ICs allows for the combination of the advantageous material properties of both semiconductor classes and opens up completely new device concepts and architectures.

For example the application of III/V material systems for n-channels with high electron mobility offers huge potential for device improvement. However, the most significant benefit of III/Vs over Si are their superior optoelectronic properties. The successful and cost-efficient integration of a reliable, long-lifetime III/V laser diode on silicon will enable chip-to-chip or even on-chip optical data transfer with all of the advantages of high bandwidth optical data communications leading to new IC functionalities.

The monolithic growth of standard III/V laser materials such as alloys of GaAs and InP leads unavoidably to the formation of a high density of misfit- and threading dislocations due to the large differences in lattice constant of GaAs and InP relative to Silicon. These threading dislocations act as non-radiative recombination channels, severely limiting laser device lifetime. The novel dilute nitride Ga(NAsP) based on GaP has been developed specifically for the lattice-matched and defect-free integration of an optically efficient direct band gap semiconductor on Si. In combination with the boron containing material systems (BGa)P and (BGa)(AsP), which form the separate confinement hetero-structure and cladding layers, a complete Ga(NAsP) laser diode has been grown using metal organic vapor phase epitaxy (MOVPE) lattice-matched to an (001) Si substrate. Transmission electron microscopy (TEM) and X-ray diffraction (XRD) confirm that this novel approach suppresses misfit formation. As a result, for the first time, electrically injected laser operation of broad area Ga(NAsP) lasers on silicon have been demonstrated at low temperature, illustrating "proof of principle" for this novel integration concept.

In this paper we also discuss the latest achievement in the development and improvement of electrically injected Ga(NAsP) laser diodes lattice matched to (001) Si. In particular, the conductivity of the cladding layers has been optimized to reduce the voltage drop across the device hetero-structures. While these first laser diodes have been deposited on 2 inch Si wafer in a horizontal MOVPE reactor, the current research activity concentrates on the process transfer to 300 mm Si wafers in a production epitaxy system. The challenges and opportunities of growing GaP based materials on 300 mm substrates in a Close-Couple-Showerhead cluster tool as well as the final device concept will be discussed in view of CMOS capability and industrial process optimisation.

978-1-61284-243-1/11 $26.00 © 2011 IEEE

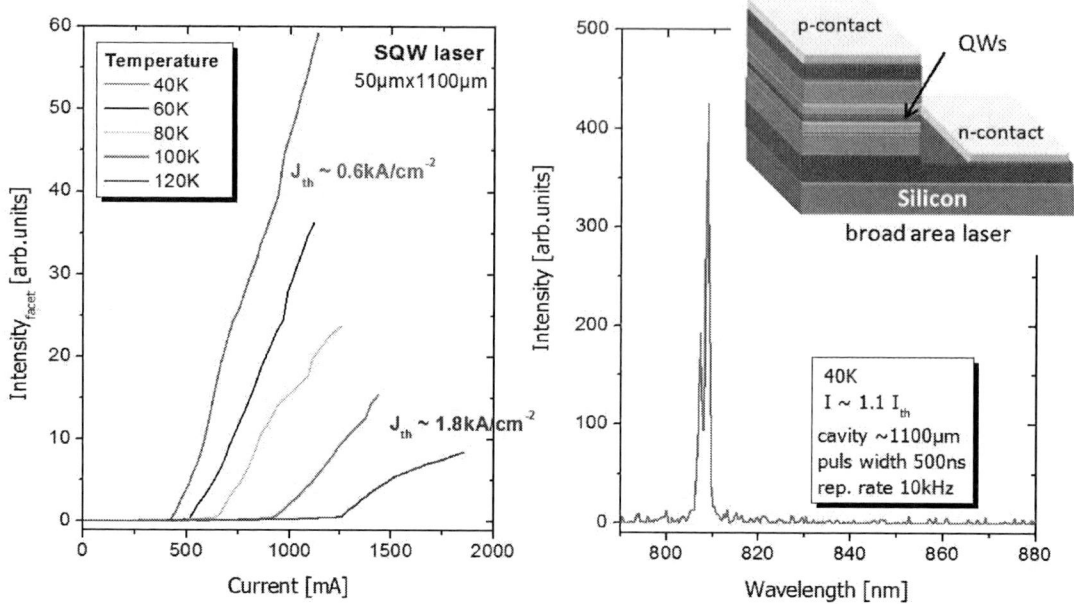

Figure 1: Facet intensity versus current for a Ga(NAsP) single quantum well laser grown on (001) Si substrate (left). Corresponding laser spectrum above threshold (right). Schematic of the processed broad area laser bars (inset right).

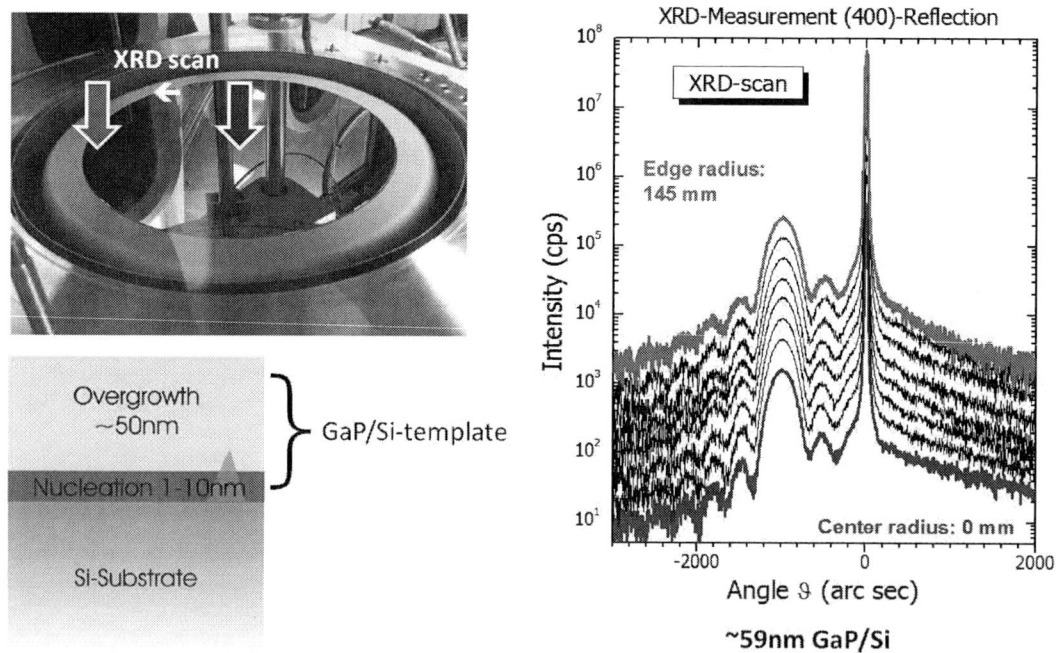

Figure 2: 59nm GaP grown on 300 mm Silicon wafer in a Close-Couple Shower head reactor. Photograph of the susceptor with a 300 mm Si wafer (top). Schematic of the GaP/Si-template (bottom). XRD scans from wafer center (blue) to the edge (red)(right).

Lateral Carrier Injection with n-type Modulation-doped Quantum Wells in VCSELs

Chin-Han Lin[1], Yan Zheng[1], Matthias Gross[3], Mark J. W. Rodwell[1], and Larry A. Coldren[1,2]

[1]Department of Electrical and Computer Engineering, University of California, Santa Barbara, CA 93106, USA
[2]Department of Materials, University of California, Santa Barbara, CA 93106, USA
[3]Ziva Corporation, San Diego, CA 92121, USA
Phone: 1-805-893-7065 Fax: 1-805-893-4500 Email: chinhan@ece.ucsb.edu

We have demonstrated a novel Field-Induced Charge-Separation Laser (FICSL) in a Vertical-Cavity Surface-Emitting Laser (VCSEL) embodiment. In addition to the initial optical modulation results that have been presented [1], we here for the first time present details on the novel lateral charge injection structure as well as the advanced bandgap engineering involved in the gate structure. These features together permit high-speed light modulation with a nearly constant injection current. The result is an entirely new concept for high-speed directly-modulated semiconductor lasers.

In conventional diode VCSELs, carriers are typically injected into the quantum wells from the perpendicular direction (Fig. 3a-b). The common epitaxial design is a SCH (Separate Confinement Heterostructure) with p-type and n-type DBR mirrors on the opposite side of the active region. On the other hand, to realize direct modulation via a gate voltage, any highly doped intra-cavity contact layer above the quantum wells such as in Fig. 3b would pin the Fermi level, screening the driving force from the gate. Hence, carriers have to be injected laterally into the active region (Fig 3c). To allow only hole movement and isolate the electrons between the active channel and the off-state well, the first period of the n-DBR has to be band-gap engineered to form a quantum barrier (Fig. 4).

Lateral injection of carriers into quantum wells has been demonstrated on lateral current injection (LCI) lasers. However, the sheet resistance in LCI lasers is typically 2-3 times higher than a vertically-injected ridge laser with the same dimensions [2]. In a VCSEL structure, the comparatively small dimensions makes the sheet resistance even more detrimental, especially when there has to be a finite setback distance d between the current aperture edge and the metal ring contact to avoid absorption loss (Fig. 1). To enhance the conductivity of lateral injection, delta-doping can be applied to the quantum well barriers to provide free carriers.

Early research showed that modulation doping was promising for quantum well performance [3]. Furthermore, both n-type and p-type doped quantum wells have been incorporated into diode lasers and demonstrated the reduction of threshold current density [4, 5]. However, doping near the optical standing wave E^2 peaks has to be treated carefully since the increase in absorption loss and dopant scattering may compromise overall performance, rendering the delta-doping benefit marginal.

To examine lateral injection with two-dimensional electron gas (2DEG) and two-dimensional hole gas (2DHG), three $In_{0.2}Ga_{0.8}As/GaAs$ quantum well samples were grown by Molecular Beam Epitaxy with different delta-doping level in the barriers; silicon was used for n-doping, and carbon via a CBr_4 source was used for p-doping. The sheet charge density and the carrier mobility of these samples were acquired by room temperature Hall measurement, and the overall sheet conductivity was calculated (Fig. 5). It is clear that due to the large difference in mobility, the conductivity of 2DEG is more than one order of magnitude higher than a 2DHG with the same sheet charge density. Material gain of the modulation-doped quantum wells was characterized by fabricating broad area lasers. It was confirmed that both n-type and p-type modulation-doping reduced the transparency current density, while p-type doping also increased the differential gain, dg/dJ (Fig. 6).

Due to the large difference in conductivity, n-type modulation-doped quantum wells are chosen to form the active channels. FICSLs were fabricated and the measurement results are shown in Fig. 7. The required gate voltage swing was higher than expected, indicating that the fabricated FICSLs had very large resistance in the n-DBR, causing a large voltage drop. The large n-DBR resistance also limited the small-signal bandwidth to be around 11GHz [1]. The differential series resistance between injector and channel was measured to be 550Ω in a 5μm (aperture diameter) device.

This work was supported by DARPA through a STTR with Ziva Corp. and by NSF through a GOALI program. A portion of this work was done in the UCSB nanofabrication facility, part of the NSF funded NNIN network.

[1] C.-H. Lin et al., International Semiconductor Laser Conference (ISLC) 2010, post deadline paper, PD2, September 2010.

[2] E.H. Sargent et al., Journal of Lightwave Technology, Vol. 16, No. 10 (1998)

[3] K. Uomi, Japanese Journal of Applied Physics, Vol. 29, No. 1, pp. 81-87 (1990).

[4] T. Mukaihara et al., Physica B, Vol. 227, Issue 1-4, pp. 400-403 (1996).

[5] N. Hatori et al., Electron. Lett., vol. 33, pp. 1096–1097 (1997).

[6] L. A. Coldren and S. W. Corzine, Diode Lasers and Photonic Integrated Circuits, Chapter 4, Wiley, NY (1995).

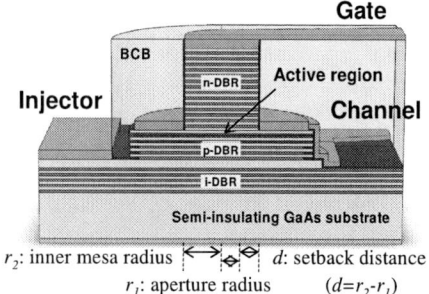

Fig. 1. Schematic of a FICSL. The injector-channel junction is DC biased, while the gate is the modulation terminal.

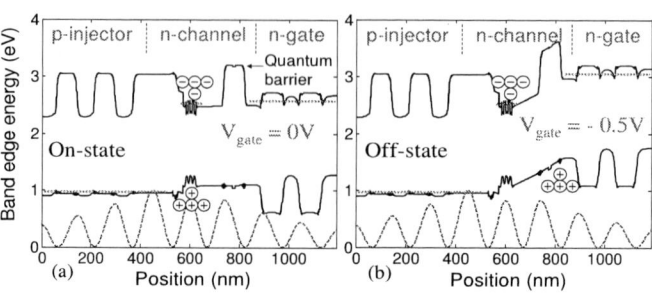

Fig. 2. Band diagrams of a FICSL (a) on-state: electrons and holes are aligned with a standing wave E^2 peak to provide gain. (b) off-state: a negative bias is applied to the gate, driving holes away from electrons and reducing output power.

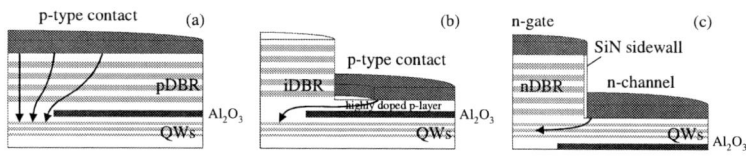

Fig. 3. Structural comparison of oxide-confined VCSELs: (a) top-contacted (b) intra-cavity contacted (c) laterally injected FICSL with top gate for modulation (this work). Arrows depict carrier injection flows. Oxide aperture layers are typically placed on the p-side due to the low spreading characteristics of holes compared to electrons.

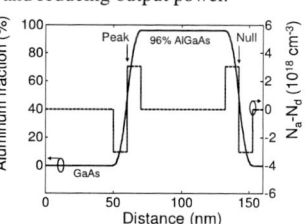

Fig. 4. Quantum barrier design to isolate electron transport.

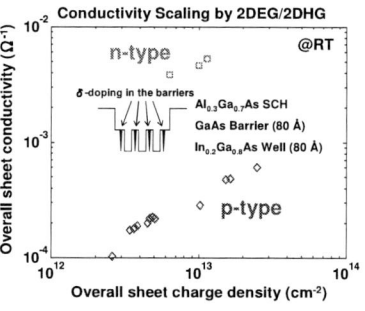

Fig. 5. Sheet charge conductivity of 2DEG and 2DHG in three modulation δ-doped quantum wells.

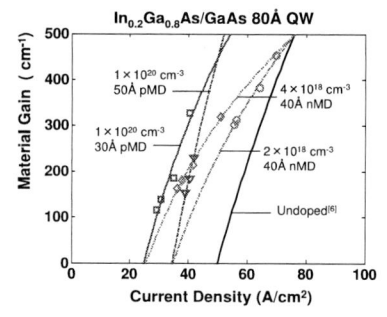

Fig. 6. Gain characteristics of modulation-doped QWs. Both n-doping and p-doping reduce transparency current density.

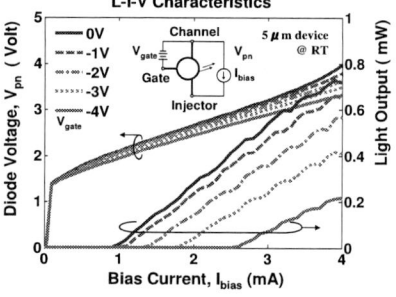

Fig. 7. DC performance of FICSL. Above threshold, negatively biasing the gate would reduce the light output given the same bias current to the injector-channel junction.

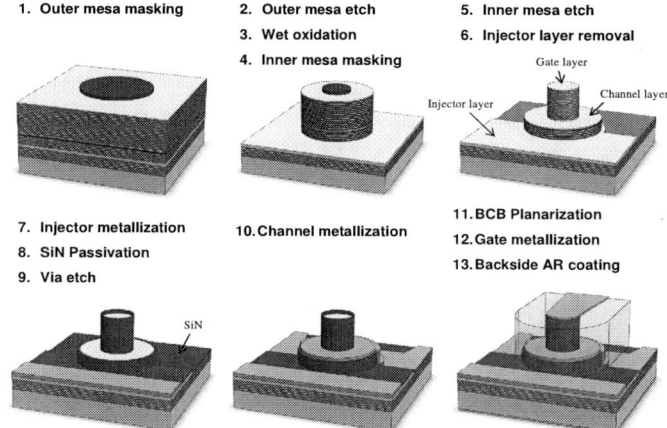

Fig. 8. Process flow for FICSLs. All steps are done with i-line lithography.

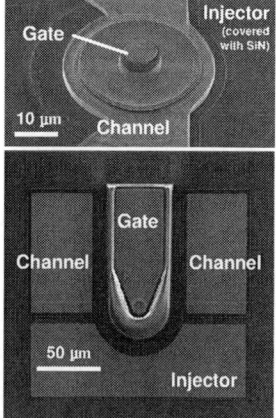

Fig. 9. SEM images of a fabricated FICSL before BCB planarization (upper) and after the completion of process (lower).

978-1-61284-243-1/11 $26.00 © 2011 IEEE

Near UV AlGaN-Cladding Free Nonpolar InGaN/GaN Laser Diodes

Daniel A. Haeger[1*], Casey Holder[1], Robert M. Farrell[1], Po Shan Hsu[1], Kathryn M. Kelchner[2], Kenji Fujito[3], Daniel A. Cohen[2], Steven P. DenBaars[1,2], James S. Speck[1], and Shuji Nakamura[1,2]

[1]*Materials Department, University of California, Santa Barbara, CA 93106, U.S.A.*
[2]*Electrical and Computer Engineering Department, University of California, Santa Barbara, CA 93106, U.S.A.*
[3]*Optoelectronic Laboratory, Mitsubishi Chemical Corporation, 1000 Higashi-Mamiana, Ushiku, Ibaraki 300-1295, Japan*

We have previously demonstrated the AlGaN-cladding-free (ACF) laser diode (LD) concept over a wide visible spectral range and various nonpolar/semipolar crystallographic orientations. The benefits of this ACF epitaxial design include lower operating voltages and higher production yields. Nonpolar LDs have been demonstrated out to 500nm[1] with semipolar reaching as long as 534nm[2]. No results have been published on nonpolar/semipolar orientations for short wavelength LDs. In this paper we report on the short wavelength limits of this epitaxial design on nonpolar bulk m-plane GaN substrates.

Structures were grown by atmosphere pressure metal organic chemical vapor deposition (AP-MOCVD) on $1°$ miscut towards the [000-1] (-c direction) m-plane GaN substrates. Instead of using low index AlGaN cladding and waveguide layers, the structures relied on thick, 8nm wide, InGaN multi-quantum-well (MQW) design to provide adequate optical confinement. InGaN waveguides such as those used for longer wavelength nonpolar LDs are unsuitable for devices near this wavelength. Following growth the wafers were processed into standard ridge waveguide structures using a self-aligned ridge etch and dielectric liftoff process. The facets were etched using a standard reactive ion etching (RIE) process using a bi-layer photoresist mask.

Lasing at 395nm was achieved at the low threshold current density of 2.68 kA/cm^2 for a 15x1800 micron device. A significantly higher threshold current density of 14.03 kA/cm^2 was measured for a 2.5x1800 micron device lasing at 386nm. The epitaxial structures are identical except for indium composition in the MQW's. Transverse confinement factors were calculated to be 0.045 and 0.036 for the 395nm and 386nm device respectively. Optical confinement decays quickly beyond 385nm. InGaN waveguides were not used due to the fact that the emission is too close to the band edge of GaN. Studies on m-plane ACF LD at UCSB have demonstrated a linear relationship between material gain and current density. This suggests that the cause of the increase in threshold current density cannot be fully explained by the reduced optical confinement, but is instead related to electrical transport. Temperature dependant measurements were performed on the 386nm device. The characteristic temperature, T_o, was calculated to be 40K. This low value is believed to be due to the increased carrier overflow from small GaN barriers. However, increasing the barrier height through use of AlGaN barriers results in rapid degradation of optical confinement. Future work will require the use of high aluminum composition AlGaN cladding, waveguide, and active region layers.

[1] Okamoto, K., J. Kashiwagi, et al. (2009). "Nonpolar m-plane InGaN multiple quantum well laser diodes with a lasing wavelength of 499.8 nm." *Applied Physics Letters* **94**(7):

[2] Adachi, M., Y. Yoshizumi, et al. (2010). "Low Threshold Current Density InGaN Based 520-530nm Green Laser Diodes on Semi-Polar {20(2)over-bar1} Free-Standing GaN Substrates." *Applied Physics Express* **3**(12):

Fig 1. LIV and Spectra for 386nm ACF LD

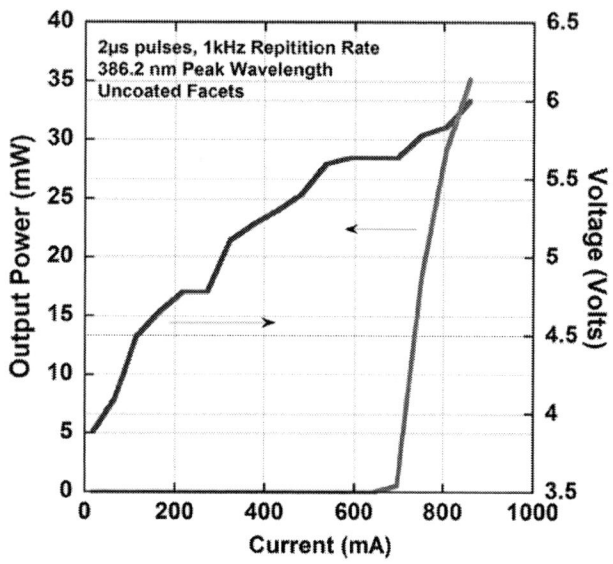
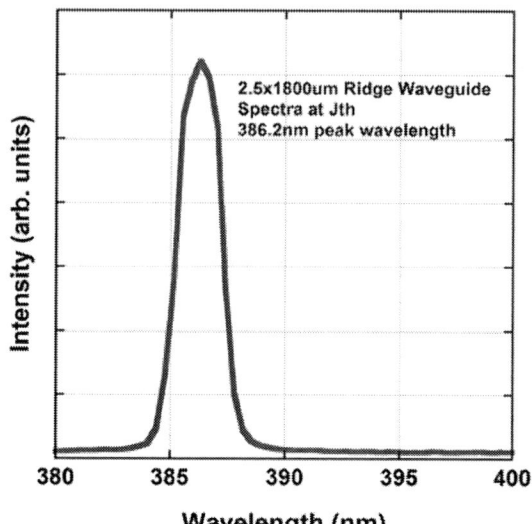

Fig 2. Temperature dependant threshold current density

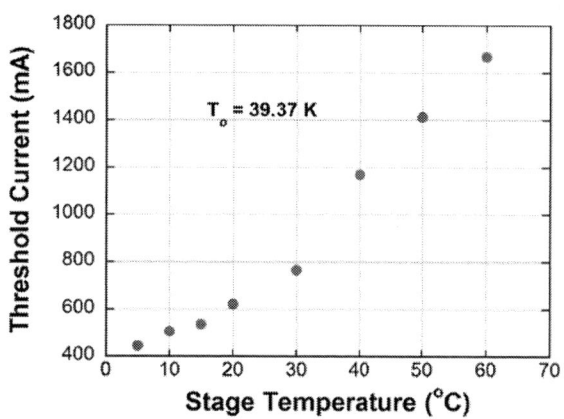

Fig. 3 Optical mode profile for ACF LD and confinement factor dependence on lasing wavelength

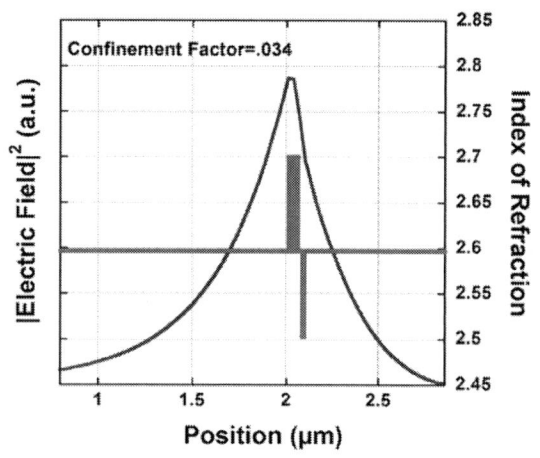
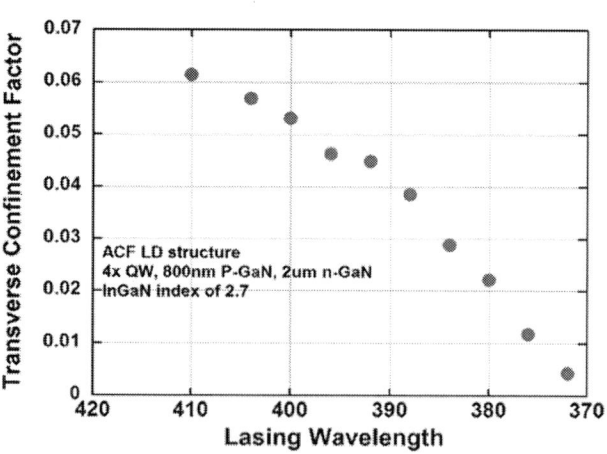

RIE Lag Directional Coupler based Integrated InGaAsP/InP Ring Mode-locked Laser

John S. Parker,[1] Pietro R.A. Binetti,[1] Yung-Jr Hung,[2] Erik J. Norberg,[1] Larry A. Coldren[1]

[1]*Electrical and Computer Engineering Department, University of California, Santa Barbara, CA 93106*
[2]*Dept. of Electronic Engineering, National Taiwan University of Science and Technology, 43 Keelung Rd., Sec. 4, Taipei 106, Taiwan*
E-mail: JParker@ece.ucsb.edu, Phone: (805)893-5955

We have demonstrated the first integrated ring mode-locked laser (MLL) with a reactive ion etch (RIE) lag coupler. The RIE lag directional coupler (RL-DC) is highly advantageous for integrated MLLs as it has an insertion loss <1 dB and can be designed to provide any coupling value. This provides the RL-DC with a much needed flexibility in large photonic systems unlike standard multimode interference (MMI) couplers, which typically provide only 3 dB power splitting.

InGaAsP/InP MLLs operating at a 1.55 μm wavelength are very stable pulsed sources, which makes them attractive components for high-speed optical fiber communication with optical-time-division-multiplexing (OTDM) [1], multi-wavelength sources for wavelength-division-multiplexing (WDM) [2], and clock distribution systems [3]. MLLs built on a highly versatile InGaAsP/InP material platform provide the capability to create monolithically integrated systems-on-chip. Previously, Y. Shi has demonstrated a single RIE lag directional coupler defined by electron-beam lithography [4]. To allow ease of fabrication of the current MLL device in large photonic integrated circuits (PICs), we have defined the entire structure using i-line stepper lithography and a single etch.

A standard offset quantum well (OQW) InGaAsP/InP integrated platform was used with 7 QWs positioned above a 300 nm tall 1.3Q waveguide with a confinement factor of 7.1%. A wet-etch removes the QWs for low loss passive waveguides followed by a single blanket p-cladding regrowth. Waveguides were defined by stepper lithography on a photoresist/Cr/SiO$_2$ three-layer mask. The patterned SiO$_2$ mask was used to mask the InGaAsP/InP in Cl$_2$/H$_2$/Ar etch chemistry with Inductively Coupled Plasma (ICP) Reactive Ion Etching (RIE). The RIE lag effect, which acts to slow the etch rate of smaller features, was used to define a 300 μm long 700 nm wide directional coupler on a deeply etched 4400 μm ring with a single etch-step, as shown in Fig. 1. The directional coupler has an etch depth of ~2.65 μm (100 nm from the bottom of the waveguide), while the deeply etched waveguides have an etch depth of 3.6 μm (below the waveguide layer by 850 nm), as shown in Fig. 2. A deeply etched directional coupler requires an extremely narrow gap <200 nm to have appreciable coupling. This typically requires more complicated Electron-Beam-Lithography (EBL), while the severe RIE lag effect from the narrow feature necessitates long etch times. This etch is difficult to make vertical and smooth, which increases scattering losses. We overcome these issues by adopting the single-etch process, which uses the RIE lag to our advantage and allows more streamlined processing of directional couplers without the need for a separate surface ridge waveguide defined by wet-etching and deeply etched waveguide defined by dry-etching.

As shown in Fig. 3, the measured cross coupling of the RIE lag directional coupler varies from 7-to-10% over the telecom C-band. The measured peak power off-chip was ~200μW (-7 dBm). The RF spectra of the fundamental and second harmonic from ESA measurements is shown in Fig. 4. The raised plateau on the RF spectra at 3 GHz is due to distortion from a low noise amplifier in the ESA and appears regardless of signal. The MLL shows stable operation over a wide range of SA biases -8 to -3 V and Semiconductor Optical Amplifier (SOA) drive currents of 170-290 mA. The mode-locked regime with RF power >25 dB above the noise is shown in Fig. 5. The pulse width variation measured by an Inrad SHG autocorrelator (AC) is shown in Fig. 6. The minimum pulse width is 1.1 ps with a spectral width of ~6 nm.

This work was supported by the Office of Naval Research (ONR). A portion of this work was done in the UCSB nanofabrication facility, part of the National Science Foundation (NSF) funded NNIN network.

[1] V. Kaman and J.E. Bowers, "120Gbit/s OTDM System Using Electroabsorption Transmitter and Demultipleter Operating at 30GHz," *Electronics Letters*, **36**[17], 1477 (2000).
[2] B.R. Koch, A.W. Fang, O. Cohen, M. Paniccia, D.J. Blumenthal, and J.E. Bowers, "Multiple Wavelength Generation from a Mode Locked Silicon Evanescent Laser," in *Proc. ISLC*, ThB3, Sorrento, Italy (Sept. 2008).
[3] P.J. Delfyett, D.H. Hartman, and S.Z. Ahmad, "Optical Clock Distribution Using a Mode-Locked Semiconductor Laser Diode System," *J. Lightwave Tech.* **9**[12], 1646 (1991).
[4] Y. Shi, S. He, and S. Anand, "Ultracompact Directional Coupler Realized in InP by Utilizing Feature Size Dependent Etching," *Optics Letters* **33**[17], 1927 (2008).

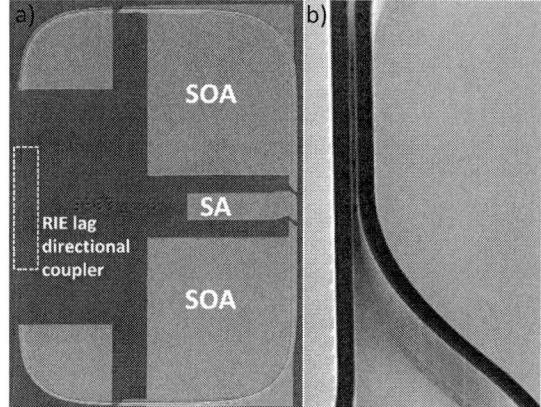

Fig. 1. Top-down SEM image of a) fabricated ring mode-locked laser and b) RIE lag directional coupler.

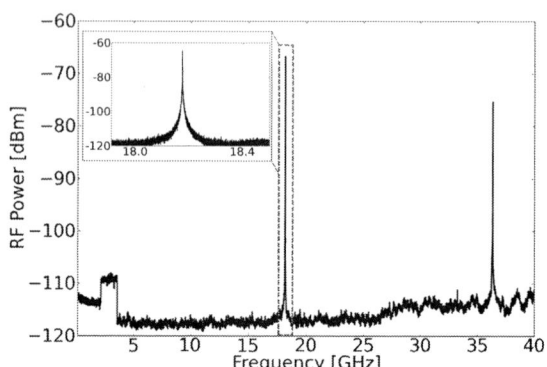

Fig. 4. ESA RF power spectrum at mode-locking showing first and second harmonic. Due to the low input power, the input electrical signal passes through a 30 dB low noise amplifier (LNA). The pedestal seen at 3 GHz is an artifact due to the LNA.

Fig. 2. SEM image of RIE lag directional coupler cross-section. The center etch depth is 200 nm into the waveguide layer.

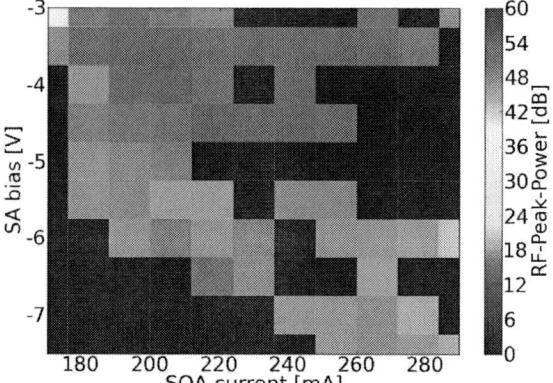

Fig. 5. Measured RF power of the MLL over the operating regime.

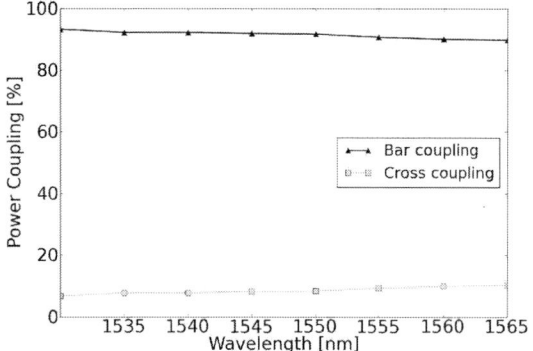

Fig. 3. Measured bar and cross coupling of 300 μm long RIE lag directional coupler. Insertion loss was measured at <1 dB.

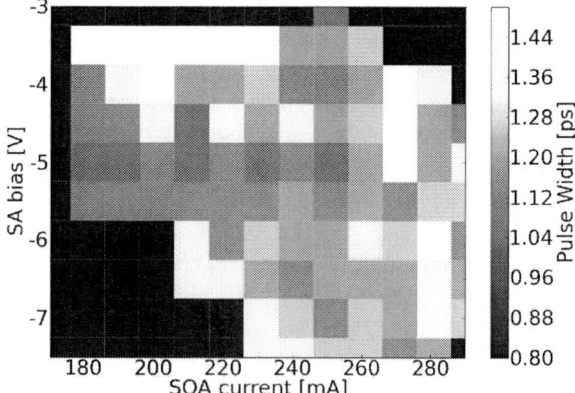

Fig. 6. Measured pulse width of the MLL over the operating regime.

Integrated Non-III-Nitride/III-Nitride Tandem Solar Cell

Nikholas G. Toledo[1], Samantha C. Cruz[2], Carl J. Neufeld[1], Jordan R. Lang[2], Michael A. Scarpulla[2, a],
Trevor Buehl[2], Arthur C. Gossard[1, 2], Steven P. Denbaars[1, 2], James S. Speck[2] and Umesh K. Mishra[1]

[1]Department of Electrical and Computer Engineering, University of California, Santa Barbara, CA 93106
[2] Materials Department, University of California, Santa Barbara, CA 93106
E-mail: nik@ece.ucsb.edu / Phone: (805) 893-3812 (x207) / Fax: (805) 893-8714

III-nitrides have recently been demonstrated as potential photovoltaic device material particularly in the high-energy portion of the solar spectrum [1-2]. The large lattice mismatch between InN and GaN however, makes it difficult to grow good quality high In-composition InGaN films for low bandgap subcells. The integration of III-N based solar cells, which have currently been demonstrated to work well above 2.0 eV, with mature IV and III-V based solar cell technologies, which work well at bandgaps ≤ 2.0 eV, has the potential to improve the efficiency of current multi-junction solar cells. In this paper, we present the first on-wafer integration of InGaN/GaN solar cells with non-III-nitride (GaAs) solar cells.

The bonded GaAs-InGaN tandem solar cell is a four-terminal device, wherein the non-III-nitride subcell is optically-coupled to but electrically-isolated from the III-nitride subcell (see Fig. 1). The subcells are not current-matched since the current contribution of low-indium composition III-nitride subcells is too low compared to the current generated by non-III-nitride subcells. ITO-coated Ga-face InGaN/GaN MQW solar cells are bonded to GaAs solar cells using a benzocyclobutene (BCB) interlayer. The process is summarized in Fig. 2 and described in [3]. Figure 3 shows a cross-sectional SEM of the bonding interface and the adjacent subcells. The resulting bond is strong and relatively void-free. The voids are primarily due to imperfections in the III-nitride surface such as hillocks and pits.

In the bonded solar cell device structure, light is transmitted through the sapphire, InGaN/GaN subcell, ITO and BCB interlayers before reaching the GaAs solar cell (see Fig. 4). The AM1.5G, one-sun intensity LIV characteristics of a bonded device show a J_{SC} = 10.6mA/cm^2 and a V_{OC} = 0.889V for the 500μm x 500μm GaAs subcell and a J_{SC} = 0.442mA/cm^2 and a V_{OC} = 1.8V for the 780μm x 780μm InGaN/GaN subcell (see Fig. 5). An unbonded GaAs solar cell device was also fabricated and characterized. The unbonded GaAs solar cell, which was processed with the same contact ring design as the bonded cell, produced a J_{SC} = 11.9mA/cm^2 and a V_{OC} = 0.955V. The drop in the J_{SC} for the bonded GaAs subcell is mainly due to the filtering of the solar spectrum as light travels through the bonded InGaN/GaN subcell and the bonding interlayer. Light transmitted through the sapphire passes through the GaN buffer first thereby absorbing any light with wavelengths < 363nm before reaching the InGaN/GaN subcell (see Fig. 6). For the GaAs subcell, only the portion of the spectrum with wavelengths longer than the absorption edge of the InGaN/GaN subcell is available for photovoltaic conversion. In this spectral region, the BCB layer does not significantly absorb photons (see Fig. 7). Reflection at the sapphire/air interface also resulted in lower J_{SC} for both subcells. The device characteristics are summarized in Table I.

The continuous improvement in III-nitride solar cell growth and design has resulted in longer wavelength absorption edges and EQEs as high as 63% [1]. With the design optimization of the current collecting layers adjacent to the bonding interface and by incorporating anti-reflection films in the bonding layer and the sapphire/air interface, the EQEs of both subcells can be greatly improved. The bonded non-III-nitride/III-nitride solar cell therefore has the potential of generating power outputs higher than what can be produced by either subcell alone.

In principle, the GaAs solar cell can be substituted by non-III-nitride multi-junction solar cells in which case, the InGaN/GaN solar cell, used as a top subcell, can potentially improve the overall efficiency of current high-efficiency multi-junction solar cells.

In summary, we demonstrated the first integrated non-III-nitride/III-nitride solar cells. By optimizing the design of each subcell for the portion of the solar spectrum available to each junction and the voltage matching necessary to combine the power of the subcells, more efficient photovoltaic devices may be achieved.

References: [1] C.J. Neufeld, et. al., Appl. Phys. Lett. **93**, 143502 (2008). [2] E. Matioli, et. al., Appl. Phys. Lett. **98**, 21102 (2011). [3] N. G. Toledo, et. al., submitted to Electronic Materials Conference 2011.

Acknowledgements: This project is funded by DARPA under Grant No. HR0011-10-1-0049 and by the Solid-State Lighting and Energy Center (SSLEC) at UC Santa Barbara.

[a] Present address: Depts. of Electrical Engineering and Materials Science & Engineering, University of Utah, Salt Lake City, UT 84112.

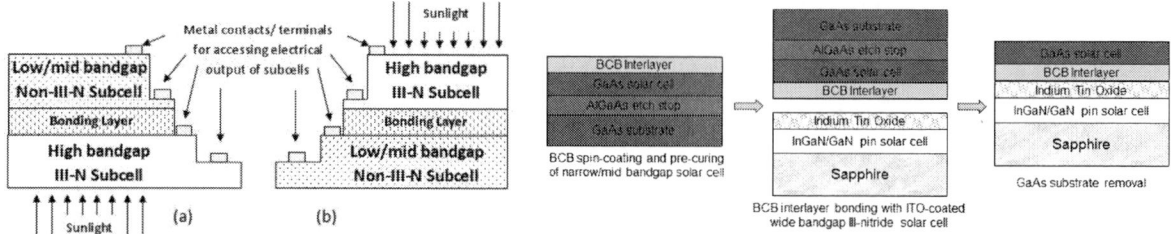

Figure 1. Cross-sectional schematic of a four-terminal wafer bonded stacked tandem multi-junction photovoltaic cell.

Figure 2. Wafer bonding process for the bonded solar cell.

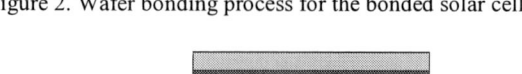

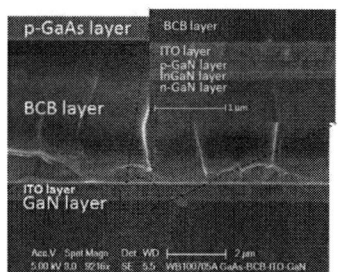

Figure 3. Cross-sectional SEM of a GaAs layer bonded to a GaN/InGaN solar cell using a BCB interlayer.

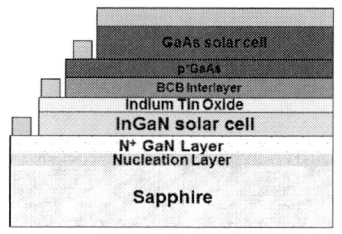

Figure 4 Bonded GaAs-InGaN/GaN device structure. The bonding interface is between the BCB interlayer and the ITO layer. Light is transmitted from the sapphire side.

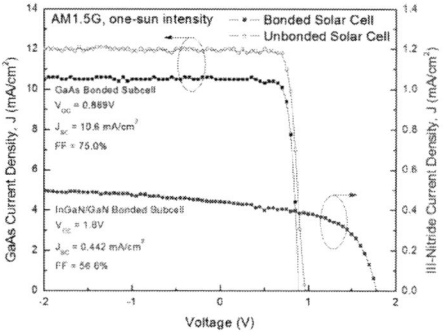

Figure 5. LIV characteristics of the bonded GaAs-InGaN/GaN solar cell and the unbonded GaAs solar cell.

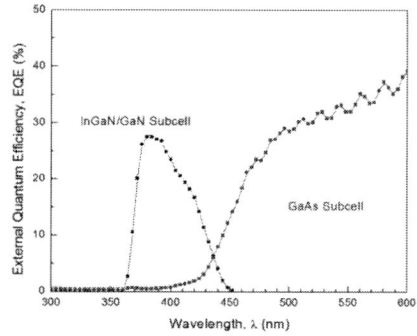

Figure 6. EQE plots of the bonded subcells.

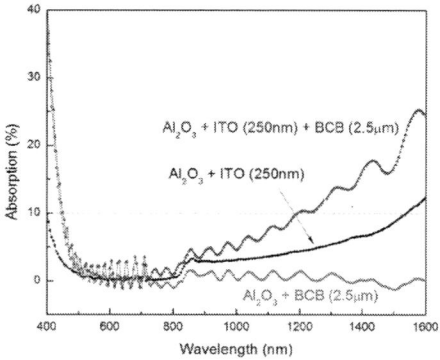

Figure 7. Absorption characteristics of ITO-Al_2O_3, BCB-Al_2O_3 and BCB-ITO-Al_2O_3. The 10% value is represented by the dashed line.

Table I. AM1.5G, one-sun (100mW/cm^2) LIV Characteristics

Solar Cell	V_{OC} (V)	J_{SC} (mA/cm^2)	Fill Factor (%)
Bonded Cell	-	-	-
GaAs subcell	0.889	10.6	75.00
InGaN/GaN subcell	1.8	0.442	56.6
Unbonded GaAs	0.955	11.9	72.6

978-1-61284-243-1/11 $26.00 © 2011 IEEE

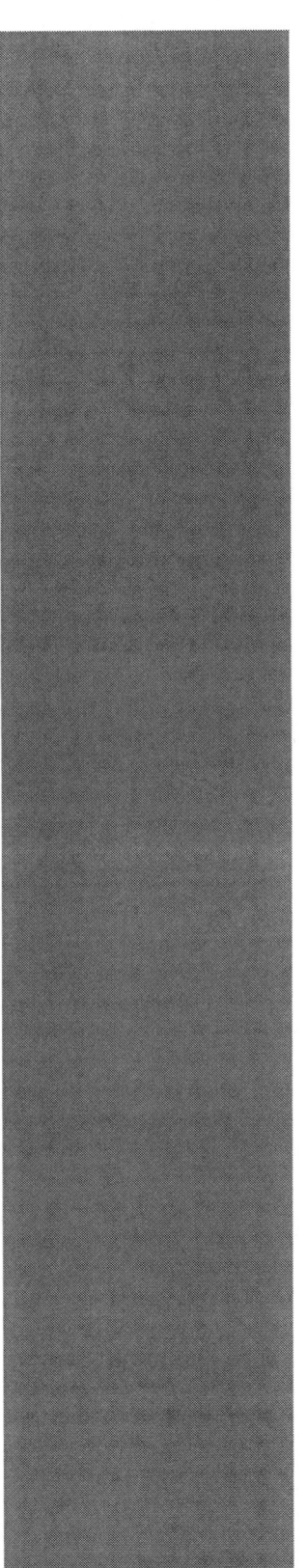

Session VII.B (Corwin Pavilion West)

High Speed Devices

Wednesday PM, June 22nd, 2011

Session Chairs: TBA

1:30 PM VII.B-1 Student Paper
N-polar GaN HEMTs with f_{max} > 300 GHz using high-aspect-ratio T-gate design
D.J. Denninghoff[1], S. Dasgupta[1], D.F. Brown[3], S. Keller[1], J. Speck[2], and U.K. Mishra[1], [1]Department of Electrical and Computer Engineering, University of California, Santa Barbara, California, USA, [2]Department of Materials, University of California, Santa Barbara, California, USA, and [3]HRL, Malibu, California, USA

1:50 PM VII.B-2 Student Paper
1.0 THz f_{max} InP DHBTs in a refractory emitter and self-aligned base process for reduced base access resistance
V. Jain[1], J. C. Rode[1], H.-W. Chiang[1], A. Baraskar[1], E. Lobisser[1], B. J. Thibeault[1], .M. Rodwell[1], M. Urteaga[2], D. Loubychev[3], A. Snyder[3], Y. Wu[3], J. M. Fastenau[3], W. K. Liu[3], [1]ECE Department, University of California, Santa Barbara, California, USA, [2]Teledyne Scientific & Imaging, Thousand Oaks, California, USA, and [3]IQE Inc., Bethlehem, Pennsylvania, USA

2:10 PM VII.B-3 Student Paper
Effect of optical phonon scattering on the performance limits of ultrafast GaN transistors
T. Fang, R. Wang, G. Li, H. Xing, S. Rajan, and D. Jena, Electrical Engineering, University of Notre Dame, Notre Dame, Indiana, USA

2:30 PM VII.B-4 Invited Paper
Device Scaling Technologies for Ultra-High-Speed GaN-HEMTs
K. Shinohara[1], D. Regan[1], I. Milosavljevic[1], A. L. Corrion[1], D. F. Brown[1], S. Burnham[1], P. J. Willadsen[1], C. Butler[1], A. Schmitz[1], S. Kim[1], V. Lee[2], A. Ohoka[2], P. M. Asbeck[2], and M. Micovic[1], [1]HRL Laboratories LLC, California, USA. and [2]Department Electrical and Computer Engineering, University of California, San Diego, California, USA

3:10 PM Break

3:30 PM VII.B-5 Student Paper
Trap-related Delay analysis of self-aligned N-polar GaN/InAlN HEMTs with record extrinsic g_m of 1105 mS/mm
Nidhi, S. Dasgupta, J. Lu, F. Wu, S. Keller, J. S. Speck and U. K. Mishra, ECE Department, University of California Santa Barbara, Santa Barbara, California, USA

3:50 PM VII.B-6
130nm InP DHBTs with f_t>0.52THz and f_{max}>1.1THz
M. Urteaga[1], R. Pierson[1], P. Rowell[1], V. Jain[2], E. Lobisser[2], M.J.W. Rodwell[2], [1]Teledyne Scientific Company, Thousand Oaks, California, USA and [2]Department of ECE, University of California, Santa Barbara, California, USA

4:10 PM VII.B-7
Late News

4:30 PM VII.B-8
Late News

4:50 PM VII.B-9
Late News

978-1-61284-243-1/11 $26.00 © 2011 IEEE

978-1-61284-243-1/11 $26.00 © 2011 IEEE

N-polar GaN HEMTs with f_{max} > 300 GHz using high-aspect-ratio T-gate design

D.J. Denninghoff,[*] S. Dasgupta, D.F. Brown[3], S. Keller, J. Speck[2], U.K. Mishra

Department of Electrical and Computer Engineering, University of California, Santa Barbara, CA, USA
[2]Department of Materials, University of California, Santa Barbara, CA, USA
[3]Currently with HRL, Malibu, CA, USA

We report measured f_{max} data of over 300 GHz on an MOCVD-grown N-polar GaN HEMT using a high-aspect-ratio T-gate. To our knowledge, this > 300-GHz f_{max} value is the highest reported to date for N-polar GaN HEMTs and is 50% higher than the previously reported value [1].

The extremely rapid performance advances of N-polar GaN HEMTs are a significant indication of their relevance and viability in modern electronic device applications [2], [3]. The advantages of HEMTs on N-polar GaN include low contact resistance, built-in back-barrier for improved electron confinement and modulation efficiency, and ability to scale the barrier thickness independently of 2DEG channel charge. The current study highlights a novel high-aspect-ratio T-gate design and its resulting state-of-the-art RF device performance on the N-face GaN material system.

The epitaxial layer structure of the device in this study is shown in Figure 1. A 5-nm MOCVD SiN gate dielectric was deposited on the 10-nm GaN channel. Selective-area 50-nm thick N+ GaN regrowth by MBE was spaced 250 nm between source and drain, and the T-gate was approximately centered therein. Standard mesa etching and probe pad metallization were also performed.

The RF data were measured using an Agilent E8361A PNA, after which the probe pads were de-embedded from the extrinsic data using established procedures [4]. Extrapolating the de-embedded h_{21} and U measured data at a 20-dB-per-decade decay, the f_T/f_{max} obtained was 115/310 GHz (Figure 2). The peak f_{max} value was obtained at Vgs = -1.5 V, Vds = 10 V. Also included in the figure is the simulated data from a standard small-signal circuit FET model, which fits the general form of h_{21} and U.

In Figure 3 is an SEM image of the high-aspect-ratio T-gate which was designed specifically for obtaining high f_{max} values. The gate length is 70 nm and the stem height is 380 nm. The purpose of the tall stem is to reduce the parasitic capacitance contribution of the top of the T-gate. The processing sequence used in the T-gate formation uniquely allows complete filling of a high-aspect-ratio stem, using only conventional electron-beam evaporation—no atomic layer deposition (ALD) was required. To reduce the gate resistance, 740 nm of Au was deposited on the 380-nm stem, resulting in a total gate metal thickness of 1.1 μm. The measured gate resistance for the 2x12.5-μm-wide device is 6 Ω.

Figure 4 shows the device dc characteristics, with 1.2 A/mm drain current and good pinch off behavior. The on-resistance of 1.93 Ohm-mm is relatively high and can easily be reduced with graded InGaN and InN regrowth [5]. The threshold voltage is -3.35 V at Vds = 5 V. The peak extrinsic transconductance is 398 mS/mm and peak intrinsic transconductance is 640 mS/mm, which was calculated using a 0.95-Ohm-mm source resistance.

The RF data at all relevant bias points are shown in Figure 5. At large drain biases, the f_{max} values are relatively constant for a wide range of drain current values. Also shown is the trend to obtain high f_{max}: large drain bias and Vgs at -1.5 V.

Future devices can achieve an even higher f_{max} by using optimized regrown contacts. By reducing the source and drain resistances to those typically achieved by fellow colleagues [5], this device design can obtain f_{max} values of over 450 GHz.

[1] Nidhi *et al.* 2010 ISCS "N-polar GaN-based double-recess gate MIS-HEMT technology with record fmax of 195: Pathway to improved Ka band performance", [2] Nidhi *et al.* 2009 IEDM "N-polar GaN-based highly scaled self-aligned MIS-HEMTs with state-of-the-art $f_T.L_G$ product of 16.8 GHz-μm", [3] see article and references in: S. Kolluri *et al.* 2011 IEEE EDL v34 "RF performance of deep-recessed N-polar GaN MIS-HEMTs using a selective etch technology without ex-situ surface passivation", [4] Koolen *et al.* 1991 IEEE Bipolar Circuits and Technology Meeting "An improved de-embedding technique for on-wafer high-frequency characterization", [5] S. Dasgupta *et al.* 2010 APL v96 "Ultralow nonalloyed ohmic contact resistance to self aligned N-polar GaN HEMTs by In(Ga)N regrowth"

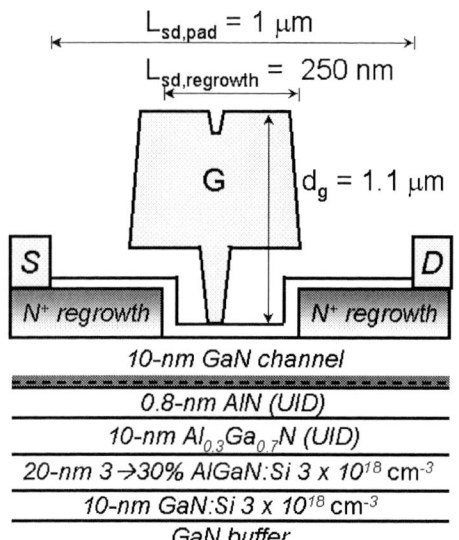

Figure 1: Epitaxial layer structure of MOCVD-grown sample used in this study, including a 5-nm MOCVD SiN gate dielectric. N+ GaN regrowth source-drain spacing is 250 nm and source-drain metal pad spacing is 1 μm.

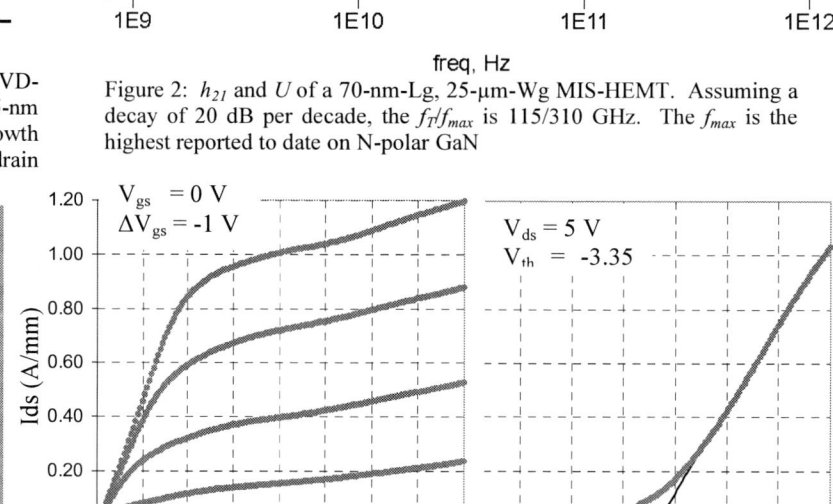

Figure 2: h_{21} and U of a 70-nm-Lg, 25-μm-Wg MIS-HEMT. Assuming a decay of 20 dB per decade, the f_T/f_{max} is 115/310 GHz. The f_{max} is the highest reported to date on N-polar GaN

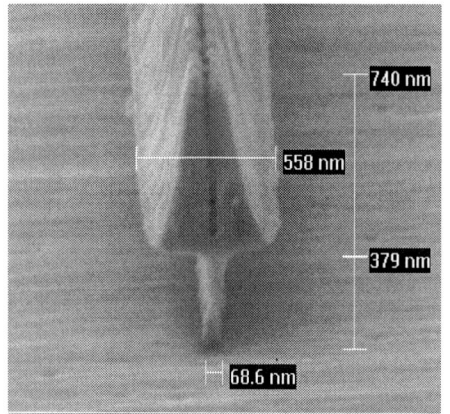

Figure 3: SEM of Ti/Au high-aspect-ratio T-gate. Lg = 70 nm, stem height = 380 nm, total gate metal thickness = 1.1 μm. Rg = 6 Ω.

Figure 4: dc characteristics of 70-nm-Lg, 25-μm-Wg device with maximum current density of 1.2 A/mm and good pinch-off behavior. The threshold voltage as extrapolated from peak transconductance bias point -3.35 V. On-resistance is 1.93 Ohm-mm. Output conductance is 2030 Ohm (at Vgs = -1 V, Vds = 8 V).

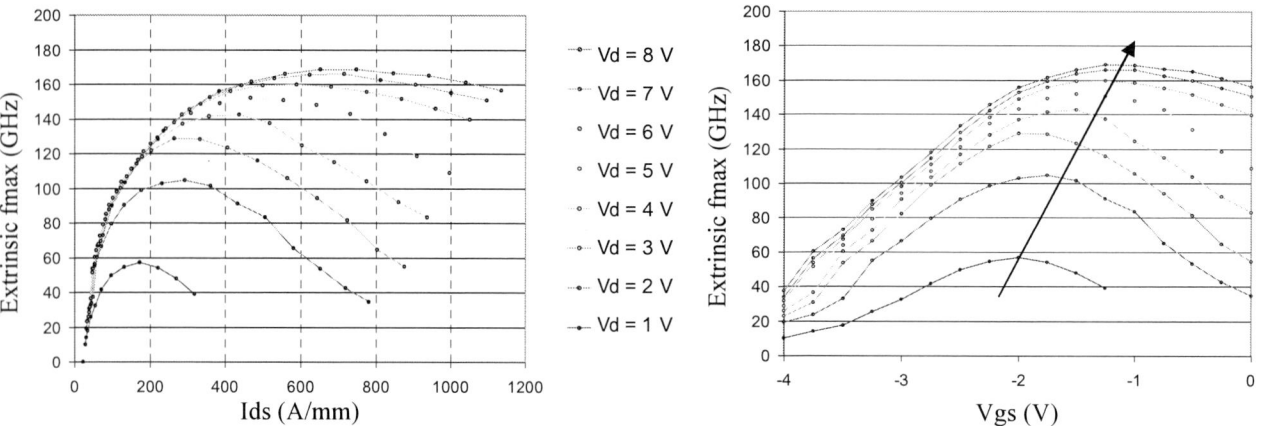

Figure 5: Device performance map of extrinsic fmax at all bias points. fmax remains high over a wide current range when drain bias is high. fmax increases with drain voltage, and the Vgs-value at peak fmax becomes less negative with increasing Vds.

978-1-61284-243-1/11 $26.00 © 2011 IEEE

1.0 THz f_{max} InP DHBTs in a refractory emitter and self-aligned base process for reduced base access resistance

Vibhor Jain[1], Johann C. Rode[1], Han-Wei Chiang[1], Ashish Baraskar[1], Evan Lobisser[1], Brian J. Thibeault[1],
Mark Rodwell[1], Miguel Urteaga[2], D. Loubychev[3], A. Snyder[3], Y. Wu[3], J. M. Fastenau[3], W.K. Liu[3]

[1]ECE Department, University of California, Santa Barbara, CA 93106-9560
Phone: 805-893-3273, Fax: 805-893-3262, Email: vibhor@ece.ucsb.edu
[2]Teledyne Scientific & Imaging, Thousand Oaks, CA 91360
[3]IQE Inc., 119 Technology Drive, Bethlehem, PA 18015

We report 220 nm InP double heterojunction bipolar transistors (DHBTs) demonstrating f_τ = 480 GHz and f_{max} = 1.0 THz. Improvements in the emitter and base processes have made it possible to achieve a 1.0 THz f_{max} even at 220 nm wide emitter-base junction with a 1.1 μm wide base-collector mesa. A vertical emitter metal etch profile, wet-etched thin InP emitter semiconductor with less than 10 nm undercut and self-aligned base contact deposition reduces the emitter semiconductor-base metal gap (W_{gap}) to ~ 10 nm, thereby significantly reducing the gap resistance term (R_{gap}) in the total base access resistance (R_{bb}), enabling a high f_{max} device. Reduction in the total collector base capacitance (C_{cb}) through undercut in the base mesa below base post further improved f_{max}. These devices employ a Mo/W/TiW refractory emitter metal contact which allows biasing the transistors at high emitter current densities (J_e) without problems of electromigration or contact diffusion under electrical stress [1].

Improved bandwidth for the HBTs can be achieved through epitaxial scaling of base (T_b) and collector (T_c) thicknesses for reduced transit delays; and lithographic scaling with lower contact resistivities for reduced RC delays [2]. For the DHBTs reported here, emitter design includes a 10 nm thick highly doped n-In$_{0.53}$Ga$_{0.47}$As cap (N$_d$ > 5 × 10^{19} cm^{-3}) for reduced contact resistance and a thin (30 nm) InP layer for controlled emitter undercut and reduced emitter depletion region resistance [3]. InGaAs base is 30 nm thick having a 9 – 5 × 10^{19} cm^{-3} doping gradient and T_c is 100 nm. The base-collector grade includes a 13.5 nm InGaAs setback and a 16.5 nm InGaAs/InAlAs chirped-superlattice grade. The epitaxial structure was grown by IQE Inc. on a 4" semi-insulating InP substrate [4]. Extrinsic base, in the emitter-base gap (W_{gap}), has a higher sheet resistance ($R_{sh,ex}$) than intrinsic base due to surface depletion from Fermi level pinning and surface damage from processing – $R_{sh,ex}$ = 920 Ω/sq and $R_{sh,int}$ = 710 Ω/sq measured from pinched and non-pinched TLM measurements. Thus, $R_{gap} = R_{sh,ex} \cdot W_{gap} / 2 \cdot L_e$ becomes a dominant component of R_{bb} for large W_{gap} [2]. For these HBTs, dry etch for Mo/W/TiW (10% Ti by weight) stack is optimized to obtain a vertical emitter metal profile; thin (30 nm) InP emitter layer is wet-etched for controlled semiconductor undercut (< 10 nm) and self-aligned base contact (Pt/Ti/Pd/Au) lift-off process is used. These features greatly reduce W_{gap}, and therefore R_{gap}, consequently increasing f_{max}. This reduction in R_{bb} made it possible to achieve 1.0 THz f_{max} even at 220 nm wide emitter with a 1.1 μm wide, misaligned base mesa. Fabrication details are as in [4].

1-67 GHz measurements of HBTs embedded in a ground-signal-ground pad structure were carried out after performing a standard line-reflect-reflect-match (LRRM) calibration on an Agilent E8361A PNA, bringing the reference planes to the probe tips. On wafer, short and open circuit pad structures identical to those used by the devices were measured after calibration to de-embed associated transistor pad parasitics. The same calibration and de-embedding procedures were used for 80-105 GHz measurements on Agilent 8510XF system. Peak RF performance for HBTs with A_{je} = 0.22 × 2.7 μm^2 was obtained at I_c = 12.1 mA and V_{ce} = 1.64 V (V_{cb} = 0.7 V, J_e = 20.4 mA/μm^2, P = 33.4 mW/μm^2). Extrapolations from single-pole fit to the measured current gain H_{21} and Mason's Unilateral gain U indicate cut off frequencies f_τ = 480 GHz and f_{max} = 1.0 THz. Total emitter access resistivity ρ_{ex} ~ 4.2 Ω-μm^2 was extracted from RF data. HBTs with an emitter junction width of 220 nm (W_e) show peak DC common emitter current gain β = 17 and common emitter breakdown voltage $V_{BR,CEO}$ = 3.7 V (J_e = 0.1 mA/um^2). The Kirk effect is observed at J_e = 23 mA/μm^2 (V_{cb} = 0.7 V) when f_τ falls to 95% of its peak value.

A linear fit to the extracted C_{cb} variation with emitter length (L_e) has ~ 0 fF intercept suggesting negligible capacitance contribution from the base post. This is also evident from weak dependence of f_τ on L_e; peak f_τ changes from 480 GHz to 465 GHz for L_e increase from 3 μm to 5 μm. For low R_{bb}, base metal resistance $R_{metal} = R_{sh,m} \cdot L_e / 6 \cdot W_{bc}$ [2] becomes a significant fraction of total R_{bb} and thus f_{max} decreases with increase in L_e. Further improvement in f_τ / f_{max} can be achieved through reduced emitter and base-collector junction areas and lower contact resistivities.

This work was supported by the DARPA THETA program under HR0011-09-C-0060. A portion of this work was done in the UCSB nanofabrication facility, part of NSF funded NNIN network and MRL Central Facilities supported by the MRSEC Program of the NSF under award No. MR05-20415

[1] Y. K. Fukai *et al.*, *Microelectronics Reliability*, vol. 49, no. 4, pp. 357-364, April 2009
[2] M. J. W. Rodwell *et al.*, *Proc. of the IEEE*, vol. 96, no. 2, pp. 271-286, Feb 2008
[3] V. Jain *et al.*, *Electron Device Letters, IEEE*, vol. 32, no.1, pp.24-26, Jan 2011
[4] V. Jain *et al.*, to be presented at *International Conference on Indium Phosphide and Related Materials (IPRM)*, May 2011

Fig. 1. Cross-sectional TEM of emitter mesa of DHBT showing the emitter profile with 220 nm wide emitter-base junction and 1.1 μm wide, non-symmetric base-collector mesa. Magnified view of the emitter mesa shows the small undercut in InP emitter and narrow base-emitter gap (W_{gap} ~10 nm)

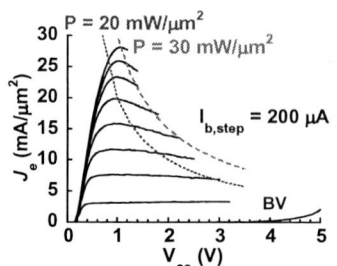

Fig. 2. Common-Emitter current density-voltage (J_e-V) characteristics of a DHBT with $A_{je} = 0.22 \times 4.7$ μm^2

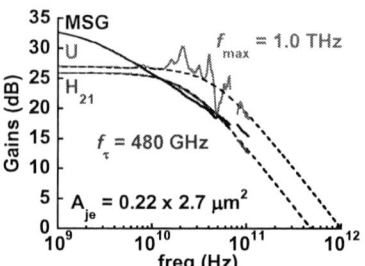

Fig. 3. Measured RF gains for the DHBT in 1 - 67 GHz and 80-105 GHz bands using off-wafer LRRM calibration in a lumped pad structure. The DHBT was biased at $I_c = 12.1$ mA, $V_{ce} = 1.64$ V

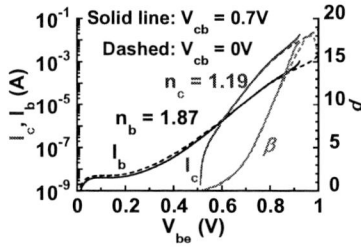

Fig. 4. Gummel Plot of a DHBT with $A_{je} = 0.22 \times 4.7$ μm^2

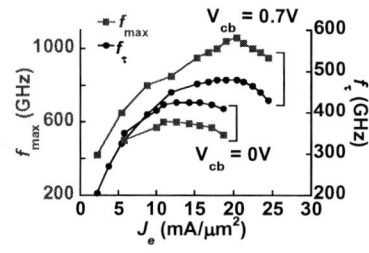

Fig 5. f_τ/f_{max} dependence on V_{cb} and J_e

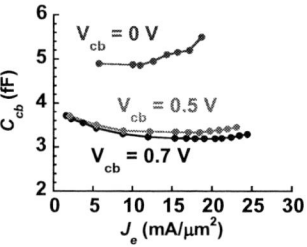

Fig 6. Variation in C_{cb} with V_{cb} and J_e

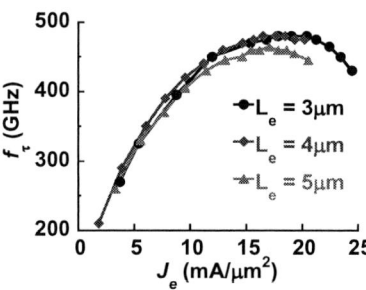

Fig 7. Variation in f_τ with J_e for different L_e at $V_{cb} = 0.7$ V for DHBTs having $W_e = 220$ nm and same base-collector mesa width

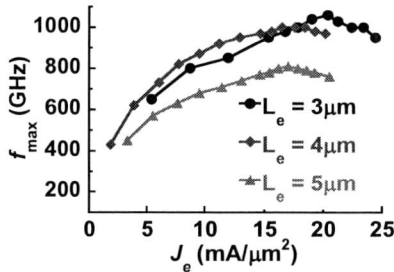

Fig 8. Variation in f_{max} with J_e for different L_e at $V_{cb} = 0.7$ V for DHBTs having $W_e = 220$ nm and same base-collector mesa width

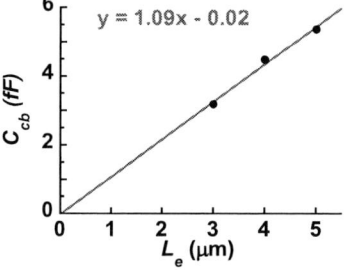

Fig 9. Variation in extracted C_{cb} with L_e for $W_e = 220$ nm and same base-collector mesa width

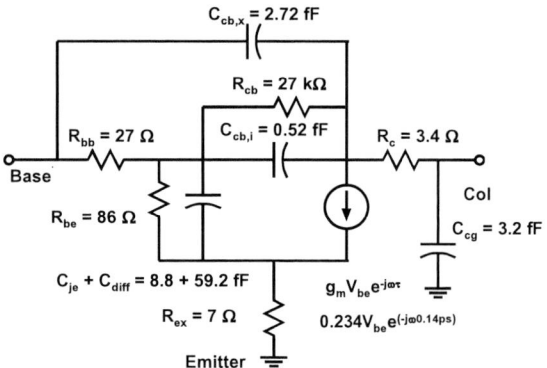

Fig 10. Hybrid-π equivalent circuit at peak RF performance from 1 – 67 GHz RF data for the bias conditions: $I_c = 12.1$ mA, $V_{ce} = 1.64$ V

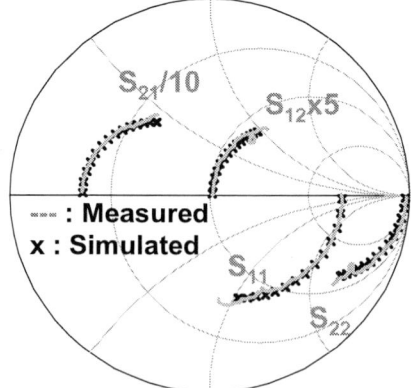

Fig 11. 1 - 67 GHz measured and simulated S-Parameters from equivalent circuit model in Fig 10

978-1-61284-243-1/11 $26.00 © 2011 IEEE

Effect of optical phonon scattering on the performance limits of ultrafast GaN transistors

Tian Fang, Ronghua Wang, Guowang Li, Huili Xing, Siddharth Rajan[+] & Debdeep Jena*

Electrical Engineering, University of Notre Dame, Notre Dame, IN 46556, USA

*Email: djena@nd.edu, *Phone: 574-631-8835

[+]Electrical and Computer Engineering, The Ohio State University, Columbus, OH 43210, USA

As GaN HEMTs are scaled down to push performance into 100's of GHz range, it is timely to investigate their performance limits. Unlike Si MOSFETs and most other III-V semiconductor based HEMTs, the electron - polar optical phonon interaction is exceptionally strong in GaN. As a result, the mean free path of hot electrons in GaN is $\lambda_{op} \sim$ 3.5nm, far shorter than typical HEMT gate lengths (L_g). Thus while Si MOSFETs and other III-V HEMTs can approach near ballistic behavior by reduction of parasitic delays and L_g, the situation is starkly different for GaN HEMTs. Here, we investigate the intrinsic performance limits of GaN HEMTs by incorporating the effect of polar optical phonon backscattering into a quasi-ballistic model. Then, we include parasitic elements and quantitatively investigate the degradation in performance. The method used is semi-analytical, and will prove very helpful in designing future generations of devices. The work not only sets a roadmap for scaling to high speeds, it also offers clear physical reasons for a number of unexplained features observed in state-of-the-art GaN HEMTs.

The device structure in Fig.1(a) shows the carrier injection from the source-injection point. The carrier density at this point is controlled by the gate voltage V_{gs}. Fig. 1(b) depicts the **k**-space electron distributions at the top of the barrier at 0 K. At low carrier concentrations (near pinchoff), electrons at the injection point cannot emit optical phonons since maximum electron kinetic energy is less than optical phonon energy $\hbar\omega_o$. Even if they absorb phonons down the channel, they cannot return to the source injection point at high fields. A critical density when the Fermi energy of right-going carriers becomes equal to $\hbar\omega_o$ marks the onset of optical phonon emission. Above this density, hot electrons are scattered back into source, and the distribution function now includes left-going carriers as in Fig 1(b). This picture repeats for the second (and higher) subbands if the carrier density increases further. By assuming the Fermi level of injection electrons is exactly one phonon energy ($\hbar\omega_o$) higher than back-scattered electrons, the ensemble injection velocity v_{inj} is calculated from the **k**-space distribution, and shown in Fig. 2(a) for 0K & 300 K. The injection velocity v_{inj} increases until the optical phonon emission occurs at an injection point 2DEG density~4×10^{12}/cm^2, at which point it reaches ~1.6×10^7cm/s at 0K, but degrades to ~1.3×10^7cm/s at 300K. This peak velocity point also leads to peak g_m and f_T. The injection velocity curve has a small bump at the onset of injection from the second subband (~2.5×10^{13}cm^{-2}), which is clear at 0K, but smoothened out at 300K. Fig 2(a) is the central result of phonon-limited high-field transport properties in GaN HEMTs, and all device characteristics (drain current, transconductance, and speed) of the HEMT can be derived from this result by combining it with device parasitics. The corresponding saturation current $J_d = env_{inj}$ is shown in Fig 2(b). In the following calculations, we ignore the second subband.

Using the model, we calculate the device characteristics of a prototype 5 nm 35% AlGaN/GaN HEMT. We ignore short channel effects, but consider parasitic elements. The carrier density n and gate capacitance C_{gs} are calculated self-consistently using Schrodinger-Possion equations, and shown in Fig. 2(c). The carrier concentration at the injection point decreases with the source resistance R_S due to the voltage drop across R_S. The instrinsic drain current $J_d = env_{in}$ and intrinsic transconductance $g_m = \partial I_d / \partial V_{gs}$ at different V_{gs} are calculated and shown in Fig. 3(a). Note the sharp drop in g_m beyond the peak, and the sub-linear increase of the drain current with gate bias. Both these experimentally observed features in GaN HEMTs remain unexplained; our model uncovers the physics behind this behavior. The phonon model predicts a second peak in the g_m-vs-V_{gs} curve at the onset of occupation of the second subband (not shown here).

The device characteristics and the effect of parasitic resistances (contact+access, S+D) are shown in Fig. 3(a). The sharply peaked 'intrinsic' curves are smoothened by the parasitic resistances. Fig. 3(b) shows the cutoff frequency f_T with L_g=50nm vs V_{gs}. The trend mimics the g_m curve. The strong degradation of the peak f_T with parasitic resistances is evident. The scaling of the peak f_T with L_g is shown in Fig. 3(c). The intrinsic peak f_T (C_{gd}=0) curve is the straight-line $f_T L_g \sim$ 23 GHz·μm. C_{gd} and R_S (and R_D) further decrease the peak f_T. Some recent experimental reports of record high performance GaN HEMTs are plotted in the figure [1-4]. The performances of these devices are rather close to the limits based on the model. Our model clearly shows the intrinsic limit of f_T that we can achieve at the respective gate lengths. The nature of ultrafast optical phonon emission forces a high degree of vertical scaling to speed up GaN HEMTs. In addition, the model quantitatively shows the significant role played by parasitic resistances and capacitances in limiting the f_T for short gate length (<50nm) GaN HEMTs.

In summary, a phonon emission model is able to explain the DC and RF behavior of ultrafast GaN HEMTs. A major departure from earlier models is the dependence of the injection velocity on the 2DEG density *and* on optical phonons in GaN. It is this injection velocity that determines both DC and RF characteristics, and in combination with parasitic elements is successful in explaining experimental data. The model should prove valuable for further improvement of GaN transistors.

[1] Shinohara et al, *IEDM*, 672, (2010),   [2] Chung et al, *IEDM*, 676, (2010),

[3] Sun et al., EDL, 31, 957 (2010),   [4] Wang et al., Submitted (2011).

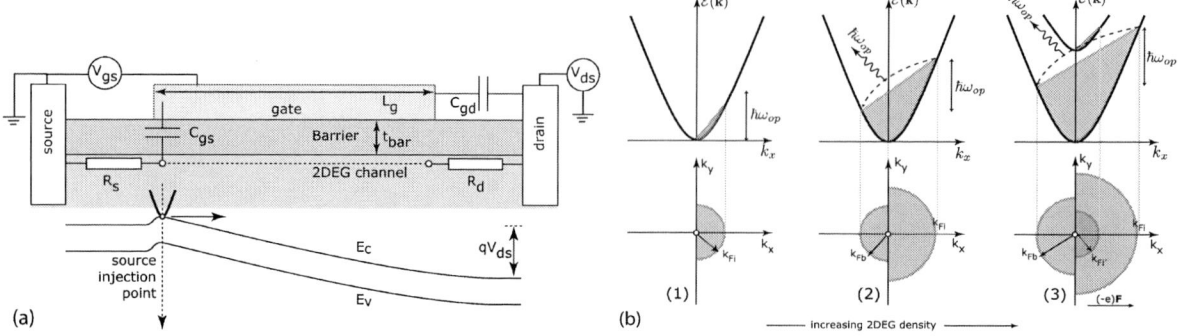

Figure 1: (a) Device geometry of a GaN HEMT and band-diagram under bias. (b) Bandstructure and distribution functions $f(k)$ at source injection point. The gate voltage V_{gs} controls the total carrier concentration at the source injection point, but at high field, phonon emission determines the shape of $f(k)$. Wavy arrows show the optical phonon emission by hot electrons.

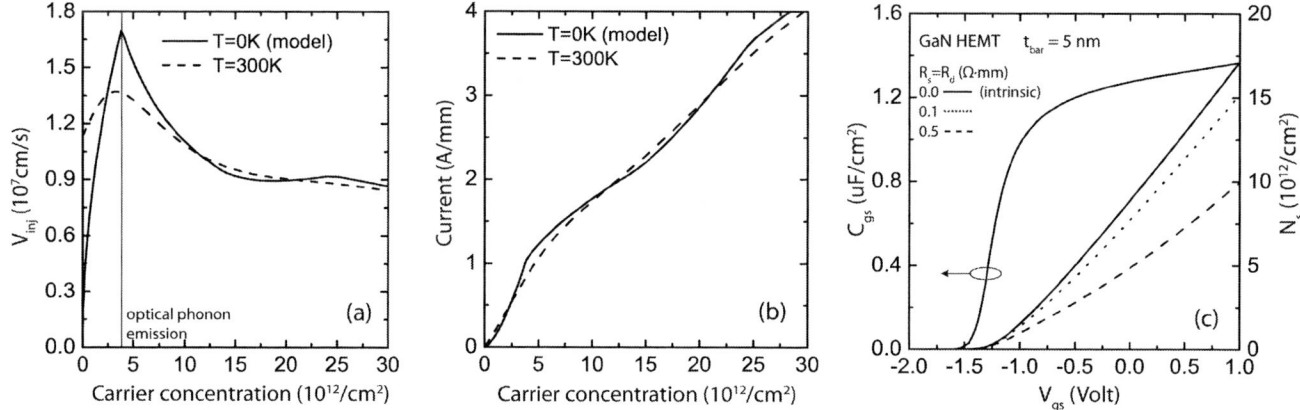

Figure 2: (a) Carrier ensemble saturation velocity as a function of 2DEG density based on the phonon model at two temperatures. (b) Current density as a function of 2DEG carrier concentration at the source-injection point. (c) Gate capacitance and carrier concentration versus gate voltages at 300K, evaluated by self-consistent solution of Schrodinger-Poisson equations.

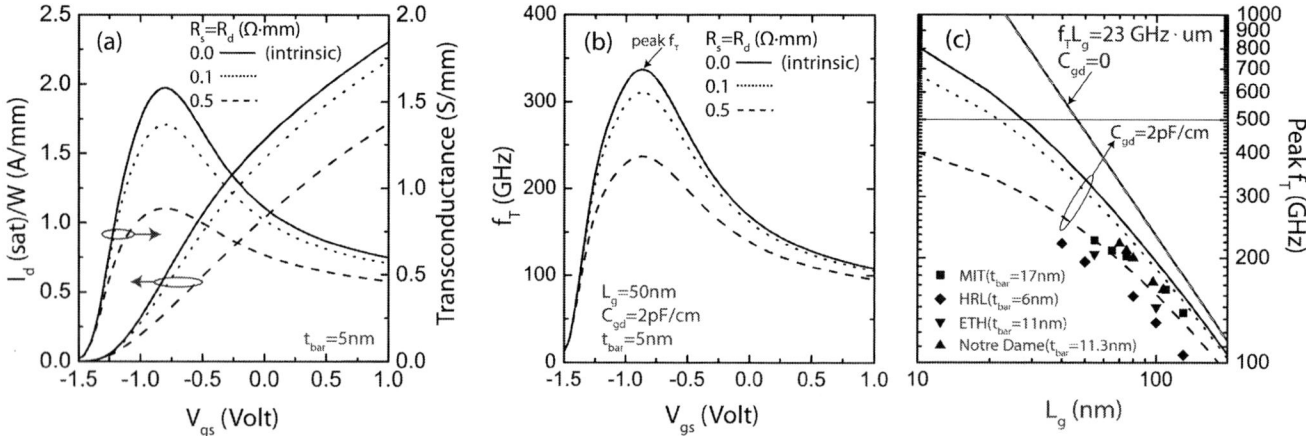

Figure 3: (a) The net drain current and transconductance versus V_{gs}. The peak in g_m occurs at the onset of optical phonon emission at the source injection point. The degradation of transconductance and saturation current with parasitic access+contact resistances are also shown. (b) The corresponding f_T versus V_{gs} plot follows the transconductance behavior, with the peak f_T near the onset of phonon emission. Parasitic resistances degrade the peak f_T severely. (c) The peak f_T is plotted against the gate length L_g. Interestingly, the intrinsic limit on the peak f_T follows the locus $f_T L_g \sim 23$ GHz.μm. The severe degradation with parasitics is also evident, and experimental record values follow the predicted trend.

978-1-61284-243-1/11 $26.00 © 2011 IEEE

Device Scaling Technologies for Ultra-High-Speed GaN-HEMTs

K. Shinohara, D. Regan, I. Milosavljevic, A. L. Corrion, D. F. Brown, S. Burnham, P. J. Willadsen,
C. Butler, A. Schmitz, S. Kim, V. Lee[†], A. Ohoka[†], P. M. Asbeck[†], and M. Micovic

HRL Laboratories LLC, 3011 Malibu Canyon Road, Malibu, CA 90263, U.S.A.

[†]Department Electrical and Computer Engineering, University of California, San Diego, CA 92093 USA

Phone: 310-317-5093, E-mail: kshinohara@hrl.com

The high frequency performance of GaN-based HEMTs has been significantly improved through innovative device scaling technologies such as AlGaN [1] or InGaN back barriers [2], thin AlN top barriers [3], lattice-matched InAlN barriers [4], ultra-short gates [5], and self-aligned gates [6]. In this paper, we review our scaling technologies for ultra-high-speed operation of GaN-HEMTs [7-10], which provide not only high yield and uniformity but also a large-scale integration of E/D-mode HEMTs for future RF and mixed-signal applications.

Figure 1 shows progress of cutoff frequency (f_T) of GaN-HEMTs with the reduction of the gate length (L_g). As the L_g is decreased below 100 nm, vertical epi scaling, lateral source-drain (S-D) scaling, and reduction of parasitic resistances become increasingly critical to obtain scaled f_T's since excess delays caused by the short channel effect, drain extension, and parasitic RC charging begin to limit the total device delay. Our projected device scaling technologies are summarized in Fig. 2. Vertically scaled AlN/GaN/Al$_{0.08}$Ga$_{0.92}$N double hetero-junction (DH) HEMT structures, shown in Fig. 3, were grown by MBE to minimize the gate-to-channel distance while maintaining a high 2DEG density. The Al$_{0.08}$Ga$_{0.92}$N back barrier increases carrier confinement and suppresses the short channel effect. E/D-mode device operation was precisely controlled by the AlN thickness (Fig. 4). High 2DEG mobilities (~1200 cm^2/V·s) were obtained for both E- (2.0 nm) and D-mode (3.5 nm) epi structures after SiN passivation. The AlN top barrier in the ohmic contact region was etched by RIE and highly Si-doped n^+-GaN (7×10^{19} cm^{-3}) was regrown by MBE to achieve a low access resistance (R_{ac}) (Fig. 5). An extremely small R_{ac}, defined as a resistance between the ohmic metal to the 2DEG, was realized for both E- (0.09 Ω·mm) and D-mode (0.08 Ω·mm) structures (Fig. 6). Figure 7 illustrates a cross-section of a fabricated D-mode HEMT in which the source-drain spacing (L_{sd}) is defined by regrown n^+-GaN contact edges. A 40-nm gate DH-HEMT with L_{sd} = 1 μm showed a very small on-resistance (R_{on}) of 0.81 Ω·mm, a high I_{dmax} of 1.62 A/mm, and off-state breakdown voltage (BV_{off}) of 42V (Fig. 8). The standard deviation of the threshold voltage, $\sigma(V_{th})$, was only 27 mV at V_{ds} = 6 V reflecting the very uniform gate-to-channel distance determined by the MBE thickness. The g_m did not degrade with reduction of L_g down to 40 nm, confirming suppressed short channel effect (Fig. 9). The device exhibited f_T/f_{max} = 220/289 GHz at V_{ds} = 2 V and 186/400 GHz V_{ds} = 6 V (Fig. 10). Small signal modeling and delay time analysis were performed on the 40-nm gate device to assess frequency limitations [9]. Figure 11 plots total delay times (= $2\pi f_T$) divided into parasitic charging, channel charging, drain delay, and gate transit time as a function of V_{ds}. The parasitic charging time was just 10% of total delay time owing to the reduction of R_{ac}, and the gate transit time was nearly independent of V_{ds}. The reduction of f_T with increasing V_{ds} was attributed to the increased drain delay resulting from extended drain depletion. Figure 12 depicts dependence of delay components at V_{ds} = 2 V on L_g and shows proportional scaling of the gate transit time with the L_g down to 40 nm, demonstrating scalability of this technology. An average electron velocity (v_{ave}) of 1.1×10^7 cm/s was deduced from the slope. Lateral S-D scaling further improved device characteristics, resulting in an extremely small R_{on} of 0.44 Ω·mm, a high I_{dmax} of 2.3 A/mm, a high peak g_m of 905 mS/mm, and a simultaneously high f_T/f_{max} = 260/394 GHz in a 45-nm gate device with L_{sd} = 170 nm (Fig. 13-15). Figure 16 compares peak f_T and f_{max} as a function of V_{ds} between the scaled device with L_g/L_{sd} = 45 nm/170 nm and the unscaled device with L_g/L_{sd} = 40 nm/1 μm. The scaled device showed a continuous increase of f_T with V_{ds} above the knee voltage until reaching a maximum value at 5 V. This result clearly illustrates that the drain extension was terminated by the reduced gate-drain spacing (L_{gd}), resulting in an enhanced electron velocity by the increased electric field under the gate at higher V_{ds}. The estimated v_{ave} reached 1.5×10^7 cm/s, which is 35% higher than the unscaled device. Figure 17 shows a SEM image of monolithically-integrated E/D-mode epi layers formed by selective area MBE regrowth. E- and D-mode devices were successfully fabricated and integrated using the regrowth technique as shown in Fig. 18.

This work was sponsored by DARPA-MTO NEXT program under DARPA/CMO Contract No. HR0011-09-C-0126. The views and conclusions contained in this document are those of the authors and should not be interpreted as representing the official policies, either expressly or implied, of the Defense Advanced Research Projects Agency or the U.S. Government.

[1] M. Micovic et al., IEDM 2004, p.807. [2] T. Palacios et al., IEEE EDL, vol. 27, no. 1, p. 13. [3] Y. Cao et al., APL vol. 90, no. 18, p. 182 112. [4] H. Sun et al., IEEE EDL, vol. 31, no. 9, p. 957. [5] M. Higashiwaki et al., DRC 2006. [6] Nidhi et al., IEDM 2009, p. 955. [7] I. Milosavljevic et al, DRC 2010, p. 159. [8] A. Corrion et al, IEEE EDL, vol. 31, no. 10, p. 1116. [9] K. Shinohara et al., IEDM 2010, p. 673. [10] D. Brown et al, IEEE TED 2011.

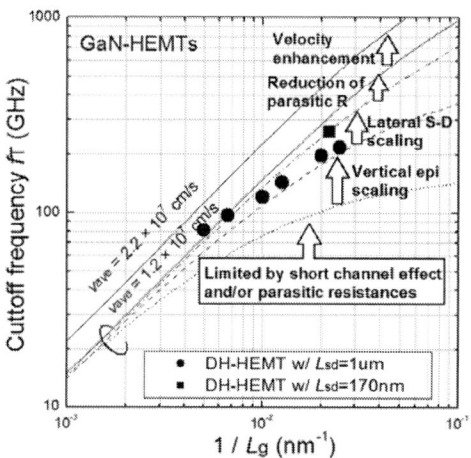

Figure 1: Progress of cutoff frequency (f_T) through device scaling. Lines represent simulated results by assuming average electron velocities (v_{ave}) of 1.2×10^7 cm/s and 2.2×10^7 cm/s.

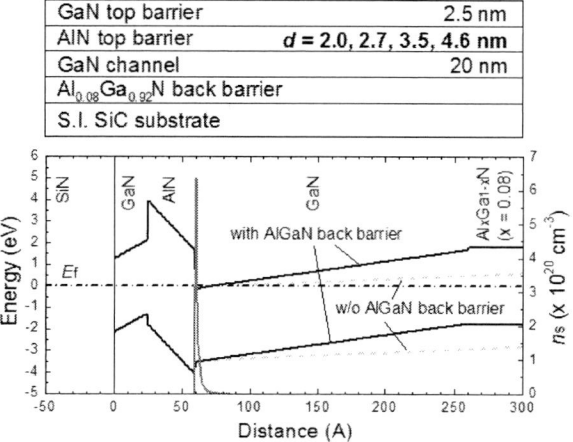

Figure 3: Vertically-scaled AlN/GaN/AlGaN DH-HEMT epitaxial structure and its band diagram with a SiN passivation layer.

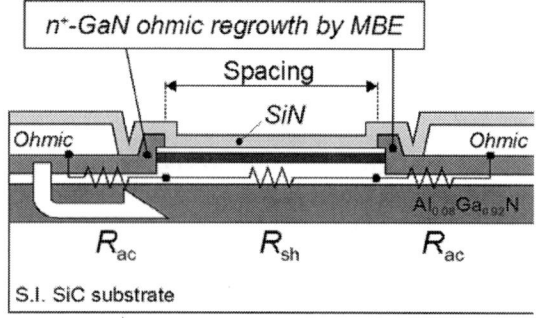

Figure 5: n⁺-GaN ohmic regrowth by MBE for reduction of device access resistance (R_{ac}). R_{ac} is defined as a resistance between the ohmic metal and the 2DEG as indicated by an arrow.

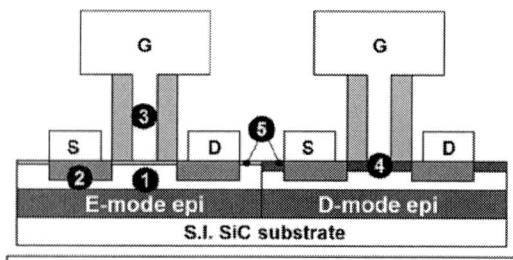

- ❶ Vertically-scaled AlN/GaN/AlGaN DH-HEMT epi
- ❷ Low resistance n⁺-GaN ohmic contact regrowth
- ❸ Latarally-scaled self-aligned gate
- ❹ No gate recess process for high V_{th} uniformity
- ❺ Monolithic E/D-mode epi integration

Figure 2: Projected device scaling technologies with monolithically integrated E/D-mode GaN-HEMTs.

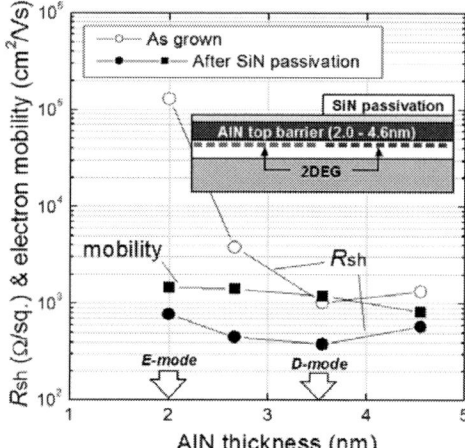

Figure 4: R_{sh} and electron mobility vs. AlN barrier thickness in AlN/GaN/AlGaN DH-HEMTs with and without SiN surface passivation layer.

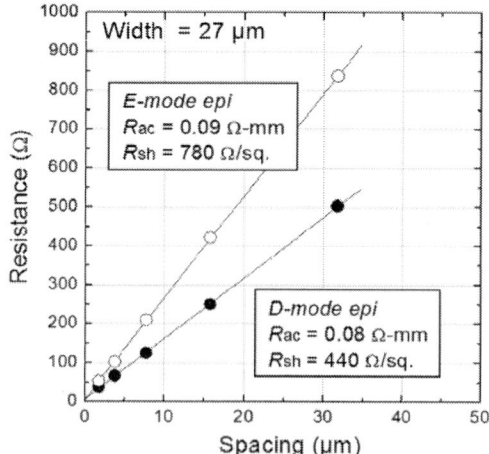

Figure 6: TLM data showing extremely small R_{ac} for both E and D-mode epi structures.

978-1-61284-243-1/11 $26.00 © 2011 IEEE

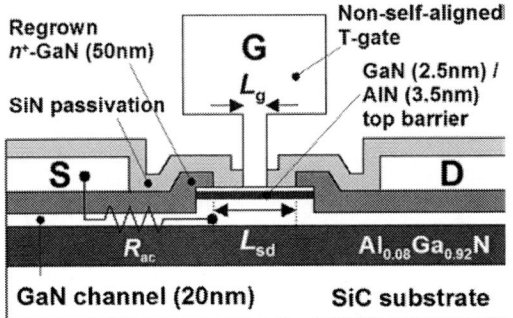

Figure 7: Technology cross-section of a deeply scaled GaN DH-HEMT with a non-self-aligned T-gate. Source-drain spacing (L_{sd}) is defined by regrown n^+-GaN contact edges.

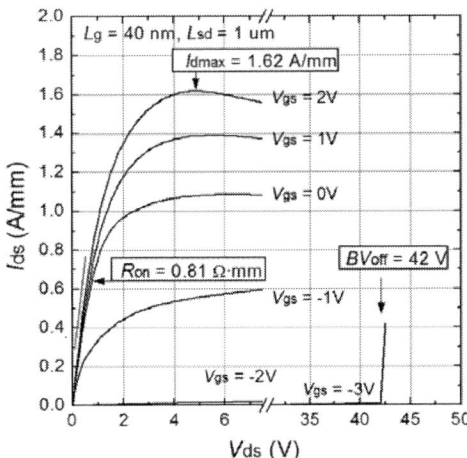

Figure 8: Output characteristic of a 40-nm gate DH-HEMT with $L_{sd} = 1$ μm.

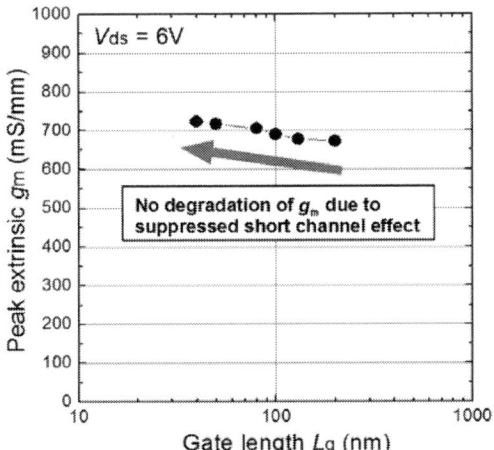

Figure 9: Peak extrinsic g_m vs. L_g of DH-HEMTs with $L_{sd} = 1$ μm.

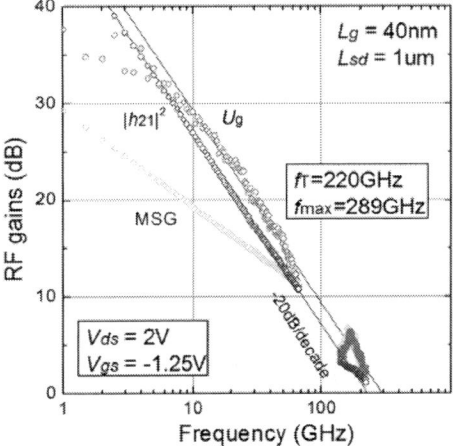

Figure 10: RF gains vs. frequency for a 40-nm gate DH-HEMT with $L_{sd} = 1$ μm measured at $V_{ds} = 2$V.

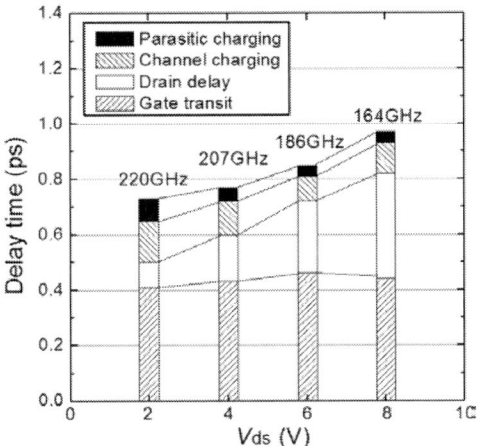

Figure 11: Delay time components as a function of V_{ds} for a 40-nm gate DH-HEMT with $L_{sd} = 1$ μm.

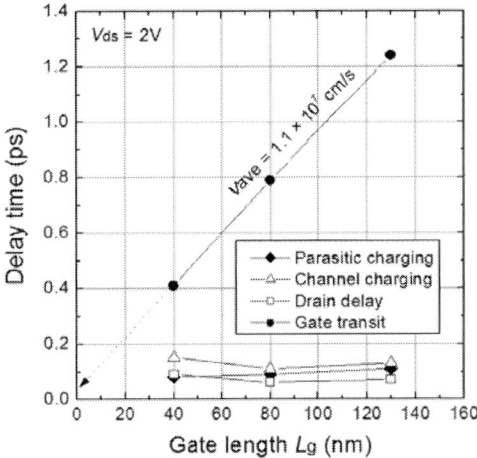

Figure 12: Delay time components as a function of L_g for DH-HEMTs with $L_{sd} = 1$ μm at $V_{ds} = 2$V.

978-1-61284-243-1/11 $26.00 © 2011 IEEE

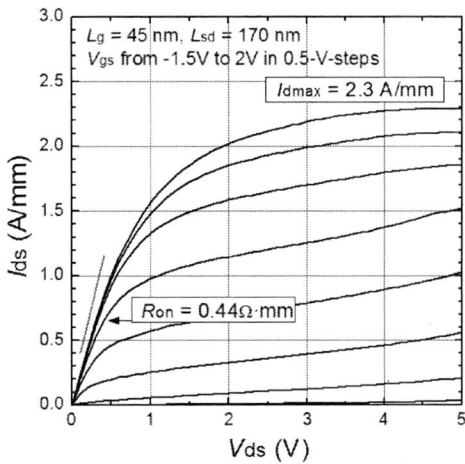

Figure 13: Output characteristics of a laterally S-D scaled 45-nm gate DH-HEMT with $L_{sd} = 170$ nm.

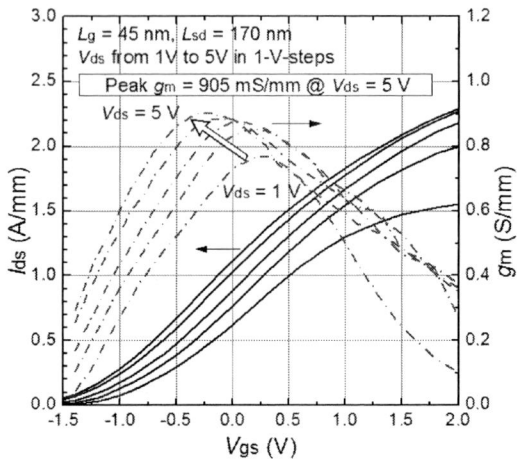

Figure 14: Transfer characteristics of a laterally S-D scaled 45-nm gate DH-HEMT with $L_{sd} = 170$ nm.

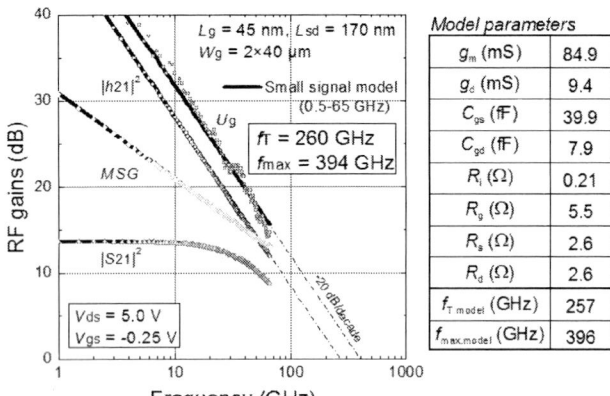

Model parameters	
g_m (mS)	84.9
g_d (mS)	9.4
C_{gs} (fF)	39.9
C_{gd} (fF)	7.9
R_i (Ω)	0.21
R_g (Ω)	5.5
R_s (Ω)	2.6
R_d (Ω)	2.6
$f_{T \, model}$ (GHz)	257
$f_{max.model}$ (GHz)	396

Figure 15: RF gains vs. frequency for a laterally S-D scaled 45-nm gate DH-HEMT.

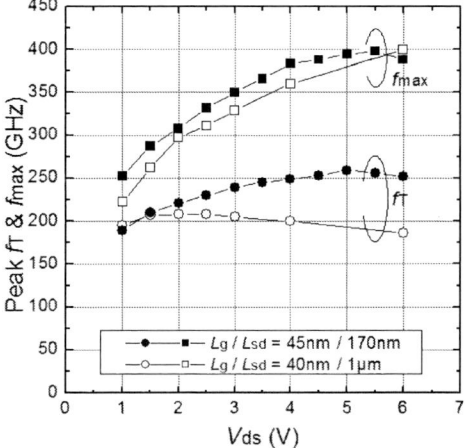

Figure 16: Comparison of dependency of f_T and f_{max} on V_{ds} between scaled HEMT with $L_{sd} = 170$ nm and unscaled HEMT with $L_{sd} = 1$ μm.

Process flow

- E-mode epi growth by MBE
- SiO₂ mask on E-mode device area
- RIE etch into SiC substrate
- Full D-mode epi regrowth by MBE
- SiO₂ mask removal

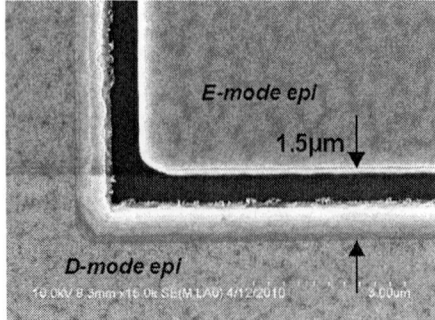

Figure 17: Monolithically integrated E/D-mode epi in vicinity by selective area MBE regrowth.

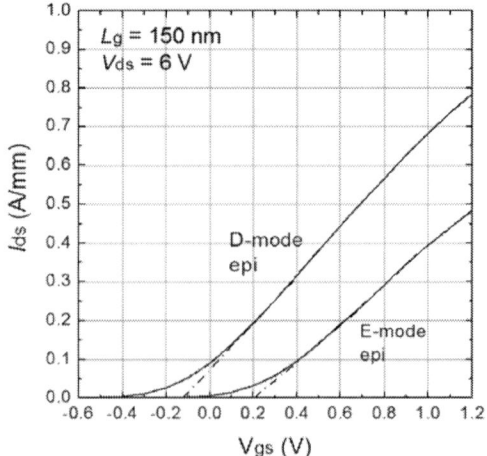

Figure 18: Transfer characteristics of monolithically integrated E/D-mode HEMTs with 150-nm L_g.

978-1-61284-243-1/11 $26.00 © 2011 IEEE

Trap-related Delay analysis of self-aligned N-polar GaN/InAlN HEMTs with record extrinsic g_m of 1105 mS/mm

Nidhi[*], S. Dasgupta, J. Lu, F. Wu, S. Keller, J. S. Speck and U. K. Mishra

ECE Department, University of California Santa Barbara, Santa Barbara, CA 93106, USA

Ga-polar InAlN-based charge-inducing barrier for HEMTs have been recently demonstrated as a viable technology for high frequency applications due to high polarization charge and hence, low resistance channels [1,2]. In this paper, we report on MBE-grown N-polar GaN/InAlN HEMTs with excellent DC and RF performance. There exists a discrepancy in the DC and RF data for N-polar MBE InAlN devices which is explained through several measurements and analysis and possible solutions are discussed.

The device layer structure (Fig. 1a) consists of 35 nm $In_{0.17}Al_{0.83}N$ barrier capped with 1 nm of GaN grown at low temperature to prevent In-desorption as the substrate-temperature is raised for the growth of subsequent layers. AlN interlayer is used to reduce alloy scattering and is capped with 10 nm GaN channel resulting in sheet resistance of 330 $\Omega/\square$. The growth details are given in [3] and band diagram is shown in Fig. 1b. MOCVD SiN_x (3 nm) is used as the gate dielectric. The device fabrication consisted of blanket-deposition of gate-stack (W/Cr/SiO$_2$/Cr), e-beam lithography for gate definition and subsequent selective-etching of the gate-stack. PECVD SiN_x-spacers are formed around the gate before MBE-regrowth of highly-doped graded InGaN/InN to get very low ohmic contact resistance of 25 Ω-μm (Fig. 2). More details on the gate-first self-aligned device fabrication process are given in [4, 5].

The DC measurements resulted in saturated drain current density ($I_{D,sat}$) of 2.13 A/mm and 2.25 A/mm at V_G = 0 V for L_G = 60 nm and 30 nm respectively and peak extrinsic transconductance (g_m) of 1105 mS/mm and 1000 mS/mm at V_{DS} = 2 V for L_G = 60 nm and 30 nm respectively (Fig. 3). This is the highest extrinsic g_m reported for III-nitride HEMTs on any orientation. The small-signal measurements resulted in f_T and f_{MAX} of 155 GHz and 20 GHz respectively for L_G = 30 nm (Fig. 4). Due to self-aligned device design and low sheet resistance, the current, peak g_m and f_T scale well up to gate-lengths of 50-60 nm (Fig. 5).

The f_T values although excellent were not commensurate with the high g_m values measured under DC conditions. To explain this, it should be noted that the MBE-grown InAlN due to low growth temperature, which results in different In and Al surface adatom mobility, shows lateral compositional fluctuation as shown in the cross-sectional TEM and HAADF images (Fig. 6) which could result in shallow traps near the conduction band edge of InAlN which could cause DC-RF dispersion, hence the discrepancy. To test this theory, pulsed-IV measurements are performed to observe the presence of traps with V_G pulsed from -3 V to 1 V with an output load-line of 25 Ω. As shown in Fig. 7, there is no dispersion up to pulse width of 120 ns (not surface traps), however dispersion is observed at 40 ns gate pulses. Also, all gate-lengths show knee walkout, however current collapse becomes more severe as the gate-length is reduced. This can be explained by the hypothesis that there are shallow traps near the conduction band which respond in the high field regions near the drain causing knee walkout and as the field penetrates under the gate towards the source for shorter gate-lengths, the devices begin to show current collapse. Therefore, there are two effects to be analyzed: knee walkout (can cause additional delay at the output) and current collapse (causing g_m-dispersion).

Total delay versus gatelength for 10 nm GaN channel is plotted for both InAlN, as well as AlGaN backbarrier (Fig. 8). The lower velocity in InAlN can be attributed to increasing g_m dispersion as gatelength is reduced resulting in change in slope. Also, L_G-independent delay in InAlN is found to be much higher than AlGaN attributed to trap-related RC-delay in the high field drain region. This drain delay can be modeled as additional capacitance reflected at the input: $\delta C_{GS} = (\delta Q_{trap}/\delta I_D)(\delta I_D/\delta V_G) = \tau_{trap} g_m$. To observe these effects, the small signal equivalent circuit was extracted using lumped element model (Fig. 9). Fig. 10 shows the change in C_{GS} from geometric capacitance ($C_{G,geom}$) and drop in g_m from its DC value versus L_G. It can be shown that by putting $C_{GS} = C_{G,geom}$ and $g_m = g_{m,DC}$ in the circuit model, much higher f_T values are achievable (Fig. 11).

To conclude, we have demonstrated excellent DC-RF performance from MBE-grown N-polar InAlN with extensive experiments to explain the DC-RF discrepancy. Similar devices with InAlN grown at higher temperature using other growth techniques should not show these traps related to lateral composition fluctuation.

[1] H. Sun et. al., IEEE EDL 31, 957 (2010), [2] H. Wang et. al., CSManTech 2010, [3] S. Dasgupta et. al., APEX, Mar 2011, [4] S. Dasgupta et. al., APL, 96, 143504 (2010), [5] Nidhi et. al., IEEE EDL 32, 33 (2011).

*Corresp. author: nidhi@ece.ucsb.edu, Phone: +1-805-893-3812-ext. 202

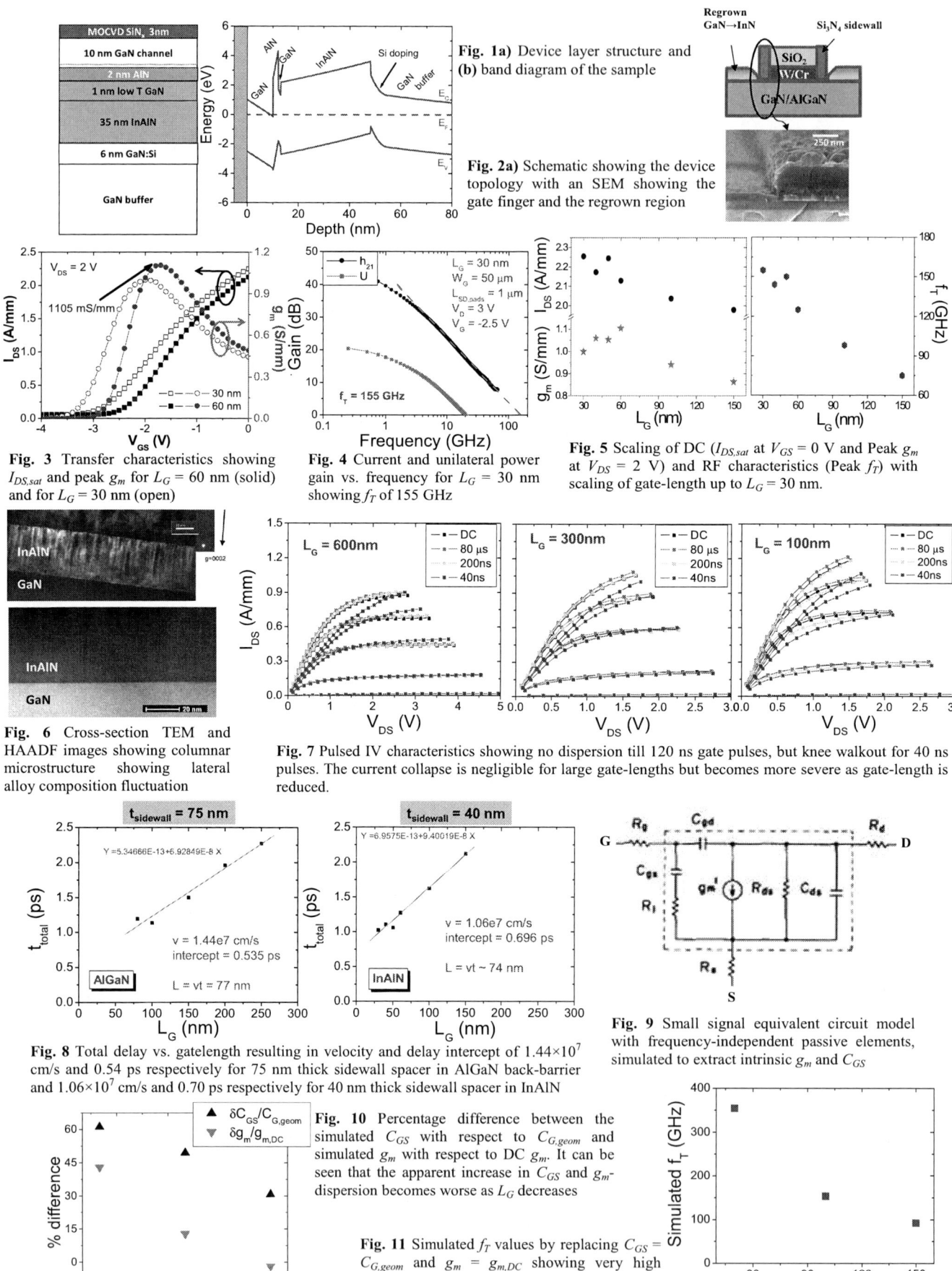

Fig. 1a) Device layer structure and **(b)** band diagram of the sample

Fig. 2a) Schematic showing the device topology with an SEM showing the gate finger and the regrown region

Fig. 3 Transfer characteristics showing $I_{DS,sat}$ and peak g_m for $L_G = 60$ nm (solid) and for $L_G = 30$ nm (open)

Fig. 4 Current and unilateral power gain vs. frequency for $L_G = 30$ nm showing f_T of 155 GHz

Fig. 5 Scaling of DC ($I_{DS,sat}$ at $V_{GS} = 0$ V and Peak g_m at $V_{DS} = 2$ V) and RF characteristics (Peak f_T) with scaling of gate-length up to $L_G = 30$ nm.

Fig. 6 Cross-section TEM and HAADF images showing columnar microstructure showing lateral alloy composition fluctuation

Fig. 7 Pulsed IV characteristics showing no dispersion till 120 ns gate pulses, but knee walkout for 40 ns pulses. The current collapse is negligible for large gate-lengths but becomes more severe as gate-length is reduced.

Fig. 8 Total delay vs. gatelength resulting in velocity and delay intercept of 1.44×10^7 cm/s and 0.54 ps respectively for 75 nm thick sidewall spacer in AlGaN back-barrier and 1.06×10^7 cm/s and 0.70 ps respectively for 40 nm thick sidewall spacer in InAlN

Fig. 9 Small signal equivalent circuit model with frequency-independent passive elements, simulated to extract intrinsic g_m and C_{GS}

Fig. 10 Percentage difference between the simulated C_{GS} with respect to $C_{G,geom}$ and simulated g_m with respect to DC g_m. It can be seen that the apparent increase in C_{GS} and g_m-dispersion becomes worse as L_G decreases

Fig. 11 Simulated f_T values by replacing $C_{GS} = C_{G,geom}$ and $g_m = g_{m,DC}$ showing very high numbers achievable by InAlN HEMTs

978-1-61284-243-1/11 $26.00 © 2011 IEEE

130nm InP DHBTs with $f_t > 0.52$ THz and $f_{max} > 1.1$ THz

M. Urteaga[1], R. Pierson[1], P. Rowell[1], V. Jain[2], E. Lobisser[2], M.J.W. Rodwell[2]

[1]Teledyne Scientific Company, Thousand Oaks, CA 93160. [2]Department of ECE, University of California, Santa Barbara, CA 93106. E-mail: murteaga@teledyne-si.com

We report results from a 130nm Indium Phosphide (InP) double heterojunction bipolar transistor (DHBT) technology. A $0.13 \times 2\mu m^2$ transistor exhibits a current gain cutoff frequency $f_t > 520$ GHz, with a simultaneous extrapolated power gain cutoff frequency $f_{max} > 1.1$ THz. The HBTs exhibit these RF figures-of-merit while maintaining a common-emitter breakdown voltage $BV_{CEO} = 3.5$V ($J_E = 10\mu A/\mu m^2$). Additionally, scaling of the emitter junction length to $2\mu m$ enables high device performance at low total power levels. Transistors in the InGaAs/InP material system have demonstrated the highest reported transistor RF figures-of-merit. Previous published results include strained-InGaAs channel high-electron mobility transistors (HEMTs) with f_{max} of >1THz [1,2], and InP DHBTs with $f_{max} > 800$ GHz [3]. High bandwidth DHBTs have applications in a number of RF and mixed-signal applications due to their high power handling and high levels of integration relative to HEMTs. The HBTs reported in this work are designed for transceiver applications at the lower end of the THz frequency band [0.3-3 THz].

General scaling laws for InP HBTs have been outlined in [4]. We perform aggressive lateral scaling of the emitter-base junction dimensions using an electroplated emitter post process with dielectric sidewall spacers. Details of the process for 250nm HBTs are reported in [5]. Critical features of the process are the use of electron beam lithography and an Au-based plating process to form an emitter post contact with a large height-to-width ratio and a vertical sidewall profile. These characteristics are advantageous for a self-aligned emitter-base HBT process flow. The 130nm process features lateral scaling of the junction and sidewall dimensions and vertical scaling of the HBT epitaxy relative to our 250nm process. The HBT epitaxy has a 25nm InGaAs base with high carbon doping ($>5 \times 10^{19} cm^{-3}$) to reduce base contact resistivity. A contact resistivity $\rho_c < 5$ $\Omega\text{-}\mu m^2$ is extracted from TLM measurements on fabricated wafers. The base-collector heterojunction grade (33nm thickness) consists of an InGaAs setback layer, a InGaAs/InAlAs chirped-superlattice, and an InP pulse doping layer. The remaining N- collector region is 67nm of InP doped at $5 \times 10^{16} cm^{-3}$, for a total collector thickness of 100nm. The emitter is InP and super-lattice grading is also used at the base-emitter heterojunction. The emitter and grade layers are designed to support high current densities with low emitter access resistance. HBTs demonstrate a forward collector ideality factor $n_C = 1.1$ and an extrinsic emitter resistance $\rho_{ex} < 4\Omega\text{-}\mu m^2$ extracted from measurements of the low-frequency (500MHz) transconductance versus collector current.

Fig. 1 shows a measurement of the HBT Gummel characteristics. The HBTs exhibit a peak current gain $\beta \sim 17$. Measurements of β versus emitter area and periphery show the current gain is limited by both bulk and surface recombination due to the high base doping level and aggressively scaled base-emitter spacing, respectively. The current gain is sufficient for our present application and we expect it could be improved with further process and epitaxy optimization. Common-emitter IV characteristics are shown in Fig. 2. The highly scaled emitter junction permits operation at high current ($J_E > 30$mA$/\mu m^2$) and power (>50mW$/\mu m^2$) densities.

On-wafer S-parameter measurements are performed in thin-film microstrip test structures in order to reduce probe-to-probe coupling and unwanted mode propagation in the InP substrate that can corrupt the measurement of S_{12} in highly scaled HBTs. The microstrip is formed using the first layer metal as a ground plane, a 7um thick spin-on-polymer (BCB, $\varepsilon_r = 2.7$) as the interlayer dielectric and an electroplated top metal level for the signal line. A multi-line Through-Reflect-Line (TRL) calibration is performed using on-wafer standards. Open and short circuit deembedding structures are used to remove the capacitive and inductive loading effects of vias through the 7um BCB layer. Total input and output capacitances of ~4fF are deembedded from the HBT measurements.

Fig. 3 shows the measured transistor gains of a $0.13 \times 2\mu m^2$ HBT at $I_C = 6.9$mA and $V_{CE} = 1.6$V. Measurements were performed from 8-50GHz and 75-105GHz. The HBT figures-of-merit f_t and f_{max} are extrapolated from least squares fits to single-pole transfer functions of the measured h_{21} and unilateral power gain (U), respectively. Fits are performed on the data to 50GHz, as measurements of U show considerable variation at high frequencies (>75GHz). The HBT exhibits an extrapolated f_t/f_{max} of 521GHz/1.15THz. Fig. 4 shows the variation of f_t and f_{max} versus collector current at varying values of V_{CE}. At a total power dissipation of 1.2mW ($I_C = 1.2$mA and $V_{CE} = 1.0$V), the HBT exhibits an f_t/f_{max} of 338GHz/639GHz. We note that the HBT f_{max} is enhanced at increasing values of V_{CE} due to collector capacitance cancellation arising from electron velocity modulation [6]. A collector-base capacitance of 1.9 fF is extracted at peak RF bias. Fig. 5 and Fig. 6 show the variation of f_t and f_{max}, respectively, versus current for varying emitter lengths. Peak f_t increases for longer device lengths due to a decrease in the relative contribution of parasitic capacitance associated with the base via contact. A peak f_t/f_{max} of 566GHz/875GHz is measured for a $4\mu m$ long HBT. f_{max} is observed to decrease for longer devices due to contributions from the base metal resistance down the length of the narrow transistor mesa. To the best of our knowledge, these results represents the first bipolar technology with a simultaneous $f_t > 500$ GHz and $f_{max} > 1$ THz.

[1] R. Lai, et. al, *IEDM* Washington DC, 2007   [2] D-H. Kim, et. al, *IEDM*, San Fransisco, CA, 2010
[3] V.Jain, et. al, *IEEE EDL* vol. 32, no.1 Jan 2011   [4] M. Rodwell, et. al, *Proc. of IEEE*, vol. 96, no.12, Feb 2008
[5] M. Urteaga, et. al. to be presented *2011 Indium Phosphide and Related Materials (IPRM) Conference,* May 2011
[6] Y. Bester, et. al, *IEEE TED* vol. 46, no.4 April 1999

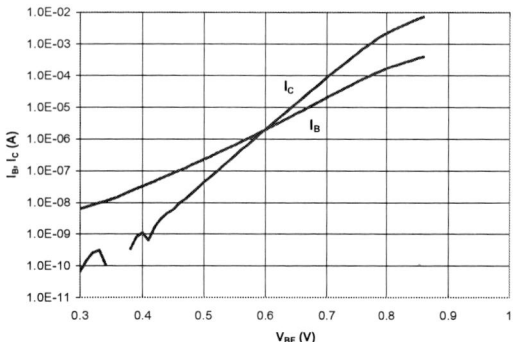

Fig. 1 Gummel characteristics of $0.13 \times 2\mu m^2$ HBT

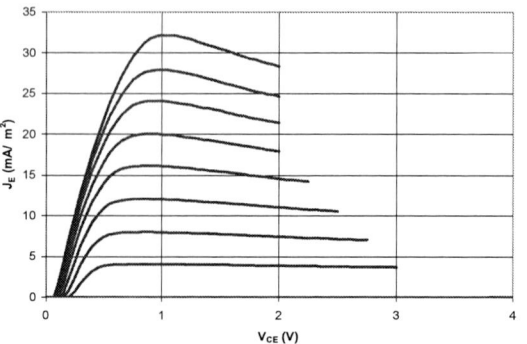

Fig. 2 Common-emitter IV characteristics of 130nm HBT normalized to emitter area

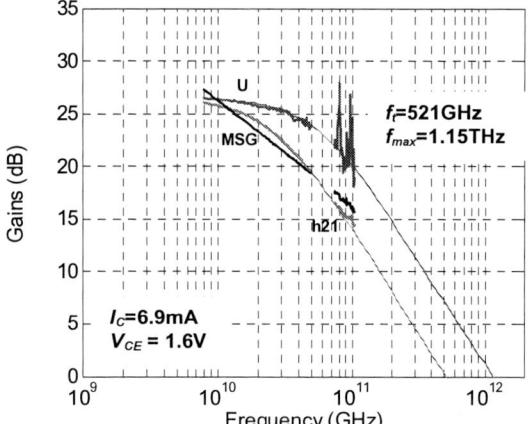

Fig. 3 RF gains of $0.13 \times 2\mu m^2$ HBT

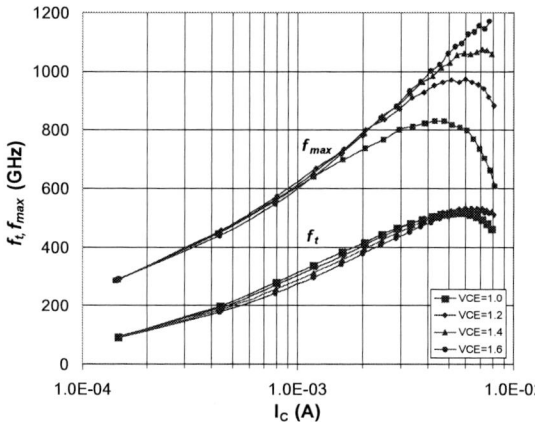

Fig. 4 f_t and f_{max} versus collector current at varying values of V_{CE} for $0.13 \times 2\mu m^2$ HBT

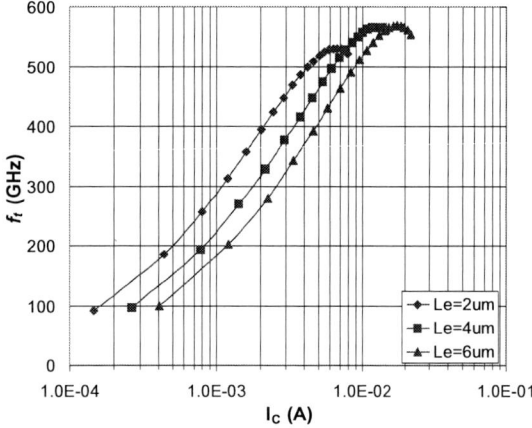

Fig. 5 Variation of f_t versus collector current for varying emitter length ($V_{CE} = 1.4V$)

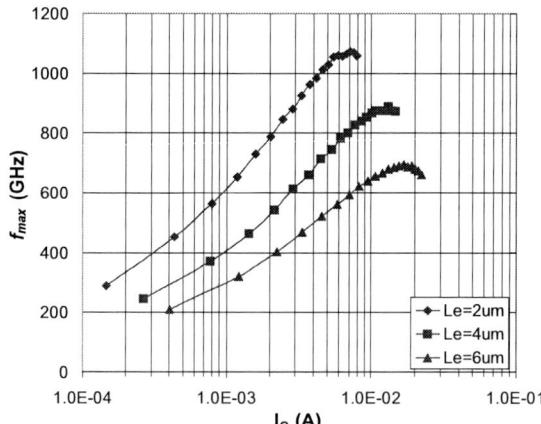

Fig. 6 Variation of f_{max} versus collector current for varying emitter length ($V_{CE} = 1.4V$)

This work was supported by DARPA CMO Contract No. HR0011-09-C-0060. The views, opinions and/or findings contained in this article are those of the authors and should not be interpreted as representing the official policies, either expressed or implied, of the Defense Advanced Research Projects Agency, or the Department of Defense.

978-1-61284-243-1/11 $26.00 © 2011 IEEE

Author Index

A

Abraham, D. W. 171
Agarwal, S. 199
Agrawal, A. 27
Ajoy, A. 113
Akinwande, D. 123
Akyol, T. 127
Ali, A. 27
Almanza-Workman, M. A. 241
Alomari, M. 73
Ancona, M. G. 143
Appelbaum, I. 159
Appenzeller, J. 33
Aroshvili, G. 149
Asbeck, P. M. 275
Assefa, S. 253
Augustine, C. 125
Auluck, K. 169

B

Baccarani, G. 201
Backes, D. 165
Bae, M.-H. 41
Baek, C.-K. 179
Balakrishnan, V. 187
Banerjee, S. K. 123, 127
Bao, D. 91
Baraskar, A. 271
Baykan, M. 237
Bayraktaroglu, B. 75, 245
Bedau, D. 165
Behin-Aein, B. 161
Behnam, A. 129
Bennett, B. R. 27, 143
Bernstin, K. 183
Bersuker, G. 117, 127
Bessire, C. D. 181
Beyer, A. 257
Bhowmik, D. 163
Bielawski, C. W. 127
Billingsley, D. 55
Binetti, P. R. 263
Björk, M. T. 119, 181
Boos, J. B. 27, 143
Borg, M. B. 21
Borgström, M. 21, 105
Boutros, K. S. 213, 221
Bouvet, D. 145
Brian, H. 221
Brown, D. F. 141, 269, 275
Buehl, T. E. 265
Burek, G. J. 21
Burke, P. J. 77, 131
Burnham, S. 213, 275
Butler, C. 275

C

Cai, J. S. 71
Campbell, P. M. 39
Cantley, K. D. 99
Cao, J. 189
Cao, Y. 139
Carlin, J.-F. 73

Carpenter, G. 123
Carter, A. D. 19
Chakrabarti, B. 99
Chan, J. 85
Chan, M. K. 155
Chapman, R. A. 99
Char, K. 95
Chen, A. 167
Chen, F. 89
Chen, H.-Y. 33
Chen, K. J. 71
Chen, M. 213
Chen, W. H. 71
Chen, Y.-S. 89
Chen, Y.-T. 117
Cheng, H.-W. 103
Chiang, H.-W. 271
Chiu, Y.-Y. 103
Chow, E. K. 129
Chu, R. 213
Chu, R. 221
Chuang, S. 13
Chung, H.-J. 31
Chung, U.-I. 31
Ciftcioiglu, B. 147
Coffie, R. 217
Cohen, D. A. 261
Colby, R. 93
Coldren, L. A. 259, 263
Colombo, L. 85, 123
Corbet, C. 35, 123
Corrion, A. L. 213, 275
Crowell, P. A. 155
Cruz, S. C. 265

D

Dasgupta, S. 141, 269, 279
Datta, S. 17, 27, 117, 161, 203, 237
De Michielis, L. 111
Delage, S. L. 73
Dellabetta, B. J. 65
DenBaars, S. P. 215, 261, 265
Deng, J. 55
Denninghoff, D. J. 269
Dery, H. 147
Dey, A. W. 21
Divakar, K. 247
Dodabalapur, A. 247
Dora, Y. 217
Dorgan, V. 41
Downer, M. 127
Dua, C. 73
Ducatteau, D. 219

E

Eddy, C. R. 39
Egawa, T. 221
Elder, R. E. 241

F

Fallahazad, B. 35
Fang, H. 13

Fang, T. 139, 273
Farrell, R. 261
Fastenau, J. M. 271
Fasth, C. 105
Fay, P. J. 139, 205
Feng, H. Z. 71
Fenk, B. 249
Ferrer, D. A. 123, 127
Fichtenbaum, N. 217
Fischetti, M. V. 193
Ford, A. C. 13
Fox, O. J. 137
Frank, D. J. 197
Fujito, K. 261
Fukata, N. 87

G

Gaidis, M. C. 171
Gallagher, W. J. 171
Ganapathi, K. 79
Ganguly, S. 121
Garcia, R. A. 241
Garlid, E. S. 155
Gaska, R. 55
Gaskill, K. D. 39
Geppert, C. 155
Ghoneim, H. 181
Gilbert, M. J. 65
Gnani, E. 201
Gnudi, A. 201
Goel, N. 117
Gossard, A. C. 19, 265
Grabinski, W. 145
Grandjean, N. 73
Green, W. M. J. 253
Gross, M. 259
Gupta, A. 123
Gupta, S. K. 67
Gutsche, C. 53

H

Habib, K. M. M. M. 109
Haeger, D. A. 261
Haensch, W. 5
Hahn, K. 249
Haley, B. 23
Hansen, N. H. 57
Happy, H. 149
Hegde, G. 23
Heo, J. 31
Hobbs, C. 237
Holder, C. 261
Holland, E. 241
Hong, B. H. 37
Hong, Y. 97
Horst, F. 253
Hossain, N. 257
Hsu, P. S. 261
Hu, C. 133
Hu, G. 171
Hu, H. 101
Hu, J. 135
Hu, Q. O. 155
Huang, J. 117
Huang, M. 147
Huang, S. 87
Hudnall, T. W. 127

Hughes, B. 213
Hung, Y. 263
Hwang, W. 39, 205

I

Ikeda, M. 57
Ionescu, A. M. 111, 145, 189
Ishibashi, K. 87
Islam, S. 41
Iwai, T. 59

J

Jackson, W. B.. 241
Jacobson, Z. A. 133
Jain, D. 77, 131
Jain, V. 271, 281
Jammy, R. 237
Javey, A. 13
Jeans, A. H. 241
Jeffrey, F. 241
Jena, D. 39, 81, 121,139, 273
Jeong, Y.-H. 179
Jin, S. 257
Johnson, J. W. 139
Joshi, T. 247

K

Kälblein, D. 177, 249
Kan, E. C. 169
Kao, C.-Y. 89, 207
Kao, K.-H. 193
Kapadia, R. 13
Karg, S. 119
Karmalkar, S. 113
Kawakami, R. 147
Kelchner, K. 261
Keller, S. 215, 269, 279
Kent, A. D. 165
Kern, K. 249
Khalil, S. 213
Khayer, M. A. 51
Kim, D. 97
Kim, H. 241
Kim, K. H. 101
Kim, K.-H. 179
Kim, S. 35
Kim, S. H. 133
Kim, S. 275
King Liu, T. 133
Kirk, W. P. 85
Klauk, H. 57, 177, 249
Klimeck, G. 23, 233
Ko, C. 187
Ko, M.-D. 179
Koester, S. J. 43, 151
Kohn, E. 73
Kolluri, S. K. 215
Kosel, T. 205
Koswatta, S. O. 81, 197, 205
Krishna, S. 13
Krivorotov, I. 147
Kshirsagar, C. 151

Kubis, T. 23
Kunert, B. 257
Kuniharu, T. 13
Kuo, J. 205
Kuwabara, H. 57
Kwak, J. 95
Kwon, O. 241

L

Laboutin, O. 139
Lake, R. K. 51, 109
Langer, J. 165
Lattanzio, L. 111
Lau, K. 107
Laux, S. 113
Law, J. 19
Lee, C. 95
Lee, C.-G. 247
Lee, C.-H. 207
Lee, H. 95
Lee, H. Y. 89
Lee, J. C. 127
Lee, J.-S. 179
Lee, K. 35
Lee, M.-H. 89, 207
Lee, S. 95
Lee, S.-H. 31
Lee, V. 275
Leedy, K. D. 75, 245
Lei, M. 127
Leonelli, D. 193
Li, F. I. 69
Li, F.-H. 103
Li, G. 121, 139, 273
Li, J. 159
Li, Q. 107
Li, R. 205
Li, X. 101, 115
Li, Y. 103
Liang, Q. 83
Liebich, S. 257
Lim, J. 95
Lim, T. 131
Lin, C.-H. 259
Lin, J. 91
Lin, J.-Y. 143
Lin, M.-R,. 167
Lind, E. 21
Liu, H. 165
Liu, L. 17
Liu, Q. 205
Liu, T. 63
Liu, W. K. 271
Lobisser, E. 19, 271, 281
Lörtscher, E. 119
Loubychev, D. 271
Lu, J. L. 211, 279
Lu, Y. 205
Luo, H. 241
Luo, J. 83
Lyon, K. 169
Lyons, A. S. 129
Lysov, A. 53

M

Madan, H. 117
Madsen, M. 13
Magnus, W. 193
Maier, D. 73
Maiorano, P. 201
Majhi, P. 237
Malatabes, J. 241
Manandhar, P. 165
May, P. W. 137
Mayer, T. 203
McLelland, H. 137
Medjdoub, F. 219
Mehrban, J. 241
Mehrotra, S. 233
Meng, N. 149
Mensch, P. 119
Miao, X. 115
Micovic, M. 275
Milosavljevic, I. 275
Mishra, U. K. 141, 211, 215, 217, 265, 269, 279
Misra, R. 27
Mitchell, W. J. 19
Mohata, D. K. 203
Mojumder, N. N. 67
Moran, D. A. 137
Moselund, K. E. 119, 181
Movva, H. P. 123
Murali, K. V. R. M. 113
Myers-Ward, R. L. 39

N

Nainani, A. 143
Najmzadeh, M. 145
Nakamura, S. 9, 261
Nakane, R. 235
Narayanan, V. 17
Neufeld, C. J. 265
Nidhi, N. 141, 279
Norberg, E. J. 263
Nowak, J. J. 171
Nylund, G. 105

O

Offer, M. 53
Ohoka, A. 275
Ok, I. 237
OSullivan, E. J. 171
Ozkan, C. 91
Ozkan, M. 91

P

Palestri, P. 111
Palmstrøm, C. J. 155
Panagopoulos, G. 125
Parikh, P. 217
Park, C.-H. 179
Park, H.-H. 23
Parker, J. S. 263
Patel, P. 133
Paul, A. 233
Pavlidis, D. 149
Perlov, C. 241
Pflaum, J. 57
Phillips, J. D. 75

Pierson, R. 281
Pillarisetty, R. 231
Poisson, M.-A. 73
Pop, E. 41, 129
Povolotskyi, M. 23
Prost, W. 53

R
Rajamohanan, B. 203, 237
Rajan, S. 273
Rajwade, S. R. 169
Ramanathan, S. 187
Ramon, M. E. 123
Regan, D. 275
Reggiani, S. 201
Regolin, I. 53
Reiber-kyle, J. 91
Richter, M. 181
Riel, H. 119
Robertazzi, R. P. 171
Rode, J. C. 271
Rodwell, M. J. 19, 23, 233, 259, 271, 281
Rogers, J. A. 3, 101
Rolland, N. 219
Rolland, P. A. 219
Rooyackers, R. 193
Rouhi, N. 77
Rowell, P. 281
Roy, K. 67, 125
Russell, S. 137
Rylyakov, A. 253
Ryu, H. 177

S
Salahuddin, S. 79, 163
Samuelson, L. 105
Saraswat, K. C. 135, 143
Saripalli, V. 17, 203
Sarkar, A. 161
Sattu, A. K. 55
Scarpulla, M. A. 265
Schiffer, P. 27
Schmid, H. 119, 181
Schmidt, O. G. 177
Schmitz, A. 275
Schow, C. 253
Scott, R. C. 245
Seabaugh, A. 39, 205
Sekitani, T. 57
Selmi, L. 111
Selvaraj, L. 221
Seo, S. 31
Sham, L. J. 147
Shannon, J. M. 61
Shaw, J. 169
Shen, L. 217
Shi, J. 147
Shin, J. 31
Shin, J. C. 101
Shin, S.-K. 87
Shinohara, K. 275
Shur, M. 55
Siddiqui, J. J. 75
Silva, S. P. 61
Simin, G. 55
Singisetti, U. 211

Snider, G. 139
Snyder, A. 271
Soga, I. 59
Sohn, C.-W. 179
Solomon, P. M. 197
Someya, T. 57
Song, Y. 83
Song, Y. 147
Soree, B. 193
Speck, J. S. 141, 211, 261, 265, 279
Sporea, R. A. 61
Srinivasan, S. 161
Stach, E. A. 93
Steiger, S. 23
Stolz, W. 257
Storm, K. 105
Subramaniam, A. 99
Suh, C. S. 217
Suh, J. 235
Sun, J. 171
Sweeney, S. J. 257

T
Tahy, K. 39
Takagi, S. 235
Takei, K. 13
Takenaka, M. 235
Takimiya, K. 57
Taoka, N. 235
Taussig, C. P. 241
Tedesco, J. L. 39
Tegude, F.-J. 53
Thelander, C. 21, 105
Thibeault, B. J. 19, 271
Toledo, N. G. 265
Trouilloud, P. L. 171
Trovinger, S. 241
Tsai, M.-J. 89
Tutuc, E. 35, 123

U
Urteaga, M. 271, 281

V
Vandenberghe, W. G. 193
Vandooren, A. 193
Vasen, T. 205
Veksler, D. 117
Venugopal, A. 85
Verhulst, A. S. 193
Verma, J. 121
Vignale, G. 63
Vignaud, D. 149
Vlasov, Y. A. 253
Vogel, E. M. 85, 99
Volz, K. 257
Vullev, V. 91

W
Wallentin, J. 105
Wang, C. 93
Wang, R. 139, 273
Wang, W. 91
Watanabe, A. 221

Welser, J. 183
Wernersson, L.-E. 21
Willadsen, P. J. 275
Wistey, M. 205
Wong, H.-S. P. 135
Wong, M. H. 211
Wu, F. 279
Wu, H. 147
Wu, Y. 271
Wu, Y.-F. 217

X
Xing, H. 39, 121, 139, 205, 273
Xu, M. 93
Xu, Q. 83

Y
Yablonovitch, E. 199
Yamaguchi, J. 59
Yamamoto, T. 57
Yang, C.-L. 89, 207
Yang, H. 31
Yang, J. 55
Yang, Z. 187
Ye, P. D. 93
Yoon, Y. 79
Yu, K. 101
Yuan, Z. 143
Yum, J. H. 127

Z
Zegaoui, M. 219
Zehnder, D. 213, 221
Zhang, Q. 205
Zhao, C. 83
Zhao, P. 81
Zheng, Y. 259
Zhong, J. 91
Zhou, G. 205
Zhou, H. C. 71
Zhou, H. 83
Zhou, Q. 71
Zhou, X. 107
Zhu, H. 205
Zhuge, J. 193
Ziegler, S. 257
Zimmermann, T. 121
Zimprich, M. 257
Zschieschang, U. 57, 177, 249
Zuo, J.-M. 101
Zutic, I. 147